Numerical Methods for Fluid Dynamics

The Institute of Mathematics and its Applications Conference Series

---

RECENT TITLES

Computational Methods and Problems in Aeronautical Fluid Dynamics, *edited by* B.L. Hewitt, C.R. Illingworth, R.C. Lock, K.W. Mangler, J.H. McDonnell, Catherine Richards and F. Walkden
Optimization in Action, *edited by* L.C.W. Dixon
The Mathematics of Finite Elements and Applications II, *edited by* J.R. Whiteman
The State of the Art in Numerical Analysis, *edited by* D.A.H. Jacobs
Fisheries Mathematics, *edited by* J.H. Steele
Numerical Software—Needs and Availability, *edited by* D.A.H. Jacobs
Recent Theoretical Developments in Control, *edited by* M.J. Gregson
The Mathematics of Hydrology and Water Resources, *edited by* E.H. Lloyd, T.O'Donnell and J.C. Wilkinson
Mathematical Aspects of Marine Traffic, *edited by* S.H. Hollingdale
Mathematical Modelling of Turbulent Diffusion in the Environment, *edited by* C.J. Harris
Mathematical Methods in Computer Graphics and Design, *edited by* K.W. Brodlie
Computational Techniques for Ordinary Differential Equations, *edited by* I. Gladwell and D.K. Sayers
Stochastic Programming, *edited by* M.A.H. Dempster
Analysis and Optimisation of Stochastic Systems, *edited by* O.L.R. Jacobs, M.H.A. Davis, M.A.H. Dempster, C.J. Harris and P.C. Parks
Numerical Methods in Applied Fluid Dynamics, *edited by* B.Hunt
Recent Developments in Markov Decision Processes, *edited by* R. Hartley, L.C. Thomas and D.J. White
Power from Sea Waves, *edited by* B. Count
Sparse Matrices and their Uses, *edited by* I.S. Duff
The Mathematical Theory of the Dynamics of Biological Populations II, *edited by* R.W. Hiorns and D. Cooke
Floods due to High Winds and Tides, *edited by* D.H. Peregrine
Third IMA Conference on Control Theory, *edited by* J.E. Marshall, W.D. Collins, C.J. Harris and D.H. Owens
Numerical Methods in Aeronautical Fluid Dynamics, *edited by* P.L. Roe
Numerical Methods for Fluid Dynamics, *edited by* K.W. Morton and M.J. Baines

IN PREPARATION

Multi-Objective Decision Making, *edited by* L.C. Thomas

Numerical Methods for Fluid Dynamics

Based on the proceedings of a conference on Numerical Methods in Fluid Dynamics, organised by The Institute of Mathematics and its Applications and held at the University of Reading, 29-31 March, 1982

Edited by

K.W. MORTON
M.J. BAINES

University of Reading
Reading, UK

1982

ACADEMIC PRESS
A Subsidiary of Harcourt Brace Jovanovich, Publishers
London New York
Paris San Diego San Francisco São Paulo
Sydney Tokyo Toronto

ACADEMIC PRESS INC. (LONDON) LTD.
24/28 Oval Road
London NW1

United States Edition published by
ACADEMIC PRESS INC.
111 Fifth Avenue
New York, New York 10003

British Library Cataloguing in Publication Data
Numerical Methods for Fluid Dynamics.
1. Fluid dynamics—Congresses
2. Numerical calculations—Congresses
I. Morton, K.W.
II. Baines, M.J.
532'.05'.015117 QA911

ISBN 0-12-508360-2

LCCCN 82-72597

Printed in Great Britain by
Whitstable Litho Ltd., Whitstable, Kent

CONTRIBUTORS

J.W. BARRETT; *(Imperial College, London) now at Department of Mathematics, Virginia Polytechnic Institute & State University, Blacksburg, Virginia, USA.*

R. BERKOWICZ; *Air Pollution Laboratory, National Agency of Environmental Protection, Risø National Laboratory, DK-4000, Roskilde, Denmark.*

A. BRANDT; *Department of Applied Mathematics, The Weizmann Institute of Science, Rehovot, Israel 76100.*

A.N. BROOKS; *(Stanford University) now Fluid Dynamics Engineer, AeroVironment Inc., Pasadena, California, USA.*

G. BROWNING; *Department of Applied Mathematics, California Institute of Technology, Pasadena, California 91125, USA.*

P. CONCUS; *Lawrence Berkeley Laboratory, University of California, Berkeley, California 94720, USA.*

A.W. CRAIG; *(University of Reading) now at Department of Civil Engineering, University College of Swansea, Singleton Park, Swansea, SA2 8PP.*

M.J.P. CULLEN; *The Meteorological Office, London Road, Bracknell, Berkshire.*

A.M. DAVIES; *Institute of Oceanographic Sciences, Bidston Observatory, Birkenhead, Merseyside, L43 7RA.*

A.R. DAVIES; *Department of Applied Mathematics, University College of Wales, Aberystwyth, Dyfed, SY23 3BZ.*

E. DETYNA; *Department of Mathematics, University of Reading, Whiteknights, Reading, RG6 2AX.*

P.P.G. DYKE; *Department of Offshore Engineering, Heriot-Watt University, Riccarton, Currie, Edinburgh EH14 4AS.*

C.M. ELLIOTT; *Department of Mathematics, Imperial College of Science and Technology, Huxley Building, 180 Queen's Gate, London SW7.*

S.A.E.G. FALLE; *Department of Mathematics, University of Leeds, Leeds, LS2 9JT.*

D.F. GRIFFITHS; *Department of Mathematical Sciences, University of Dundee, Dundee, DD1 4HN.*

W. HERRMANN; *Sandia National Laboratories, Albuquerque, USA.*

H. HOLSTEIN; *Department of Computer Science, University College of Wales, Aberystwyth, Dyfed, SY23 3BZ.*

B.J. HOSKINS; *Department of Meteorology, University of Reading, Earley Gate, Whiteknights, Reading, RG6 2AU.*

N.E. HOSKIN; *AWRE, Aldermaston, Reading, RG7 4PR.*

T.J.R. HUGHES; *Division of Applied Mechanics, Durand Building, Stanford University, Stanford, California 94305, USA.*

I.P. JONES; *Applied Mathematics Group, Computer Science Division, AERE, Harwell, Oxon., OX11 ORA.*

H.O. KREISS; *Department of Applied Mathematics, California Institute of Technology, Pasadena, California 91125, USA.*

D.C. LESLIE; *Department of Nuclear Engineering, Queen Mary College, University of London, Mile End Road, London E1 4NS.*

J.N. LILLINGTON; *UKAEA, AEE Winfrith, Dorchester, Dorset, DT2 8DH.*

J.D. MACDOUGALL; *UKAEA, AEE Winfrith, Dorchester, Dorset, DT2 8DH.*

N.C. MARKATOS; *Concentration Heat & Momentum Ltd. (CHAM), Bakery House, 40 High Street, Wimbledon, London SW19 5AU.*

K.W. MORTON; *Department of Mathematics, University of Reading, Whiteknights, Reading, RG6 2AX.*

J. NITTMANN; *Department of Applied Mathematical Studies, University of Leeds, Leeds, LS2 9JT.*

S. OSHER; *Department of Mathematics, University of California, Los Angeles, California 90024, USA.*

D.J. PADDON; *School of Mathematics, University of Bristol, Bristol, BS8 1TW.*

G.D. PHELPS; *Department of Offshore Engineering, Heriot-Watt University, Riccarton, Currie, Edinburgh EH14 4AS.*

P.L. PRAHM; *Air Pollution Laboratory, National Agency of Environmental Protection, Risø National Laboratory, DK-4000, Roskilde, Denmark.*

J. RAE; *Theoretical Physics Division, AERE, Harwell, Oxon., OX11 ORA.*

P.L. ROE; *Fluid Mechanics Division, RAE, Bedford, MK41 6AE.*

N. RHODES; *Concentration Heat & Momentum Ltd. (CHAM), Bakery House, 40 High Street, Wimbledon, London SW19 5AU.*

D.G. TATCHELL; *Concentration Heat & Momentum Ltd. (CHAM), Bakery House, 40 High Street, Wimbledon, London SW19 5AU.*

T.E. TEZDUYAR; *Division of Applied Mechanics, Durand Building, Stanford University, Stanford, California 94305, USA.*

W.L. WOOD; *Department of Mathematics, University of Reading, Whiteknights, Reading, RG6 2AX.*

D.L. YOUNGS; *AWRE, Aldermaston, Reading, RG7 4PR.*

Z. ZLATEV; *Air Pollution Laboratory, National Agency of Environmental Protection, Risø National Laboratory, DK-4000, Roskilde, Denmark.*

PREFACE

From the earliest developments of electronic computers, von Neumann foresaw them as having a major impact on the understanding and prediction of complicated and economically important fluid flows. He was greatly over-optimistic in predicting how soon this would happen and in the early stages progress in changing engineering design practice and decision processes was slow. Indeed, even in the 1970's major computer manufacturers were seriously underestimating future needs for large "number-crunchers", as the computers needed for major practical fluid flow computations were somewhat derisorily called. Now, however, there can be no doubting the importance of fluid flow computations in many fields and computer manufacturers are responding to the requirements.

In aircraft and aero-engine design, for instance, several authoritative accounts have recently paid tribute to the increasingly important role of large scale flow computations - see e.g. D.R. Chapman "Trends and pacing items in computational aerodynamics" at the 7th International Conference on Numerical Methods in Fluid Dynamics, Stanford, 1980, and W.F. Ballhaus "Computational aerodynamics and design" at the corresponding 8th Conference, Aachen, 1982. Aerospace is perhaps now the most active and economically successful field of application but in many others such computations are either already established practice or are becoming increasingly common: weather forecasting, flood and tidal prediction, hydraulic engineering, heat transfer, reactor safety, oil reservoir modelling are just a few examples.

In such fields it is not just the rapidly increasing power and flexibility of computers that has led to the greater use of flow computations. It is in roughly equal measure the improvements in numerical algorithms that is responsible. Thus the conference held at the University of Reading in March, 1982, of which this book is the proceedings, was organised to present and discuss the numerical methods which are used in fluid dynamics and, in particular, to bring together numerical analysts, with primary interests in the development and analysis of methods, and fluid dynamicists, principally concerned with applying such methods in various applications areas.

The whole development of numerical methods for approximating differential equations has always been strongly influenced and motivated by their application to the equations of fluid dynamics. Not only the work of von Neumann but that of Courant, Richardson, Southwell and many more recent contributors bears ample testimony to this fact. Increasingly, however, there is a trend towards the development of methods specifically for fluid dynamics: hence the small but significant change in the title of this book from the less specific title of the conference.

The choice of invited speakers at the conference was guided by the objective of obtaining a fair balance in three different directions: between the numerical techniques of finite differences, finite elements, spectral methods and other more specialised methods; between such differing fluid flow models and equations as potential flow, Euler equations, shallow water equations, Navier-Stokes equations, turbulence models etc; and between differing applications areas, as already referred to. These invited papers are collected at the beginning of the book in the order in which they were presented so that the structure of the conference can be largely deduced from them. Similarly the contributed papers are arranged in the order presented. One point regarding the balance between applications areas should be noted: the conference held a year earlier at the University of Reading on "Numerical Methods in Aeronautical Fluid Dynamics", the proceedings of which have been edited by P.L. Roe and form a companion to the present volume, has amply covered developments in this field; we have therefore been able to pay more attention to other fields and emphasise the common problems and widespread relevance of recent advances in techniques and understanding.

The readers of this volume will rapidly recognise that a proper understanding of modern computational fluid dynamics requires a sound background in numerical analysis, fluid dynamics and some aspects of computer science. This is unlikely to be acquired on most undergraduate courses in universities or polytechnics and only with great difficulty through subsequent industrial experience. Thus we anticipate that post-graduate courses of various lengths will be increasingly used for this purpose and we hope that this volume will be of help in the planning and teaching of these courses.

The rapid publication of this volume, in the same year as the conference, has been possible only with the ready co-operation of all the authors who observed the several deadlines set them, the industry and care of the staff at the IMA who retyped all the papers and the enthusiastic help of the publishers. Our thanks are due to all of them.

October, 1982

K.W. Morton
M.J. Baines

ACKNOWLEDGEMENTS

The Institute thanks the authors of the papers, the editors Professor K.W. Morton (University of Reading) and Dr. M.J. Baines (University of Reading), and also Mrs S. Hockett, Mrs. J. Parsons and Miss D. Wright for typing the manuscript.

CONTENTS

GENERALISED GALERKIN METHODS FOR STEADY AND UNSTEADY PROBLEMS

K.W. Morton

(Department of Mathematics, University of Reading)

1. INTRODUCTION

It is almost certainly true that the majority of practical fluid flow calculations are presently carried out using finite difference methods. In meteorology, aerodynamics, hydraulics, heat transfer and many other fields there is a large investment of experience, effort and expense in their use and they usually perform well enough. Finite element methods have as yet made a practical impact only in relatively few instances, see for example (Hirsch and Warzée, 1976), (Kawahara, 1978) and (Jameson, 1982). On the other hand, there is a very large literature covering their theory and their development for model problems. They have inherent advantages for equilibrium problems governed by quadratic extremal principles, that is, where the equations of motion are linear, elliptic and self-adjoint. But in fluid flow problems it is generally true that at least one of these properties is far from being satisfied.

In this paper we shall consider two particular developments of finite element methods to enable them to deal successfully with the wider classes of problems occurring in fluid flow. One is directed towards the solution of steady, linear diffusion-convection problems, which epitomise the effects of losing self-adjointness and whose successful solution is a necessary preliminary to tackling the Navier-Stokes equations at moderate to high Reynolds numbers. The other is concerned with evolutionary problems governed by hyperbolic equations. Both involve generalisations to the Galerkin formulation, from which the finite element method has drawn many of its advantageous properties.

Consider the following extremal problem for functions v defined in a region Ω;

$$\text{minimise} \quad \{\|Tv\|^2 - 2\langle f,v\rangle\}, \qquad (1.1a)$$

the solution u satisfying the equation

$$T^*Tu = f, \tag{1.1b}$$

where T is a linear differential operator of order m, T* is its adjoint, f is a given function and $\langle\cdot,\ \cdot\rangle$, $\|\cdot\|$ denote, respectively, the L^2 inner product and norm over Ω: in the minimisation v is to lie in $H^m(\Omega)$, the space of functions with square integrable mth derivatives. Leaving aside the important but rather technical issues of how the boundary and boundary conditions are approximated - for which we refer the reader to standard texts such as (Strang and Fix, 1973) - suppose u is approximated from the conforming finite element space $S^h \subset H^m(\Omega)$, spanned by basis functions $\phi_j(x)$, that is,

$$S^h = \{V \in H^m(\Omega) \,|\, V(x) = \Sigma_{(j)}\ V_j\phi_j(x)\}. \tag{1.2}$$

Then carrying out the minimisation in (1.1a) over S^h gives the approximation U which satisfies the Galerkin equations

$$\langle TU,\ T\phi_i\rangle = \langle f,\phi_i\rangle \qquad \forall\ \phi_i \in S^h. \tag{1.3}$$

Since u also satisfies these equations we have

$$\langle T(u-U),T\phi_i\rangle = 0, \tag{1.4}$$

and from this it follows that

$$\|\ T(u-U)\ \| = \min_{V\in S^h} \|\ T(u-V)\ \|. \tag{1.5}$$

This equation expresses the crucial optimal approximation property of the Galerkin method: U is the best approximation to u from the trial space S^h in the energy norm determined by T. Both the theoretical error analysis and the practically important superconvergence properties follow from this equation.

Similarly, for the hyperbolic evolutionary problem

$$\frac{\partial u}{\partial t} + Lu = 0, \tag{1.6}$$

where L is a first order spatial operator, suppose u at each time t is approximated from S^h. Then the ordinary differential equations for the nodal parameters $U_j(t)$ may be determined by Galerkin equations

$$<\frac{\partial U}{\partial t} + LU, \phi_i> = 0 \qquad \forall \phi_i \in S^h. \tag{1.7}$$

In many cases these again have important superconvergence properties: for example, with piecewise linear elements on a uniform mesh one obtains fourth order accuracy. Moreover, by multiplying each equation by U_i and summing, one obtains

$$\frac{d}{dt} \tfrac{1}{2}\|U\|^2 + <LU,U> = 0. \tag{1.8}$$

Thus, just as for the exact solution u, the L^2 energy of the approximation is conserved or dissipated according to whether L is conservative or dissipative, i.e. $<Lv,v> = \text{or} \geq 0$. By similar arguments one can show that the method has valuable non-linear stability properties.

In the next section we consider diffusion-convection problems and the Petrov-Galerkin methods which have been developed over recent years for their solution. We shall show that the widely used upwind schemes, including the streamline diffusion method, can be placed in a unified framework which provides a useful basis of comparison and sharp error bounds. This framework is based on symmetrizing the bilinear form associated with each problem, which can be done in two natural and distinct ways and yields approximations which are therefore optimal in alternative norms. Upwind schemes can generally be regarded as approximations to that symmetrization which leads to optimality in the Dirichlet norm, i.e. that arising from Poisson's equation: the alternative corresponds to that used by Barrett and Morton (1980) which leads to optimal approximations in a near least squares sense.

Petrov-Galerkin methods have also been developed by several authors for hyperbolic equations. However, in section 3 we present a generalisation of the Galerkin method which is based more directly on the use of characteristics. With piecewise linear basis functions it is shown to be extremely accurate for smooth advection problems and to lead to several practically useful approximate schemes both in one and two space dimensions: these require little more computation than Galerkin methods and give large gains in stability and accuracy. With piecewise constant basis functions, the approach leads to methods for shock problems which are closely related to the difference methods of Engquist and Osher (1981).

2. DIFFUSION-CONVECTION PROBLEMS

Consider the following problem in two or three dimensions:

$$-\underline{\nabla}\cdot(a\underline{\nabla}u - \underline{b}u) = f \text{ in } \Omega \tag{2.1a}$$

$$u = g \text{ on } \Gamma_D, \quad \partial u/\partial n = 0 \text{ on } \Gamma_N. \tag{2.1b}$$

Here a is a positive diffusion coefficient, $\underline{b}$ a convective velocity and f a given source. We shall assume that the convective medium is incompressible, $\underline{\nabla}.\underline{b} = 0$, and that, of the boundary $\Gamma_D \cup \Gamma_N$ of Ω, the Dirichlet part Γ_D includes all the inflow boundary; that is, if $\underline{n}$ is the outward normal then $\underline{b}\cdot\underline{n} \geq 0$ on Γ_N. A typical configuration is indicated in Fig. 1 and the non-dimensional Peclet number bL/a, where L is a characteristic length, may have values ranging from 10^2 for pollutant dispersal in a river to 10^4 and higher in heat transfer problems.

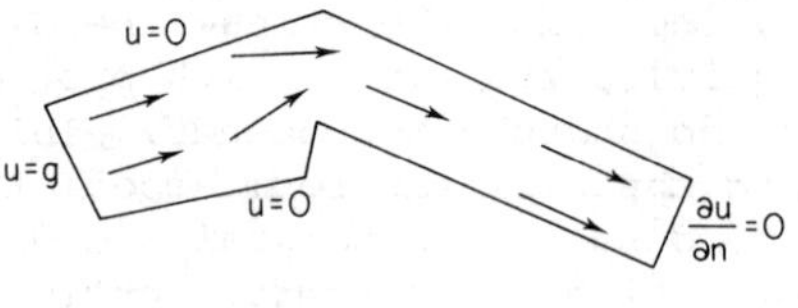

Fig. 1 A typical diffusion-convection problem in 2D.

As is well-known, the Galerkin method in these situations often leads to wildly oscillatory solutions: in effect, central difference approximations to the convective term are generated and the discrete equations become almost singular. The remedy with difference methods has long been to use upwind differencing for this term and, in order to avoid the excessive damping associated with a wholly upwind scheme, to adopt a mixed strategy such as that advocated by Allen and Southwell (1955): in one dimension, for $-au'' + bu' = 0$ on a uniform mesh, this gives

$$-a\delta^2 U_j + bh[(1-\xi)\Delta_0 + \xi\Delta_-]U_j = 0 \tag{2.2a}$$

$$\text{i.e. } -(a + \tfrac{1}{2}\xi bh)\delta^2 U_j + bh\Delta_0 U_j = 0, \tag{2.2b}$$

where h is the mesh spacing, $\Delta_0 U_j := \frac{1}{2}(U_{j+1} - U_{j-1})$ and $\delta^2 U_j := U_{j+1} - 2U_j + U_{j-1}$; with the exponentially fitted choice of mixing parameter,

$$\xi = \coth(\tfrac{1}{2}bh/a) - (\tfrac{1}{2}bh/a)^{-1}, \tag{2.3}$$

this scheme gives exact nodal values for this simple model equation. Problems of accuracy still occur in practical situations through excessive crosswind diffusion and with variable flow fields and source terms.

Zienkiewicz (1975) seems to have been the first to recognise that various upwind schemes could be generated with finite elements if the weighting or test functions ϕ_i in (1.3) were modified. There is now a large literature on such Petrov-Galerkin methods and the reader is referred to the review by Heinrich and Zienkiewicz (1979) together with the other articles in the conference proceedings edited by Hughes (1979). Most methods aim to reduce to the Allen and Southwell scheme in the form (2.2a) by an appropriate choice of parameters: Hughes and Brooks (1979, 1981) on the other hand develop their streamline diffusion method from the form (2.2b) by adding an extra tensor diffusivity to the problem before using the Galerkin method, though this can also be regarded as a Petrov-Galerkin method.

A seemingly alternative approach was taken by Barrett and Morton (1980, 1981). Their aim was to choose a test space in such a way that the bilinear form associated with (2.1) was symmetrized and thus the optimal approximation property (1.5) restored to the method. However, as Morton (1981) has pointed out, both classes of method can be viewed from this objective of symmetrization, the difference being in the resulting symmetric form which is aimed at.

2.1 Alternative symmetrizations

Introducing the first order operators

$$T_1 v := a^{\frac{1}{2}}\underline{\nabla} v, \quad T_2 v := a^{\frac{1}{2}}\underline{\nabla} v - (\underline{b}/a^{\frac{1}{2}})v, \tag{2.4}$$

equation (2.1a) may be written as

$$T_1^* T_2 u = f \quad \text{in } \Omega, \tag{2.5}$$

demonstrating the lack of self-adjointness. Similarly, by introducing the bilinear form

$$B(v,w) := \langle a\underline{\nabla}v,\ \underline{\nabla}w\rangle + \langle \underline{\nabla}\cdot(\underline{b}v),\ w\rangle \tag{2.6}$$

we obtain the weak form of (2.1) for $u \in H_E^1$ as

$$B(u,w) = \langle f,w\rangle \qquad \forall w \in H_0^1, \tag{2.7}$$

where

$$H_E^1 := \{v \in H^1(\Omega) \mid v = g \text{ on } \Gamma_D\} \tag{2.8a}$$

and

$$H_0^1 := \{w \in H^1(\Omega) \mid w = 0 \text{ on } \Gamma_D\}; \tag{2.8b}$$

then this may be written as

$$\langle T_2 u,\ T_1 w\rangle + \int_{\Gamma_N} \underline{b}.\underline{n}uw\,dS = \langle f,w\rangle \qquad \forall w \in H_0^1. \tag{2.9}$$

There are two obvious symmetric forms related to this: one is the symmetric part of $B(\cdot,\cdot)$ and can be written in terms of T_1 as

$$B_1(v,w) := \langle T_1 v, T_1 w\rangle + \tfrac{1}{2}\int_{\Gamma_N} \underline{b}\cdot\underline{n}vw\,dS \tag{2.10a}$$

$$= \tfrac{1}{2}[B(v,w) + B(w,v)]; \tag{2.10b}$$

the other is that used by Barrett and Morton (1980) and based on T_2,

$$B_2(v,w) := \langle T_2 v, T_2 w\rangle + \int_{\Gamma_N} \underline{b}\cdot\underline{n}vw\,dS \tag{2.11a}$$

$$= \langle a\underline{\nabla}v,\ \underline{\nabla}w\rangle + \langle (|\underline{b}|^2/a)v,w\rangle. \tag{2.11b}$$

(Barrett and Morton actually introduce a weighting function ρ in their definition of B_S which we have taken as a^{-1}.) The objective of approximately symmetrizing $B(\cdot,\cdot)$ can be sought through either form.

Since $B(v,w)$ is continuous on $H_0^1 \times H_0^1$ and an equivalent norm on H_0^1 may be defined from either $B_1(v,v)$ or $B_2(v,v)$, one may deduce from the Riesz representation theorem that operators R_1 and R_2 exist such that

$$\text{for } m = 1, 2 \qquad B(v,w) = B_m(v, R_m w) \qquad \forall v, w \in H_0^1 . \tag{2.12}$$

Leaving aside for the moment the problem of explicitly representing R_m, or its inverse, consider the Petrov-Galerkin method for U in a trial space S_E^h with basis functions ϕ_i, the subscript denoting that the essential boundary condition $U = g$ on Γ_D is satisfied, and based on test functions ψ_i spanning a test space $T_0^h \subset H_0^1$:

$$B(U,\psi_i) = \langle f,\psi_i \rangle \qquad \forall \psi_i \in T_0^h. \tag{2.13}$$

Since u satisfies the same equations we have the projection property for the error

$$B(u-U,\psi_i) = 0 \qquad \forall \psi_i \in T_0^h. \tag{2.14}$$

Now suppose we were able to choose the test functions ψ_i^* so that, for $m = 1$ or 2,

$$\text{span } \{R_m \psi_i^*\} = \text{span } \{\phi_i \in H_0^1\} = S_0^h := S^h \cap H_0^1. \tag{2.16}$$

Then denoting the corresponding Petrov-Galerkin approximation by U_m^* and noting that $u - U_m^* \in H_0^1$, we have from (2.12), (2.14)

$$B_m(u-U_m^*,\phi_i) = 0 \qquad \forall\phi_i \in S_0^h \tag{2.17}$$

and hence the optimal approximation property holds,

$$\|u-U_m^*\|_{B_m} = \min_{V \in S_E^h} \|u-V\|_{B_m} . \tag{2.18}$$

In fact consider any test space T_0^h which has the same dimension as S_0^h, and for which the positive definiteness of $B(v,v)$ ensures the non-singularity of the stiffness matrix in (2.13): and suppose the closeness with which S_0^h can be approximated by $R_m T_0^h$ is described by the constant Δ such that

$$\min_{W \in T_0^h} \|V-R_m W\|_{B_m} \leq \Delta\|V\|_{B_m} \qquad \forall\, V \in S_0^h. \tag{2.19}$$

Then one can show for the corresponding Petrov-Galerkin approximation U,

$$\|u-U\|_{B_m} \leq (1-\Delta^2)^{\frac{1}{2}} \|u-U_m^*\|_{B_m} . \tag{2.20}$$

In particular, one can deduce that the Galerkin approximation falls short of the optimal approximation by a factor dominated by the mesh Peclet number $|\underline{b}|h/a$.

2.2 *Schemes derivable from $B_1(v,w)$ symmetrization*

Now let us consider how closely we can in practice approximate either of the ideal test spaces given by (2.16). For $m = 1$, the relation (2.12) can be regarded as an equation for w with $R_1 w$ given, which takes the form

$$\langle a\underline{\nabla}(w-R_1 w) + \underline{b}w, \underline{\nabla}v\rangle - \tfrac{1}{2}\int_{\Gamma_N} \underline{b}\cdot\underline{n}(R_1 w)v\,ds = 0 \qquad \forall v \in H_0^1. \tag{2.21}$$

In one dimension, on the unit interval with a and b positive constants and Dirichlet boundary conditions at $x = 1$ as well as

$x = 0$, it becomes

$$aw' + bw = a(R_1 w)' + \text{const.}, \quad w(0) = w(1) = 0, \tag{2.22a}$$

the constant being determined by the two boundary conditions; and with a Neumann boundary condition at $x = 1$ it becomes

$$aw' + bw = a(R_1 w)' + \tfrac{1}{2}b(R_1 w)(1), \quad w(0) = 0. \tag{2.22b}$$

When the trial functions are piecewise linear, one obtains from these equations as the ideal test functions the negative exponential functions used successfully by Hemker (1977): on a uniform mesh, they are proportional to $\psi_j(x) = \psi(\frac{x}{h} - j)$ where

$$\psi(t) = \begin{cases} 1 - e^{-\beta(t+1)} & -1 \le t \le 0 \\ e^{-\beta t} - e^{-\beta} & 0 \le t \le 1, \end{cases} \tag{2.23}$$

and $\beta = bh/a$. These not only give the Allen and Southwell scheme but for $-au'' + bu' = f$ they give <u>exact nodal values for any source function f</u>. This is consistent with (2.18) for $m = 1$ because optimality in this norm for piecewise linear approximations corresponds to linear interpolation between the nodes.

For large values of β the exponentials in (2.23) are awkward to deal with and in any case it is difficult to see how (2.21) could be solved in two dimensions. However, simpler functions can be used which reproduce the Allen and Southwell scheme and approximate (2.23) sufficiently well as regards modelling the effect of the source function. They include the quadratic functions used by Christie et al. (1976) and Heinrich et al. (1977) which, again on a uniform mesh, become

$$\psi(t) = \phi(t) + \alpha\sigma(t) \tag{2.24a}$$

with

$$\sigma(t) = \begin{cases} -3t(1-|t|) & |t| \le 1 \\ 0 & |t| > 1 : \end{cases} \tag{2.24b}$$

choosing the parameter α so that $\alpha > 1 - 2/\beta$ ensures no oscillation in the solution, while taking $\alpha = \xi$, given by (2.3), gives the Allen and Southwell scheme. Moreover it is easy to extend this scheme into two dimensions using bilinear elements on rectangles. The trial basis functions are given by the product $\phi_{ij}(x,y) = \phi_i(x)\phi_j(y)$ and the test functions can be taken as

$$\psi_{ij}(x,y) = [\phi_i(x) + \alpha_1\sigma_i(x)]\ [\phi_j(y) + \alpha_2\sigma_j(y)] \tag{2.25}$$

with α_1, α_2 based on the two components of $\underline{b}$.

As noted above, the methods of Hughes and Brooks (1979, 1981) are prompted by the form (2.2b): the oscillations produced by the central differencing for the convective term $\underline{b}\cdot\underline{\nabla}u$ are damped by introducing extra diffusion, while still retaining the Galerkin test functions. In two dimensions the convective differencing is approximately in the streamline direction and thus they add the extra diffusion only in this direction: that is, the diffusion a in the term $-\underline{\nabla}\cdot(a\underline{\nabla}u)$ is replaced by a tensor diffusivity,

$$-\underline{\nabla}\cdot(\underline{\underline{A}}\nabla u), \text{ where } A_{\ell m} = a\delta_{\ell m} + \tilde{a}\ b_\ell b_m/|\underline{b}|^2. \tag{2.26}$$

On a uniform rectangular mesh with spacings h_1,h_2 the suggested choice of the parameter $\tilde{a}$ is

$$\tilde{a} = \tfrac{1}{2}(\xi_1 b_1 h_1 + \xi_2 b_2 h_2) \tag{2.27}$$

$$\text{with} \quad \xi_m = \coth(\tfrac{1}{2}b_m h_m/a) - (\tfrac{1}{2}b_m h_m/a)^{-1}, \quad m = 1, 2.$$

This scheme seems to work well in many practical situations. To place it in our general framework, we note that in their more recent paper Hughes and Brooks (1981) show that it is, in part at least, a Petrov-Galerkin method since

$$<\underline{\underline{A}}\nabla v, \underline{\nabla}\phi> = <a\underline{\nabla}v, \underline{\nabla}\phi> + <\underline{b}\cdot\underline{\nabla}v, (\tilde{a}/|b|^2)\underline{b}\cdot\underline{\nabla}\phi>. \tag{2.28}$$

That is, it is equivalent to using test functions

$$\psi_{ij} = \phi_{ij} + (\tilde{a}/|\underline{b}|^2)\underline{b}\cdot\underline{\nabla}\phi_{ij}, \tag{2.29}$$

on just the convection term. For linear or bilinear trial functions these are discontinuous test functions and lie outside our theoretical framework: but if that part of $B(U,\psi_{ij})$ arising from the diffusion term and the $\underline{b}\cdot\underline{\nabla}\phi_{ij}$ term in (2.29) is evaluated element by element, it gives no contribution since $\nabla^2 U = 0$ on each element. In this way (2.29) can be regarded as a test function used in the whole of the bilinear form. Although Hughes and Brooks did not originally use the test function (2.29) on the source term, Johnson and Nävert (1981) in their analysis of the streamline diffusion method for the singular case $a = 0$ do in fact modify the source function in a way that is consistent with applying (2.29) and Hughes and Brooks (1981) now apply (2.29) consistently on source and time derivative terms.

Thus to summarise the $m = 1$ case, in one dimension the exact test functions (2.23) of Hemker, those advocated by Heinrich et al. (1977) and typified by (2.24) and those in effect used by Hughes and Brooks (1981) and given by (2.29) all reproduce the Allen and Southwell difference operator but differ in the way in which they sample the source function. These three test functions ψ_i for typical values of the mesh Peclet number β are plotted in Fig. 2a. The last two appear to differ significantly from the first: yet from Fig. 2b, which shows in each case the extent to which $R_1\psi_i$ can reproduce the trial basis function ϕ_i, we see that either of them is very effective. These figures give a rough pictorial representation of the approximation properties of each scheme as defined by (2.19) and (2.20). Actual calculations of the parameter Δ in each case and in the Galerkin case give the results in Table I. Though there is little to choose between the two main practical methods in this simple case, they differ much more markedly, and of course much more importantly, in the way in which they extend to 2D.

Table I

Ratios of Petrov-Galerkin error to optimal error, given by $(1-\Delta^2)^{-\frac{1}{2}}$ from (2.19), (2.20) with $m = 1$.

β	Galerkin	Heinrich et al.	Hughes and Brooks
2	1.1547	1.0060	1.0924
5	1.7559	1.0468	1.1509
50	14.468	1.2022	1.1547
500	144.43	1.2344	1.1547
10^5	28868	1.2383	1.1547

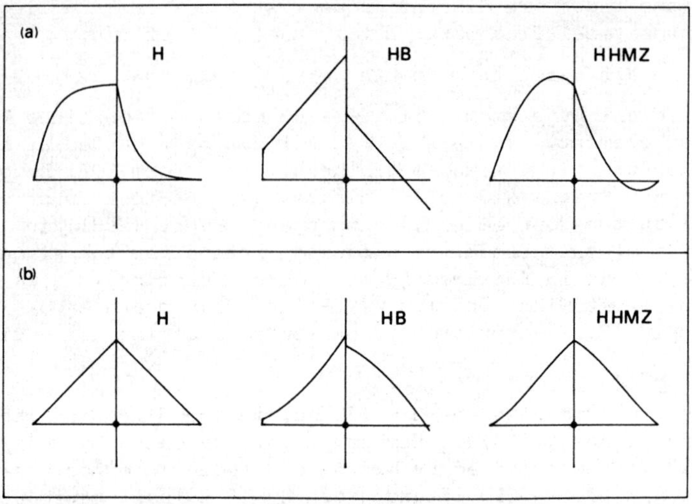

Fig. 2 (a) Test functions ψ_i used by Hemker (H), Hughes and Brooks (HB) and Heinrich et al. (HHMZ), and (b) corresponding approximations to the trial function ϕ_i constructed from $R_1\psi_i$. The mesh Peclet number $\beta = 5$.

2.3 Symmetrization using $B_2(v,w)$

Turning now to the case m = 2 adopted by Barrett and Morton (1980, 1981), the equation corresponding to (2.12) and (2.21) becomes

$$<\underline{\nabla}w - [\underline{\nabla}(R_2w) - a^{-1}\underline{b}(R_2w)], a\underline{\nabla}v - \underline{b}v> + \int_{\Gamma_N} \underline{b}\cdot\underline{n}(w - R_2w)\, vdS = 0 \tag{2.31}$$

$$\forall v \in H_0^1.$$

In one dimension, on the unit interval with a and b positive constants and Dirichlet boundary conditions, this becomes

$$w' = (R_2w)' - a^{-1}b(R_2w) + \text{const.}e^{-bx/a}, \quad w(0) = w(1) = 0, \tag{2.32}$$

the constant again being determined by the boundary conditions. Clearly the negative exponential plays a less important role in this case and it is a straightforward matter to compute the ideal test functions defined by (2.16) for any choice of trial functions. It is, however, unnecessary to do this computation and it is preferable to move directly to the symmetrized form of the equations. Assuming for simplicity that the Dirichlet boundary conditions are homogeneous, the Petrov-Galerkin equations (2.13) with the test functions (2.16) can be written, using (2.12) as

$$B_2(U_2^*,\phi_i) = \langle f,\psi_i^*\rangle \qquad \forall\ \phi_i \in S_0^h. \tag{2.33}$$

Now suppose that instead of solving (2.16) we solve

$$R_2^* \tilde{f} = f \tag{2.34}$$

where R_2^* is the adjoint operator to R_2. Then (2.33) becomes

$$B_2(U_2^*,\phi_i) = \langle \tilde{f},\phi_i\rangle \qquad \forall\ \phi_i \in S_0^h, \tag{2.35}$$

<u>a symmetric system of Galerkin equations</u> using a transformed source function. For the simple one-dimensional problem we obtain from the equation corresponding to (2.32)

$$\tilde{f}(x) = f(x) + a^{-1}b\,[F(x) - \bar{F}] \tag{2.36}$$

where

$$F(x) = \int_0^x f(y)\,dy, \quad \bar{F} = \int_0^1 e^{-bx/a}F(x)\,dx \Big/ \int_0^1 e^{-bx/a}\,dx.$$

This is easily generalised to variable a, b and to general boundary conditions. The main two points to notice are that, first, the discrete operator on the left of (2.35) which one obtains with linear basis functions is no longer the Allen and Southwell operator but, instead,

$$-a\delta^2 U + bh\beta(1 + \tfrac{1}{6}\delta^2)U, \tag{2.37}$$

which corresponds to the self-adjoint differential form $-au'' + a^{-1}b^2u$; secondly, approximations to this ideal scheme are obtained, just as in the m = 1 case, by approximating the source function term - for instance, by omitting $\bar{F}$ in (2.36).

The most important difference, however, of this m = 2 case from the m = 1 case is that U_2^* is a best fit in the norm given by (2.11b), which for large Peclet numbers becomes the L^2 norm, while U_1^* was exact at the nodes. This means that in a sharp boundary layer U_2^* exhibits oscillations, but of a controlled kind: indeed the extent of the overshoot is a valuable measure of the thickness of the boundary layer. The general problem of recovering information about u given its L^2 best fit has now been studied by many authors - see, for instance, (Barrett, Moore and Morton, 1982) and the references therein. The formulae correspond in some sense to the interpolation formulae that are needed to recover u from its nodal values: one of the best known gives accurate nodal values from the nodal parameters of a best linear fit on a uniform mesh,

$$\left|u(jh) - \frac{1}{12}(U_{j-1} + 10U_j + U_{j+1})\right| \leq \frac{1}{360} h^4 \|u^{(iv)}\|_\infty . \qquad (2.38)$$

We shall see that L^2 best fits play an important rôle in the next section too. Thus although it is inappropriate to dwell at length on the recovery problem here, it is important to note that there are few disadvantages and often some advantage (as in the boundary layer case) in a method yielding an L^2 best fit rather than nodal values.

To continue the discussion of the m = 2 case, there are several ways in which two dimensional problems may be treated. In (Morton and Barrett, 1980) tensor products of the one-dimensional ideal test functions were used, as with the method of Heinrich and Zienkiewicz (1979). These are both somewhat awkward to use and less successful with strongly curved streamlines than the mixed method used in (Barrett and Morton, 1981) and forming the natural extension of the form (2.35). Without attempting to solve the equation for f, we introduce the flux function

$$\underline{v} = \underline{b}u - a\nabla \underline{u} \qquad (2.39)$$

and can then write the equation (2.7) for u as

$$B_2(u,w) = \langle f,w\rangle + \langle a^{-1}\underline{b}\cdot\underline{v},w\rangle \qquad \forall\, w \in H_0^1 . \qquad (2.40)$$

This is approximated by

$$B_2(U,\phi_i) = \langle f,\phi_i\rangle + \langle a^{-1}\underline{b}\cdot\underline{V},\phi_i\rangle \qquad \forall\ \phi_i \in S_0^h \tag{2.41}$$

with $\underline{V}$ obtained by approximating the equation $\underline{\nabla}\cdot\underline{v} = f$. Then one finds that, if S^* is the best L^2 fit to $a^{-1}\underline{b}\cdot\underline{v}$ from S_0^h, one has

$$\| U-U_2^* \|_{B_2} \le \| \, |\underline{b}|^{-1}(a^{-1}\underline{b}\cdot\underline{V} - S^*) \| . \tag{2.42}$$

Although it is far from clear that the best procedures for approximating $\underline{V}$, or rather $\underline{b}\cdot\underline{V}$, have yet been found the results obtained so far are encouraging.

2.4 A test problem

We end this section with a few numerical results for a test problem which is a modification of one put forward by Hutton (1981). The flow field $\underline{b}$ is indicated in Fig. 3 and is derived from a stream function $(1-x^2)(1-y^2)$. In Hutton's test problem a tanh input profile for u was specified on $y = 0$, $-1 \le x \le 0$ with Dirichlet conditions on the tangential boundary consistent with pure convection: the main test was for the output profile for various values of the Peclet number. We have tested the Heinrich et al. scheme using (2.25), the Hughes and Brooks scheme using (2.29) and the Barrett and Morton (1981) scheme and all performed reasonably well on this problem with the Hughes and Brooks scheme giving the best results, presumably because of its small crosswind diffusion: a two dimensional version of the Allen and Southwell scheme by contrast gave very poor results. Our modification to the problem is to specify $u = 0$ on the input and all the tangential boundaries except $x = 1$, where we put $u = 100$: this models a situation where a cold fluid is channelled past a hot plate.

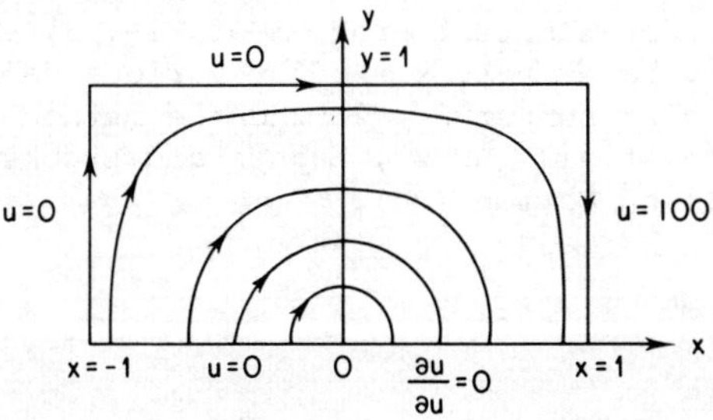

Fig. 3 A test problem modified from Hutton (1981).

The interesting profiles are those for fixed values of y. In Fig. 4 we show the profiles for each scheme at y = 0.9, y = 0.5 and y = 0 when the mesh Peclet number $\beta = 20$; Fig. 5 shows corresponding results for $\beta = 100$. Despite their objective of non-oscillatory solutions both the Heinrich et al. scheme and that of Hughes and Brooks show considerable oscillation, particularly the latter. The Barrett and Morton scheme, on the other hand, is aimed at a best fit in the norm defined by (2.11b) and would be expected to be oscillatory for these values of β. Indeed the variation of the boundary layer thickness with y, which is most obvious in Fig. 4, can be calculated from these results and each thickness is within a few per cent of that calculated from an asymptotic analysis.

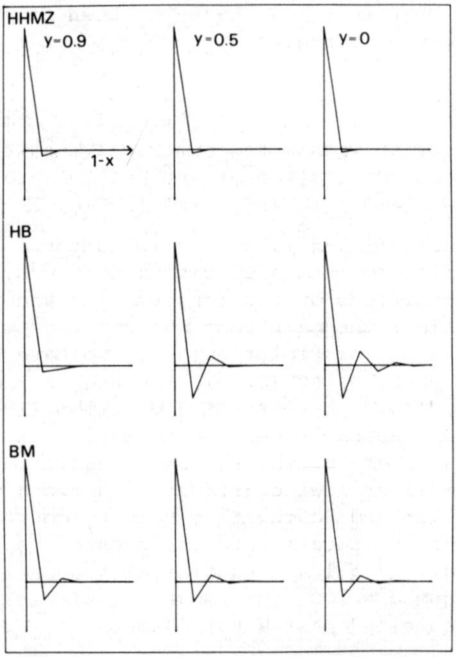

Fig. 4 Results for the test problem of Fig. 3, showing the boundary layer near x = 1 for various values of y. The mesh Peclet number $\beta = 20$ and the methods used are Heinrich et al. (HHMZ), Hughes and Brooks (HB) and Barrett and Morton (BM).

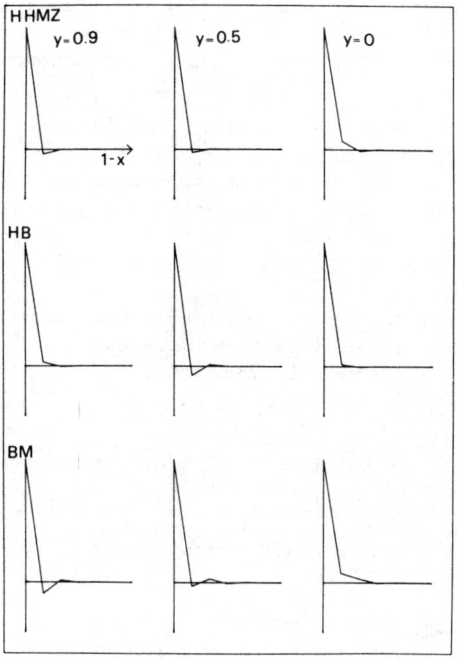

Fig. 5 Similar to Fig. 4 for $\beta = 100$.

More analysis is clearly required to explain fully the behaviour in this two-dimensional example of the two schemes which were motivated by the Allen and Southwell difference method and which we have associated with the B_1 symmetrization. However, when the results are combined with those for the Barrett and Morton scheme and viewed in the context of the general analysis given above, they add to the growing evidence that generalised Galerkin methods can successfully handle a wide class of diffusion-convection problems: the important point is that their output must be correctly analysed and not viewed as if it came from just another, rather complicated, difference scheme.

3. HYPERBOLIC EQUATIONS

Any method for approximating hyperbolic equations sacrifices a good deal if it takes no account of the presence of characteristics. The semi-discrete Galerkin equations (1.7) yield such methods: thus as soon as a standard time discretisation is introduced, disadvantages to the Galerkin formulation appear.

First of all, a reduced stability range for explicit schemes is generally obtained. For example that for the leapfrog method is reduced by a factor $\sqrt{3}$: while Euler's method applied to $u_t + au_x = 0$ gives the central difference scheme which is well-known to be stable only for $\Delta t = O(h^2)$. Indeed the phenomenon in this latter case is very similar to that in the previous section and some form of upwinding is strongly indicated. One of the consequences of this loss of stability is that the schemes cannot be used when the CFL number is unity, while most common difference schemes are exact in this limit, a fact which improves their accuracy over the whole stability range.

Many authors have sought means to remedy these defects and several of them are based on a Petrov-Galerkin approach. Thus suppose a one-step method in time is used and $u(x,t)$ is approximated at $n\Delta t$ by

$$U^n(x) = \Sigma_{(j)} U_j^n \phi_j(x). \tag{3.1}$$

Then test functions $\psi_i(x)$ are sought for the equations

$$< \frac{U^{n+1} - U^n}{\Delta t} + L(\theta U^{n+1} + \overline{1-\theta} U^n), \psi_i > = 0. \tag{3.2}$$

Morton and Parrott (1980) introduced special test functions $\chi_i(x)$, corresponding to the use of linear trial functions ϕ_j for the model scalar problem $u_t + au_x = 0$, which have the property of giving exact results when the CFL number $\mu = a\Delta t/h$ is unity (the unit CFL property): because the Galerkin method is highly accurate for small μ they therefore used in the general case

$$\psi_i = (1 - \nu)\phi_i + \nu\chi_i \tag{3.3}$$

with $\nu = \mu$ or $\nu = \mu^2$, determined by a Fourier analysis. Highly accurate schemes which are closely related to well-known finite difference schemes result for either Euler's method, $\theta = 0$, or Crank-Nicolson, $\theta = \frac{1}{2}$. A similar scheme was given for leap-frog time differencing and both this and the Crank-Nicolson scheme retain conservation of U though not of U^2: the leap-frog scheme is 4th order accurate in both Δt and h. Convenient generalisations were given for hyperbolic systems and the use of bilinear elements allowed Morton and Stokes (1981) to extend the methods to two dimensions. However, limitations were found with the

Petrov-Galerkin formulation when triangular elements were used and an approach even more closely based on the characteristics was introduced.

3.1 Euler characteristic Galerkin method (ECG Method) in one dimension

Consider the scalar conservation law in one dimension

$$\partial_t u + \partial_x f(u) = 0, \tag{3.4a}$$

or

$$\partial_t u + a(u)\partial_x u = 0, \tag{3.4b}$$

where $a(u) = \partial f/\partial u$. Then u is constant along the characteristics $dx/dt = a$ so that, if we write $u^n(x)$ for $u(x,n\Delta t)$ and similarly for a and f, we have for smooth flows

$$u^{n+1}(y) = u^n(x) \text{ where } y = x + a^n(x)\Delta t. \tag{3.5}$$

Thus the L^2 projection of u^{n+1} onto the trial space S^h spanned by $\{\phi_j\}$ is related to that of u^n by

$$\begin{aligned} \langle u^{n+1} - u^n, \phi_j\rangle &= \int_{-\infty}^{\infty} u^{n+1}(y)\phi_j(y)dy - \int_{-\infty}^{\infty} u^n(x)\phi_j(x)dx \\ &= \int_{-\infty}^{\infty} u^n(x)\ [\phi_j(y)\ \frac{dy}{dx} - \phi_j(x)]dx \\ &= \int_{-\infty}^{\infty} u^n(x)\ [\frac{d}{dx}\int_x^y \phi_j(z)dz]dx \\ &= -\int_{-\infty}^{\infty} \partial_x u^n(x)\ [\int_x^y \phi_j(z)dz]dx. \end{aligned} \tag{3.6}$$

On a uniform mesh with $\phi_j(x) = \phi(\frac{x}{h} - j)$, we introduce the upwind-averaged test function

$$\Phi(s,\mu) = \frac{1}{\mu}\int_s^{s+\mu} \phi(\sigma)\,d\sigma \qquad (3.7)$$

and set

$$\Phi_j^n(x) = \Phi\left(\frac{x}{h} - j,\ \frac{a^n(x)\Delta t}{h}\right)$$

$$= \frac{1}{a^n(x)\Delta t}\int_x^{x+a^n(x)\Delta t} \phi_j(z)\,dz\,. \qquad (3.8)$$

This test function $\Phi(s,\mu)$ is plotted for various values of μ in Fig. 6, where it is seen to have its maximum at $-\frac{1}{2}\mu$: this corresponds to Φ_j^n peaking midway between jh and the foot of the characteristic drawn back from $(jh, \overline{n+1}\Delta t)$ to time level n. From (3.6) and (3.8) we then get

$$<u^{n+1} - u^n, \phi_j> + \Delta t <a^n \partial_x u^n,\ \Phi_j^n> = 0. \qquad (3.9)$$

This exact relationship does not of course allow complete tracking of the evolution of u^n, as does (3.5), since only the projection quantities $<u^n,\phi_j>$ are obtained at each level and these are insufficient to calculate the second term of (3.9). But several approximation schemes can be based on this relation.

We refer to the following as the (exact) Euler Characteristic Galerkin (ECG) method.

$$<U^{n+1} - U^n, \phi_j> + \Delta t\ <a(U^n)\partial_x U^n,\ \Phi_j^n> = 0, \qquad (3.10)$$

where U^n is given by (3.1) and is assumed continuous and Φ_j^n is given by (3.8) with $a^n(x)$ taken as $a(U^n)$. We leave the second term in the form $a(U)\partial_x U$ because $a(U)$ will cancel: but the evaluation of this inner product still involves considerable computation and various approximate schemes will be considered below. The merit of (3.10), however, is that the only error involved is that due to the projection at each time step: if the initial data are projected into S^h then this is carried forward

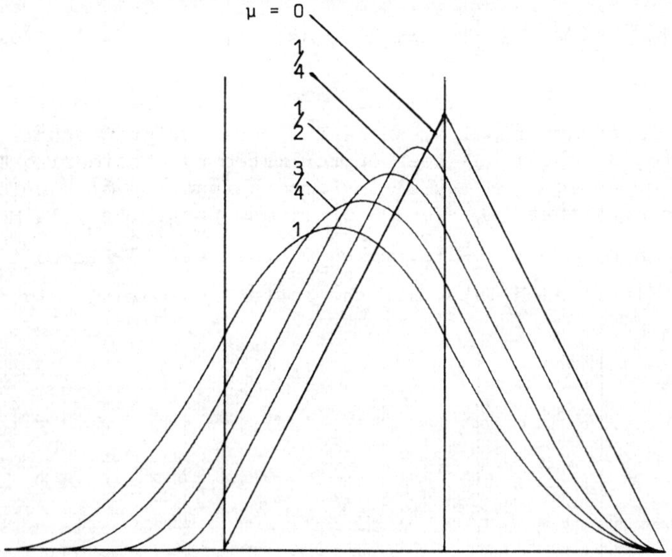

Fig. 6 Upwind-average test function $\Phi(s,\mu)$ for = 0, ¼, ½, ¾, 1.

exactly through the first time step before being projected again, and so on. Thus, if the objective is to carry forward the L^2 projection of $u(t)$ onto S^h, this is the very best that can be achieved by a one-step algorithm using the time step Δt, unless more information is deduced regarding the form of u and used in a recovery procedure.

3.2 Approximate ECG schemes

We confine ourselves here to piecewise linear basis functions ϕ_j and to schemes which are exact when $a(U)$ is constant and the CFL number $\mu = a\Delta t/h$ lies in $(0,1)$. In this case, (3.10) involves only three neighbouring nodes and their coefficients can be correctly reproduced by either of the following replacements for Φ_j:

$$\Phi_j \approx \Phi_j^T := (1-\tfrac{1}{2}\mu)\phi_j + \tfrac{1}{2}\mu\phi_{j-1} + \frac{1}{12}\mu(3-2\mu)(\phi_j'-\phi_{j-1}') + M[(\phi_j-\phi_{j-1}) + \tfrac{1}{2}(\phi_j'-\phi_{j-1}')]; \qquad (3.11)$$

$$\Phi_j \approx \Phi_j^S(x) := \frac{1}{6}[\phi_j(x) + 4\phi_j(x + \tfrac{1}{2}\mu h) + \phi_j(x + \mu h)]. \quad (3.12)$$

In the first alternative, where the superscript T denotes that the integral (3.7) has been approximated by considering Taylor series expansions, M may be any smooth function of μ which tends to zero at either limit $\mu \to 0$ or 1: one possibility is $M = 0$, and another, $M = \frac{1}{2}\mu(1-\mu)^2$, makes Φ_j^T the best L^2 fit to Φ_j by a linear fit in each interval; the latter is shown in Fig. 7.

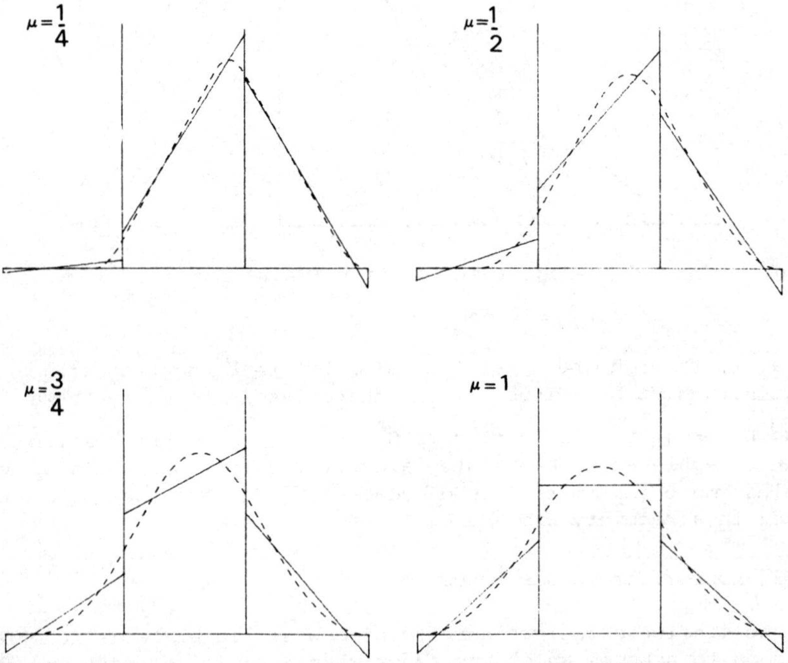

Fig. 7 The approximate test function $\Phi^T(s,\mu)$, with $M = \frac{1}{2}\mu(1-\mu)^2$, compared with the exact $\Phi(s,\mu)$.

For Φ^T only one extra set of inner products needs to be evaluated as compared with the Galerkin method, $\langle\phi_i',\phi_j'\rangle$ as well as $\langle\phi_i',\phi_j\rangle$. In the second alternative, given by (3.12) where the superscript S stands for "shifted", or for Schoombie (1982) who introduced such test functions in a Petrov-Galerkin setting, two extra sets of inner products need to be evaluated: moreover, these also depend on the value of μ so that in more general cases they cannot be evaluated once and for all.

There is a third possibility besides (3.11) and (3.12), which approximates directly the inner product $\langle\phi_i',\Phi_j\rangle$ needed in the linear case of (3.10):

$$\langle\phi_i',\Phi_j\rangle \approx (1-\mu)^2 \langle\phi_i',\phi_j\rangle - \mu(1-\mu)\langle\phi_i',\phi_{j-1}\rangle \tag{3.13a}$$

$$+ \mu(3-2\mu)\langle\phi_i,\phi_j - \phi_{j-1}\rangle.$$

Here there are no extra inner products needed at all. Moreover, integrating by parts we see that this scheme is equivalent to using a test function

$$\Phi_j \approx \Phi_j^I(x) := (1-\mu)^2\phi_j - \mu(1-\mu)\phi_{j-1} + \mu(3-2\mu)\int_{-\infty}^{x}(\phi_{j-1}-\phi_j)dy, \tag{3.13b}$$

which is shown in Fig. 8 where we see that it is an exact match at $\mu = 1$ and extremely accurate at $\mu = \frac{3}{4}$.

There is clearly no stability limit on (3.10) while there will be such limits on the approximate schemes given above. For (3.11) this is independent of M and is actually $-\frac{1}{2} \le \mu \le \frac{3}{2}$ although for reasons of accuracy one would wish to keep μ in $(0,1)$.

The application of either (3.11), (3.12) or (3.13) to a general scalar conservation law is straightforward: one merely replaces μ in the formulae by the local CFL number $a(U_j^n)\Delta t/h$. However, because $a(U^n)$ does not cancel as it does in (3.10), either some of the speed advantage has been lost through having to integrate inner products which contain $a(U^n)$ explicitly, or some accuracy is lost by replacing $a(U^n)$ by the constant $a(U_j^n)$ and absorbing it into the coefficients in (3.11), (3.12) and (3.13). An alternative is to use the product approximation strongly advocated by Christie et al. (1981): that is, to use the approximation

$$\partial_x f(u) \approx \Sigma_{(j)} f(U_j^n)\phi_j'. \tag{3.14}$$

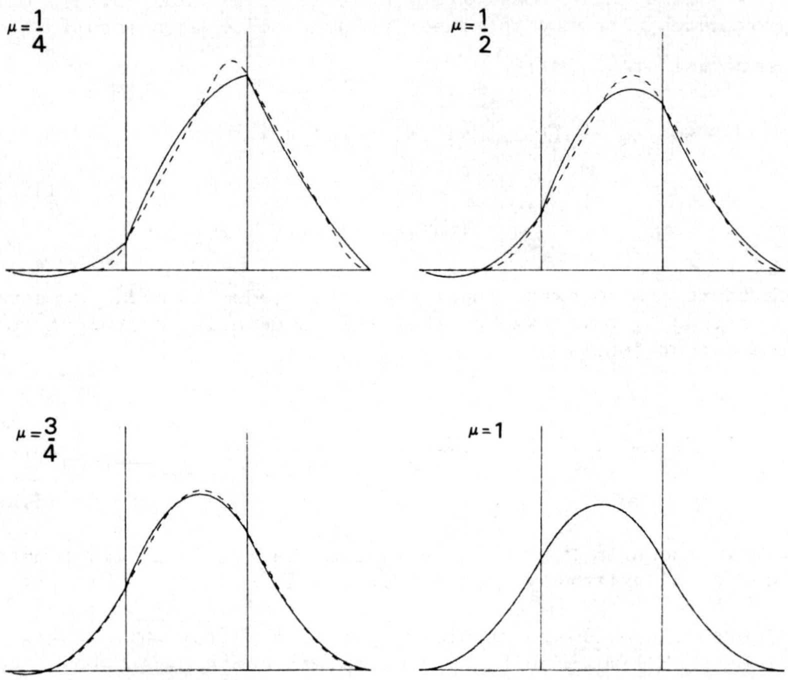

Fig. 8 The approximate test function $\Phi^I(s,\mu)$ compared with the exact $\Phi(s,\mu)$.

Such an approximation is of particular advantage when coupled to (3.13a), the amount of computation then differing little from that for the Galerkin method applied to the simple advection equation $u_t + au_x = 0$.

3.3 Test results in one dimension

We will show a sample of results obtained with these schemes: a more complete account will be found in (Morton and Stokes, 1982). The first tests are for pure advection. Fig. 9 shows the results for a Gaussian profile and for a ramp function as compared with Gadd's modification of the Lax-Wendroff method (Gadd, 1978). A Fourier analysis shows that the error generated in the L^2 projection at each time step is given by

$$\text{error} \sim \frac{1}{24}\mu^2(1-\mu^2)\xi^4 \tag{3.15}$$

where $\xi = kh$ and k is the wave number.

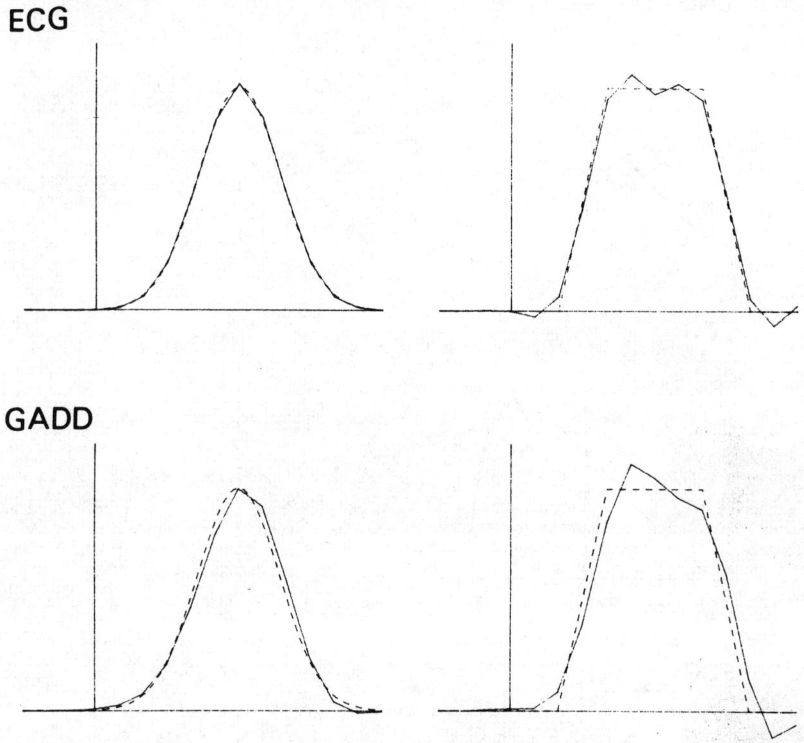

Fig. 9 Advection of a Gaussian profile and a ramp function by the ECG scheme and Gadd's scheme, for $\mu = 0.8$.

Fig. 10 gives results obtained for the non-linear advection equation $u_t + uu_x = 0$ with an initial isolated cosine wave. Each picture shows the leading edge of the wave at $t = \frac{1}{2}$ where $t = 2/\pi$ is the time to first breaking: on the left are results for Crank-Nicolson-Galerkin (the most reliable second order accurate time-stepping for Galerkin), the exact ECG scheme (3.10) and the approximate ECG scheme (3.11); on the right the same schemes are used but coupled with the product approximation (3.14).

Taken together these two sets of results demonstrate impressive accuracy for the ECG methods and this is confirmed by results obtained by other authors experimenting with similar schemes - Bercovier and Pironneau (1981) and Benqué and Ronat (1982). Note that all the results presented for the finite

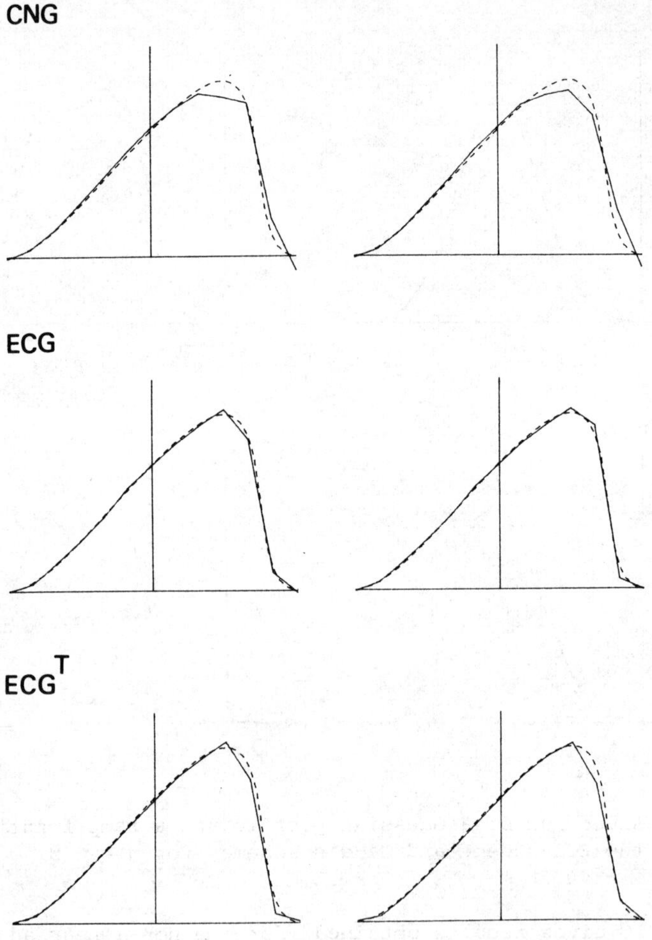

Fig. 10 Non-linear advection by Crank-Nicolson-Galerkin (CNG), exact ECG and approximate ECG (ECG^T) schemes: product approximation is used in the right-hand set.

element schemes should be interpreted as approximations to best L^2 fits. The particular interpretation that is used is unimportant for linear constant coefficient problems but is very important for non-linear problems - see Cullen and Morton (1980). Finally we should point out that similar test functions can be developed for other basis functions and for other time-differencing schemes and these will be found in Morton and Stokes (1982). In particular, though one has to generalise the development of the algorithm to deal with discontinuous basis functions, it can then lead to methods related to the upwind schemes presently used in shock modelling: thus piecewise constant elements plus an exact Riemann solver to replace (3.5) gives the method of Godunov (1959) while simpler extensions to (3.5) can lead to the basic method of Engquist and Osher (1981) - see (Morton, 1982).

3.4 ECG schemes in two dimensions

For the linear advection equation

$$\partial_t u + \underline{a}\cdot\underline{\nabla} u = 0 \tag{3.16}$$

the exact Euler Characteristic Galerkin Method uses an upwind-averaged test function completely analogous to (3.7) and (3.8). However, for piecewise linear elements over triangles, the computation of the test function, or of its inner products with $\underline{a}\cdot\underline{\nabla}U$, is clearly a considerable task. Fortunately either (3.11) or (3.13) extends in a natural and economic manner to give very accurate results. The vector $-\underline{a}\Delta t$ extends from the node j into a triangle for which this is one vertex and defines the foot of the characteristic drawn back from node j at time level n + 1 to time level n: the approximation to Φ_j uses the basis functions, and with (3.11) their gradients, corresponding to all the vertices of this triangle - see Fig. 11. Using the notation of this figure, the generalisation of (3.11) is most simply given in the local coordinate system based on node j in which the triangle becomes a canonical right triangle as shown in Fig. 11 and $\underline{a}\Delta t$ becomes (μ_1, μ_2). No choice of coefficients gives an exact match to the perfect test function but there is a two parameter family of methods corresponding to (3.11) which give third order accuracy: the simplest, corresponding to M = 0 in (3.11), takes the form

$$\begin{aligned}\Phi_A \simeq \Phi_A^T := {} & (1-\tfrac{1}{2}\mu_1-\tfrac{1}{2}\mu_2)\phi_A + \tfrac{1}{2}\mu_1\phi_B + \tfrac{1}{2}\mu_2\phi_C \\ & + \tfrac{1}{2}\mu_1[(\tfrac{1}{2} - \tfrac{1}{3}\mu_1)\partial_\xi(\phi_A-\phi_B) + \tfrac{1}{3}\mu_2\partial_\xi\phi_C] \\ & + \tfrac{1}{2}\mu_2[(\tfrac{1}{2} - \tfrac{1}{3}\mu_2)\partial_\eta(\phi_A-\phi_C) + \tfrac{1}{3}\mu_1\partial_\eta{}_B].\end{aligned} \tag{3.17}$$

Stability also depends on the choice of parameters and the stability region of (3.17) has been shown to include $\mu_1^2 + \mu_2^2 \leq 1$, and therefore all cases where $-\underline{a}\Delta t$ lies in a triangle with node j as a vertex.

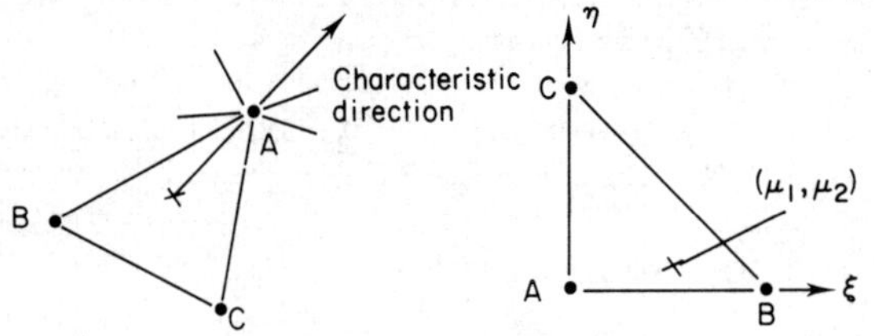

Fig. 11 Layout of triangle for calculation of approximate ECG test function from (3.17).

Numerical tests on this and related schemes have so far been carried out with advection of a Gaussian, both on straight line tracks and around a circle, with excellent accuracy. Fig. 12 shows radial cross sections for a Gaussian carried round a circular trajectory after travelling a quarter, a half, three-quarters and a complete revolution. The problem is solved on $(-1,1)\times(-1,1)$ with right-triangular elements, $\Delta x = \Delta y = \frac{1}{16}$, $\Delta t = 0.8\Delta x$, circle radius $\frac{1}{2}$ and standard deviation $1/2\sqrt{2}$: after 160 time-steps the phase error is about one mesh length.

The present objective is to apply the method to the shallow water equations. The test function (3.17) is applied to just the advection terms and the Galerkin formulation used for the remainder: it may be desirable however to use the alternative time-stepping schemes referred to briefly at the end of the previous section. The immediate aim is to improve the stability range and accuracy reported for model problems in Cullen and Morton (1980) and obtained with schemes which generalised the purely Galerkin approach only in regard to modelling the non-linear terms by a two-stage Galerkin procedure.

4. CONCLUSIONS

Diffusion-convection problems and hyperbolic equations provide two related but distinct problem areas where the deficiencies of the Galerkin derivation of finite element methods are most apparent. We have shown that generalised Galerkin methods can be formulated that can, in the one case completely and in the other very largely, restore the optimal approximation properties that make the Galerkin approach so successful with self-adjoint equilibrium problems. These ideal methods are not completely practical but we have also shown how a number of existing methods can be viewed as approximations to them and also how they provide guidelines to the development of new practical algorithms.

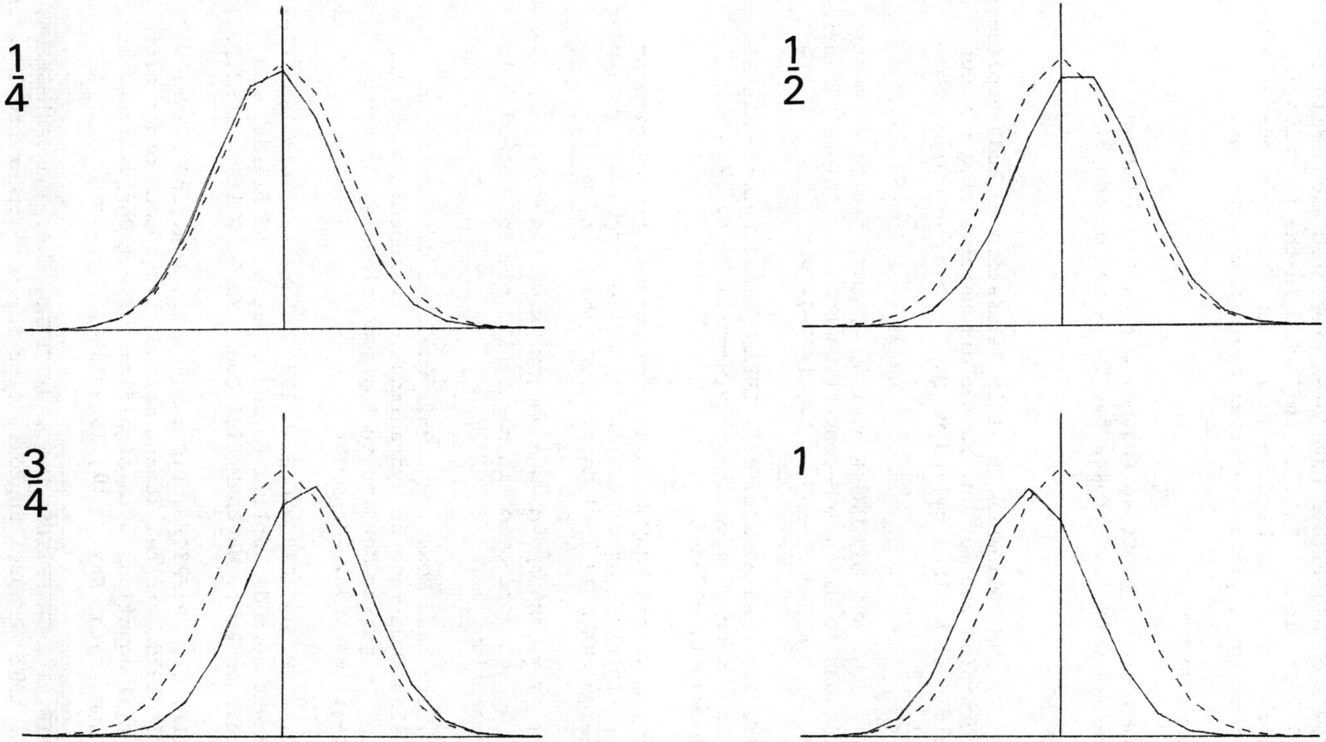

Fig. 12 Convection of a Gaussian after ¼, ½, ¾ and 1 revolution with approximate ECG scheme.

A few such algorithms have been presented along with results for model problems. But, as indicated in the introduction, much work has yet to be done to develop them into competitive techniques for typical practical fluid flow problems.

ACKNOWLEDGEMENTS

I am indebted to Bryan Scotney for the computations in Section 2 and to Alan Stokes for those in Section 3.

REFERENCES

ALLEN, D., and SOUTHWELL, R. 1955 Relaxation methods applied to determining the motion, in two dimensions, of a viscous fluid past a fixed cylinder. *Q. J. Mech. and Appl. Math.* VIII, 129-145.

BARRETT, J.W. and MORTON, K.W. 1980 Optimal finite element solutions to diffusion-convection problems in one dimension. *Int. J. Num. Meth. Eng.* **15**, 1457-1474.

BARRETT, J.W. and MORTON, K.W. 1981 Optimal Petrov-Galerkin methods through approximate symmetrization. *IMA J. Num. Analysis* **1**, 439-468.

BARRETT, J.W. and MORTON, K.W. 1981 Optimal finite element approximation for diffusion convection problems. To appear in Proc. MAFELAP 1981 Conf. (J.R. Whiteman, Ed.).

BARRETT, J.W., MOORE, G. and MORTON, K.W. 1982 Optimal recovery and defect correction in the finite element method. In preparation.

BENQUÉ, J.P. and RONAT, J. 1982 Quelques difficultes des modeles numerique en hydraulique. Presented at 5th Int. Symp. Computing Methods in Applied Sciences and Engng., Versailles 1981. To appear.

BERCOVIER, M. and PIRONNEAU, J. 1981 Characteristics and finite element methods applied to the equation of fluids. To appear in Proc. MAFELAP 1981 Conf. (J.R. Whiteman, Ed.).

CHRISTIE, I., GRIFFITHS, D.F., MITCHELL, A.R. and ZIENKIEWICZ, O.C. 1976 Finite element methods for second order differential equations with significant first derivatives. *Int. J. Num. Meth. Eng.* **10**, 1389-1396.

CHRISTIE, I., GRIFFITHS, D.F., MITCHELL, A.R. and SANZ-SERNA, J.M. 1981 Product approximation for non-linear problems in the finite element problem. *IMA J. Num.Analysis* **1**, 253-266.

CULLEN, M.J.P. and MORTON, K.W. 1980 Analysis of evolutionary error in finite element and other methods. *J.Comput. Phys.* **34**, 245-268.

ENGQUIST, B. and OSHER, S. 1981 One sided difference equations for non-linear conservation laws. *Maths. Comput.* **36**, 321-352.

GADD, A.J. 1978 A numerical advection scheme with small phase speed errors. *Q. J. R. Met. Soc.* **104**, 583-594.

GODUNOV, S.K. 1959 A finite difference method for the numerical computation of discontinuous solutions of the equations of fluid dynamics. *Mat. Sb.* **47**, 271-290.

HEINRICH, J.C., HUYAKORN, P.S., MITCHELL, A.R. and ZIENKIEWICZ, O.C. 1977 An upwind finite element scheme for two-dimensional convective transport equation. *Int. J. Num. Meth. Eng.* **11**, 131-143.

HEINRICH, J.C. and ZIENKIEWICZ, O.C. 1979 The finite element method and 'upwinding' techniques in the numerical solution of convection dominated flow problems. In Finite Element Methods for Convection Dominated Flows (T.J.R. Hughes, Ed.) *AMD* Vol. **34**, *Am. Soc. Mech. Engng.* (New York), 105-136.

HEMKER, P.W. 1977 A numerical study of stiff two-point boundary problems. Thesis, Mathematisch Centrum, Amsterdam.

HIRSCH, Ch. and WARZEE, G. 1976 A finite element method for through-flow calculations in turbomachines. *J. Fluids Engng. Trans. ASME* **98**, 403-421.

HUGHES, T. (Ed.) 1979 Finite Element Methods for Convection Dominated Flows, *AMD* Vol. **34**, *Am. Soc. of Mech. Engng.* (New York).

HUGHES, T.J.R. and BROOKS, A. 1979 A multi dimensional upwind scheme with no crosswind diffusion. In Finite Element Methods for Convection Dominated Flows (T.J.R. Hughes, Ed.) *AMD* Vol. **34**, *Am. Soc. Mech. Engng.* (New York), 19-35.

HUGHES, T.J.R. and BROOKS, A. 1981 A theoretical framework for Petrov-Galerkin methods with discontinuous weighting functions: application to the streamline-upwind procedure. To appear in Finite Elements in Fluids Vol. **4** (R.H. Gallagher, Ed.) J. Wiley and Sons (New York).

HUTTON, A.G. 1981 The numerical representation of convection. IAHR Working Group Meeting, May 1981.

JAMESON, A. 1982 Transonic aerofoil calculations using the Euler equations. In Numerical Methods in Aeronautical Fluid Dynamics (P.L. Roe, Ed.), Academic Press.

JOHNSON, C. and NÄVERT, U. 1981 An analysis of some finite element methods for advection-diffusion problems. In Conf. on Analytical and Numerical Approaches to Asymptotic Problems in Analysis (O. Axelsson, L.S. Frank and A. van der Sluis, Eds.) North-Holland.

KAWAHARA, M. 1978 Steady and unsteady finite element analysis of incompressible viscous fluid. In Finite Elements in Fluids 3 (R.H. Gallagher et al., Eds.) Wiley and Sons (New York), 23-54.

MORTON, K.W. 1981 Finite element methods for non-self-adjoint problems. Univ. of Reading, Num. Anal. Rep. 3/81.

MORTON, K.W. 1982 Shock capturing, fitting and recovery.In Proc. VIIIth Int. Conf. on Num. Meth. in Fluid Dynamics, to appear.

MORTON, K.W. and BARRETT, J.W. 1980 Optimal finite element methods for diffusion-convection problems. In Proc. Conf. Boundary and Interior Layers - Computational and Asymptotic Methods (J.J.H. Miller, Ed.) Boole Press, Dublin, 134-148.

MORTON, K.W. and PARROTT, A.K. 1980 Generalised Galerkin methods for first order hyperbolic equations. *J.Comput. Phys.* **36**, 249-270.

MORTON, K.W. and STOKES, A. 1981 Generalised Galerkin methods for hyperbolic equations. To appear in Proc. MAFELAP 1981 Conf. (J.R. Whiteman, Ed.).

MORTON, K.W. and STOKES, A. 1982 Characteristic Galerkin methods for hyperbolic equations. In preparation.

SCHOOMBIE, S.W. 1982 Spline Petrov-Galerkin methods for the numerical solution of The Kortweg-de Vries equation. *IMA J. Num.Analysis* **2**, 95-109.

ZIENKIEWICZ, O.C., GALLAGHER, R.H. and HOOD, P. 1975 Newtonian and non-Newtonian viscous incompressible flow, temperature induced flows: finite element solutions. In 2nd IMA Conf. Mathematics of Finite Elements and Applications (J.R. Whiteman, Ed.), Academic Press (London).

SCALING AND COMPUTATION OF ATMOSPHERIC MOTIONS

G. Browning and H.O. Kreiss

(California Institute of Technology)

1. INTRODUCTION

In many applications the underlying partial differential equations allow solutions on different time scales, a "slow" and a number of "fast" scales. Often one is not interested in the fast scales. In recent years (Kreiss, 1979, 1980 and Browning et al., 1980) a mathematical theory for symmetric hyperbolic partial differential equations with different time scales has been developed. This theory allows us to derive refined reduced systems and provides also an easier way to initialize the data in limited areas. As an illustration of the theory we shall apply it to atmospheric and oceanographic motions.

Any hyperbolic system which fully describes atmospheric and oceanographic motions must contain solutions with more than one time scale. For example, the shallow water equations govern two classes of motions with different time scales. The "slow" Rossby-type motions and the "fast" inertia-gravity motions. Meteorologically the main interest is in the first type of motions. Therefore methods have been developed to filter out the fast waves. There are two different approaches to accomplish this. In the first, one keeps the underlying equations but one prepares the initial data - called initialisation - such that the fast waves are not excited. In the second, one changes the equations to a set of so-called reduced equations which allow only the slow motions.

Currently, in most large scale meteorological calculations, one uses the primitive equations and initialises the data - using nonlinear normal mode analysis such that the gravity waves are filtered out (Baer, 1977 and Machenhauer, 1977). This can be considered as a combination of the two above methods: the primitive equations are obtained from the full gas dynamic equations by making the hydrostatic assumption thus eliminating the "fast" vertical sound waves; then one initialises the data such that the remaining "fast" gravity waves are not excited.

Experience shows that the primitive equations are not totally satisfactory. The calculation of the ultra long waves and of the vertical velocity is often not very accurate. Also, troubles with the boundary conditions appear when one uses the primitive equations in limited areas. We want to use our theory to re-examine the scaling of atmospheric motions and derive accurate reduced systems which can be used in limited areas without any problems at the boundaries. Also, we believe, these models lead to a more accurate evaluation of the vertical velocity and the ultra long waves.

2. MATHEMATICAL BACKGROUND

The gas dynamic equations, describing large scale atmospheric flows, can be rewritten as a hyperbolic system of partial differential equations of the form

$$D(x,t,u,\varepsilon)\,\partial u/\partial t = \varepsilon^{-\beta} P_0(\varepsilon,\partial/\partial x)u + P_2(x,t,u,\varepsilon)u. \tag{2.1}$$

Here $u = (u_1,\ldots,u_n)^T$ is a vector function depending on $x = (x_1,x_2,x_3)$, t and

$$D = \begin{pmatrix} d_1 & 0 & \cdot & \cdot & \cdot & 0 \\ 0 & d_2 & 0 & \cdot & \cdot & 0 \\ \cdot & \cdot & \cdot & \cdot & \cdot & \cdot \\ 0 & \cdot & & \cdot & 0 & d_n \end{pmatrix}, \quad d_j \equiv d_j(x,t,u,\varepsilon) > 0, \tag{2.2}$$

is a positive definite diagonal matrix which depends smoothly on u,x,t,ε: $\varepsilon > 0$ is a small constant, $\beta \geq 1$ is an integer and

$$P_0 \equiv \sum_{j=1}^{3} A_j\,\partial/\partial x_j + B, \quad A_j = A_j^*, \quad B = -B^* \tag{2.3}$$

is a first order antisymmetric differential operator with matrix coefficients which depend smoothly on ε but not on x,t. Finally $P_1(x,t,u,\varepsilon,\partial/\partial x)$ is a quasilinear antisymmetric first order differential operator whose coefficients depend smoothly on x,t,u and ε.

Problems of this kind have solutions on different time scales: fast scales of order $O(\varepsilon^{-\alpha})$, $\alpha \le \beta$ and a slow scale of order $O(1)$.

We are interested only in the solutions on the slow scale. In a number of papers (Kreiss, 1979, 1980 and Browning et al., 1980) we have shown that these slow solutions can be obtained using a simple principle:

Choose the initial data in such a way that at $t = 0$ a number $p > 0$ of time derivatives are bounded independently of ε.

We have proved that this principle leads to solutions which only vary on the slow scale on a time interval $0 \le t \le T$, T independent of ε. The size of T will in general depend on p. The larger the number of bounded derivatives at $t = 0$ the longer it takes for the fast waves to appear and to pollute the slow solution. However, the number of linear independent constraints which the initial data have to satisfy such that the time derivatives are bounded independently of ε does not depend on p. The constraints just become more refined with increasing p.

The constraints consist of a number of elliptic differential equations which the initial data have to satisfy. As long as the solution stays on the slow scale it satisfies these elliptic equations also at later times because these constraints are satisfied whenever the time derivatives are bounded independently of ε. This enables us to derive accurate reduced systems by replacing part of the original hyperbolic system by these elliptic equations.

3. SCALING OF THE BASIC EQUATIONS FOR LARGE SCALE ATMOSPHERIC FLOW

In Cartesian coordinates x,y,z directed eastward, northward and upward, respectively, the Eulerian equations can be written as

$$\begin{aligned} &ds/dt = 0 \\ &d\underline{V}/dt + \rho^{-1}\underline{\nabla}p + f(\underline{k} \times \underline{V}) + g\,\underline{k} = \underline{0}, \\ &dp/dt + \gamma p\, \underline{\nabla}\cdot\underline{V} = 0 . \end{aligned} \tag{3.1}$$

Here t is the time, $\underline{V} = (u,v,w)^T$ is the velocity, ρ the density, p the pressure and $s = \rho p^{-1/\gamma}$ is the entropy. Also, $f = f(y)$ is the Coriolis parameter, g is the constant gravitational acceleration, $\gamma = 1.4$ and $\underline{k} = (0,0,1)^T$ is the unit vector in the vertical direction. The total differential operator d/dt is given by

$$d/dt \equiv \partial/\partial t + \underline{V}\cdot\underline{\nabla} \equiv \partial/\partial t + u\partial/\partial x + v\partial/\partial y + w\partial/\partial z.$$

We assume that f is given by the tangent plane approximation $f = 2\Omega[\sin\omega_0 + (y/r)\cos\omega_0]$ where Ω is the angular speed of the earth, ω_0 is the latitude of the coordinate origin and r is the radius of the earth.

We shall now introduce dimensionless variables to identify the relative magnitude of all terms in the equation. We change the independent variables via the relations

$$x = L_1 x', \ y = L_2 y', \ z = Dz', \ t = Tt'. \tag{3.2}$$

Here L_1, L_2, D, T are the representative horizontal lengths, vertical depth and time scale, respectively. For the dependent variables we introduce the relations

$$\begin{aligned}
&u = Uu', \quad v = Vv', \quad w = Ww', \\
&\rho = R_0\rho_0(z) + R\rho', \quad p = P_0 p_0(z) + Pp', \\
&R/R_0 = P/P_0, \quad P_0(p_0(z))_z + R_0 g\rho_0(z) = 0, \\
&s = (R_0\rho_0(z) + R\rho')(P_0 p_0(z) + Pp')^{-1/\gamma} = S_0 s_0(z) + Ss', \\
&S_0 = R_0 P_0^{-1/\gamma}, \quad s_0(z) = \rho_0(z)(p_0(z))^{-1/\gamma}, \quad S = RP_0^{-1/\gamma}, \\
&f = 2\Omega f' = 2\Omega[\sin\omega_0 + L_2(y/r)\cos\omega_0], \ 2\Omega = 10^{-4}s^{-2}, \\
&g = Gg' = 10ms^{-2}.
\end{aligned} \tag{3.3}$$

Density and pressure depend on two parameters because there are basic horizontal mean values

$$R_0 = 1kgm^{-3}, \quad P_0 = 10^5 kgm^{-1}s^{-2} \tag{3.4}$$

due to the presence of gravity.

We assume also that

$$T = L_1 U^{-1} = L_2 V^{-1}, \tag{3.5}$$

so that for each equation the dimensionless time derivative will be of the same order of magnitude as the horizontal advection terms. Then substituting the relations (3.2) and (3.3) into the system (3.1) gives us

$$\begin{aligned}
&ds'/dt + S_1^{-1} S_2 (s_0(z'))_{z'} w' = 0, \\
&dp'/dt + S_1^{-1} [S_2 (p_0(z'))_{z'} w' + \gamma (p_0(z') + S_1 p')d] = 0, \\
&du'/dt + S_3 (\rho_0(z') + S_1 \rho')^{-1} p'_{x'} - S_4 f' v' = 0, \\
&dv'/dt + \tilde{S}_3 (\rho_0(z') + S_1 \rho')^{-1} p'_{y'} + \tilde{S}_4 f' u' = 0, \\
&dw'/dt + S_5 (\rho_0(z') + S_1 \rho')^{-1} (p'_{z'} + S_6 g' \rho') = 0,
\end{aligned} \tag{3.6}$$

where

$$d = u'_{x'} + v'_{y'} + S_2 w'_{z'}, \quad dg/dt \equiv g_{t'} + u' g_{x'} + v' g_{y'} + S_2 w' g_{z'}.$$

The S_i are dimensionless scaling parameters defined by

$$\begin{aligned}
&S_1 = R/R_0 = P/P_0, \ S_2 = TW/D, \ S_3 = P/(R_0 U^2), \ \tilde{S}_3 = P/(R_0 V^2), \\
&S_4 = 2\Omega T L_2/L_1, \ \tilde{S}_4 = 2\Omega T L_1/L_2, \ S_5 = TP/(DW), \ S_6 = DGR/P.
\end{aligned} \tag{3.7}$$

Finally we can express ρ' in terms of s', p'

$$\rho' = \rho_0 (\gamma p_0)^{-1} p' - p_0^{1/\gamma} s' + O(PP_0^{-1}).$$

We neglect the term $O(PP_0^{-1})$.

4. LARGE SCALE MOTIONS

There are quite a number of different scales which govern meteorological motions. In particular if fronts and jetstreams are present then one has to allow $L_1 \neq L_2$. All these different possibilities will be discussed in a forthcoming paper. Here we restrict ourselves to the large scale for which

$$\begin{aligned} &L_1 = L_2 = 10^6 m,\ D = 10^4 m,\ U = V = 10 m/sec, \\ &P = 10^3 kg/msec^2,\ P/R = 10^5 m^2/sec^2. \end{aligned} \tag{4.1}$$

We leave $W = \varepsilon^{-\alpha}$, $\varepsilon = 10^{-1}$, free and shall show that α is determined by the other parameters. Neglecting the term $S_1\rho' = \varepsilon^2\rho'$ and dropping the primes, we obtain for (3.6)

$$\begin{aligned} &ds/dt + a(z)w = 0,\ a(z) = \varepsilon^{\alpha-3} s_{0z} > 0,\ \varepsilon = 10^{-1}, \\ &\varepsilon^2 dp/dt + \gamma p_0(\varepsilon^{\alpha-1} Lw + u_x + v_y) = 0,\ Lw \equiv w_z - (g' \rho_{0z}/\gamma p_0)w, \\ &\varepsilon du/dt + \rho_0^{-1} p_x - f' v = 0, \\ &\varepsilon dv/dt + \rho_0^{-1} p_y + f' u = 0, \\ &\varepsilon^{\alpha+4} dw/dt - \rho_0^{-1}(L^* p + g' p_0^{1/\gamma} s) = 0,\ - L^* p \equiv p_z + (g' \rho_{0z}/\gamma p_0)p, \\ &d/dt \equiv \partial/\partial t + u\partial/\partial x + v\partial/\partial y + \varepsilon^{\alpha-1} w\partial/\partial z. \end{aligned} \tag{4.2}$$

Here a(z) describes the stability of the atmosphere. It is well known that $a(z) > 0$, otherwise rapid exponential growth would occur. We assume that $a(z) \sim 1$ which is consistent with $ds/dt = O(1)$. We consider only the midlatitude where $f' = 1$. Therefore du/dt, dv/dt are of order O(1) if and only if the so called geostrophic approximations

$$\rho_0^{-1} p_x - fv = O(\varepsilon),\ \rho_0^{-1} p_y + fu = O(\varepsilon) \tag{4.3}$$

hold. Also dp/dt and dw/dt are of order O(1) if

$$\varepsilon^{\alpha-1} Lw + u_x + v_y = -(\varepsilon^2/\gamma p_0)\, dp/dt = O(\varepsilon^2) \tag{4.4}$$

and the hydrostatic approximation

$$L^{*}p + g' p_{o}^{1/\gamma} s = \varepsilon^{\alpha+4} \rho_{o} dw/dt = O(\varepsilon^{\alpha+4}) \tag{4.5}$$

are valid. The β-plane approximation becomes

$$f' = f_{o} + \varepsilon f_{1} y. \tag{4.6}$$

Cross-differentiation of (4.3) gives us

$$\rho_{o}^{-1} p_{xy} - f' v_{y} = O(\varepsilon), \quad \rho_{o}^{-1} p_{xy} + f' u_{x} = O(\varepsilon).$$

Therefore

$$f' (u_{x} + v_{y}) = O(\varepsilon), \tag{4.7}$$

and (4.4) implies necessarily $\alpha \geq 2$. We want to show that $\alpha = 2$. For that reason we introduce new functions by

$$\rho_{o}^{-1} p_{x} - f' v = \varepsilon \tilde{p}_{x} + \varepsilon H, \quad \rho_{o}^{-1} p_{y} + f' u = \varepsilon \tilde{p}_{y} + \varepsilon G,$$

$$H_{x} + G_{y} = 0 . \tag{4.8}$$

Assuming that u,v,p are known we can find G,H,p satisfying the above relations if and only if

$$H_{y} - G_{x} = \varepsilon^{-1} [(\rho_{o}^{-1} p_{x} - f' v)_{y} - (\rho_{o}^{-1} p_{y} + f' u)_{x}] =$$

$$- \varepsilon^{-1} (u_{x} + v_{y}) - f_{1} v = -f_{1} v + \varepsilon^{\alpha-2} L w + (\varepsilon/\gamma p) dp/dt. \tag{4.9}$$

With this notation the horizontal momentum equations become

$$du/dt + \tilde{p}_{x} + H = 0, \quad dv/dt + p_{y} + G = 0. \tag{4.10}$$

Assume now that $\alpha > 2$. Then to first approximation

$$H_{x} + G_{y} = 0, \quad H_{y} - G_{x} = -f_{1} v$$

and by (4.4)

$$u_x + v_y = 0. \tag{4.11}$$

Also by (4.2)

$$d/dt = d_H/dt \equiv \partial/\partial t + u\partial/\partial x + v\partial/\partial y. \tag{4.12}$$

Thus to first approximation there is no coupling between the planes z = constant. Therefore the z-derivatives due to the shear would become large and the scaling assumption would be violated. This proves that $\alpha = 2$ is the right scaling.

We shall now derive the primitive equations. We have $d^2w/dt^2 = O(1)$ if and only if

$$O(\varepsilon^8) = \varepsilon^2[d(L^*p)/dt + g' d(p_0^{1/\gamma}s)/dt] = \varepsilon^2[L^* dp/dt + g' p_0^{1/\gamma} ds/dt]$$

$$+ O(\varepsilon^3) = -\gamma p_0 L^*(u_x + v_y) - \varepsilon\gamma p_0 L^* Lw - \varepsilon^2 g' p_0^{1/\gamma} a(z)w + O(\varepsilon^3). \tag{4.13}$$

Thus if we neglect terms of order $O(\varepsilon^3)$ we obtain from (4.5) and (4.13) the so-called primitive equations

$$\begin{aligned}
&ds/dt + a(z)w = 0,\\
&\varepsilon du/dt + \rho_0^{-1}p_x - f'v = 0,\\
&\varepsilon dv/dt + \rho_0^{-1}p_y + fu = 0, \qquad (4.14)\\
&L^*p + g' p_0^{1/\gamma}s = 0,\\
&L^*(u_x + v_y) + \varepsilon L^* Lw + \varepsilon^2 g' p_0^{1/\gamma} a(z)w = 0.
\end{aligned}$$

Numerically one has to solve the system in the following way. Assume we know all variables at time t. Then we use the first three equations to determine s,u,v at time $t + \Delta t$. Then we can determine p and w from the last two equations. We shall now discuss the difficulties with this approach.

(1) We need one boundary condition to determine p from the hydrostatic equation. One should use p at the surface (z = 0).

However this value is not known. Therefore one assumes that p is given at the top of the atmosphere and uses that as a boundary condition. Mathematically this is not satisfactory. The operator L^* is such that the solutions of $L^* p = 0$ decay exponentially with height and therefore $p \to 0$ for $z \to \infty$ is no boundary condition.

(2) The calculation of w from Richardson's equation requires that we have to calculate the horizontal divergence $u_x + v_y$ and divide it by ε.

By (4.14)

$$-\varepsilon^{-1} d(u_x + v_y)/dt = -\varepsilon^{-1}[(du/dt)_x + (dv/dt)_y] + \varepsilon^{-1} j_0(u,v) +$$

$$j_1(u,v,w) = \varepsilon^{-2}(\rho_0^{-1} p_x - f' v)_x + (\rho_0^{-1} p_y + f' u)_y +$$

$$\varepsilon^{-1} j_0(u,v) + j_1(u,v,w), \tag{4.15}$$

where

$$j_0 = ((u_x)^2 + 2u_y u_x + (v_y)^2), \quad j_1 = w_x u_z + w_y v_z.$$

Therefore, if we want to calculate w with two accurate figures we have to calculate the balance expression

$$(\rho_0^{-1} p_x - f' v)_x + (\rho_0^{-1} p_y + f' u)_y$$

accurate to four figures. Remembering that the first two figures of the total pressure are constant we have to know the total pressure accurate to six digits.

We obtain the primitive equations by demanding that dw/dt and d^2w/dt^2 are of order O(1). We shall now use the condition that also the time derivatives of the other variables are of order O(1) to derive a better conditioned system. The system is an improved quasi-geostrophic model.

For clarity we consider a simplified system by assuming that f', ρ_0, p_0 are constant and that also the convection operator

$$d/dt \equiv \partial/\partial t + U_0\partial/\partial x + V_0\partial/\partial y + \varepsilon W_0\partial/\partial z$$

has constant coefficients. Then d/dt commutes with partial derivatives. For example

$$d(u_x)dt = (du/dt)_x, \quad d(Lw)/dt = L(dw/dt).$$

The general case differs from the simplified case by lower order terms only. We obtain

$$-(\varepsilon^2/\gamma p_0)d^2p/dt^2 = \varepsilon d(Lw)/dt + (du/dt)_x + (dv/dt)_y =$$

$$\varepsilon d(Lw)/dt - \varepsilon^{-1}(\rho_0^{-1}\Delta p - f'\xi), \ \Delta p \equiv p_{xx} + p_{yy}, \ \xi \equiv u_y - v_x.$$

Thus $d^2p/dt^2 = O(1)$ if and only if the balance relation

$$\rho_0^{-1}\Delta p + f\xi = O(\varepsilon^2) \tag{4.16}$$

holds.

For the third derivative we obtain

$$-(\varepsilon^3/\gamma p_0)d^3p/dt^3 = \varepsilon^2 d^2Lw/dt^2 - \rho_0\Delta(dp/dt) - fd\xi/dt,$$

i.e., $d^3p/dt^3 = O(1)$ if

$$\rho_0^{-1}\Delta(dp/dt) + fd\xi/dt = O(\varepsilon^2).$$

Now

$$d\xi/dt = \varepsilon^{-1}f(u_x + v_y) = -fLw + (\varepsilon/\gamma p_0)dp/dt.$$

Therefore

$$\rho_0^{-1}\Delta(dp/dt) - f^2Lw + (\varepsilon/\gamma p_0)dp/dt = O(\varepsilon^2),$$

and using the hydrostatic relation,

$$\rho_0^{-1}\Delta(dL^*p/dt) - f^2L^*Lw + (\varepsilon/\gamma p_0)fd(L^*p)dt =$$

$$- g'\rho_0^{-1}p_0^{1/\gamma}\Delta(ds/dt) - f^2L^*Lw - (\varepsilon/\gamma p_0)fg'p_0^{1/\gamma}ds/dt.$$

The entropy equation gives us an elliptic equation for w

$$Pw \equiv a(z)\Delta w - \rho_0(f^2/g'p_0^{1/\gamma})L^*Lw - (\varepsilon/\gamma p_0)a(z)w = O(\varepsilon^2). \tag{4.17}$$

In the general nonlinear situation, (4.16) and (4.17) become

$$\begin{aligned} \rho_0^{-1}\Delta p + f\xi &= F_1(u,v) + O(\varepsilon^2) \\ Pw &= F_2(u,v,p) + O(\varepsilon^2). \end{aligned} \tag{4.18}$$

Neglecting the $O(\varepsilon^2)$ terms we can combine these equations with

$$\begin{aligned} du/dt + \tilde{p}_x + H = 0, \quad dv/dt + \tilde{p}_y + G = 0 \\ u_x + v_y + Lw = 0 \\ H_x + G_y = 0,\ H_y - G_x = -f_1v + Lw + (\varepsilon/\gamma p)dp/dt, \end{aligned} \tag{4.19}$$

to obtain a system which is mathematically completely satisfactory.

REFERENCES

BAER, F. 1977 Adjustments of initial conditions required to suppress gravity oscillations in nonlinear flows. *Contr. Atmos. Phys.* **50**.

BROWNING, G., KASAHARA, A. and KREISS, H.O. 1980 Initialization of the primitive equations by the bounded derivative method. *J. Atmos Sci. Vol.* **37**.

KREISS, H.O. 1979 Problems with different time scales for ordinary differential equations. *SIAM J. Num. Analysis* Vol. **16**.

KREISS, H.O. 1980 Problems with different time scales for partial differential equations. *Commun. Pure Appl. Math.* Vol. **33**.

MACHENHAUER, B. 1977 On the dynamics of gravity oscillation in a shallow water model with applications to normal mode initialization. *Contr. Atmos. Phys.* **50**.

SPECTRAL METHODS IN METEOROLOGY

B.J. Hoskins

(Department of Meteorology, University of Reading)

1. INTRODUCTION

Until 1976, all the major numerical models of the atmosphere for use both in forecasts and in modelling the general circulation were based on finite difference techniques in all three space dimensions. Since that time there has been a move towards global models using a spectral representation in the horizontal, a move whose speed has been remarkable considering the huge investment in time represented by these models and the desire not to change them too frequently. In this chapter, the subject will be viewed largely from a historical perspective so that the reasons for the sudden popularity of the method are apparent. For a thorough review of the state of the art in 1974, the reader is referred to GARP (1974), and for more detailed descriptions of the spectral method to Bourke et al. (1977), and Machenhauer (1979).

2. THE BAROTROPIC VORTICITY EQUATION

The earliest quantitative models of the atmosphere which met with some success were those based on the conservation of the vertical component of absolute vorticity in two-dimensional non-divergent flow. If the motion on the surface of a sphere is described by a streamfunction ψ, then the relative vorticity is

$$\xi = \nabla^2\psi \ , \qquad (2.1a)$$

and the absolute vorticity

$$\zeta = 2\Omega\mu + \xi \ , \qquad (2.1b)$$

where μ is the sine of the latitude and Ω is the rotation rate of the underlying planet and of the coordinate system. The conservation of vorticity may then be written

$$\frac{\partial}{\partial t}\zeta = - J(\psi,\zeta) \ , \tag{2.2}$$

or

$$\frac{\partial}{\partial t}\nabla\psi = -(2\Omega/a^2)\frac{\partial\psi}{\partial\lambda} - J(\psi,\nabla^2\psi), \tag{2.3}$$

where J is the Jacobian, λ is the longitude and a the planetary radius: (2.2) and (2.3) are often known as the barotropic vorticity equation.

Following work in planar geometry, Haurwitz (1940) studied solutions of (2.3) which are perturbations about a state of no motion. In this case the Jacobian term is neglected and the solutions are the spherical harmonics:

$$Y_n^m(\mu,\lambda) = P_n^m(\mu)\exp(im\lambda) \ . \tag{2.4}$$

The order m is the longitudinal wavenumber, $n - |m|$ is the number of zeros between the north and south poles and the degree n is the total wavenumber since

$$\nabla^2 Y_n^m = - n(n+1)a^{-2} Y_n^m . \tag{2.5}$$

From (2.4) and (2.5) the perturbation solutions to (2.3) are of the form

$$\psi = A\,Y_n^m(\mu,\lambda)\exp im(\lambda - ct) \ , \tag{2.6}$$

where $c = -2\Omega/[n(n+1)]$.

These modes are usually termed Rossby-Haurwitz waves.

In 1946 Neamtan noted that, since these Rossby-Haurwitz waves have $\nabla^2\psi \propto \psi$, the Jacobian term in (2.3) would be zero even for finite amplitude waves. Thus Rossby-Haurwitz waves are solutions of the nonlinear barotropic vorticity equation.

Spherical harmonics form a complete and orthogonal system of basis functions on the sphere:

$$\iint Y_q^p\,(Y_s^r)^*\,d\mu\,d\lambda = 2\pi\,\delta_{pr}\,\delta_{qs} \ , \tag{2.7}$$

the constant depending on the normalisation applied. Thus it is natural to pose a representation of any vorticity or stream-function distribution in terms of them, e.g.

$$\xi = \sum_{n \geq |m|} \xi_n^m (t) \, Y_n^m (\mu, \lambda) \, . \tag{2.8}$$

From (2.1),

$$\psi_n^m = - a^2 n^{-1} (n+1)^{-1} \xi_n^m \, ,$$

$$\zeta_1^0 = \xi_1^0 + (.375)^{-\frac{1}{2}} \text{ and } \zeta_n^m = \xi_n^m \quad \text{otherwise.} \tag{2.9}$$

Substituting in (2.3), multiplying by $\left(Y_n^m\right)^*$ and integrating over the sphere gives

$$\frac{d}{dt} \xi_n^m = \frac{2im\Omega}{n(n+1)} \xi_n^m + \sum_{n_1 \geq m_1} \sum_{n_2 \geq m_2} I_{n,n_1,n_2}^{m,m_1,m_2} \xi_{n_1}^{m_1} \xi_{n_2}^{m_2} \, . \tag{2.10}$$

The first term on the right-hand side corresponds to the Rossby-Haurwitz wave motion. The representation of the non-linear term includes the interaction coefficients I. This approach to the solution of (2.3) was first discussed by Silberman (1954).

Clearly, for application of this interaction coefficient method, the representation (2.8) must be truncated in some manner. Two ways of performing this truncation have been widely used (see Fig. 1). In the rhomboidal truncation, for each zonal wavenumber m the same number of latitudinal modes is retained: mathematically,

$$|m| \leqslant M \, , \quad |m| \leqslant n \leqslant |m| + J \, . \tag{2.11}$$

The triangular truncation retains all modes whose total scale is greater than a certain length: in mathematical symbols,

$$n \leqslant M \, . \tag{2.12}$$

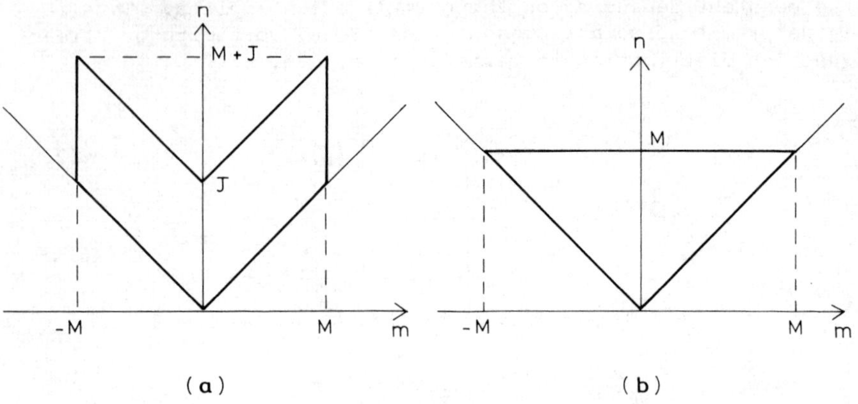

Fig. 1 Two widely used spectral truncations named after their shapes in graphs using as axes the order (m) and index (n) of the spherical harmonic functions:

(a) rhomboidal, (b) triangular.

For severe truncations it appears possible that the rhomboidal truncation gives a more efficient representation of atmospheric flow which is dominated by large zonal wavelengths embedded in an eastward flow. Otherwise it would appear that the isotropic triangular truncation is preferable. This is clearly so in models containing dissipative processes dependent on the local scale. It is worthy of note that a spectral model with triangular truncation is entirely independent of the coordinate system used: every point and every direction is treated identically. This is a tremendous bonus compared with the polar singularity that has troubled finite difference modellers for many years (see, e.g., Arakawa, 1977).

The application of the method of interaction coefficients (2.10) is possible for very severe truncations, but generally suffers from the overwhelming difficulty that the amount of storage required and the number of calculations per time step both rise as the fifth power of M. This led to only limited use of the spectral method for well over a decade following the work of Silberman (1954). In this time the emphasis of large-scale atmospheric modelling shifted towards the use of the so-called primitive equations instead of the barotropic vorticity equation and the inclusion of physical processes such as rainfall. Robert (1966) discussed the possible application of the spectral method to the primitive equations, but the inclusion of physical processes, which are of their essence local in physical space, appeared to be an insurmountable problem.

3. THE SPECTRAL-TRANSFORM METHOD

The introduction of the spectral-transform method by Orszag (1970) and Eliasen et al. (1970) removed the basic difficulties associated with storage, computation and local physical processes. The method is most simply described by considering the optimum way of obtaining the Fourier coefficients, a_k, of

$$a = b \times c \tag{3.1}$$

where a, b and c are periodic functions in the range $0 \leq x \leq 2\pi$, defined by the truncated Fourier series:

$$a = \sum_{|k| \leq K} a_k e^{ikx} , \quad b = \sum_{|k| \leq K} b_k e^{ikx} , \quad c = \sum_{|k| \leq K} c_k e^{ikx} . \tag{3.2}$$

Substitution of (3.2) in (3.1), multiplication by e^{-ikx} and integration over the domain gives the convolution sum

$$a_k = \sum_{p+q=k} b_p c_q . \tag{3.3}$$

This is the equivalent of the interaction coefficient method, and for all the coefficients a_k involves of order K^2 multiplications.

An alternative is to obtain grid point values of b and c by Fourier synthesis and then determine a_k by Fourier analysis:

$$a_k = \frac{1}{2\pi} \int_0^{2\pi} b \times c \, e^{-ikx} dx . \tag{3.4}$$

Orszag (1970) observed that the product of b and c gives waves in the range $|k| \leq 2K$ so that the integrand in (3.4) can only contain modes with $|k| \leq 4K$. It may thus be determined exactly by finite Fourier analysis on at least 4K + 1 points. However, as noted by Orszag (1971), the coefficients a_k are required only for $|k| \leq K$. For this range, the integrand in (3.4) contains modes with $|k| \leqslant 3K$ so that it may be performed exactly on a grid of $M_g \geqslant 3K + 1$ equally spaced points in the range 0 to 2π. The Fourier synthesis and analysis may be executed using fast Fourier transform algorithms so that the whole operation takes of order $K \log_2 K$ operations. For large enough truncations, this spectral-transform method must therefore be more efficient than evaluation of the convolution sum.

To apply this method to the barotropic vorticity equation (2.2), we write it in the form

$$\frac{\partial \zeta}{\partial t} = -\frac{1}{a(1-\mu^2)} \frac{\partial}{\partial \lambda}(U\zeta) - \frac{1}{a}\frac{\partial}{\partial \mu}(V\zeta)\,, \tag{3.5}$$

where

$$(U,V) = \left(-\frac{1}{a}(1-\mu^2)\frac{\partial \psi}{\partial \mu}\,,\ \frac{1}{a}\frac{\partial \psi}{\partial \lambda}\right) \tag{3.6}$$

is the velocity multiplied by the cosine of the latitude. Multiplication by $\left(Y_n^m\right)^*$ and integration over the surface of the sphere gives

$$2\pi \frac{d\zeta_n^m}{dt} = \int_{-1}^{1} -\frac{P_n^m(\mu)}{a(1-\mu^2)} \left[\int_0^{2\pi} \frac{\partial}{\partial \lambda}(U\zeta)\, e^{-im\lambda}\, d\lambda\right] d\mu$$

$$- \int_0^{2\pi} \frac{e^{-im\lambda}}{a} \left[\int_{-1}^{1} \frac{\partial}{\partial \mu}(V\zeta)\, P_n^m(\mu)\, d\mu\right] d\lambda.$$

Integration by parts gives

$$\frac{d\zeta_n^m}{dt} = \frac{1}{2\pi a}\int_{-1}^{1} \left[\frac{P_n^m(\mu)}{(1-\mu^2)}(-\,im)\int_0^{2\pi} U\zeta e^{-im\lambda} d\lambda + \frac{dP_n^m(\mu)}{d\mu}\int_0^{2\pi} V\zeta e^{-im\lambda} d\lambda\right] d\mu. \tag{3.7}$$

The Fourier integrals may be evaluated exactly using $M_g \geqslant 3M+1$ equally spaced longitudinal points. Eliasen et al. (1970) found that the integrands in the μ integrals were polynomials in μ and could therefore be performed exactly using Gaussian quadrature. The maximum degree of these polynomials is $2M + 3J - 1$ for rhomboidal truncation and $3M - 1$ for triangular truncation. Therefore the number of "Gaussian" latitudes must be $J_g \geqslant \frac{2M + 3J}{2}$ and $\frac{3M}{2}$, respectively. In practice, these latitudes are very nearly equally spaced.

Thus in the spectral transform method, values of ζ_n^m at time t give grid point values of ζ, U and V:

$$\zeta(\mu,\lambda,t) = \Sigma\ \zeta_n^m(t)\ P_n^m(\mu)\ e^{im\lambda} \quad , \tag{3.8}$$

$$U(\mu,\lambda,t) = a\ \Sigma\ \frac{\zeta_n^m}{n(n+1)}\ (1 - \mu^2)\ \frac{dP_n^m}{d\mu}\ e^{im\lambda} \quad , \tag{3.9}$$

and
$$V(\mu,\lambda,t) = -ia\ \Sigma\ \frac{m\ \zeta_n^m}{n(n+1)}\ P_n^m\ e^{im\lambda} \quad : \tag{3.10}$$

(3.9) and (3.10) are derived from (2.9) and (3.6). The grid point array is composed of M_g equally spaced points in longitude and J_g Gaussian latitudes. The evaluation of (3.8)-(3.10) requires simple products and Fourier transforms. In grid-point space, the products $U \times \zeta$ and $V \times \zeta$ are evaluated. The Fourier integrals in (3.7) are then performed exactly using finite Fourier series and the μ integration by Gaussian quadrature. Note that the contribution from latitude circles to the spectral tendencies may be accumulated sequentially. The evaluation of $d\zeta_n^m/dt$ at time t then allows the prediction of ζ_n^m at time $t + \Delta t$ using finite differences in time. Full spectral-transform models were first produced by Machenhauer and Rasmussen (1972) and Bourke (1972).

The computation per time-step is of order M^3 in the limit of large M (and $J \sim M$ rhomboidal truncation) which demonstrates the efficiency of the spectral-transform method as opposed to the interaction coefficient method. Even for a comparatively severe rhomboidal truncation with $M = 16$, $J = 15$ Bourke (1972) found the former to be about ten times faster for a model based on equations slightly more complicated than the barotropic vorticity equation.

The storage required for a spectral-transform model is not punitive and the grid-point stage allows the calculation and inclusion of parameterizations of local physical processes. In fact some meteorologists prefer to consider spectral-transform models as grid-point models in which derivatives are determined in spectral space.

4. THE PRIMITIVE EQUATIONS IN σ-COORDINATES

In the first attempts to solve the primitive equations using spectral methods, Robert (1966) and Eliasen et al. (1970) tackled the momentum equations directly. The fact that the variables under discussion were not true vectors or scalars led to some difficulties. Bourke (1972) pointed out that the horizontal motion could be described uniquely by its vorticity and divergence (D), that these were true scalars, and that the momentum equations could be turned into predictive equations for them. They are thus the natural dependent variables to describe the horizontal component of the motion in a spectral model.

In the three-dimensional meteorological equations, it is usual to make the lower boundary a coordinate surface. The most common way of achieving this is to use as vertical coordinate

$$\sigma = p/p_* , \tag{4.1}$$

where p is the pressure and p_* its surface value. The boundary conditions are then

$$\dot{\sigma} = 0 \text{ at } \sigma = 0,1 . \tag{4.2}$$

Using this coordinate, the equations of motion for an inviscid, adiabatic, hydrostatic, perfect gas in a thin shell surrounding a rotating spherical planet may be written (Bourke, 1974; Hoskins and Simmons, 1975):

$$\frac{\partial \zeta}{\partial t} = \frac{1}{a(1-\mu^2)} \frac{\partial}{\partial \lambda} F_V - \frac{1}{a} \frac{\partial}{\partial \mu} F_U , \tag{4.3}$$

$$\frac{\partial D}{\partial t} = \frac{1}{a(1-\mu^2)} \frac{\partial}{\partial \mu} F_U + \frac{1}{a} \frac{\partial}{\partial \mu} F_V - \nabla^2 \left[\frac{U^2+V^2}{2(1-\mu^2)} + \phi + R\,\bar{T}\,\ell n p_* \right] , \tag{4.4}$$

$$\frac{\partial T'}{\partial t} = - \frac{1}{a(1-\mu^2)} \frac{\partial}{\partial \lambda} (UT') - \frac{1}{a} \frac{\partial}{\partial \mu} (VT') + DT' - \dot{\sigma} \frac{\partial T}{\partial \sigma} + K\, T \frac{D \ln p}{Dt} ,$$

(4.5)

$$\frac{\partial \ln p_*}{\partial t} = - \frac{U}{a(1-\mu^2)} \frac{\partial \ln p_*}{\partial \lambda} - \frac{V}{a} \frac{\partial \ln p_*}{\partial \mu} - D - \frac{\partial \dot{\sigma}}{\partial \sigma} , \quad (4.6)$$

$$\frac{\partial \phi}{\partial \ln \sigma} = - R\, T , \quad (4.7)$$

where

$$T = \bar{T}(\sigma) + T' , \quad (4.8)$$

$$F_U = V\zeta - \dot{\sigma} \frac{\partial U}{\partial \sigma} - T' \frac{\partial \ln p_*}{\partial \lambda} , \quad (4.9)$$

and
$$F_V = - U\zeta - \dot{\sigma} \frac{\partial V}{\partial \sigma} - T' (1-\mu^2) \frac{\partial \ln p_*}{\partial \mu} : \quad (4.10)$$

ϕ is the geopotential, $\zeta = 2\Omega\mu + \nabla^2\psi$, $D = \nabla^2\alpha$, and

$$U = - \frac{1-\mu^2}{a} \frac{\partial \psi}{\partial \mu} + \frac{1}{a} \frac{\partial \alpha}{\partial \lambda} , \quad V = \frac{1}{a} \frac{\partial \psi}{\partial \lambda} + \frac{1-\mu^2}{a} \frac{\partial \alpha}{\partial \mu} .$$

(4.11),(4.12)

Using finite differences in σ, the form of (4.3)-(4.12) is similar to that of the barotropic vorticity equation (3.5)-(3.6), involving simple horizontal derivatives and products. This is true except for the vertical advection terms. $\dot{\sigma}$ is obtained from a vertical integral of (4.6) and therefore involves products, so that terms like $\dot{\sigma}\ \partial U/\partial \sigma$ include triple products. The spectral-transform technique may be used in a quite straightforward manner for the horizontal representation in primitive equation models (Bourke, 1974; Hoskins and Simmons, 1975). Usually a grid fine enough to treat only the simple product terms is adequate since the aliasing associated with the very small triple product appears to be unimportant.

The time-step usable in a simple leap-frog scheme for primitive equation models is limited by the CFL condition

$$C \frac{[M(M+1)]^{\frac{1}{2}}}{a} \Delta t \lesssim 1 \quad , \tag{4.13}$$

where C is the external gravity wave speed ($\sim$ 300 m s^{-1}). The role of external and internal gravity modes is generally thought of as providing rapid adjustment towards the slowly evolving, meteorologically interesting, modes for which C in (4.13) would be the fastest flow speed in the model ($\sim$ 75 m s^{-1}). Therefore a semi-implicit time scheme is often employed (Robert, 1969; Kwizak and Robert, 1971), in which the terms dominant in gravity wave motion are treated implicitly while elsewhere the simple leap-frog scheme is retained. For example, the last two terms in (4.4) and (4.6) are averaged in time. A time-step at least a factor of four larger is possible, but at the expense of solving a second order elliptic equation. For a spectral model this equation becomes algebraic and its solution is trivial. More details are given in (Hoskins and Simmons, 1975).

Attempts have also been made to produce vertical schemes superior to second order finite differences. Machenhauer and Daley (1972) used the spectral-transform method in this direction also but encountered problems due to vertically propagating waves. Finite element techniques in the vertical along with spectral-transform in the horizontal would provide an attractive three-dimensional Galerkin model. Such a vertical scheme has been formulated by Staniforth and Daley (1977) and is being actively considered elsewhere (Burridge, pers. comm.).

5. THE USE OF SPECTRAL-TRANSFORM MODELS

Spectral-transform models have proved a valuable tool for meteorological research in the last decade. As an example, Fig. 2 shows a result achieved with a 5-layer T42 (triangular truncation with M = 42) spectral-transform model using a transform grid of 64 Gaussian latitudes and 128 equally-spaced longitudes. The model contained no forcing except for a small internal dissipation of the form $- K\nabla^4$ on vorticity, divergence and temperature. The initial state was symmetric about the equator and composed of an atmosphere warm at the equator and cold at the pole balanced by a westerly wind increasing with height, with a small, local perturbation in surface pressure. The basic flow is unstable to so-called baroclinic waves whose structures are very similar to the mobile depressions of middle latitudes. However this instability is triggered by the initial perturbation in an orderly manner with successive surface pressure lows and highs appearing downstream. At the day shown, it is noticeable that despite the use of global basic functions

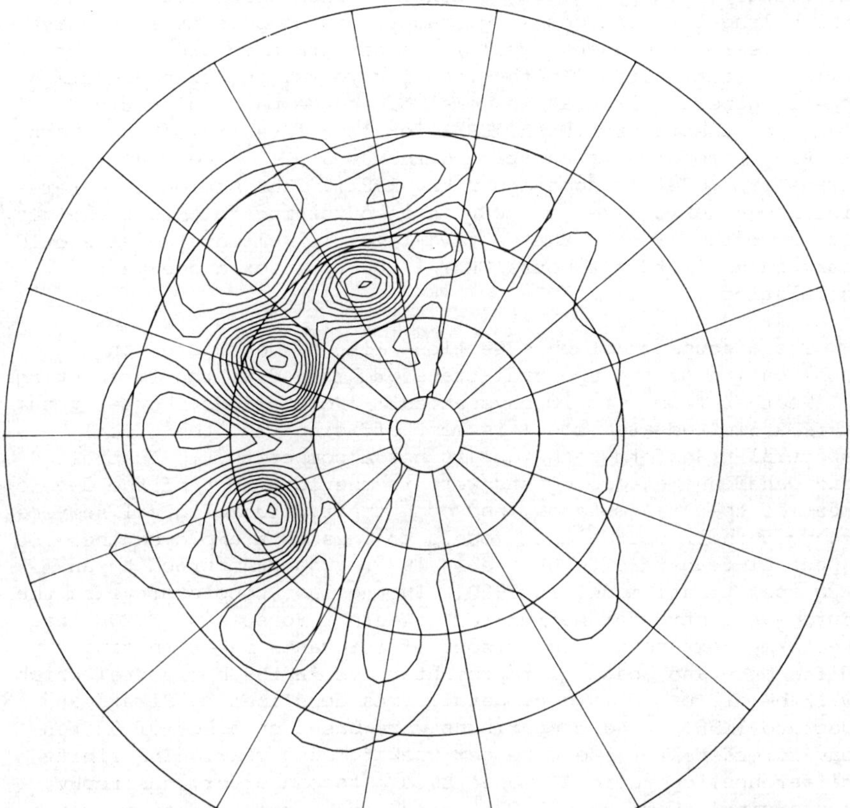

Fig. 2 The surface pressure map at day 13 for the experiment described in the text. Shown is the downstream development of mid-latitude systems from an initial perturbation. The experiment and relevant meteorological theory are discussed in detail in Simmons and Hoskins (1979).

the hemisphere is divided into a region of significant disturbances and one of almost zero disturbance.

Whereas finite difference models with coarse resolution are unable to represent correctly the growth and movement of the baroclinic waves discussed above, quite severely truncated spectral models perform quite reasonably. This means that such models are ideally suited to general circulation studies in which long integrations (e.g. many years of perpetual January or 50 years with annual cycle) are performed in order to obtain climate statistics. In the last few years, the standard atmospheric general circulation model of the Geophysical Fluid Dynamics Laboratory (NOAA, USA) for long integrations has been an R15 (rhomboidal truncation with M = J = 15) spectral-transform model (Manabe and Hahn, 1981). The National Center for Atmospheric Research, USA is also turning to this technique. It is worthy of note that a five-layer T21 climate model should take only of order 2 hours CRAY 1 CPU time for a one-year simulation.

The discussion of this section will now be centred on the application of the spectral-transform method to the forecasting of weather from 1 to 10 days ahead. (For periods shorter than this a limited area model is most efficient and the global spectral-transform technique is not a competitor.) Australian and Canadian meteorologists were in the forefront of the development of spectral methods, and both their meteorological services changed their operational models to ones that employed the spectral-transform technique in 1976. The USA turned to an R30 operational model in 1980. In the last couple of years the European Centre for Medium-range Weather Forecasts (ECMWF) has performed extensive comparisons of forecasts based on finite difference and spectral representations in the horizontal which will be discussed in some detail. As described by Girard and Jarraud (1982), the comparisons were based on a set of 53 ten-day forecasts for one case per week for one year. The finite difference, operational model used a second order, enstrophy conserving scheme on a 1.875° grid and is referred to as N48 because of the number of latitudinal points between equator and pole. This was compared with a T63 spectral-transform model utilising a very similar grid and very similar CPU time. The two models employed the same vertical discretisation involving 15 levels and the same physical parameterisations. The difficulties of comparing two forecast models are considerable, and the efforts made at ECMWF are detailed by Girard and Jarraud (1982).

The first problem is that the crucial package of physical parameterisations is always developed with one model in mind and may tend to compensate for that model's errors. Likewise the analysis of the meteorological data at the initial time and its preparation for the model is also developed in a particular context. Both these factors presumably tend to favour the existing model.

The other problems are all associated with how to compare the forecasts from the two models. One way is to consider a specific region and look for occasions on which the forecast from one model is clearly superior to that from the other. Such occasions are rare and, though Girard and Jarraud (1982) give an example when the spectral forecast was markedly better than the finite difference one, the evidence as to the total behaviour of the models is small. One of the major errors frequently present in forecast models is the incorrect movement of surface low pressure centres. Consequently, Girard and Jarraud (1982) looked at the average error for the models for lows that, between days 1 and 2, moved less than 5°, between 5° and 10° and so on. The slow moving systems were similarly treated by the two models but for those that moved more than 15° in one day, the underestimate of the speed was about 40% less for the T63 model.

Objective measures of forecasts such as those based on the root mean square error of the height of a pressure surface reward models that under-estimate growth and tend to the climatological mean state rather quickly. One objective measure that has some virtues is a height anomaly correlation. If h_c, h_f and h_o are the climatological mean, forecast and observed values of the height of a particular pressure surface then the height anomaly correlation is the average correlation of $(h_f - h_c)$ and $(h_o - c)$ expressed as a percentage. Fig. 3 shows a typical example of the 1000 mb height anomaly correlation for the region 20°N - 82.5°N for the ECMWF forecast model.

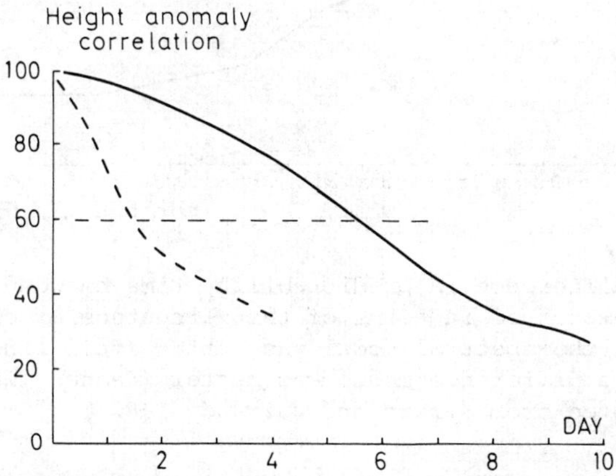

Fig. 3 1000 mb height anomaly correlations for the region 20°N - 82.5°N obtained by the ECMWF operational model. The 60% level may be used as a definition of predictability time. The result from a zero model of persistence is shown by heavy dashes. (Adapted from Bengtsson and Simmons, 1983.)

Experience has suggested that a value greater than 60% usually provides a useful forecast. Here the 60% line is crossed at 5.5 days. For comparison, persistence of the initial data typically provides a useful forecast for less than 1½ days. For the sequence of comparison forecasts, Fig. 4 shows a summary of the number of occasions on which each model crossed the 60% anomaly correlation line a certain number of hours after the other one. As measured by this index, the T63 model is clearly superior, though the increase in the length of the useful forecast is only of the order of ¼ day in 5 days.

Based on these experiences, ECMWF have decided to switch to a T63 spectral model.

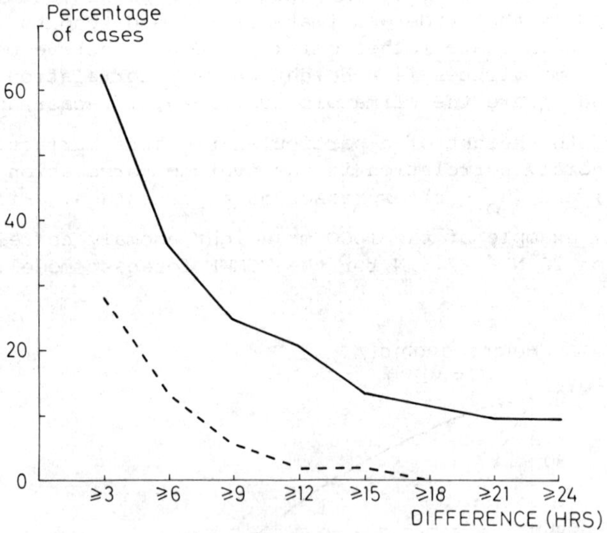

Fig. 4 The difference in predictability time between T63 and N48 expressed in terms of the percentage of cases for which the spectral model was better (full line) and the finite difference model was better (dashed line). (Adapted from Girard and Jarraud, 1982.)

6. CONCLUSION

The global domain is perhaps second only to the periodic box as being suitable for the spectral technique and so it is hardly surprising that, with the advent of the transform method it is proving so attractive to meteorological modellers. The improvement over present day finite difference methods for the same CPU time is small but probably significant. Finite element methods show some promise (Cullen and Hall, 1979) but do not presently appear competitive. In the vertical dimension, the almost universal use of second order finite differences could well be changed in the next decade, with finite elements appearing to be a strong contender.

REFERENCES

ARAKAWA, A. 1977 Computational aspects of numerical models for weather prediction and climate simulation. In Methods in Computational Physics, Vol. **17**: General Circulation Models of the Atmosphere (J. Chang, Ed.), Academic Press.

BENGTSSON, L. and SIMMONS, A.J. 1983 Medium range weather prediction - operational experience at ECMWF. In Large Scale Dynamical Processes in the Atmosphere (B.J. Hoskins and R.P. Pearce, Eds), Academic Press.

BOURKE, W. 1972 An efficient one-level, primitive-equation spectral model. *Mon. Weath.Rev.*, **100**, 683.

BOURKE, W. 1974 A multi-level spectral model. I. Formulation and hemispheric integrations. *Mon.Weath. Rev.*, **102**, 687.

BOURKE, W., McAVENEY, B., PURI, K. and THURLING, R. 1977 Global modelling of atmospheric flow by spectral methods. In Methods in Computational Physics, Vol. **17**: General Circulation Models of the Atmosphere (J. Chang, Ed.), Academic Press.

CULLEN, M.J.P. and HALL, C.D. 1979 Forecasting and general circulation results from finite element models. *Q.J. R. Met. Soc.*, **105**, 571.

ELIASEN, E., MACHENHAUER, B. and RASMUSSEN, E. 1970 On a numerical method for integration of the hydrodynamical equations with a spectral representation of the horizontal fields. Report No. 2, Institut for Teoretisk Meteorologi, Univ. of Copenhagen.

GARP, 1974 Proceedings of the International Symposium on Spectral Methods for Numerical Weather Prediction, GARP WGNE Report No. 7, World Meteorological Organisation, Geneva.

GIRARD, C. and JARRAUD, M. 1982 Short and medium range forecast differences between a spectral and grid-point model. An extensive quasi-operational comparison. ECMWF Technical Report, Reading.

HAURWITZ, B. 1940 The motion of atmospheric disturbances on the spherical earth. *J. Marine Res.*, **5**, 254.

HOSKINS, B.J. and SIMMONS, A.J. 1975 A multi-layer spectral model and the semi-implicit model. *Q.J. R. Met. Soc.*, **101**, 637.

KWIZAK, M. and ROBERT, A. 1971 A semi-implicit scheme for grid-point atmospheric models of the primitive equations. *Mon. Weath.Rev.*, **99**, 32.

MACHENHAUER, B. 1979 The spectral method. In Numerical Methods used in Atmospheric Models. GARP Pub. Series No. 17, World Meteorological Organization, Geneva.

MACHENHAUER, B. and DALEY, R. 1972 A baroclinic primitive equation model with a spectral representation in three dimensions. Report No. 4, Institut for Teoretisk Meteorologi, Univ. of Copenhagen.

MACHENHAUER, B. and RASMUSSEN, E. 1972 On the integration of the spectral hydrodynamical equations by a transform method. Report No. 3, Institut for Teoretisk Meteorology, Univ. of Copenhagen.

MANABE, S. and HAHN, D.G. 1981 Simulation of atmospheric variability. *Mon.Weath.Rev.*, **109**, 2260.

NEAMTAN, S.M. 1946 The motion of harmonic waves in the atmosphere. *J. Met.*, **3**, 53.

ORSZAG, S.A. 1970 Transform method for calculation of vector-coupled sums: Application to the spectral form of the vorticity equation. *J. Atmos. Sci.*, **27**, 890.

ORSZAG, S.A. 1971 Numerical simulation of incompressible flows within simple boundaries. I. Galerkin (spectral) representation. *Stud. Appl. Math.*, **50**, 293.

ROBERT, A. 1966 The integration of a low order spectral form of the primitive meteorological equations. *J. Met. Soc. Japan*, **44**, 237.

ROBERT, A. 1969 The integration of a spectral model of the atmosphere by the implicit method. In Proceedings of WMO/IUGG Symposium, Japn Met. Agency, 1968, VII, 19.

SILBERMAN, I. 1954 Planetary waves in the atmosphere. *J. Met.*, **11**, 27.

SIMMONS, A.J. and HOSKINS, B.J. 1979 The downstream and upstream development of unstable baroclinic waves. *J. Atmos. Sci.*, **36**, 1239.

STANIFORTH, A.N. and DALEY, R. 1977 A finite-element formulation for the vertical discretization of sigma-coordinate primitive equation models. *Mon. Weath. Rev.*, **105**, 1108.

SIMULATION METHODS FOR TURBULENT FLOWS

D.C. Leslie

(Queen Mary College, London)

1. THE ROLE OF LARGE EDDY SIMULATION IN TURBULENCE RESEARCH

At any level above the pure correlation of experimental data, a calculation on turbulent flow involves some input from the Navier-Stokes equations. These equations contain more information than one could ever use, but the more information one rejects (in the sense of not attempting to reproduce it in the calculation) the more one must model, invent, guess.

The k-ε model, which is now standard for advanced engineering investigations, is a good example. One retains only the mean values of the velocities and the turbulent stresses, and one has to:-

Assume the stresses can be calculated with an eddy viscosity representation;
Make a model of the diffusion of k;
Invent an ε equation, which is only weakly connected to the Navier-Stokes equations.

One then introduces modelling constants, which are supposed to be universal but cannot be so. They are spectral averages across the big eddies, and depend on the structure of these eddies, which must vary from one flow to another. Without denigrating the very real achievements of this model, one must have doubts about its universality and therefore about its ability to predict a radically new flow.

If k-ε is insufficient - and that is of course a matter for judgement and debate - where should one turn. There are three essentially different approaches:-

I More detailed time-averaged closures,

II Classical closures such as Direct Interaction,

III Detailed computation of the time evolution of the turbulent field - the Monte Carlo approach.

1.1 More detailed time-average closures

These are attractive in that they sidestep the eddy viscosity assumption, but they do not avoid the more fundamental difficulty of specifying the length scale. It is for judgement whether their complexity, which can be quite formidable in a meteorological problem, repays the better accuracy. As of today, the collective view seems to be that it does not.

1.2 Classical closures

These retain far more information than the time-averaged closures. For example they need no length scale assumption, and in that sense they are a real competitor to Monte Carlo (the term is defined below). However, it seems that they are not competitive in their use of computer time. They deal with pair correlations rather than with the primitive field, and in all but the simplest problems this higher dimensionality (1 time + 2x3 space vs 1 time + 3 space if the problem is stationary) more than outweighs the greater smoothness of the object functions. For example, a Direct Interaction calculation on the simplest real flows (e.g. fully developed shear-driven flow in a channel) can only be accommodated on today's computers by using some form of subgrid model.

Taking into account the computational disadvantage and the stronger assumption needed in the classical closures (namely, expansion on a Gaussian basis) it is hard to see them competing with the Monte Carlo methods for engineering work. Of course this does not diminish their value for theoretical studies.

2. MONTE CARLO METHODS

The phrase implies that the solution of a stochastic equation is approximated by tracing particular realisations forward in time. To implement the method for the Navier-Stokes equations, one sets up a discrete representation appropriate to the geometry, supplies an appropriate realisation of the initial field and advances the solution step by step. The discretisation may be either by spectral representation or on a finite difference mesh: the more advanced codes mix the two. The volume balance approach of Schumann(1975) is usually regarded as falling into the finite difference category, but it is closer in spirit to the finite element method. Fourier functions are the natural basis for homogeneous problems, and Chebyshev polynomials are the preferred choice for the cross-stream direction in a channel simulation (Orszag and Kells, 1980). They are not optimal, but they can be handled by a small modification of the Fast Fourier transform technique. Also they concentrate representative points near the walls, so that the field is represented best in the places where it is changing fastest.

Unless the simulation includes fine detail near a boundary, the viscous stability limit is not very pressing and explicit methods can be used for time-advancement. Schumann's group at Karlsruhe used leapfrog in their channel simulations, in which the wall is replaced by synthetic boundary conditions, the viscous layers not being represented at all. Both the Stanford group under Reynolds and Ferziger (Ferziger, 1981) and the Turbulence Unit at QMC (Antonopoulos-Domis, 1981a) have used Adams-Bashforth for simulating homogeneous flows with an occasional Euler step to stabilise the computation. Because the sampling error is relatively large, there is no need to require high accuracy from the numerical scheme.

These explicit methods seem to be satisfactory in such applications, but it will not be easy to detect a non-disastrous instability when the solution is so chaotic. The codes are of course tested against problems with known (smooth) solutions, and one can only hope that this is enough.

When the viscous layers near the wall are being simulated in detail, some degree of implicitness is needed. It is customary to use a semi-implicit scheme, with Crank-Nicolson for the most unstable terms (viscous, normal pressure gradient and perhaps some subgrid terms) and Adams-Bashforth for the rest (Moin and Kim, 1981).

The main problem in generating the initial fields is to decide what one wants. For example, one could require that the starting field for a homogeneous simulation should have zero mean, a Gaussian distribution and a specified three-dimensional scalar spectrum: it must of course satisfy continuity. There is no problem about producing such a field from the output of a standard random number generator, but when one has done this one discovers that the field is under-specified. Because it is Gaussian, the triple correlations are zero; the inertial transfer is insufficient until these have built up, and one can actually see this happening. Should one have specified the triple correlations in the starting field? It would be hard to do, and it does not seem to be necessary, in that later stages of the simulation do not seem to be affected by the "starting transient".

The problem is much harder when the flow is inhomogeneous, and the QMC team has given considerable attention to it. The aim is that the starting field should conform to the leading features of the flow, as determined by experiment. This is not cheating, since if this input does not correspond to the natural state of the simulated flow, then the simulated flow will diverge from it. This strategy can save a lot of computing time: if one starts from a really bad initial field, it can take far longer than one can afford to get to a stationary state. The early stages of the brilliant Moin-Kim (1981) channel simulation reinforce this warning. Their basic difficulty was an inadequate subgrid model, but this was concealed by the unrealistic starting field.

3. THE LIMITATIONS OF FULL SIMULATION

There is no fundamental reason why one should not be able to simulate a turbulent flow in any desired detail. The Navier-Stokes equations are believed to be a sufficient description, discretisation errors can in principle be made as small as one wishes, and existing algorithms are probably sufficient*. (It is not so clear that this is equally true of the software.) The reason why one cannot do so in practice has been understood for at least 20 years (Corrsin, 1961). In a high Reynolds number engineering flow, the largest scales (eddies) will be 10^3 times as big as the smallest ones: in an oceanic or atmospheric flow, the ratio might be 10^5:1. Thus even for an engineering flow one might need 10^{10} mesh points, implying the manipulation of up to 10^{12} bytes of data at each time step. We have no computer capable of such operations, and it is commonly said that we cannot foresee the time when we shall have one.

It seems that there are at present only three types of computer to be considered for simulations on the largest possible scale. These are the CRAY1, the ICL-DAP and CYBER 205: I have no experience of the last, and my remarks are confined to the first two. A 64^3 simulation packs comfortably into both CRAY and DAP†. This is quite inadequate for the simulation of real flows, but interesting work can be done on the decay of homogeneous turbulence. The (Taylor microscale) Reynolds number does not exceed 70, and is more usually around 40.

---

* Orszag (lecture at Nice, September 1981) thinks that the improvement in algorithms has been no less rapid than that in computers.

† The recent 256^3 simulation of the Taylor-Green vortex by Orszag and Morf takes advantage of the high symmetry of this vortex. Its computational demands are roughly the same as those of an unrestricted 64^3 simulation.

4. RATIONALE OF LES

If computer limitations prevent one simulating all scales in a real turbulent flow, then the remedy is to simulate the large scales only, accepting that the small ones will not be properly represented. This idea, due to Smagorinsky (1963) and Lilly (1966), has succeeded brilliantly because it accords with the fundamental nature of the flow. It is the large eddies which determine the characteristics of the individual flow - the heat transfer in a pipe, the motion of a hurricane. These large eddies are characteristic of the individual flow, vary from one flow to another and are hard to model in any universal way. In the absence of analytical solutions, computation must be the preferred way of studying their properties.

In contrast the small eddies are fairly universal and moreover they have little effect on the large-scale properties of the flow. Their function is simply to dissipate into heat the energy which is injected into the turbulence by the large eddies. They are not modelled directly in Large Eddy Simulation. What one represents is the inertial drain of energy out of the grid scales (the eddies or wave numbers which are represented in the simulation) into the subgrid scales (the wave numbers which cannot be represented for lack of resolution). The immediate transfer is from the top end of the represented range into the bottom end of the non-represented range, and this is all that is modelled. The subsequent fate of the transferred energy is irrelevant and is not modelled. (In fact this energy will be transferred into the dissipation range and there converted into heat.) Provided the cut between the represented and the non-represented wave numbers is well into the inertial range, one universal subgrid model (representation of the transfer across the cut) should apply to all flows.

This conventional picture is not realised in any actual Large Eddy Simulation. The method is often used when the (local) Reynolds number is too small for there to be any inertial range, and even when there is such a range, one would never waste precious computational points on a feature whose properties are relatively well understood. In practice the cut is made towards the top end of the production range, and it is not clear why the universal subgrid model works as well as it does. In Kraichnan's phrase, it suffers from an embarrassment of success.

Having extolled the advantages of LES, it is important to stress its disadvantages too. First, engineering turbulence is inherently three-dimensional. Since LES is seeking to simulate it properly, the simulation must also be three-dimensional even though the problem is of lower dimensionality, and the size of the computation is inevitably large*. The other side of this coin is that, in principle at least, the computational disparity between LES and (e.g.) k-ε should diminish as the geometry of the problem gets more complex.

There is another drawback, perhaps not so obvious as the first, and this is the unforgiving nature of the technology. One is attempting to reproduce the flow rather than to model it: there is therefore almost nothing to adjust. One cannot improve the results of an unsatisfactory simulation by adjusting model constants, and one must therefore go back to the drawing board.

5. FORMULATION OF THE LES EQUATIONS

The first step is to write

$$u = \bar{u} + u'. \tag{5.1}$$

Here the overbar denotes an operation which picks out the grid scales, and the equation simply says that whatever is not a grid scale must be a subgrid scale. At this stage, the separating operation is not defined: it is not a time average. The formal procedure is to substitute (5.1) into the Navier-Stokes equations, and then to perform the separating operation. For constant property flows and in kinematic units, the result is

$$\frac{\partial \bar{u}_i}{\partial t} + \frac{\partial}{\partial x_j}\overline{\bar{u}_i\bar{u}_j} = -\frac{\partial P}{\partial x_j} + \nu\nabla^2\bar{u}_i - \frac{\partial}{\partial x_j}\tau_{ij} \tag{5.2}$$

where

$$\tau_{ij} = R_{ij} - \frac{1}{3}R_{kk}\,\delta_{ij} \tag{5.3}$$

$$P = \bar{p} + \frac{1}{3}R_{kk} \tag{5.4}$$

with

$$R_{ij} = \overline{u_i'\bar{u}_j} + \overline{\bar{u}_i u_j'} + \overline{u_i'u_j'}. \tag{5.5}$$

R_{ij} is the raw subgrid stress, the effect on the grid scale equation of motion of the incompleteness of the simulation. This interaction between the seen and the unseen operates through the inertial terms (including the pressure). There would be no subgrid stresses if the Navier-Stokes equations were linear.

---

*Schumann (1973, 1975) has done successful simulations with meshes as small as 8^3, but this reduction is achieved by using synthetic boundary conditions of questionable validity.

The equations are rearranged so that the subgrid stress is trace-free and therefore has the same character as the Reynolds stress: the price is the incorporation of an unknown component into the grid scale pressure*. Note also the rather curious form of the grid scale inertial term: this will be discussed later.

In equation (5.5), the terms $\overline{u'\bar{u}}$ represent the "forward scattering" of energy from the grid scales (GS) into the subgrid scales (SGS), while $\overline{u'u'}$ represents backscattering from the SGS into the GS. The backscattering is surprisingly large: studies by Leslie and Quarini (1979) show that it is typically one-half of the net forward scattering.

It is clear from equation (5.2) that, over and above the problems common to all Monte Carlo simulations of the Navier-Stokes equations, there are two problems special to LES. These are the definition of the separation process and the representation of the subgrid stress τ_{ij} in terms of the grid scale velocity only. This representation is known as subgrid modelling (SGM). It is a closure approximation, but it is much more forgiving than those made in formulating e.g. the k-ε equations.

6. THE SEPARATION PROCESS

The process is more or less self-evident for a spectral calculation, in which

$$u = u(\underline{n})$$

$\underline{n}$ being a three-dimensional modal index. The separation process is then defined by

$$\left.\begin{aligned} u(\underline{n}) &= \bar{u}(\underline{n}) \text{ if } \underline{n} \text{ is a represented mode} \\ &= u'(\underline{n}) \text{ if not} \end{aligned}\right\} \qquad (6.1)$$

a represented mode being one which is included in the calculation. The simplest example of this process is the simulation of a homogeneous flow. The represented modes will usually be contained within the box

$$- K \leq k_1, k_2, k_3 \leq + K \qquad (6.2)$$

*The QMC team has studied the importance of this term by simulating the same homogeneous flow on meshes of 8^3, 16^3 and 32^3. $\langle P^2 \rangle$ changed substantially when the mesh was increased from 8^3 to 16^3, but the further increase to 32^3 did not produce much further change in the pressure.

which defines the process . Even in this simple case there are complications, since interest will be focused on the scalar spectrum E(k), where $k = |\underset{\sim}{k}|$. Quarini (1977) has shown that so far as the scalar spectrum is concerned, equation (6.1) is equivalent to

$$\bar{u}(k) = G(k)u(k) \tag{6.3}$$

and has given the form of G(k).

When we come to finite difference representations, it is no longer obvious how the separation process should be defined. Leonard (1974) transforms equation (6.3) into configuration space and defines

$$\bar{u}(\underset{\sim}{x}) = \int_{-\infty}^{\infty} G(\underset{\sim}{x}-\underset{\sim}{x}')u(\underset{\sim}{x}')d\underset{\sim}{x}' \tag{6.4}$$

$u(\underset{\sim}{x}')$ being the primitive or unfiltered field. His preferred form of filter is the Gaussian

$$G(\underset{\sim}{x}-\underset{\sim}{x}') = \left(\frac{6}{\pi\Delta^2}\right)^{3/2} \exp\{-\frac{6}{\Delta^2}|\underset{\sim}{x}-\underset{\sim}{x}'|^2\} \tag{6.5}$$

Δ being the equivalent width of the filter. This operation smooths out the small eddies while leaving the large ones almost untouched, which is just what is wanted.

The sound relation of Leonard's proposal to the nature of the separation process is attractive. Moreover it deals simply and naturally with the inertial term

$$\frac{\partial}{\partial x_j}\overline{\bar{u}_i\bar{u}_j}$$

in the fundamental equation (5.2), which is not simply deducible from $\bar{u}_i$. Following Leonard, one writes

$$\overline{\bar{u}_i\bar{u}_j} = \bar{u}_i\bar{u}_j + L_{ij} \tag{6.6}$$

and the problem is to determine the functional relationship between the "Leonard stress" L_{ij} and the GS velocity field $\bar{u}_i$. With the filter procedure

$$L_{ij} = \gamma\Delta^2\nabla^2(\bar{u}_i\bar{u}_j) + O(\Delta^4) \tag{6.7}$$

where γ is related to the second moment of the filter. For the

Gaussian filter, $\gamma = \frac{1}{24}$. Thus the left-hand side of equation (5.2) is approximated by

$$\frac{\partial \bar{u}_i}{\partial t} + \frac{\partial}{\partial x_j}(\bar{u}_i \bar{u}_j) + \frac{1}{24}\frac{\partial}{\partial x_j}\{\Delta^2 \nabla^2 (\bar{u}_i \bar{u}_j)\} = \ldots\ldots$$

Investigations at Stanford suggest that the optimum value of Δ is about 2h, h being the mesh spacing.

Despite the attractions of filtering, we at QMC now believe that it is not the best procedure. There are three objections to it.

(i) The operation (6.5) implies that the field is moved through the filter, and this operation seems to be undefinable when the field is inhomogeneous.

(ii) How does the field know that the filter is Gaussian?

(iii) Careful studies by Antonopoulos-Domis (1981b) show that, at least at low Reynolds numbers, the effect of the Leonard stress is to backscatter energy from the higher wave number parts of the represented range into the biggest eddies, and thus to distort the spectrum in the represented range.

(i) is generally agreed. For example, although Stanford is the home of filtering, Moin and Kim (1981) did not use it in the direction normal to the wall in their recent highly successful simulation of fully developed channel flow. (ii) is a purely philosophical point: if the simulation cannot know the nature of the filter, is the operation well defined? (iii) is still being debated: if true, it seems decisive.

The QMC team therefore agrees with Schumann that it is better not to filter, and to seek schemes that make the Leonard stress identically zero. The separation process (6.1) achieves this automatically for spectral codes (Leslie and Quarini, 1979). For finite difference codes we use the volume-balance method of Deardroff (1970) and Schumann (1973), in which the equation of motion is integrated across each elementary computational volume. Provided the velocities are regarded as constant across each such volume, L_{ij} is automatically zero.

7. SUBGRID MODELLING

Smagorinsky (1963) suggested that the subgrid stress τ_{ij} should be related to the GS velocity field by an eddy viscosity model

$$\tau_{ij} = \nu_n S_{ij} \tag{7.1}$$

where

$$S_{ij} = \frac{\partial \bar{u}_i}{\partial x_j} + \frac{\partial \bar{u}_j}{\partial x_i} \tag{7.2}$$

in the GS strain tensor. Townsend's justification of the mixing length model (Townsend, 1961) is much better for the small eddies than for the whole of the velocity field, since diffusion is a large-eddy phenomenon which will be small in the subgrid field. It therefore seems likely that (7.1) - plus - (7.2) is a good approximation. Smagorinsky also proposed

$$\nu_n = C^2 h^2 \bar{s} \tag{7.3}$$

where

$$\bar{s} = \left(\frac{1}{2} S_{ij} S_{ji}\right)^{1/2} \tag{7.4}$$

while h is the mesh spacing and C is a non-dimensional constant (~ 0.18). Subsequently, Lilly (1966) justified the model in terms of turbulence theory, showed how it could be adapted to the finite difference structure of actual codes and evaluated the constant C.

This model has stood the test of time remarkably well. The only important modification is that of Schumann (1975), and this is an extension rather than a replacement. Lilly showed that, since the subgrid eddy viscosity in (7.1) is isotropic, this equation must imply isotropy of the subgrid scales. Schumann realised that near the wall there would be significant anisotropy of the subgrid scales (his "inhomogeneous part") and that this must be represented if the simulation was to succeed. This was later confirmed by Moin and Kin (1981), who were unable to generate enough energy to sustain the flow without this additional term.

Schumann separated out the inhomogeneous or anisotropic part by averaging over planes parallel to the wall. We shall denote this operation by <>: note that this is <u>not</u> a realisation average. (His procedure is slightly more complex, and the simplified version which we shall describe is due to Moin and Kim.) He writes

$$S_{ij} = (S_{ij} - \langle S_{ij}\rangle) + \langle S_{ij}\rangle \tag{7.5}$$

the first term being the isotropic part, which is adequately represented by the Smagorinsky-Lilly model. In detail Schumann's model is

$$\tau_{ij} = -\nu_n(S_{ij} - \langle S_{ij}\rangle) - \nu_n^* \langle S_{ij}\rangle \tag{7.6}$$

$$\nu_n = C^2h^2\left\{\frac{1}{2}(S_{ij} - \langle S_{ij}\rangle)(S_{ji} - \langle S_{ji}\rangle)\right\}^{1/2} \tag{7.7}$$

$$\nu_n^* = L^2\left\{\frac{1}{2} \langle S_{ij}\rangle\langle S_{ji}\rangle\right\}^{1/2} \tag{7.8}$$

$$L = \min\,(C_1h,\ \kappa x_2). \tag{7.9}$$

In (7.9), C_1 is a computational constant (~ 0.1) while $\kappa \sim 0.4$ is the von Karman constant. This model reduces to Smagorinsky-Lilly when the flow is isotropic, i.e. $\langle S_{ij}\rangle = 0$.

Schumann (1973) did not attempt to simulate the viscous layers immediately adjacent to the wall. Moin and Kim (1981) have extended his model to these layers by introducing appropriate van Driest damping factors.

8. TYPICAL RESULTS

Much effort has been expended on the Large Eddy Simulation of homogeneous flows, a simple situation in which it is relatively easy to judge the adequacy of the simulation. No experimental flow is truly homogeneous, but the decaying turbulence generated by a grid in a wind-tunnel is usually regarded as being so. The agreement between the simulations and the experiments suggests that this is reasonable.

Figs. 1 and 2 show simulations by Antonopoulos-Domis (1981a) of measurements of this type by Comte-Bellot and Corrsin (1971) and Yeh and Van Atta (1973) on decaying velocity fields. Fig. 1 gives the decay of the total energy with distance from the grid, while Fig. 2 gives the three-dimensional scalar spectrum at a particular downstream station. In both cases, the agreement between simulation and experiment is extremely good.

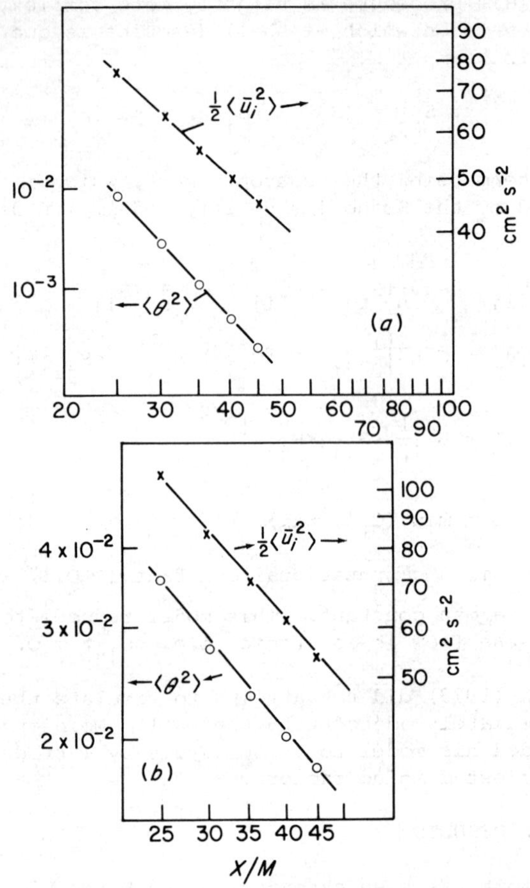

Fig. 1. Decay of $\frac{1}{2}\langle \bar{u}_i^2 \rangle$ and $\langle \theta^2 \rangle$, (a) 16^3 runs, h = 1.5 cm, (b) 32^3 runs. Leonard term not included. $C_u = 0.23$, $C_O = 2.0$. ——, filtered experimental. Simulation × × ×, $\frac{1}{2}\langle u_i^2 \rangle$; o o o, $\langle \theta^2 \rangle$*.

---

*Figs.1 and 2 are reproduced with the kind permission of the author and editor from Antonopoulos-Domis, 1981a, *JFM*, **104**, pp. 63 and 66, Figs. 2 and 5 respectively.

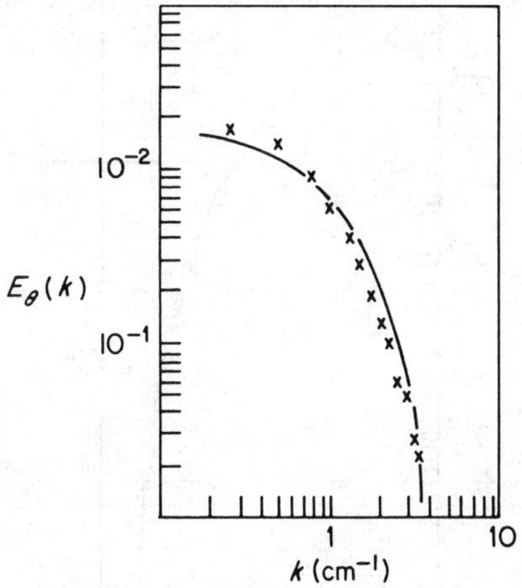

Fig. 2 Three-dimensional temperature spectra, 16^3 runs, h = 1.5 cm. No Leonard term, C_O = 2.0, X/M = 46.5. ——, filtered experimental; × × × , simulation.*

The inputs to the calculation are a value for the constant C in equation (7.3) and a spectrum at a starting station sufficiently far from the grid for local effects due to grid structure to have died away. The value of C may be regarded as the only degree of freedom in the calculation; the input spectrum is simply the initial data. With C = 0.23 the energy decays at the right rate in all simulations of both experiments, so that the constant is universal: moreover this value is in precise agreement with that calculated by Lilly (1966). The excellence of the calculated spectrum is even more striking, since this shows that energy transfers are being correctly represented. (Results at other downstream stations are equally satisfactory.)

Antonopoulos-Domis (1981a) has also used the technique to simulate the Yeh and Van Atta (1973) experiments on the decay of a passive temperature field, and his results for the decay of the "temperature energy" are also shown on Fig. 1. The agreement with experiment is again extremely good. Fig. 3 shows a temperature spectrum. The agreement is adequate, though not as good as for the velocity spectrum. The fascinating feature of this figure is that the data points with large sampling error are from the computation: the scatter in the experimental points is much less. This figure illustrates a real difficulty of LES, namely that it is very expensive to reduce the statistical scatter to an acceptably low level. Fig. 4 compares the measured and computed development of the integral scales of

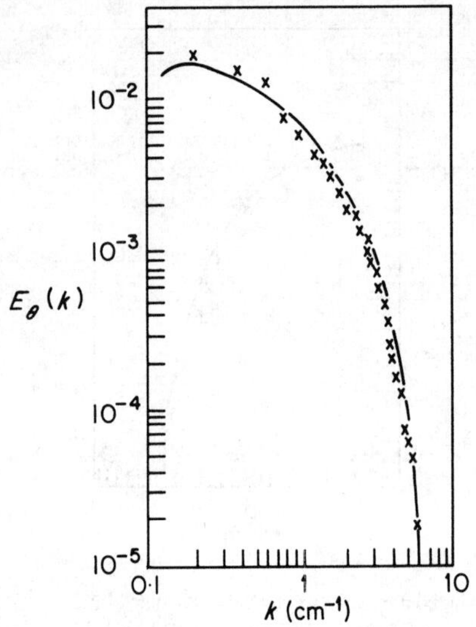

Fig. 3 Three-dimensional temperature spectra, 32^3 runs, h = 1 cm. No Leonard terms, C_0 = 2.0, X/M = 46.5. ——, filtered experimental; × × ×, simulation.*

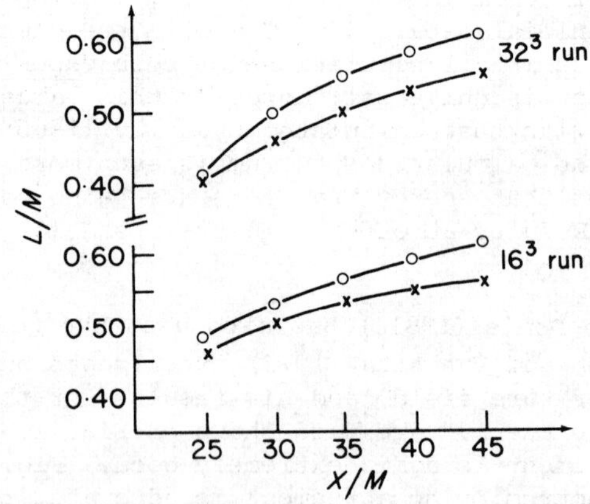

Fig. 4 Integral scales of velocity L_u and temperature L_θ. ——, filtered experimental. Simulation: × ×, L_u; O O, L_θ.*

*Figs. 3 and 4 are reproduced with the kind permission of the author and the editor from Antonopoulos-Domis, 1981a, *JFM*, **104**, pp. 68 and 70, Figs. 7 and 10 respectively.

both spectra: again the agreement is very good.

The simulation of real, inhomogeneous flows is naturally much less developed. In the writer's opinion, the outstanding achievement to date in this area is the low Reynolds number channel simulation of Moin and Kim (1981). We select two examples from the wealth of data which they present. Fig. 5 compares the computed mean velocity profile with the measurements of Hussain and Reynolds (1975): the agreement is impressive.

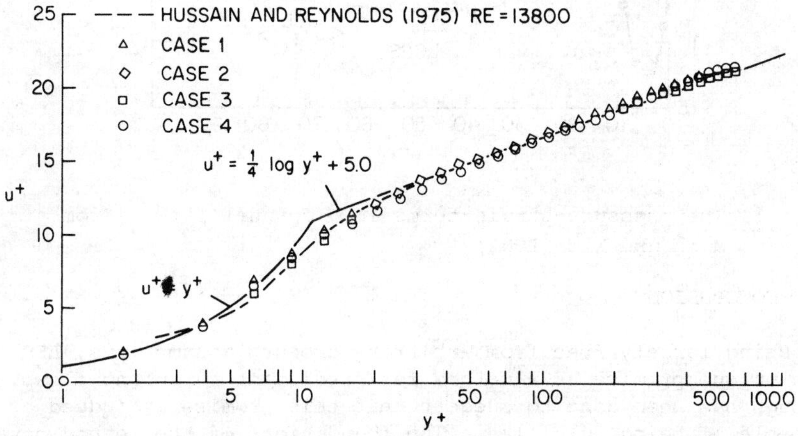

Fig. 5 Mean velocity profiles from four computed cases and comparison with experimental data.*

This figure is of great importance, since it represents the first truly non-experimental prediction of the von Karman constant. (One hesitates to call the prediction theoretical.) In contrast, Fig. 6 is purely computational: it shows the variation of the three diagonal components of the pressure-strain tensor with distance from the wall. The interest of this figure is that although these quantities are significant, there is no way of measuring them. Here we see LES being used to supplement experiment, by providing values for quantities which are not susceptible to experimental determination.

Full details of the Moin-Kim simulation are given in their paper, while Ferziger (1981) surveys other simulations of homogeneous flows.

---

*Figs. 5 and 6 are reproduced with the kind permission of the authors from Moin and Kim, 1982, NASA TM 81309.

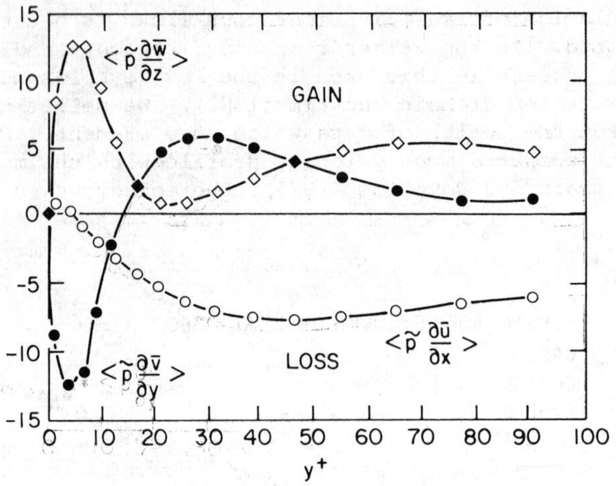

Fig. 6 The pressure-strain terms in a channel flow. From Moin and Kim (1982).

9. CONCLUSIONS

Being largely free from arbitrary closure assumptions, LES has unique promise as a method for predicting turbulent flows. Enough has been done to suggest that this promise is indeed capable of being fulfilled. The foundations of the method have been firmly established, and the agreement between the available predictions and experiment is encouraging. The principal drawback of the method is its insatiable appetite for time on the largest possible computers.

ACKNOWLEDGEMENTS

This paper rests on the work of past and present members of the QMC Turbulence Unit, particularly that of Professor M. Antonopoulos-Domis, Dr. M.D. Love, Dr. S.T.B. Young: we have been much influenced by Professor J.H. Ferziger.

I am grateful for the financial support of SERC and the Aerodynamics Division of RAE Farnborough, and for Dr. Young's help in the preparation of this paper.

REFERENCES

ANTONOPOULOS-DOMIS, M. 1981a Large-eddy simulation of a passive scalar in isotropic turbulence, *J. Fluid Mech.*, **104**, pp. 55.

ANTONOPOULOS-DOMIS, M. 1981b Aspects of large eddy simulation of homogeneous isotropic turbulence, *Int. J. Num. Meth. Fluids*, **1**, pp. 273.

COMTE-BELLOT, G. and CORRSIN, S. 1971 Simple Eulerian time correlation of full- and narrow-band velocity signals in grid-generated isotropic turbulence, *J. Fluid Mech.*, **48**, pp. 273.

CORRSIN, S. 1961 Turbulent flow, *American Scientist*, **49**, pp. 300.

DEARDORFF, J.W. 1970 A numerical study of three-dimensional turbulent channel flow at large Reynolds numbers, *J. Fluid Mech.*, **41**, pp. 453.

FERZIGER, J.H. 1981 Higher-level simulations of turbulent flow, Stanford University Report TF-16.

HUSSAIN, A.K.M.F and REYNOLDS, W.C. 1975 Measurements in fully developed turbulent channel flow, *J. Fluids Eng.*, **97**, pp. 568-578.

LEONARD, A. 1974 Energy cascade in large-eddy simulations of turbulent fluid flows, *Adv. in Geophys.*, **18A**, pp. 237.

LESLIE, D.C. and QUARINI, G.L. 1979 The application of classical closures to the formulation of subgrid modelling procedures, *J. Fluid Mech.*, **91**, pp. 65.

LILLY, D.K. 1966 On the application of the eddy viscosity concept in the inertial sub-range of turbulence, NCAR Report No. 123.

LOVE, M.D. 1980 Subgrid modelling studies with Burgers' equation, *J. Fluid Mech.*, **100**, pp. 87.

MOIN, P. and KIM, J. 1982 Numerical investigation of turbulent channel flow, *J. Fluid Mech.* **118**, 341.

ORSZAG, S.A. and KELLS, L.C. 1980 Transition to turbulence in plane Poiseuille and plane Couette flow, *J. Fluid Mech.*, **96**, pp. 159.

QUARINI, G.L. 1977 Applications of closed turbulence theories at high wave numbers, PhD Thesis, University of London.

SCHUMANN, U. 1973 Ein verfahren zur direkten numerischen simulation turbulenter strömungen in platten - und ringspaltkanälen and über seine anwendung zur untersuchung von turbulenzmodellen, PhD thesis, Karlsruhe University, Available as KFK 1854.

SCHUMANN, U. 1975 Subgrid scale model for finite difference simulations of turbulent flows in plane channels and annuli, *J. Comput. Phys.*, **18**, pp. 376.

SMAGORINSKY, J.S. 1963 General circulation experiments with the primitive equations I: the basic experiment, *Mon. Weath. Rev.*, **91**, pp. 99.

TOWNSEND, A.A. 1961 Equilibrium layers and wall turbulence, *J. Fluid Mech.*, **11**, pp. 97.

YEH, T.T. and VAN ATTA, C.W. 1973 Spectral transfer of scalar and velocity fields in heated grid turbulence, *J. Fluid Mech.*, **58**, pp. 2.

FINITE ELEMENT SOLUTIONS FOR NAVIER-STOKES EQUATIONS

J. Rae

(AERE Harwell, Didcot, Oxon)

1. INTRODUCTION

This paper reviews briefly several topics which have arisen recently in the application of finite element methods to Navier-Stokes and related equations. The points examined mostly cropped up in practical work and are not much discussed in text-books. After a brief section reminding readers of the finite element method and the Navier-Stokes equations the following four sections treat in turn the topics:

(1) the placing of a pressure reference point in a flow region.

(2) spurious modes in solutions due to unfortunate choices of elements or integration schemes.

(3) Benard problems and uniqueness of solutions.

(4) problems with fluxes on boundaries with Dirichlet boundary conditions.

The Navier-Stokes equations describing fluid flow can be written in the non-dimensional form

$$\frac{\partial \underline{u}}{\partial t} + \underline{u}.\ \nabla \underline{u} + \nabla P - \frac{1}{Re} \nabla^2 \underline{u} = 0 \qquad (1.1)$$

$$\nabla .\ \underline{u} = 0 \qquad (1.2)$$

where $\underline{u}$ is the velocity field, P the pressure and Re the Reynolds number. To specify a physical problem mathematically, boundary conditions have to be given and these are usually specified velocity values (e.g., zero on stationary walls) or specified fluxes, normal derivatives of velocity, all round the flow region. For time-dependent problems initial values also have to be given. Since only gradients of pressure appear, the equations determine pressure only up to an additive constant which has to be fixed by specifying a reference pressure at some point initially.

Finite element representations of the equations (1.1) and (1.2) can be set up in many ways and have been described by many authors (Connor and Brebbia (1976) Chung (1978); Thomasset (1981)). Here we simply outline the most common method, the Galerkin finite element method. A weak form of the equations is formed by multiplying each of (1.1) and (1.2) by suitable test functions $\underline{g}$ and h and integrating over the flow region Ω (with boundary $\partial\Omega$). This gives equations

$$\int_{\Omega} \frac{\partial \underline{u}}{\partial t} \cdot \underline{g} + \int_{\Omega} \underline{u} \cdot \nabla \underline{u} \cdot \underline{g} + \int_{\Omega} \nabla P \cdot \underline{g}$$

$$+ \frac{1}{Re} \int_{\Omega} \nabla \underline{u} \cdot \nabla \underline{g} - \frac{1}{Re} \int_{\partial\Omega} \frac{\partial \underline{u}}{\partial n} \cdot \underline{g} = 0 \tag{1.3}$$

$$\int_{\Omega} (\nabla \cdot \underline{u}) h = 0 \;. \tag{1.4}$$

The idea is that if equations (1.3) and (1.4) hold for enough choices of test functions then for sufficiently smooth functions they are equivalent to the original equations together with their natural boundary conditions. However, since lower order derivatives appear, the weak form can admit solutions which are less smooth than required by (1.1) and (1.2). Test functions are generally chosen to vanish on parts of $\partial\Omega$ where Dirichlet boundary conditions have been specified.

The spatial discretisation of the weak equations is made by dividing the region Ω into a finite number of elements, of simple geometric shape and approximating the velocities, pressures and test functions by piecewise continuous polynomials defined on each element by the values at a number of nodes. This is equivalent to representing the functions by a linear combination of basis functions ϕ_i associated with the nodes $i = 1, \ldots, N$ with ϕ_i non zero only on elements containing node i. The Galerkin method uses the same set of basis functions for the test functions and the unknown fields and so gives as many equations as there are nodal values to be found. It is not desirable to use the same basis functions to model both $\underline{u}$ and P (Jackson and Cliffe (1980), Sani et al. (1980)). The continuity equation (1.4) is then treated as the "pressure equation", although P does not appear in it, in the sense that it is the equation tested with the basis functions which have been used for P.

The simplest and commonest way of implementing Dirichlet boundary conditions is to discard the finite element equations related to the corresponding boundary nodes and use instead the specified value as the nodal value. This keeps the number of equations the same as the number of unknowns. The same treatment is usually applied to the pressure specification. The continuity equation for some pressure node is dropped from the system of equations and the pressure given a specified value there. Neumann boundary conditions other than the natural ones are satisfied by adding to equation (1.3) a boundary term of the form shown but integrated only along the appropriate part of the boundary and using the known value of the normal derivative to evaluate the integral.

2. CHOICE OF THE PRESSURE REFERENCE POINT

When the pressure is specified at a node J the continuity equation at that node is, as we already observed, deleted from the set of equations. However, it is simple to show that this equation is implied by the other nodal continuity equations. We have

$$\int_\Omega \phi_i \; \nabla . \underline{\hat{u}} = 0 \quad i \neq J \tag{2.1}$$

where $\underline{\hat{u}}$ is the finite element representation of $\underline{u}$.
For all the usual sets of basis functions, the constant 1 is in the space of test functions and normally

$$\sum_{i=1}^{N} \phi_i = 1 \tag{2.2}$$

If we now sum (2.1) over all values of $i \neq J$ we get

$$\int_\Omega (1-\phi_J) \; \nabla . \underline{\hat{u}} = 0 \tag{2.3}$$

or equivalently

$$\int_\Omega \phi_J \; \nabla . \underline{\hat{u}} = \int_\Omega \nabla . \underline{\hat{u}} = \int_{\partial\Omega} \underline{\hat{u}} . \underline{n} \tag{2.4}$$

where $\underline{n}$ is the normal to the boundary $\partial\Omega$. The last expression in (2.4) is just the net flux through the boundary and when no sources or sinks are present this should be zero in a well formulated model. This however requires some care as it is a condition on the finite element approximation to $\underline{u}$. If the modelling is not careful the term may not be zero.

The above argument also assumed exact arithmetic which is not, of course, what a computer does. Finite precision arithmetic gives nodal values of the fields $\underline{u}'_i$, p' different from the exactly calculated $\hat{\underline{u}}_i$, $\hat{p}$.

For $\underline{u}' = \sum_i \phi_i \underline{u}'_i$ we now obtain

$$\int_\Omega \phi_i \nabla.\underline{u}' = \varepsilon_i \tag{2.5}$$

with (small) non zero ε_i. The actual value of ε_i arises during the calculation and depends on the program. The line of argument used before now leads to

$$\int_\Omega (1-\phi_J) \nabla.\underline{u}' = \sum_{i\neq J} \varepsilon_i \tag{2.6}$$

and so to

$$\int_\Omega \phi_J \nabla.\underline{u}' = \int_{\partial\Omega} \underline{u}'.\underline{n} - \sum_{i\neq J} \varepsilon_i \tag{2.7}$$

In a problem where the boundary conditions are under good control we will have $\int_{\partial\Omega} \underline{u}'.\underline{n} = 0$ because, for example, the values of $\underline{u}'$ are so specified. The ε term will in general be different from zero and enters the equations in just the way that a source or sink would if placed at J. Rough estimates on (2.7) show that the velocity at J will be perturbed by a magnitude

$$(h_j/A_j) \sum_{i\neq J} \varepsilon_i$$

where A_J is the area of elements which have node J and h_J the mesh spacing at J. The velocity equation (1.3) implies there is a corresponding jump in $\hat{p}_J$. If the boundary conditions have not been carefully done the effective source can be very much bigger and so will the jumps in $\hat{\underline{u}}_j$ and $\hat{p}_j$. In both cases a local refinement near J, which will not affect most ε_i, will increase the jump as A_J decreases. Thus local refinement near the pressure specification is bad.

Numerical experiments have been made on these effects and we quote just one brief example (see Jackson (1981a) for more details). The driven cavity problem, a square fluid filled box with sliding lid, was calculated on a 17 x 17 uniform grid with IBM single precision arithmetic. The pressure was specified first at the centre of the cavity and then at the lower left corner. The pressures in the two cases differed by a constant to better than 1.10^{-4} in about 0.32 except at the lower left corner where the difference was about 2×10^{-3}. Regular grid refinement had just the effect predicted. Also on a basic 21 x 21 grid local grid refinement in the corner increased the effect. In fact a local refinement by a factor of 16 gave a pressure jump comparable to the pressure values over most of the cavity.

The conclusions are these:

(i) use high precision arithmetic

(ii) do not place the pressure reference point in a region of high refinement

(iii) if a noticeable pressure jump remains suspect the modelling of boundary conditions.

3. SPURIOUS MODES PRODUCED BY NUMERICAL INTEGRATION

One of the early problems in solving the Navier-Stokes equations by the finite element method was the occurrence of spurious pressure fluctuations for certain choices of elements (Taylor and Hood (1974); Olson and Tuann (1976); Jackson and Cliffe (1980); Sani et al. (1980)). The "mixed interpolation" rule of thumb provided an escape from the practical difficulties but one cannot say the problem was fully understood until quite recently. Spurious oscillations can arise for other reasons which, though mentioned briefly in (Zienkiewicz, 1977) seem to have had rather less study. It is one of these I wish to discuss here.

As used in practice the finite element method always employs some numerical integration scheme in the assembly of the equations. For high order or curved elements the integration is usually not exact and should therefore properly be considered as part of the discretisation. Choice of too low an order of integration can lead to trouble as we now show.

A matrix which arises in time dependent finite element analysis, for example in eq. (1.3), is the mass or capacity matrix

$$M_{ij} = \int_{\Omega} \phi_i \phi_j \tag{3.1}$$

where ϕ_i i = 1, ..., N are the basis functions. As Ω is divided into elements we also have

$$M_{ij} = \sum_{\text{elements}} \int_{\text{element}} \phi_i \phi_j = \sum_{el} M_{ij}^{el} \tag{3.2}$$

Because the integrations are done approximately each M_{ij}^{el} is replaced by its approximation D_{ij}^{el} given by

$$D_{ij}^{el} = \sum_{\sigma=1}^{n} \omega_\sigma \phi_i(x_\sigma) \phi_j(x_\sigma) \tag{3.3}$$

where x_σ, ω_σ are the n integration points and weights for the element in question. If the number of independent basis functions on this element is m and m is greater than n, then the element matrix D_{ij}^{el} has at least m - n zero eigenvalues. If we renumber nodes so that i,j on the element run from 1 to m the condition for a zero eigenvalue is

$$\sum_{j=1}^{m} D_{ij}^{el} u_j = 0 \quad i=1, \ldots, m \tag{3.4}$$

for some eigenvector u_j. This can be rewritten

$$\sum_{=1}^{n} \omega_\sigma \phi_i(x_\sigma) \sum_{j=1}^{m} u_j \phi_j(x_\sigma) = 0 \quad i=1, \ldots, m \tag{3.5}$$

so it is enough to show the existence of a function

$$\Phi = \sum_{j=1}^{m} u_j \phi_j(x) \tag{3.6}$$

such that

$$\Phi\ (x_\sigma) = 0 \quad \text{for } \sigma=1, \ldots, n. \tag{3.7}$$

But the mapping $T : R^m \to R^n$ defined by

$$T(u_1, \ldots, u_m) = \left(\sum_{j=1}^{m} u_j\phi(x_1), \ldots, \sum_{j=1}^{m} u_j\phi(x_n)\right) \tag{3.8}$$

is linear and has a nullspace of dimension at least m-n by standard linear algebra (Halmos (1958)). A constructive proof of this result can be found in (Jackson, 1981b).

The element eigenvectors found above can be combined to form an eigenmode of the complete problem if they are compatible across elements and with the boundary conditions. For example, the eight noded Serendipity quadrilateral element with a seven point Gauss scheme has, by the above, at least one zero eigenvalue. The gauss points are arranged asymmetrically on the element as in Fig. 1 so the element has an orientation given by the line in Fig. 2. The latter figure also illustrates the form of the eigenvector. It is now easy to see that an arrangement of elements as in Fig. 3 would allow a global oscillation whereas the arrangement in Fig. 4 would suppress it. Also a boundary condition fixing the left hand side of Fig. 3 would kill the oscillation except possibly on very large grids. This, and other examples, are described in (Jackson, 1981b).

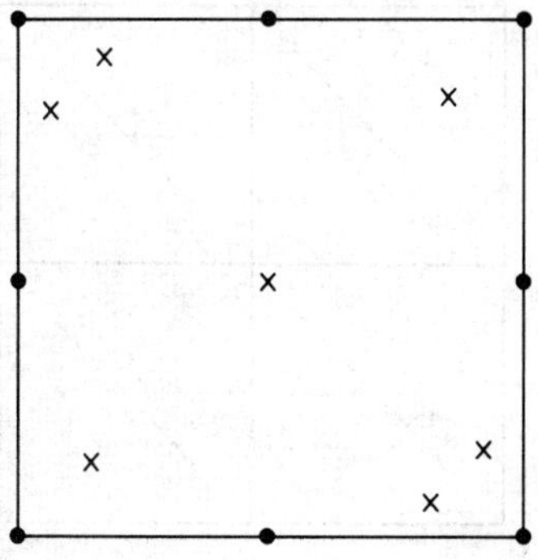

Fig. 1 The Gauss Points on the Eight-noded Serendipity Element.

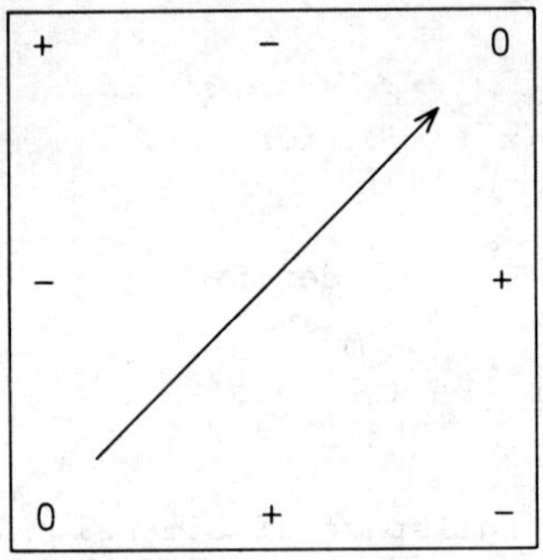

Fig. 2 The Handedness of the Element and Schematic of its Zero Energy Mode.

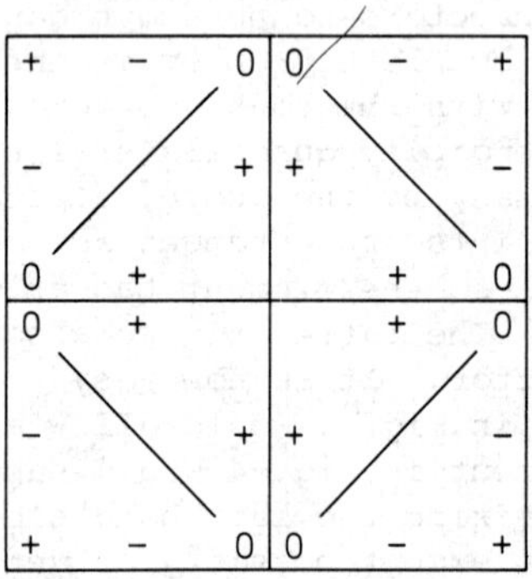

Fig. 3 A Layout of Elements Permitting the Mode to Propagate across the Mesh.

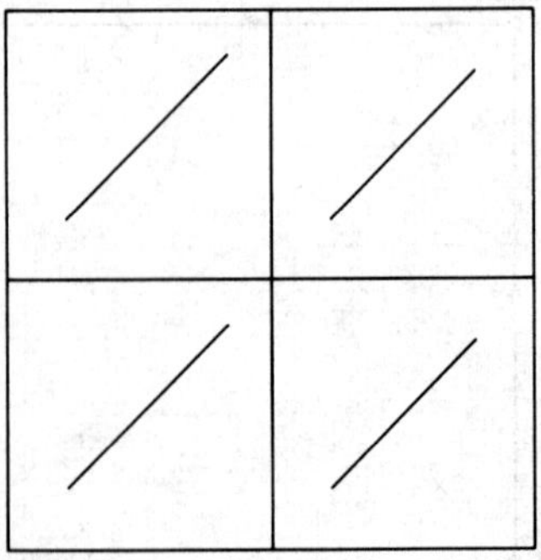

Fig. 4 A Layout of Elements which will Suppress the Mode.

Although we have used the capacity matrix to illustrate these points similar problems can arise with other terms in equations such as the matrix

$$K_{ij} = \int_{\Omega} \nabla\psi_i \cdot \nabla\psi_j \tag{3.9}$$

arising from the fourth term in equation 1.3. A dramatic illustration is given in Figs. 5 and 6. They illustrate contours of the solution of the Poisson equation on a circular region with a point source and point sink. Fig. 5 used a grid of 6-noded quadratic triangles with a 3-point Gauss scheme: Fig. 6 was the same in all respects except it used a 1-point scheme.

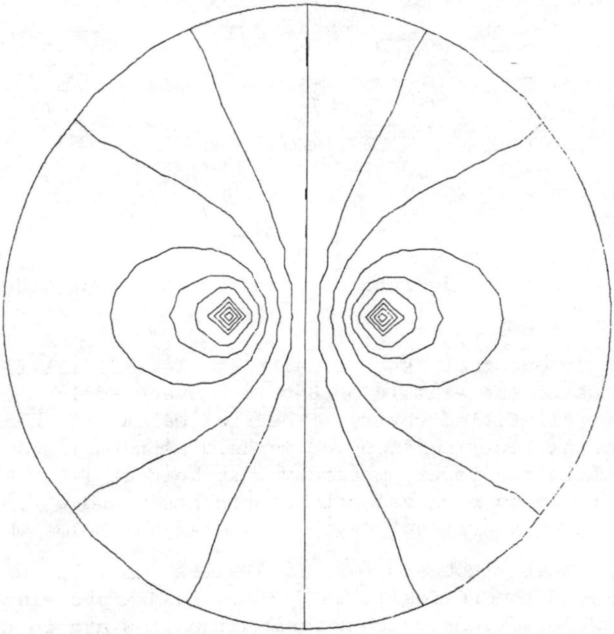

Fig. 5 Contours of the Solution of the Poisson Equation on a Circular Region with Source and Sink. (6-noded Quadratic Triangles with a 3-pt Gauss Scheme).

4. BÉNARD CONVECTION AND UNIQUENESS OF SOLUTION

It is perhaps unnecessary to point out that nonlinear equations can often have more than one solution but the topic is still found difficult by many physicists and engineers who feel that well formulated problems should have only one solution or

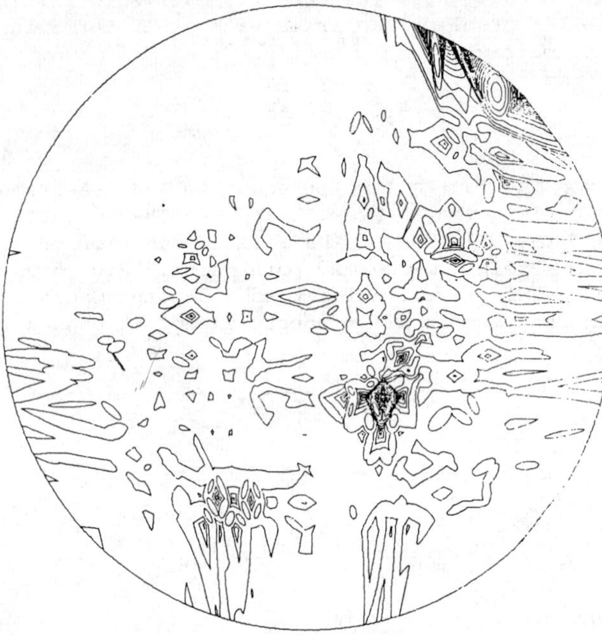

Fig. 6 Effect on Figure 5 of Changing to a 1-pt Gauss Scheme.

at least only one that is physically stable. I will illustrate the point using the example of Bénard buoyancy-driven convection in a square air-filled cavity. The side walls x=0, 1 are non-conducting, the lower surface y=0 is held at normalised temperature T=1 while the upper surface y=1 is held at T=0. All four walls are rigid so both velocity components u and v are zero there. We have a critical Rayleigh number Ra_c below which no convection takes place and for rigid walls its value was only recently found theoretically as 2585 by Luijkx and Platten (1981). The equations which described the flow are in the dimensionless form of Mallinson and De Vahl Davis (1977)

$$u \frac{\partial u}{\partial x} + v \frac{\partial u}{\partial y} + \frac{\partial p}{\partial x} - Pr \nabla^2 u = 0 \qquad (4.1)$$

$$u \frac{\partial v}{\partial x} + v \frac{\partial v}{\partial y} + \frac{\partial p}{\partial y} - Pr \nabla^2 v - Ra\ Pr\ T = 0 \qquad (4.2)$$

$$\frac{\partial u}{\partial x} + \frac{\partial v}{\partial y} = 0 \qquad (4.3)$$

and

$$u \frac{\partial T}{\partial x} + v \frac{\partial T}{\partial y} - \nabla^2 T = 0. \qquad (4.4)$$

The first three of these equations are the time-independent Navier-Stokes equations with a buoyancy term added and the fourth equation describes the heat transport. Ra is the Rayleigh number based on the width of the cavity and the Prandtl number Pr is taken as 0.71 for air.

For every solution

$$\{ u(x),\ v(x),\ p(x),\ T(x) \} \text{ then the fields}$$

$\{ -u(1-x),\ v(1-x),\ p(1-x),\ 1-T(1-x) \}$ are also a solution since they satisfy both equations and boundary conditions. The new solution corresponds to a reflection of the first solution about the line $x = 0.5$. It is known from numerical work (Upson et al. (1980)) that the solution in the literature does not have this reflection symmetry (see Fig. 7). It follows that a second solution exists. In fact there must also be the third trivial solution, $u = v = 0$, but this is known to be physically unstable for $Ra > Ra_c$.

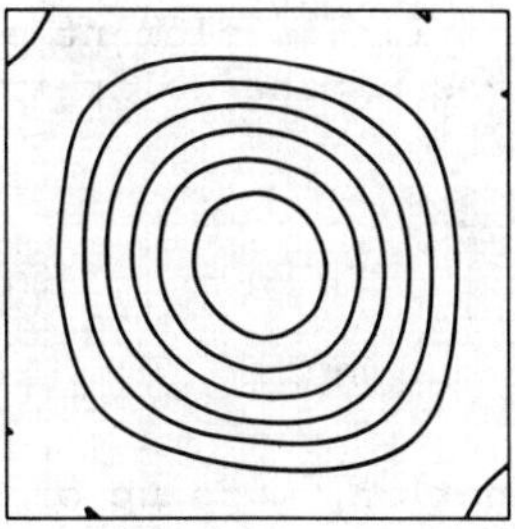

Fig. 7 Streamlines for one Steady Solution of the Bénard Problem at Rayleigh Number 10^4, above the Critical Value.

Winters (1981) has studied this problem using the finite element method and had no difficulty in generating all three solutions. He also estimated the value of Ra_c from his computations to be 2610, in good agreement with the value quoted above.

As the Rayleigh number is increased in this sort of problem the results get progressively more complicated with further bifurcations. Much remains to be done to disentangle the true solutions from those which are physically unstable or due to inadequate mesh refinement. It is also possible that in some parameter ranges there are no stable steady solutions and time dependent methods have to be used. Something of this kind may be happening in the "double-glazing problem" where the buoyancy cavity is long and thin rather than square. A sequence of complicated solutions has been found by a number of authors (Jones (1979); Winters (1981); Gresho (1981)) up to Rayleigh numbers near 3.10^5. They have all found difficulty in going beyond this with steady state codes and it may be there is no longer a stable solution.

5. BOUNDARY FLUXES

When the finite element method is used to model field variables, difficulties of various kinds can arise in the calculation of fluxes at the boundaries. The problem is present in many cases including Navier Stokes equations but we will use the following simple equation, a relative of equation 4.4, to illustrate the points.

$$\frac{\partial T}{\partial t} - \nabla^2 T = Q \qquad (5.1)$$

in a 2-dimensional region Ω with Neumann boundary conditions $\partial T/\partial n = f$, where f is a known function on the entire boundary $\partial\Omega$. If $\hat{T}$ represents the finite element approximation to T then the usual procedure gives as the Galerkin equation

$$\frac{\partial}{\partial t}\int_\Omega \hat{T}g + \int_\Omega \nabla\hat{T}.\,\nabla g - \int_{\partial\Omega} fg = \int_\Omega Qg \qquad (5.2)$$

where g is the test function, made up of element basis functions. Now all standard sets of basis functions allow the constant 1 to be a test function: in general $\Sigma\,\phi_i = 1$ when the sum is taken over all nodes. Taking this choice for g we get the conservation law

$$\frac{\partial}{\partial t}\int_\Omega \hat{T} - \int_{\partial\Omega} f = \int_\Omega Q \qquad (5.3)$$

The important point to note is that the prescribed flux f and not the finite element flux $\partial\hat{T}/\partial n$ appears here. In fact $\partial\hat{T}/\partial n \neq f$ and in general one does not even have $\int_{\partial\Omega} f = \int_{\partial\Omega} \partial\hat{T}/\partial n$. To see this it is enough to think of Ω as a single square element with linear shape function. Both $\partial\hat{T}/\partial x$ and $\partial\hat{T}/\partial y$ are constant on this and $\int_{\partial\Omega} \partial\hat{T}/\partial n = 0$ independently of how f is chosen.

In the case when Dirichlet boundary conditions are used equations are "thrown out" at nodes on the boundary where values are prescribed. The argument using $\Sigma\,\phi_i = 1$ can no longer be used to obtain a conservation law and $\partial\hat{T}/\partial n$ is just as bad a flux as above. The way to proceed (Gresho and Lee (1981)) is to solve the standard equations, that is

$$\frac{\partial}{\partial t}\int_{\Omega} \hat{T}\phi_i + \int_{\Omega} \nabla\hat{T}.\,\nabla\phi_i = \int_{\Omega} Q\phi_i \qquad (5.4)$$

for each node i including on the boundary together with equations defining a consistent flux $\hat{C}$

$$\int_{\partial\Omega} \hat{C}\phi_j = -\int_{\partial\Omega} \nabla\hat{T}.\,\nabla\phi_j \qquad (5.5)$$

for each boundary node j. Since the equations now use <u>all</u> ϕ's the $\Sigma\phi_i = 1$ argument can be used to obtain a conservation law in the form of (5.3) but with the f replaced by the consistent flux $\hat{C}$. Thus the flux defined by (5.5) is a meaningful and useful quantity defined on the boundary.

An alternative approach is to ignore the problem and use very thin elements at the boundary. The equations which are formed are now

$$\frac{\partial}{\partial t}\int_{\Omega} \hat{T}\,\phi_i + \int_{\Omega} \nabla\hat{T}.\,\nabla\phi_i = \int_{\Omega} Q\phi_i \qquad (5.6)$$

for all i <u>not</u> on the boundary. If we now think of Ω made up of the thin boundary region Ω' and the interior Ω'' the nodes i cover Ω'' including its boundary. Summing over these gives as before the conservation law

$$\frac{\partial}{\partial t}\int_{\Omega''}\hat{T} - \int_{\Omega''} Q + \sum_i \left\{ \frac{\partial}{\partial t}\int_{\Omega'} \hat{T}\,\phi_i + \int_{\Omega'} \nabla\hat{T}.\,\nabla\phi_i - \int_{\Omega'} Q\phi_i \right\} = 0 \qquad (5.7)$$

Now if the thickness ε of Ω' is taken very small the $\partial/\partial t$ and Q terms in brackets vanish like ε. In the term with $\nabla\hat{T}.\ \nabla\phi_i$ only the component with $\partial\phi_i/\partial n$ remains non zero as $\varepsilon \to 0$ because $\partial\phi_i/\partial n$ behaves like $(-1/\varepsilon)$. In each of the thin elements the contribution is to lowest order $-\partial\hat{T}/\partial n$ and the sum over i takes us right round the boundary to give

$$\frac{\partial}{\partial t}\int_{\Omega}\hat{T} - \int_{\partial\Omega}\frac{\partial\hat{T}}{\partial n} = \int_{\Omega} Q$$

to lowest order in ε.

Several difficulties remain. The consistent flux method has problems at sharp corners where it gives a continuous flux when the true flux is discontinuous. The thin element method can lead to singularities if the boundary is sufficiently curved. A recently proposed method, related to the above, appears to get round the problem (Cliffe and Jackson (1982)).

REFERENCES

CHUNG, T.J. 1978 Finite Element Analysis in Fluid Dynamics. New York: McGraw-Hill.

CLIFFE, K.A. and JACKSON, C.P. 1982 Private Communication.

CONNOR, J.J. and BREBBIA, C.A. 1976 Finite Element Techniques for Fluid Flow. London, Newnes-Butterworth.

GRESHO, P.M. 1981 Private Communication.

GRESHO, P.M. and LEE, R.L. 1981 The Consistent Method for Computing Desired Boundary Quantities when the Galerkin FEM is used to Solve Thermal and/or Fluid Problems. In Proceedings of 2nd International Conference on Numerical Methods in Thermal Problems, Venice, Italy.

HALMOS, P.R. 1958 Finite-dimensional Vector Spaces, Princeton, Van Nostrand.

JACKSON, C.P. and CLIFFE, K.A. 1981 Mixed Interpolation in Primitive Variable Finite Element Formulations for Incompressible Flow. *Int. J. Num. Meth. Eng.* **17**, 1659-1688.

JACKSON, C.P. 1981a The Effect of the Choice of the Reference Location in Numerical Modelling of Incompressible Flow. AERE Harwell Report TP 925.

JACKSON, C.P. 1981b Singular Capacity Matrices Produced by Low-order Gaussian Integration in the Finite Element Method. *Int. J. Num. Meth. Eng.* **17**, 871-877.

JONES, I.P. 1979 A Numerical Study of Natural Convection in an Air-Filled Cavity: Comparison with Experiment, Numerical Heat Transfer **2**, 193-213.

LUIJKX, J.M. and PLATTEN, J.K. 1980 On the Onset of Free Convection in a Rectangular Channel. *J. Non-Equilib. Thermodyn.* **6**, 141-158.

MALLINSON, G.D. and DeVAHL DAVIS, G., 1977 Three Dimensional Natural Convection in a Box: A Numerical Study. *J. Fluid Mech.* **83**, 1-31.

OLSON, M.D. and TUANN, S.Y. 1976 Primitive Variables versus Stream Function Finite Element Solutions of the Navier Stokes Equations. In Proceedings of 2nd International Symposium on Finite Element Methods in Flow Problems. S. Margherita Ligure, Italy.

SANI, R.L., GRESHO, P.M. and LEE, R.L. 1980 On the Spurious Pressures Generated by Certain GFEM Solutions of the Incompressible Navier-Stokes Equations In Proceedings of the 3rd International Conference on Finite Elements in Flow Problems, Banff, Canada.

TAYLOR, C. and HOOD, P. 1974 Navier-Stokes Equations using Mixed Interpolation. In *Finite Element Methods in Flow Problems* (J.T. Oden, O.C. Zienkiewicz, R.H. Gallagher and C. Taylor, Eds.) Huntsville, University of Alabama Press.

THOMASSET, F. 1981 Implementation of Finite Element Methods for Navier-Stokes Equations. New York: Springer-Verlag.

UPSON, C.D., GRESHO, P.M. and LEE, R.L. 1980 Finite Element Simulation of Thermally Induced Convection in an Enclosed Cavity, Lawrence Livermore Laboratory Report UCID 18602.

WINTERS, K.H. 1981 The Finite Element Simulation of Buoyancy Driven Flows. In 2nd International Conference on Numerical Methods in Thermal Problems, Venice, Italy.

ZIENKIEWICZ, O.C. 1977 The Finite Element Method. London: McGraw-Hill.

A PETROV-GALERKIN FINITE ELEMENT FORMULATION FOR SYSTEMS OF CONSERVATION LAWS WITH SPECIAL REFERENCE TO THE COMPRESSIBLE EULER EQUATIONS

T.J.R. Hughes, T.E. Tezduyar and A.N. Brooks

(Division of Applied Mechanics, Stanford University, USA)

1. INTRODUCTION

In this paper we are concerned with certain classes of Petrov-Galerkin finite element methods for convection dominated flows. Petrov-Galerkin methods are distinguished from the more widely known (Bubnov-) Galerkin method in that weighting functions and trial solutions are selected from <u>different</u> classes of functions. This computational framework opens the way to the development of numerical methods which are <u>optimal</u>, or nearly so, according to various measures of accuracy.

This may be contrasted with the behaviour of Galerkin methods for this class of problems, which suffer from both accuracy and stability deficiencies. Dupont (1973) was the first to point out that the Galerkin method did not generally achieve optimal asymptotic error estimates for even smooth solutions of the simplest one-dimensional, first-order hyperbolic model problem. When solutions are rough, Galerkin methods are known to produce highly oscillatory error components. Thus the "best approximation property", and its stability and optimal-accuracy by-products, which the Galerkin method achieves in typical elliptic operator problems, is lost in the context of highly nonsymmetrical convection-type operators.

Due to the potential of Petrov-Galerkin methods for flow problems, a large literature is already accumulating. The first reference that we are aware of on this topic was the paper by Dendy (1974). Works which analyzed, further developed, and generalized Dendy's approach were those of Wahlbin (1974), Raymond and Garder (1976) and Baker (1979). Dendy's formulation applied to first-order hyperbolic problems. The introduction of second-order terms into the equations posed technical problems unless interpolations with C^1-continuity were introduced. These are, of course, very inconvenient in practice and this, perhaps, was the main reason why this approach did not have a more significant impact initially.

A separate school of thought was later originated by the Dundee and Swansea research groups (see, e.g., Christie et al., 1976, Heinrich et al., 1977, Griffiths and Mitchell, 1979, and Heinrich and Zienkiewicz, 1979). In these works various perturbations of the normal Galerkin weighting functions were introduced ostensibly to account for upwind influence.* Because these perturbations could be continuous, there were no impediments to applying them to equations including second-order spatial derivatives (e.g. advection-diffusion equations). Morton and colleagues (see, e.g., Barrett and Morton, 1980, 1981, Morton, 1981, Morton and Barrett, 1980, Morton and Parrott, 1980) extended the development of these ideas to time-dependent cases and defined methods achieving optimality conditions in various norms.

Our work in this area began with the development of finite element methods for the advection-diffusion and incompressible Navier-Stokes equations which we called the "streamline-upwind method" (Hughes and Brooks, 1979, Brooks and Hughes, 1980). The method was similar in intent to the skew-differencing procedure of Raithby (1976). In attempting to generalize the procedure to a Petrov-Galerkin method we encountered the same difficulty that faced Dendy's method. This was finally overcome in (Hughes and Brooks, 1982) for the advection-diffusion equation and in (Brooks and Hughes, 1982a,b) for the incompressible Navier-Stokes equations.

We have solved a wide variety of steady and time-dependent linear and nonlinear problems to demonstrate the accuracy and stability of this formulation.

Claes Johnson (1981) has analyzed the streamline method and established optimal convergence rates and a strong discontinuity capturing property, even when the discontinuity is skew to the mesh.

In this paper we present a general Petrov-Galerkin formulation of streamline type which accommodates systems of conservation laws. The formulation can specialize to advection-diffusion, compressible Euler, compressible Navier-Stokes equations,

*Frequently, Petrov-Galerkin finite element methods are described as "upwind element methods". Fully upwind finite difference methods are so inaccurate that the terminology "upwind" has taken on a perjorative connotation. It is to be emphasized in no uncertain terms that "upwind Petrov-Galerkin finite element methods" are very different from such upwind difference methods and do not suffer the pathological, excess numerical diffusion of the latter.

etc. In this work we emphasize the advection-diffusion and, in particular, the compressible Euler equations. The former are simple and delineate the methodology, and the latter are of considerable contemporary interest in computational aerodynamics. An outline of the remainder of the paper follows.

After describing notational preliminaries in Section 2 we introduce general systems of conservation laws in Section 3. In Section 4 we propose a weighted residual formulation of Petrov-Galerkin type. Particular attention is paid to formulating a variety of physically important boundary conditions. In our opinion this is a subject which is often not adequately discussed in computational fluid dynamics. Some possible simplifications of the basic formulation are described. In Section 5 we present further details concerning the selection of optimization parameters and describe a family of new methods prompted by Jean Donea's recently developed Taylor-Galerkin schemes (Donea, 1982). Some sample calculations are presented in Section 6 and conclusions are drawn in Section 7.

To keep the paper to a reasonable size we have not discussed solution algorithms herein. The interested reader may consult our other works which contain detailed treatments of this subject (see, e.g., Brooks and Hughes, 1982a,b and Tezduyar and Hughes, 1982).

2. PRELIMINARIES

Let n_{sd} denote the number of space dimensions. Let Ω be an open region in $\mathbb{R}^{n_{sd}}$ with piecewise smooth boundary Γ. Let $\underline{x} = \{x_i\}$, $i = 1,2,\ldots,n_{sd}$, denote a general point in Ω and let $\underline{n} = \{n_i\}$ be the unit outward normal vector to Γ. We assume Γ admits the following decomposition

$$\Gamma = \overline{\Gamma_G \cup \Gamma_H} \tag{2.1}$$

$$\emptyset = \Gamma_G \cap \Gamma_H \tag{2.2}$$

where Γ_G and Γ_H are subsets of Γ. The superposed bar in (2.1) represents set closure and $\emptyset$, in (2.2), denotes the empty set. The significance of Γ_G and Γ_H will be made apparent in the sequel.

The summation convention on repeated indices is assumed in force and the subscript n is used to denote the normal component of a vector (e.g., if $n_{sd} = 3$, then $\underset{\sim}{F}_n = \underset{\sim}{F}_j n_j = \underset{\sim}{F}_1 n_1 + \underset{\sim}{F}_2 n_2 + \underset{\sim}{F}_3 F_3$). A comma is used to denote partial differentiation (e.g., $\underset{\sim}{F}_{j,j} = \partial \underset{\sim}{F}_j/\partial x_j$) and t denotes time. The Kronecker delta is denoted by δ_{ij}; if i = j, then $\delta_{ij} = 1$, otherwise $\delta_{ij} = 0$.

Consider a discretization of Ω into element subdomains Ω^e, $e = 1, 2, \ldots, n_{e\ell}$, where $n_{e}\ell$ is the number of elements. Each Ω^e is taken to be an open set and its boundary is denoted Γ^e. We assume

$$\bar{\Omega} = \bigcup_e \bar{\Omega}^e \tag{2.3}$$

$$\Gamma \subset \bigcup_e \Gamma^e \tag{2.4}$$

The set $\bigcup_e \Omega^e$ will be referred to as the element interiors. The element boundaries, modulo Γ, play an important role in what follows. We call this set the interior boundary, viz.

$$\Gamma_{int} = \bigcup_e \Gamma^e - \Gamma. \tag{2.5}$$

Two classes of functions are important in the developments which follow. The classes are distinguished by their continuity properties across Γ_{int}. It suffices to assume that all functions considered herein are smooth on the element interior. Functions of the first class are continuous across Γ_{int}. These functions are denoted by C^0. Functions of the second class are allowed to be discontinuous across Γ_{int} and are denoted by C^{-1}. The C^0 functions may be recognized as containing the standard finite element interpolations.

3. SYSTEMS OF CONSERVATION LAWS

We consider the following system of m partial differential equations:

$$\underset{\sim}{U}_{,t} + \underset{\sim}{F}_{j,j} + \underset{\sim}{G} = \underset{\sim}{0} \text{ on } \Omega, \tag{3.1}$$

where

$$\underset{\sim}{U} = \underset{\sim}{U}(\underset{\sim}{x},t) \tag{3.2}$$

$$\underset{\sim}{F}_j = \underset{\sim}{F}_j(\underset{\sim}{U},\nabla\underset{\sim}{U},\underset{\sim}{x},t) \tag{3.3}$$

$$\underset{\sim}{G} = \underset{\sim}{G}(\underset{\sim}{U},\underset{\sim}{x},t). \tag{3.4}$$

The vector $\underset{\sim}{F}_j$ is referred to as a flux vector and $\underset{\sim}{G}$ is a source term. We assume that for each $\underset{\sim}{k} = \{k_i\} \in \mathbb{R}^{n_{sd}}$ there exists a matrix $\underset{\sim}{S}$ such that

$$\underset{\sim}{S}^{-1}(k_j \underset{\sim}{A}_j)\underset{\sim}{S} = \underset{\sim}{\Lambda}, \tag{3.5}$$

where

$$\underset{\sim}{A}_j = D_1 \underset{\sim}{F}_j = \partial \underset{\sim}{F}_j / \partial \underset{\sim}{U} \tag{3.6}$$

and $\underset{\sim}{\Lambda}$ is a real, diagonal matrix. If, in addition

$$D_2 \underset{\sim}{F}_j = \partial \underset{\sim}{F}_j / \partial(\nabla \underset{\sim}{U}) = \underset{\sim}{0}, \tag{3.7}$$

then (3.1) is said to be a <u>first-order hyperbolic system</u>.

Let the total flux be decomposed into partial fluxes, $\underset{\sim}{F}_j^{(1)}$ and $\underset{\sim}{F}_j^{(2)}$, as follows:

$$\underset{\sim}{F}_j(\underset{\sim}{U}, \nabla\underset{\sim}{U}, \underset{\sim}{x}, t) = \underset{\sim}{F}_j^{(1)}(\underset{\sim}{U}, \underset{\sim}{x}, t) + \underset{\sim}{F}_j^{(2)}(\underset{\sim}{U}, \nabla\underset{\sim}{U}, \underset{\sim}{x}, t). \tag{3.8}$$

Thus, if dissipative mechanisms are present, as evidenced by the appearance of the argument $\nabla\underset{\sim}{U}$, we assume they are confined to the partial flux $\underset{\sim}{F}_j^{(2)}$.

The initial/boundary-value problem for (3.1) consists of finding a function $\underset{\sim}{U}$ which satisfies (3.1), the initial condition

$$\underset{\sim}{U}(\underset{\sim}{x}, 0) = \underset{\sim}{U}_0(\underset{\sim}{x}), \tag{3.9}$$

where $\underset{\sim}{U}_0$ is a given function of $\underset{\sim}{x} \in \Omega$, and appropriately specified boundary conditions. In this paper, for sake of simplicity, we shall limit attention to boundary conditions of the following form:

<u>Dirichlet type</u>

In this case we assume

$$\underset{\sim}{\partial}\underset{\sim}{U} = \underset{\sim}{G} \text{ on } \Gamma_G \tag{3.10}$$

where $\underset{\sim}{\partial}$ is a boundary operator and $\underset{\sim}{G}$ is a prescribed function.

<u>Neumann type</u>

In this case we assume

$$-\underset{\sim}{F}_n^{(2)} = \underset{\sim}{H} \text{ on } \Gamma_H \tag{3.11}$$

where $\underset{\sim}{H}$ is a prescribed function on Γ_H. This can be interpreted as a partial flux boundary condition or, if $\underset{\sim}{F}_j^{(1)} = \underset{\sim}{0}$, a total flux boundary condition.

Finally, we allow for no boundary condition on Γ_H. This is viewed as a sepcial case of (3.11) in which both $\underset{\sim}{F}_j^{(2)}$ and $\underset{\sim}{H}$ are assumed to be identically zero. Clearly, various combinations of the above boundary conditions may also be specified on portions of Γ, although this is not explicitly spelled out in the sequel.

4. WEIGHTED RESIDUAL FORMULATION

Consider a point $\underset{\sim}{x}$ in Γ_{int}. Designate (arbitrarily) one side of Γ_{int} to be the "plus side" and the other to be the "minus side". Let $\underset{\sim}{n}^+$ and $\underset{\sim}{n}^-$ be unit normal vectors to Γ_{int} at $\underset{\sim}{x}$ which point in the plus and minus directions, respectively. Clearly, $\underset{\sim}{n}^- = -\underset{\sim}{n}^+$. Let $\underset{\sim}{F}_j^+$ and $\underset{\sim}{F}_j^-$ denote the values of $\underset{\sim}{F}_j$ obtained by approaching $\underset{\sim}{x}$ from the positive and negative sides, respectively. The "jump" in $\underset{\sim}{F}_n$ at $\underset{\sim}{x}$ is defined to be

$$\begin{aligned} [\underset{\sim}{F}_n] &= (\underset{\sim}{F}_j^+ - \underset{\sim}{F}_j^-)n_i^+ \\ &= \underset{\sim}{F}_j^+ n_i^+ + \underset{\sim}{F}_j^- n_i^- . \end{aligned} \tag{4.1}$$

As may be readily verified from (4.1), the jump is invariant with respect to reversing the plus and minus designations.

Throughout, we shall assume that trial solutions, $\underset{\sim}{U}$, satisfy $\partial\underset{\sim}{U} = \underset{\sim}{G}$ on Γ_G and weighting functions, $\underset{\sim}{W}$, satisfy $\partial\underset{\sim}{W} = \underset{\sim}{0}$ on Γ_G. Thus all Dirichlet type boundary conditions are treated as <u>essential boundary conditions</u> in the present formulation.

The variational equation is assumed to take the form

$$\begin{aligned} &\int_\Omega \{\underset{\sim}{W}^T(\underset{\sim}{U}_{,t} + \underset{\sim}{F}^{(1)}_{j,j} + \underset{\sim}{G}) - \underset{\sim}{W}^T_{,j}\underset{\sim}{F}^{(2)}_j\}d\Omega \\ &+ \sum_{e=1}^{n_{e\ell}} \int_{\Omega^e} \tilde{\underset{\sim}{P}}^T(\underset{\sim}{U}_{,t} + \underset{\sim}{F}_{j,j} + \underset{\sim}{G})\,d\Omega = \int_{\Gamma_H} \underset{\sim}{W}^T \underset{\sim}{H}\,d\Gamma . \end{aligned} \tag{4.2}$$

In (4.2), $\underset{\sim}{U}$ and $\underset{\sim}{W}$ are assumed to be taken from the same class of typical C^0 finite element interpolations; $\underset{\sim}{\tilde{P}}$ is a C^{-1} perturbation to the weighting function.

The Euler-Lagrange conditions emanating from (4.2) may be deduced by way of integration by parts:

$$0 = \sum_{e=1}^{n_{e\ell}} \int_{\Omega^e} \underset{\sim}{\tilde{W}}^T (\underset{\sim}{U}_{,t} + \underset{\sim}{F}_{j,j} + \underset{\sim}{G})\, d\Omega$$

$$- \int_{\Gamma_H} \underset{\sim}{W}^T (\underset{\sim}{F}_n^{(2)} + \underset{\sim}{H})\, d\Gamma - \int_{\Gamma_{int}} \underset{\sim}{W}^T [\underset{\sim}{F}_n^{(2)}]\, d\Gamma. \tag{4.3}$$

From (4.3) we see that the Euler-Lagrange equations are (3.1) restricted to the element interiors, (3.11), and the (partial) flux continuity condition across inter-element boundaries, namely

$$[\underset{\sim}{F}_n^{(2)}] = \underset{\sim}{0}\ . \tag{4.4}$$

Note that (3.11) is a <u>natural boundary condition</u>.

<u>Remarks</u>

1) Note that if $\underset{\sim}{\tilde{P}} = \underset{\sim}{0}$ we have a <u>Galerkin</u> weighted residual formulation; if $\underset{\sim}{\tilde{P}} \neq 0$ we have a <u>Petrov-Galerkin</u> formulation. The modified weighting function, that is

$$\underset{\sim}{\tilde{W}} = \underset{\sim}{W} + \underset{\sim}{\tilde{P}}, \tag{4.5}$$

is confined to the element interiors and thus does not affect boundary or continuity conditions.

2) The preceding formulation generalizes (Hughes and Brooks, 1982) (which was restricted to the linear advection-diffusion equation) to systems of conservation laws. A related but somewhat different formulation for the incompressible Navier-Stokes equations is described in (Brooks and Hughes, 1982a,b). The case in which $\underset{\sim}{F}_j^{(2)}$ and $\underset{\sim}{H}$ are $\underset{\sim}{0}$ has been employed in (Tezduyar and Hughes, 1982) for the compressible Euler equations. See also (Griffiths, 1981) for a related formulation.

The following examples will illustrate potential applications of the preceding formulation.

Examples

1. The scalar linear advection-diffusion equation

In this case we have

$$\phi_{,t} + \sigma_{j,j} - \mathcal{f} = 0 \tag{4.6}$$

where

$$\sigma_j = \sigma_j^a + \sigma_j^d \quad \text{(total flux)} \tag{4.7}$$

$$\sigma_j^a = \sigma_j^a(\phi) = u_j\phi \quad \text{(advective flux)} \tag{4.8}$$

$$\sigma_j^d = \sigma_j^d(\phi) = -k_{jk}\phi_{,k} \quad \text{(diffusive flux).} \tag{4.9}$$

In the above, $\mathcal{f}$ is a source term, u_j is the flow velocity, and k_{jk} is diffusivity. Each of $\mathcal{f}$, u_j and k_{jk} is assumed to be a given function of $\tilde{x}$ and t.

If Γ_g and Γ_h decompose the boundary such that

$$\Gamma = \Gamma_g \cup \Gamma_h \tag{4.10}$$

$$\emptyset = \Gamma_g \cap \Gamma_h \tag{4.11}$$

then we shall assume Dirichlet conditions on Γ_g, that is

$$\phi = g \text{ on } \Gamma_g \tag{4.12}$$

where g is a given function defined on Γ_g. The three possibilities on Γ_h are:

total flux boundary condition

$$-\sigma_n = h \text{ on } \Gamma_h \tag{4.13}$$

where h is a given function defined on Γ_h;

diffusive flux boundary condition

$$-\sigma_n^d = h \text{ on } \Gamma_h; \tag{4.14}$$

and, finally,

no boundary condition on Γ_h.

This last condition occurs in cases of pure advection in which Γ_h is defined to be that part of Γ on which $u_n \geq 0$.

The initial condition is

$$\phi(\underset{\sim}{x},0) = \phi_0(\underset{\sim}{x}) \tag{4.15}$$

where ϕ_0 is a given function of $\underset{\sim}{x} \in \Omega$.

The linear advection-diffusion equation may be brought within the general framework by the following conditions:

$$m = 1 \tag{4.16}$$

$$\underset{\sim}{U} = \phi \tag{4.17}$$

$$\underset{\sim}{W} = w \tag{4.18}$$

$$\underset{\sim}{\tilde{P}} = \tilde{p} \tag{4.19}$$

$$\underset{\sim}{F}_j = \sigma_j \tag{4.20}$$

$$\underset{\sim}{G} = -\ell \tag{4.21}$$

$$\underset{\sim}{\partial} = 1 \tag{4.22}$$

$$\underset{\sim}{\mathcal{G}} = g \tag{4.23}$$

$$\underset{\sim}{H} = h \tag{4.24}$$

$$\underset{\sim}{U}_0 = \phi_0 \tag{4.25}$$

To attain the various conditions, $\underset{\sim}{F}_j^{(1)}$ and $\underset{\sim}{F}_j^{(2)}$ need to be set as follows:

total flux boundary condition

$$\underset{\sim}{F}_j^{(1)} = \underset{\sim}{0};\ \underset{\sim}{F}_j^{(2)} = \underset{\sim}{F}_j = \sigma_j \tag{4.26}$$

diffusive flux boundary condition

$$\underset{\sim}{F}_j^{(1)} = \sigma_j^a\ ;\ \underset{\sim}{F}_j^{(2)} = \sigma_j^d \tag{4.27}$$

no boundary condition on Γ_h

$$\underset{\sim}{F}_j^{(1)} = \underset{\sim}{F}_j = \sigma_j = \sigma_j^a ; \ \underset{\sim}{F}_j^{(2)} = \sigma_j^d = 0; \ \underset{\sim}{H} = h = 0. \tag{4.28}$$

We have found the diffusive flux boundary condition to be very effective in numerically simulating outflow conditions (see, e.g., Hughes and Brooks, 1982, Brooks and Hughes, 1982a,b). Cohen (1982) has employed the total flux condition successfully in certain problems of oil reservoir simulation.

The perturbation to the weighting function, $\tilde{p}$, may be selected in various ways. In the streamline-upwind Petrov-Galerkin formulation, originated in (Hughes and Brooks, 1979, 1982) and (Brooks and Hughes, 1980), it take the form

$$\tilde{p} = \tau_i \sigma_{i,i}^a (w) = \tau (u_i w)_{,i} \tag{4.29}$$

where τ is an optimization parameter. Selection of τ is discussed in Section 5. Kelly et al. (1980) later independently proposed a similar idea (see also (Nakazawa 1982) for further developments).

2. The compressible Euler equations (inviscid gas dynamics)

The equations are

$$\rho_{,t} + (\rho u_j)_{,j} = 0 \qquad \text{(continuity)} \tag{4.30}$$

$$(\rho u_i)_{,t} + (\rho u_j u_i)_{,j} = \sigma_{ij,j} + f_i \qquad \text{(momentum)} \tag{4.31}$$

$$(\rho e)_{,t} + (\rho u_j e)_{,j} = r + f_i u_i + (\sigma_{ij} u_i)_{,j} \qquad \text{(energy)} \tag{4.32}$$

where

$$\sigma_{ij} = -p\delta_{ij}, \ p = \hat{p}(\rho, \varepsilon) \tag{4.33}$$

$$e = \varepsilon + u^2/2, \ u = \|\underset{\sim}{u}\| = (u_i u_i)^{\frac{1}{2}} \tag{4.34}$$

in which ρ is the density, u_j is the velocity, σ_{ij} is the Cauchy stress tensor, f_i is the prescribed body force (per unit volume), e is the total energy density, r is the heat supply (per unit volume), p is the pressure, and ε is the internal energy density.

These equations may be put into the format of a hyperbolic system of conservation laws by employing the following definitions:

$$m = n_{sd} + 2 \tag{4.35}$$

$$U_1 = \rho \tag{4.36}$$

$$U_{j+1} = \rho u_j, \quad 1 \le j \le n_{sd} \tag{4.37}$$

$$U_m = \rho e \tag{4.38}$$

$$\underset{\sim}{F}_j = (U_{j+1}/U_1)\underset{\sim}{U} + \begin{Bmatrix} 0 \\ p\underset{\sim}{\delta}_j \\ p\,U_{j+1}/U_1 \end{Bmatrix} \tag{4.39}$$

$$p = \hat{p}\left(U_1,\ U_m/U_1 - \tfrac{1}{2}\left(\sum_{i=1}^{n_{sd}} U_{i+1}^2\right)/U_1^2\right) \tag{4.40}$$

$$\underset{\sim}{\delta}_j = \begin{Bmatrix} \delta_{j1} \\ \cdot \\ \cdot \\ \cdot \\ \delta_{jn_{sd}} \end{Bmatrix} \tag{4.41}$$

$$G = -\begin{Bmatrix} 0 \\ \underset{\sim}{f} \\ r + f_i u_i \end{Bmatrix} \tag{4.42}$$

Consideration of the characteristics and particular physical situation leads to appropriate boundary condition specifications. A variety of Dirichlet type conditions (3.10) may be envisaged. Three potentially useful boundary conditions on Γ_H may be set up as follows:

total flux boundary condition

$$\underset{\sim}{F}_j^{(1)} = \underset{\sim}{O} \; ; \; \underset{\sim}{F}_j^{(2)} = \underset{\sim}{F}_j \tag{4.43}$$

pressure ("traction") boundary condition

Pressure may be specified as a natural boundary condition by selecting

$$\underset{\sim}{F}_j^{(1)} = (U_{j+1}/U_1)\underset{\sim}{U} , \tag{4.44}$$

$$\underset{\sim}{F}_j^{(2)} = \begin{Bmatrix} O \\ p \, \underset{\sim}{\delta}_j \\ p \, U_{j+1}/U_1 \end{Bmatrix} , \tag{4.45}$$

and

$$\underset{\sim}{H} = - \begin{Bmatrix} O \\ h \, \underset{\sim}{n} \\ h \, u_n \end{Bmatrix} , \tag{4.46}$$

where h is the prescribed value of pressure defined on Γ_h.

no boundary condition on Γ_H

$$\underset{\sim}{F}_j^{(1)} = \underset{\sim}{F}_j \; ; \; \underset{\sim}{F}_j^{(2)} = \underset{\sim}{O} \; ; \; \underset{\sim}{H} = \underset{\sim}{O}. \tag{4.47}$$

In (Tezduyar and Hughes, 1982) we have employed the formulation defined by (4.47). In modelling outflow conditions, (4.47) appears to be effective although pressure specification may be more appropriate in some circumstances.

As in the previous example, there are many possibilities for defining the weighting function perturbation, $\underset{\sim}{\tilde{P}}$. In (Tezduyar and Hughes, 1982) we tested the following definition

$$\tilde{P} = \underset{\sim}{T}_i \; \underset{\sim}{W}_{,i} \tag{4.48}$$

where

$$\underset{\sim}{T}_i = \tau_i \; \underset{\sim}{A}_i \quad \text{(no sum)} \tag{4.49}$$

or

$$\underset{\sim}{T}_i = \tau_i \; \underset{\sim}{A}_i^T \quad \text{(no sum).} \tag{4.50}$$

(Recall that $\underset{\sim}{A}_i = \partial \underset{\sim}{F}_i / \partial \underset{\sim}{U}$.)   Each of (4.49) and (4.50) leads to interesting formulations. We have experimented with both, and also with the definition of τ_i. At present we favour (4.50) over (4.49) due to its superior behaviour on nonlinear problems. This is discussed more fully in (Tezduyar and Hughes, 1982). A sketch of the procedures used to define τ_i is contained in Section 5.

It is interesting to observe that if $\hat{p}(\rho, \; \varepsilon) = \rho \; f(\varepsilon)$, where f is an arbitrary function,[*] then $\underset{\sim}{F}_j(\underset{\sim}{U})$ is a homogeneous function of degree 1, that is $\underset{\sim}{F}_j(\alpha \; \underset{\sim}{U}) = \alpha \; \underset{\sim}{F}_j(\underset{\sim}{U})$ for all $\alpha \in \mathbb{R}$. In this case it follows that $\underset{\sim}{F}_j = \underset{\sim}{A}_j \; \underset{\sim}{U}$. The individual partial fluxes defined by (4.44) and (4.45) also are homogeneous functions of degree 1.

Remarks

1. The full, compressible Navier-Stokes equations, including thermal effects, can also be subsumed by the general weighted residual format of (4.2). Likewise, various natural boundary conditions can be built into the formulation. However, this lies outside the scope of the present paper.

2. Discretization of (4.2) is carried out by expanding $\underset{\sim}{U}$ and $\underset{\sim}{W}$ in terms of a set of finite element basis, or shape, functions. For example, the expression for $\underset{\sim}{U}$ might take the form[†]

$$\underset{\sim}{U}(\underset{\sim}{x}, \; t) = \sum_B N_B(\underset{\sim}{x}) \underset{\sim}{U}_B(t) \tag{4.51}$$

---

[*]For example, this is the case for a perfect gas defined by $\hat{p}(\rho, \varepsilon) = (\gamma - 1)\rho\varepsilon, \; \gamma \in \mathbb{R}$

[†] One could also employ different shape functions for the individual components of $\underset{\sim}{U}$. This idea becomes important in "constrained" cases, such as incompressibility (see, e.g., Thomasset, 1981). Similar considerations need to be made if compressible flow algorithms are to be exercised at very low Mach numbers.

where B is a nodal index, N_B is the shape function associated with node B, and $\mathbf{U}_B$ is the value of $\mathbf{U}$ at node B. This leads to a weighted residual formulation in the strict sense. Certain modifications of the basic formulation--such as reduced integration or use of lower-order interpolants--are interesting from the standpoints of efficiency and, occasionally, lead to improved accuracy.* We have been fond of the use of reduced integration techniques in our finite element work (see e.g. Hughes et al., 1979) and have experimented with them in the present context in (Tezduyar and Hughes, 1982).

The use of interpolants of $\mathbf{F}_j$ and $\mathbf{G}$ has an interesting consequence. It produces schemes which are reminiscent of, and at the same time generalize, classical "conservative differencing schemes". The basic idea is to approximate $\mathbf{F}_j$ and $\mathbf{G}$ by expansions in terms of shape functions and the nodal values of $\mathbf{F}_j$ and $\mathbf{G}$, respectively. For example, the expansion for $\mathbf{F}_j$ might take the following form:

$$\mathbf{F}_j(\mathbf{x}) = \sum_B N_B(\mathbf{x})\,\mathbf{F}_{jB}(t) \tag{4.52}$$

where

$$\mathbf{F}_{jB}(t) = \mathbf{F}_j(\mathbf{U}_B(t),\ (\nabla\mathbf{U})_B(t),\ \mathbf{x}_B\ ,\ t)\ , \tag{4.53}$$

$$(\nabla\mathbf{U})_B(t) = \sum_C \nabla N_C(\mathbf{x}_B)\,\mathbf{U}_C(t) \tag{4.54}$$

and likewise for $\mathbf{G}$. Christie et al. (1981) have proposed and examined schemes of this kind. They term them "product approximations". Spradley et al. (1980) have also adopted this idea in their "general interpolants method". Various other interesting finite element and finite difference concepts are synthesized in their approach.

The use of reduced integration and/or the use of interpolants simplifies element calculations. Their relative merits from the standpoints of accuracy, stability, shock structure, etc., do not seem certain at this point and thus further investigations are warranted.

---

* These violations of the weighted residual recipe are often referred to as "variational crimes", a terminology coined by Strang (see Strang and Fix, 1973).

3. The development of so-called finite volume techniques also emanates from integral forms of the conservation equations (see, e.g. Caughey and Jameson, 1979). These techniques, which are often thought of as finite difference methods, have essential features in common with the integral difference methods originated at Lawrence Livermore Laboratory (see e.g. Wilkins, 1969). More recently it has been shown how to derive the latter class of methods via finite element/reduced integration concepts (see Goudreau and Hallquist, 1982 for a survey of these ideas).

We thus anticipate that the finite volume method will also find a place within the finite element hierarchy.

5. PETROV-GALERKIN METHODS OF CHARACTERISTIC TYPE

In this Section we discuss further details of the class of Petrov-Galerkin methods described in Section 4 and propose some further generalizations.

Selection of τ_i

We begin with a presentation of the formulae we have used for the optimization parameters. As yet no universal scheme has been formulated. The following criteria have been investigated so far.

Temporal criterion

In conjunction with certain time-stepping algorithms, a criterion based upon the time step, Δt, proves effective:

$$\tau_i = \tau = F\,\alpha\,\Delta t, \qquad 1 \le i \le n_{sd}. \tag{5.1}$$

In (5.1), α is an algorithmic parameter and F is a parameter which enables us to adjust the magnitude of τ_i for various purposes. The value of α is usually set to ½ in cases in which we are interested in time accuracy, whereas it is set to 1 in cases in which we wish to rapidly obtain a steady flow. An interesting explicit algorithm which attains second-order time accuracy can be constructed by setting $\alpha = ½$ and $F = 1$ (see Dukowitz and Ramshaw, 1979 and Tezduyar and Hughes, 1982 for further details).

Since (5.1) pertains to all elements in a mesh it is a <u>global</u> criterion.

Spatial criteria

Two spatial criteria have been employed:

(a) $\tau_i = \tau = F\,\alpha\,\tilde{\xi}\,h/\lambda, \qquad 1 \le i \le n_{sd}$ (5.2)

(b) $\tau_i = F\,\alpha\,\tilde{\xi}\,h_i/\lambda_i, \qquad \text{(no sum)}, \qquad 1 \le i \le n_{sd}$ (5.3)

where λ_i is the spectral radius of $\underset{\sim}{A}_i$,

$$\lambda = \|\underset{\sim}{\lambda}\| = (\lambda_i \lambda_i)^{\frac{1}{2}} \tag{5.4}$$

$$h_i = 2\|\underset{\sim}{\nabla}\, x_i\|^{*} \tag{5.5}$$

$$h = h_i \lambda_i/\lambda \qquad (= 0 \text{ if } \lambda = 0) \tag{5.6}$$

$$\tilde{\xi} = (\coth \alpha) - 1/\alpha \tag{5.7}$$

$$\alpha = \lambda h/(2d). \tag{5.8}$$

In (5.5), $\underset{\sim}{\nabla}$ denotes the gradient operator with respect to the canonical isoparametric coordinates. The factor of 2 appears since the isoparametric parent domain is usually scaled to have a length of 2 (Zienkiewicz, 1977). The following examples should make the definition clear:

$$h = 2|\partial x/\partial \xi| \qquad \text{(one dimension)} \tag{5.9}$$

$$h_i = 2[(\partial x_i/\partial \xi)^2 + (\partial x_i/\partial \eta)^2]^{\frac{1}{2}} \qquad \text{(two dimensions)}. \tag{5.10}$$

In (5.8), d is a diffusivity coefficient which takes on the following definitions:

$$d = k_{ij}\lambda_i \lambda_j/\lambda^2 \qquad \text{(advection-diffusion eq.)} \tag{5.11}$$

$$d = 0 \qquad \text{(Euler eqs.)}. \tag{5.12}$$

Note that $d = 0$ implies $\tilde{\xi} = 1$.

*This definition was prompted by the work of Nakazawa (1982).

The values of τ_i given by (5.2) and (5.3) depend upon the geometry and state of an individual element. In this sense (5.2) and (5.3) are <u>local</u> criteria.

Various model equations have been employed to select $F\alpha$ in (5.2) and (5.3). The value $F\alpha = \frac{1}{2}$ has been determined by considering the one-dimensional, steady advection-diffusion equation (Hughes and Brooks, 1979). A boundary-layer investigation also points to this value (Hughes and Brooks, 1980). An analysis of the semi-discrete, time-dependent, advection equation in one dimension yields a value $F\alpha = 1/\sqrt{15}$ (Raymond and Garder, 1976). Fourth-order phase accuracy is thereby attained for linear elements. We have employed this value in transient advection-diffusion and Navier-Stokes calculations such as the ones described in Section 6. Values of $F\alpha \geq \frac{1}{2}$ have been experimented with to achieve effective resolution of shocks governed by first-order hyperbolic systems.

<u>Accuracy, stability and shock structure</u>

In the context of the advection-diffusion equation, the streamline-upwind/Petrov-Galerkin method described in Section 4, and above, leads to difference stencils which are skewed with respect to streamlines and are centred about points upwind of the nodal point to which they are associated. In this way the accuracy of centred, skew schemes is attained while, at the same time, the stability of upwind procedures is achieved. Claes Johnson (1981) has proved that this formulation leads to optimal error estimates* and furthermore that effects are propagated in the discrete problem approximately as in the continuous problem. In particular, Johnson (1981) has established that discontinuities of exact solutions across characteristics (i.e., streamlines in the present case) will be numerically captured in a very narrow region around the characteristic, even when the characteristics are arbitrarily skew to the mesh.

The methods proposed for hyperbolic systems possess similar interesting structure and our experience in non-linear cases with shocks is very good so far. However, it should be mentioned that, for this class of problems, a strong case can be made for monotone methods (see Harten et al., 1976 and the works of Osher and collaborators, e.g., Osher, 1980, 1981a,b, 1982, Engquist and Osher, 1980, Goorjian and Van Buskirk, 1981). We are still hopeful that within the framework that we have delineated, by appropriate local selection of the optimization parameter, we

---

* In the Galerkin method, optimal asymptotic error estimates fail to hold in general, even when the exact solution is smooth (Dupont, 1973). Rough exact solutions create particular havoc, causing spurious oscillations which pollute the solution globally.

will achieve the attributes of monotone methods, concomitantly with the desirable features delineated above. Admittedly, this is an area requiring further investigation.

We wish to now consider a definition of $\tilde{P}$ which generalizes the forms that we have previously considered. This definition is motivated by Jean Donea's development of the class of Taylor-Galerkin methods (Donea, 1982) which possess excellent phase-accuracy properties.

Let
$$\tilde{P} = \mathcal{L}W \tag{5.13}$$

where $\mathcal{L}$ is a linear operator defined by the expansion

$$\mathcal{L} = \mathcal{L}(n, L) = L + \tfrac{1}{2}L^2 + \ldots + \frac{1}{n!} L^n \tag{5.14}$$

In (5.14), L is also a linear operator. Clearly, if $n = 1$, $\mathcal{L} = L$. The definitions of the perturbation that we have used in our previous work are special cases of (5.13), (5.14) with $n = 1$. This can be seen as follows:

advection-diffusion equation; cf. (4.29)

$$\tilde{p} = \mathcal{L}w = Lw = \tau(u_i w)_{,i} \tag{5.15}$$

compressible Euler equations; cf. (4.48)

$$\tilde{P} = \mathcal{L}W = LW = T_i W_{,i} \tag{5.16}$$

Formally generalizing (5.16) to the case of $n \geq 1$ leads to the expression

$$\tilde{P} = \mathcal{L}W = \left(T_i \frac{\partial}{\partial x_i} + \ldots + \left(T_i \frac{\partial}{\partial x_i}\right)^n\right) W \tag{5.17}$$

If in (5.17)

$$T_i = \frac{\Delta t}{2} A_i^T \tag{5.18}$$

we arrive at a family of methods similar to Donea's. By using, say

$$T_i = \tau_i A_i^T \quad \text{(no sum)} \tag{5.19}$$

with τ_i selected according to spatial criteria, we generate other interesting possibilities. The spatial criteria have the advantage that steady flows are independent of the size of the time step employed, and seem better suited to meshes involving significant variation in element size. Clearly these ideas are in a speculative phase and require further investigation.

6. NUMERICAL EXAMPLES

The rotating cone (advection-diffusion equation)

An elevation plot for the rotating-cone problem is shown in Fig. 1. The cone is shown after having rotated 360°. The trapezoidal-rule algorithm was used for temporal discretization of the streamline-upwind Petrov-Galerkin formulation. A 30×30 mesh of bilinear elements was employed.

For additional numerical examples and comparisons with different approaches, see (Hughes and Brooks, 1982).

Karman vortex street*(incompressible Navier-Stokes equations)*

Our procedures for solving the incompressible Navier-Stokes equations are described in (Brooks and Hughes, 1982a,b) which may be consulted for further details.

Stationary streamlines are shown in Fig. 2 for a computation of the Karmen vortex street at a Reynolds number of 100. The elements employed assumed bilinear velocity and constant pressure. The mesh consisted of 1436 elements. The time-stepping algorithm treats velocity degrees-of-freedom explicitly and pressures implicitly.

Transonic nozzle flow (compressible Euler equations)

Shock profiles are shown in Fig. 3. The exact solution (shown solid) has been provided by Lomax et al. (1981). This problem is one of steady flow, consequently a value of $\alpha = 1$ (backward differences) was employed. Various values of F were tested and their effects on the shock profile may be seen in Fig. 3. At F = 1, the shock front is captured across one element (40 linear elements were employed). As F is increased the shock tends to be smeared somewhat. In the calculation shown $\underset{\sim}{T} = \underset{\sim}{A}^T$ and the local criterion, (5.2), was employed.

Shocks structure/entropy condition test problem

This problem was suggested to us by S. Osher. The equation is

$$u_{,t} + \left(\frac{u^2}{2}\right)_{,x} = 0 \qquad (6.1)$$

and the initial condition is shown in frame 0 of Fig. 4. Note that the initial condition involves two stable shocks and one

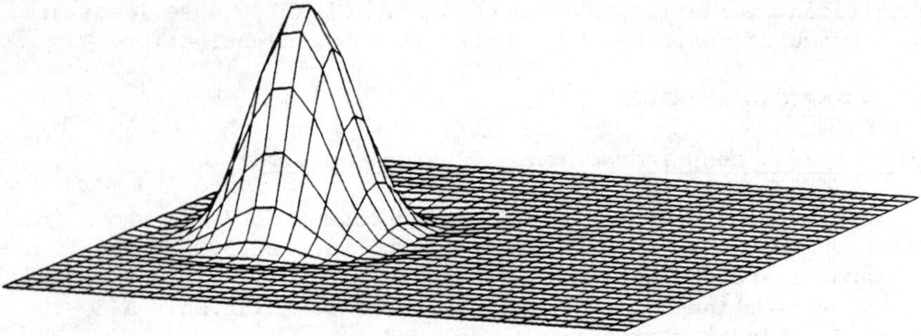

Fig. 1 Advection of a cone.

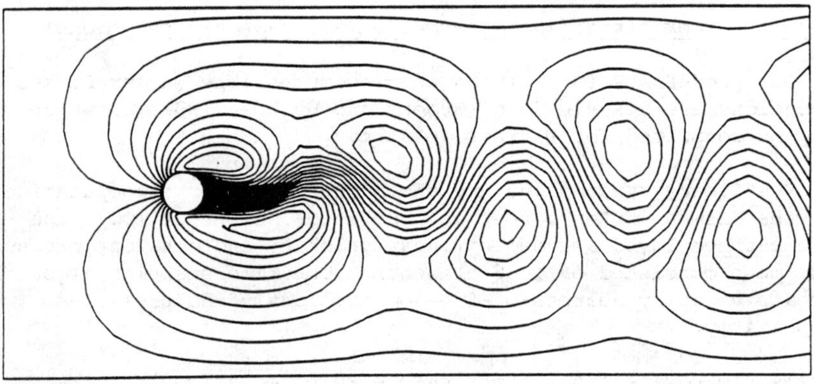

Fig. 2 Karman vortex street with Re = 100.

unstable shock. Each successive frame corresponds to a time step ($\Delta t = 2.174$). As may be seen, the unstable shock immediately breaks down and eventually the stable shocks coalesce to form a steady shock (frame 23). The finite scheme nodally interpolates the exact steady state.

Equation (6.1) is a special case of the hyperbolic system formulation described in Section 3. In the calculation shown in Fig. 4, 40 linear elements were employed, the spatial criterion (5.2) was employed, $F = 1$ and $\alpha = \frac{1}{2}$.

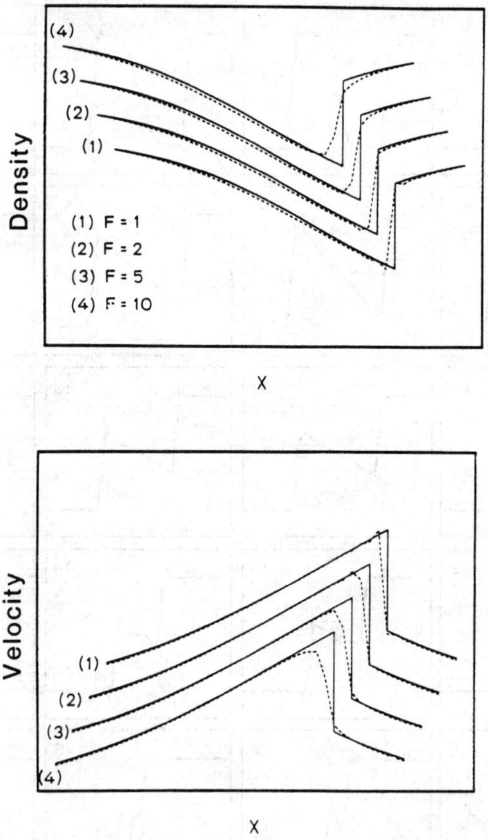

Fig. 3 Transonic nozzle flow problem.

Thin biconvex airfoil (compressible Euler equations)

Sample results are presented for a transonic flow about a thin biconvex airfoil. Details about the problem and additional results may be found in (Tezduyar and Hughes, 1982). "Medium" and "fine" meshes are illustrated in Fig. 5. In the medium mesh 16 elements are used to discretize the airfoil whereas 32 elements are used in the fine mesh. (The fine mesh represents a bisection of the medium mesh.) For this problem the weighting function perturbation was defined by (4.50), the temporal criterion (5.1) was employed, $F = 1$ and $\alpha = 1$. Pressure coefficient plots are presented in Fig. 6 for channel and free-stream boundary conditions along the upper edge. The results show a shift in the shock location of about 6% for the two boundary condition cases which is to be expected. Mach number contours are illustrated in Fig. 7 for the channel boundary condition case.

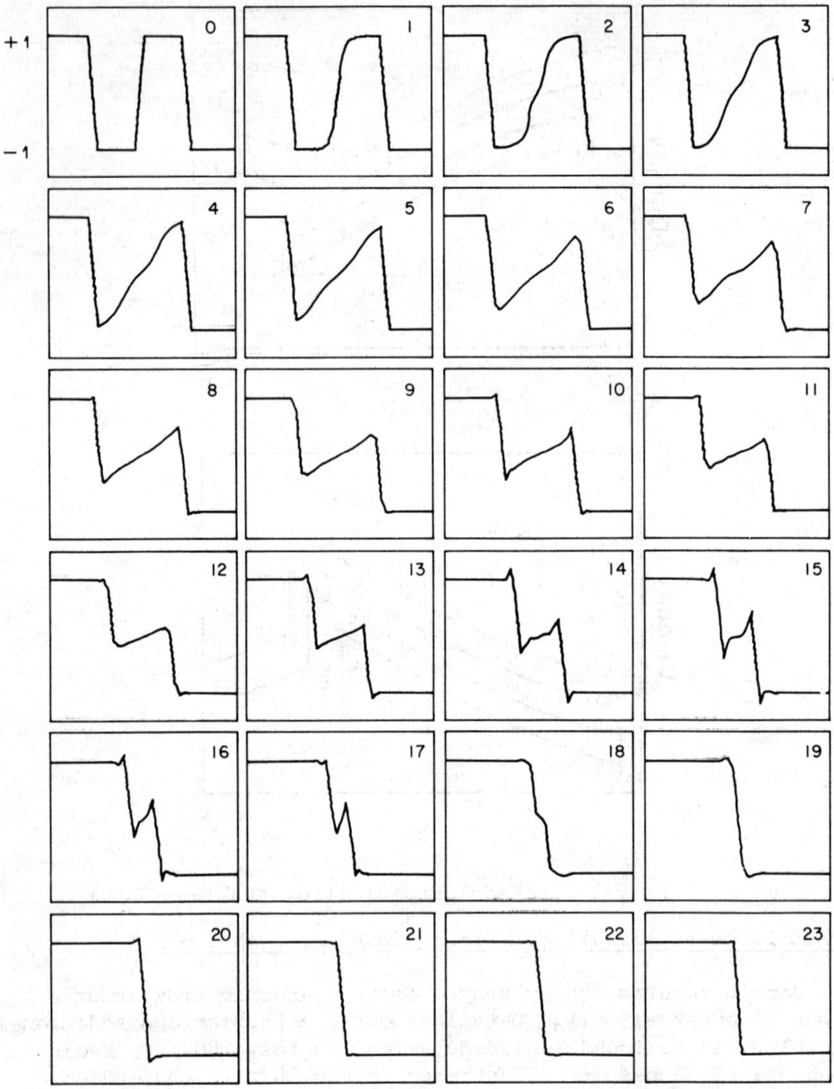

Fig. 4 Burger's equation test problem.

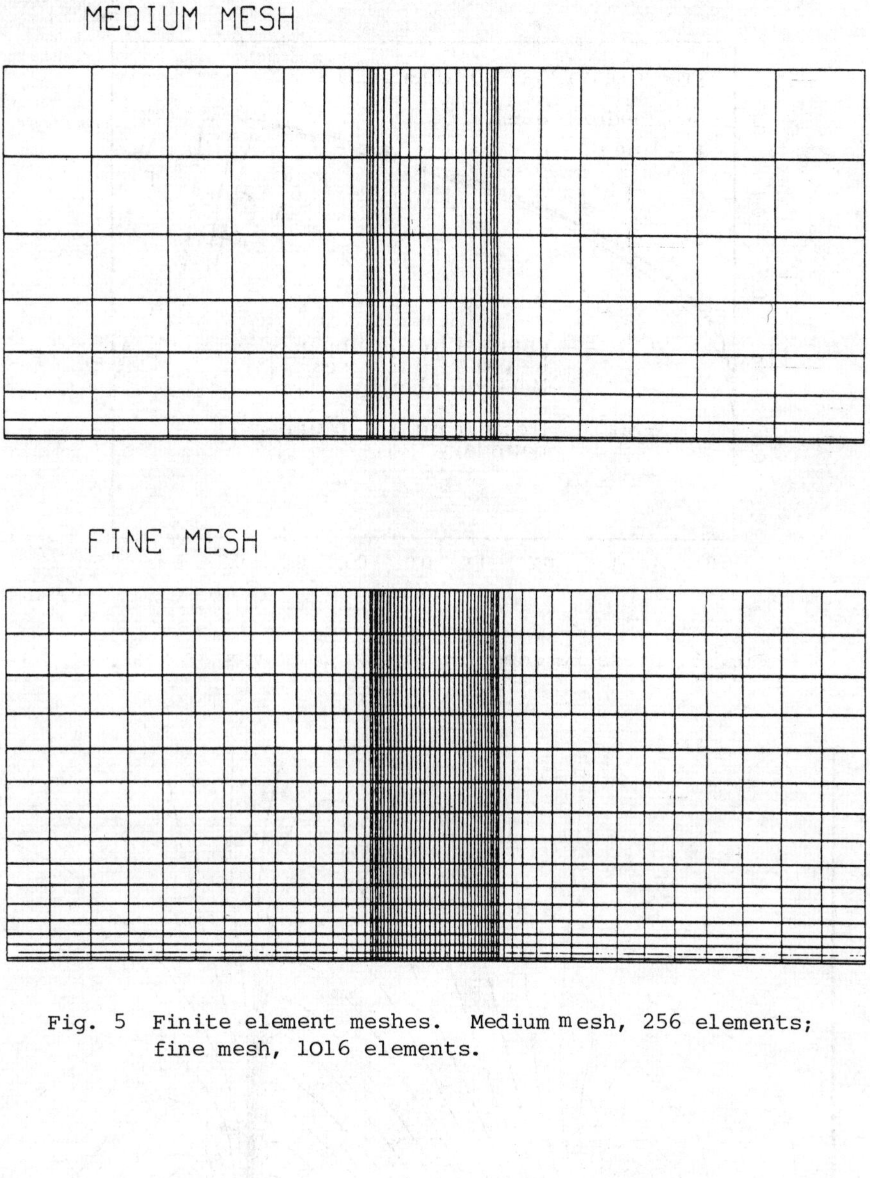

Fig. 5 Finite element meshes. Medium mesh, 256 elements; fine mesh, 1016 elements.

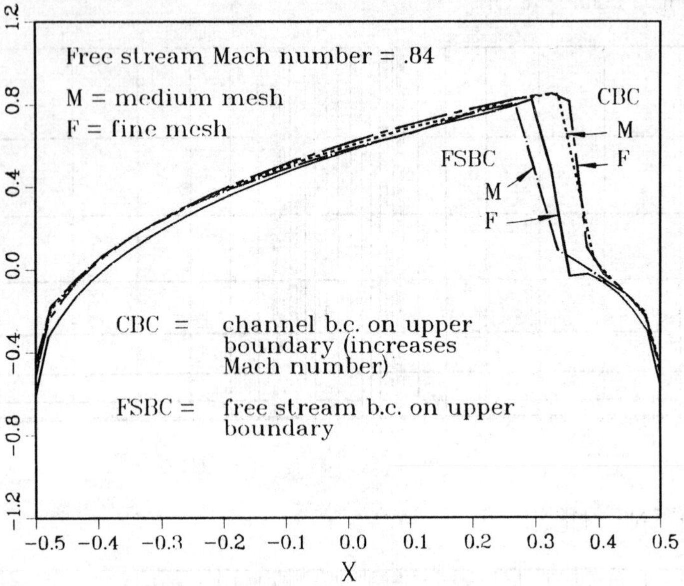

Fig. 6 Pressure coefficient for biconvex airfoil.

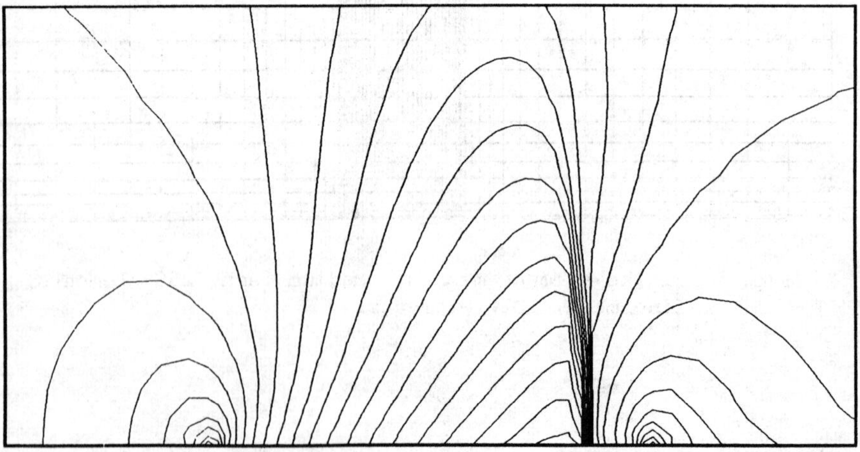

Fig. 7 Transonic case, fine mesh, isomachs
Mach number contours for the airfoil problem.

7. CONCLUSIONS

In this paper we have presented a Petrov-Galerkin formulation which generalizes the streamline upwind concept to systems of conservation laws. Emphasis has been placed on the compressible Euler equations in the present work. Numerical examples indicate the good behaviour of the method on a variety of problems, including ones with shocks.

Future work needs to focus on optimizing the formulation with respect to shock structure and entropy conditions. Furthermore, it is hoped to tackle a number of aerodynamic flows of a complex nature. Finally, the development of element-by-element, and subdomain-by-subdomain, implicit algorithms (see, e.g., Hughes et al., 1982, 1983) for the present class of problems will hopefully open the way to much more efficient solution of large problems, particularly three-dimensional ones.

ACKNOWLEDGEMENT

We would like to thank the Computational Fluid Dynamics Branch of the NASA Ames Research Center for supporting our research, and especially H. Lomax for guidance and encouragement.

REFERENCES

BAKER, A.J. 1979 Research on numerical algorithms for the three-dimensional Navier-Stokes equations, 1. Accuracy, Convergence and Efficiency, Technical Report AFFDL-TR-79-3141, Wright-Patterson Air Force Base, Ohio, USA.

BARRETT, J.W. and MORTON, K.W. 1980 Optimal finite element solutions to diffusion convection problems in one dimension, *Int. J. Num. Meth. Eng.*, **15**, pp. 1457-1474.

BARRETT, J.W. and MORTON, K.W. 1981 Optimal Petrov-Galerkin Methods through approximate symmetrization, *IMA J. Num. Analysis*, **1**, pp. 439-468.

BROOKS, A. and HUGHES, T.J.R. 1980 Streamline-upwind/Petrov-Galerkin methods for advection dominated flows. In Proceedings of the Third International Conference on Finite Element Methods in Fluid Flow, Banff.

BROOKS, A. and HUGHES, T.J.R. 1982a Streamline-upwind/Petrov-Galerkin formulation for convection dominated flows with particular emphasis on the incompressible Navier-Stokes equations. To appear in the proceedings of FENOMECH '81 and Computer Methods in Applied Mechanics and Engineering.

BROOKS, A. and HUGHES, T.J.R. 1982b An algorithm for solving the Navier-Stokes equations based upon the streamline-upwind Petrov-Galerkin formulation. To appear in Numerical Methods for Coupled Problems, London, J. Wiley & Sons.

CAUGHEY, D.A. and JAMESON, A. 1979 Numerical calculation of transonic potential flow about wing-body combinations, *AIAA J.* **17**, pp. 175-181.

CHRISTIE, I., GRIFFITHS, D.F., MITCHELL, A.R. and SANZ-SERNA, J.M. 1981 Product approximation for non-linear problems in the finite element method, *IMA J. Num. Analysis,* **1**, pp. 253-266.

CHRISTIE, I., GRIFFITHS, D.F., MITCHELL, A.R. And ZIENKIEWICZ, O.C. 1976 Finite element methods for second order differential equations with significant first derivatives, *Int. J. Num. Meth. Eng.,* **10**, pp. 1389-1396.

COHEN, M. 1982 Private Communication.

DENDY, J.E. 1974 Two methods of Galerkin type achieving optimum L^2 rates of convergence for first order hyperbolics, *SIAM J. Num. Analysis,* **11**, pp. 637-653.

DONEA, J. 1982 A Taylor-Galerkin method for convective transport problems, preprint.

DUKOWICZ, J.K. and RAMSHAW, J.D. 1979 Tensor viscosity method for convection in numerical fluid dynamics, *J. Comput. Phys.,* **32**, pp. 71-79.

DUPONT, T. 1973 Galerkin methods for first-order hyperbolics: an example, *SIAM J. Num. Analysis,* **10**, pp. 890-899.

ENGQUIST, B. and OSHER, S. 1980 One-sided difference schemes and transonic flow, *Proc. Natl. Acad. Sci. USA,* **77**, pp. 3071-3074.

GOORJIAN, P.M. and VAN BUSKIRK, R. 1981 Implicit calculations of transonic flows using monotone methods, AIAA-81-0331, AIAA 19th Aerospace Sciences Meeting, St. Louis, Missouri.

GOUDREAU, G.L. and HALQUIST, J. O. 1982 Recent developments in large scale finite element Lagrangian hydrocode technology, *Comp. Meth. Appl. Mech. Eng.,* to appear.

GRIFFITHS, D.F. 1981 A Petrov-Galerkin method for hyperbolic equations. To appear in Proceedings of MAFELAP '81 (J. Whiteman, Ed.), Brunel University.

GRIFFITHS, D.F. and MITCHELL, A.R. 1979 On generating upwind finite element methods, *AMD* **34**. In Finite Element Methods for Convection Dominated Flows (T.J.R. Hughes, Ed.), New York, ASME.

HARTEN, A., HYMAN, J.M. and LAX, P.D. 1979 On finite difference approximations and entropy conditions for shocks, *Comm. Pure Appl. Maths.*, **XXIX**, pp. 297-322.

HEINRICH, J.C., HUYAKORN, P.S., ZIENKIEWICZ, O.C. and MITCHELL, A.R. 1977 An 'upwind' finite element scheme for two-dimensional convective transport equations, *Int. J. Num. Meth. Eng.*, **11**, pp. 134-143.

HEINRICH, J. and ZIENKIEWICZ, O.C. 1979 The finite element method and 'upwinding' techniques in the numerical solution of convection dominated flow problems, *AMD* **34**. In Finite Element Methods for Convection Dominated Flows (T.J.R. Hughes, Ed.), New York, ASME.

HUGHES, T.J.R. and BROOKS, A. 1979 A multidimensional upwind scheme with no crosswind diffusion, *AMD* **34**. In Finite Element Methods for Convection Dominated Flows (T.J.R. Hughes, Ed.), New York, ASME.

HUGHES, T.J.R. and BROOKS, A. 1980 Galerkin/upwind finite element mesh partitions in fluid mechanics. In Boundary and Interior Layers - Computational and Asymptotic Methods, (J.J.H. Miller, Ed.), Dublin, Boole Press.

HUGHES, T.J.R. and BROOKS, A. 1982 A theoretical framework for Petrov-Galerkin methods with discontinuous weighting functions: application to the streamline upwind procedure. To appear in Finite Elements in Fluids, **4** (R.H. Gallagher, Ed.), London, J. Wiley & Sons.

HUGHES, T.J.R., LEVIT, I. and WINGET, J. 1982 An implicit unconditionally stable element-by-element algorithm for heat conduction analysis. To appear in *J. Eng. Mech. Div.*, ASCE.

HUGHES, T.J.R., LEVIT, I. and WINGET, J. 1983 An element-by-element solution algorithm for problems of structural and solid mechanics. To appear in *Comput. Meth. Appl. Mech. Eng.*

HUGHES, T.J.R., LIU, W.K. and BROOKS, A. 1979 Review of finite element analysis of incompressible viscous flows by the penalty function formulation, *J. Comput. Phys.*, **30**, pp. 1-60.

JOHNSON, C. 1981 Finite element methods for convection-diffusion problems, Fifth International Symposium on Computing Methods in Engineering and Applied Sciences, INRIA, Versailles.

KELLY, D.W., NAKAZAWA, S., ZIENKIEWICZ, O.C. and HEINRICH, J.C. 1980 A note on upwinding and anisotropic balancing dissipation in finite element approximations to convective diffusion problems, *Int. J. Num. Meth. Eng.*, **15**, pp. 1705-1711.

LOMAX, H. et al. 1981 Private Communication.

MORTON, K.W. 1981 Finite element methods for non-self-adjoint problems, Numerical Analysis Report 3/81, University of Reading.

MORTON, K.W. and BARRATT, J.W. 1980 Optimal finite element methods for diffusion-convection problems. In Boundary and Interior Layers- Computational and Asymptotic Methods, (J.J.H. Miller, Ed.), Dublin, Boole Press.

MORTON, K.W. and PARROTT, A.K. 1980 Generalized Galerkin methods for first-order hyperbolic equations, *J. Comput. Phys.*, **36**, pp. 249-270.

NAKAZAWA, S. 1982 Finite element analysis applied to polymer processing, PhD Thesis, University of Swansea.

OSHER, S. 1980 Approximation par éléments finis avec décentrage pour des lois de conservation hyperboliques non linéaires multi-dimensionelles, *C.R. Acad. Sc. Paris*, **t.290**, Série A., pp. 819-821.

OSHER, S. 1981a Approximation par eléménts finis avec décentrage de problèmes de perturbations singulières quasi linéaires et multi dimensionnels, *C.R. Acad. Sc. Paris*, **t.292**, Série I, pp. 99-101.

OSHER, S. 1981b Nonlinear singular perturbation problems and one sided difference schemes, *SIAM J. Num. Analysis*, **18**, pp. 129-144.

OSHER, S. 1982 Numerical solution of singular perturbation problems and hyperbolic systems of conservation laws. North Holland Math. Studies # 47, (Axelsson et al. Eds.).

RAITHBY, G.D. 1976 Skew upstream differencing schemes for problems involving fluid flow, *Comput. Meth. Appl. Mech. Eng.*, **9**, pp. 153-164.

RAYMOND, W.H. and GARDER, A. 1976 Selective damping in a Galerkin method for solving wave problems with variable grids, *Mon. Weath. Rev.*, **104**, pp. 1583-1590.

SPRADLEY, L.W., STALNAKER, J.F. and RATLIFF, A.W. 1980 Computation of three-dimensional viscous flows with the Navier-Stokes equations, AIAA-80-1348, AIAA 13th Fluid and Plasma Dynamics Conference, Snowmass, Colorado, USA.

STRANG, G. and FIX, G. 1973 An analysis of the finite element method, Englewood Cliffs, New Jersey, Prentice Hall.

TEZDUYAR, T.E. and HUGHES, T.J.R. 1982 Development of time-accurate finite element techniques for first-order hyperbolic systems with particular emphasis on the compressible Euler equations, Report prepared under NASA-Ames University Consortium Interchange No. NCA2-OR745-104.

THOMASSET, F. 1981 Implementation of finite element methods for Navier-Stokes equations, New York, Springer-Verlag.

WAHLBIN, L.B. 1974 A dissipative Galerkin method for the numerical solution of first order hyperbolic equations. In Mathematical Aspects of Finite Elements in Partial Differential Equations, (C. de Boor, Ed.), New York, Academic Press.

WILKINS, M.L. 1969 Calculation of elastic-plastic flows, Lawrence Livermore Laboratory, Rept. UCRL-7322, Rev. 1, Livermore, California, USA.

ZIENKIEWICZ, O.C. 1977 The finite element method, London, McGraw-Hill.

INTRODUCTORY REMARKS ON MULTIGRID METHODS*

A. Brandt

(The Weizmann Institute of Science, Israel)

1. WHERE AND WHY MULTIGRID CAN HELP

The starting point of the multigrid method (or, more generally, the Multi-Level Adaptive Technique - MLAT), and indeed also its ultimate upshot, is the following "*golden rule*":

The amount of computational work should be proportional to the amount of real physical changes in the computed system.

Stalling numerical processes must be wrong.

That is, whenever the computer grinds very hard for very small or slow real physical effect, there must be a better computational way to achieve the same goal. Common examples of such grinding are the usual iterative processes for solving the algebraic equations arising from discretizing partial-differential, or integro-differential, boundary-value (steady-state) problems, in which the error has relatively small changes from one iteration to the next. Another example is the solution of time-dependent problems with time-steps (dictated by numerical stability requirements) much smaller than the real scale of change in the solution. Or, more generally, the use of too fine a discretization grid, where in large parts of the computational domain the mesh-size and/or the time-step are much smaller than, again, the real scale of solution changes.

If you have such a problem, multi-level techniques may help. Your trouble is usually related to some "stiffness" in your problem; i.e., to the existence of several solution components with different scales, which conflict with each other. For example, smooth components, which are efficiently approximated on coarse grids but are slow to converge in fine-grid processes, conflict with high-frequency components which must be approxi-

*This research is sponsored by the Air Force Wright Aeronautical Laboratories, Air Force Systems Command, United States Air Force, under AFOSR 82-0063.

mated on fine grids. By employing interactively several scales of discretization, multi-level techniques resolve such conflicts, avoid stalling and do away with the computational waste.

In fully developed MLAT processes the amount of computation should be determined only by the amount of real physical information.

The main development of multi-level techniques has so far been as fast solvers of the algebraic equations arising in discretizing boundary-value problems (steady-state problems or implicit steps in evolution problems). Discretizing such problems on several levels (several grids with geometrically decreasing mesh-sizes), the multigrid solution process involves relaxation sweeps on each level, coarse-grid-to-fine-grid interpolation of corrections, and fine-to-coarse transfers of residuals. Since relaxation sweeps on each level can be *very* efficient in liquidating those error components whose wavelengths (or scale of change) are comparable to that level's mesh-size, the combined process can be extremely efficient in reducing all errors. Indeed, when properly designed, the multigrid solution process requires just a few (four to ten) work units, where a work unit is the amount of computational work involved in *expressing* the algebraic equations. This efficiency is obtained for all problems on which sufficient research has been made, from simple model problems to complicated nonlinear systems on general domains, including diffusion problems with strongly discontinuous coefficients, integral equations, and minimization problems with constraints; from regular elliptic to singular-perturbation and non-elliptic boundary-value problems. Due to the iterative nature of the method, nonlinear problems require no more work than the corresponding linearized problems: no outer Newton-like steps, and, in fact, no global linearizations at all, are needed when an advanced multigrid version, called the Full Approximation Scheme (FAS) is used. Problems with global constraints are solved as fast as the corresponding unconstrained difference equations, using a technique of enforcing the constraints only at the coarse-grid stages of the algorithm. A few work units are also all the work required in calculating each eigen-function of discretized eigenproblems (Brandt et al., 1981).

Beyond the fast algebraic solvers, multi-level techniques can be very useful in other ways related to stiffness. They can provide very efficient grid-adaptation procedures for problems (either steady-state or evolution problems) in which different scales of discretization are needed in different parts of the domain. The FAS multigrid version gives a convenient way to create non-uniform adaptable structures which are based on uniform grids and hence are very flexible, allowing quick *local* refinements and *local* coordinate transformations (to fit curved boundaries, stream directions, etc.). The resulting discrete equations are still solved with the usual multigrid efficiency (a few work units). Moreover, this grid adaptation can naturally be governed by quantities which are by-products of the FAS multigrid processing. The grid adaptation can naturally be integrated into the multigrid algorithm, yielding an unsaturated process where the error in approximating the true *differential* solution is decreased at a fast, nearly optimal rate.

Multi-level techniques can also enormously reduce the number of discrete relations employed in solving chains of many similar boundary-value problems (as in processes of continuation, in design and optimization search procedures, and in implicit solutions to time-dependent problems). They can automatically exploit the fact that often, even when the solutions themselves are not very smooth, the *changes* in the solution from one problem to the next are very smooth, hence well represented on coarse grids, at least in most parts of the domain. For the heat equation in an infinite space, for example, the amount of work required for marching in this way (still maintaining finest-grid accuracy) from initial state to 90% steady state is less than the work of 10 explicit time steps (on that finest grid). Design and optimization can be fully integrated into the multigrid solution process (using the finest grid only for local optimizations), allowing complete optimization which costs only modestly more than the solution of one boundary-value problem.

Another use of multigridding, both in steady-state and in evolution problems, is to obtain solutions with high order of approximation which cost only a fraction more than the low-order solutions. Here one exploits the fact that the higher order approximation is needed only for the smooth components of the solution (especially in non-elliptic and singular perturbation problems), hence it can inexpensively be obtained through coarse-grid defect corrections, naturally fitting into the multigrid processing.

Multigrid techniques effectively separate local from global processing. They can therefore also be used to cut vastly the required computer storage, far below the storage needed for the finest grid, without employing external storage and without severe loss of efficiency.

In solving integral equations, the kernel of integration K(x,y) usually becomes very smooth as the distance between the points x and y increases. This can be used by a multi-level discretization-and-solution process, whose number of operations is almost proportional to the number of points on the finest grid.

Finally, multi-level processes can also introduce new dimensions of efficiency in solving or treating some large systems which do not originate from partial-differential or integral equations. The common feature in those systems is that they involve many unknowns related in a low-dimensional space; i.e., each unknown u_P is defined at a point $P = (x_1, \ldots x_d)$ of a low-dimensional space (d is usually 2 or 3), and the equations are given in terms of these coordinates x_j. Moreover, the coupling between two values u_P and u_Q generally becomes weaker or smoother as the distance between P and Q increases, except perhaps for a small number of particular pairs (P,Q). Examples are the equations of multivariate interpolation of scattered data (Meinguet, 1979), geodetic problems of finding the locations of many stations that best fit a large set of local observations (Meissle, 1980), and pattern-recognition systems (Narayanan et al., 1981 and Tanimoto, 1981).

2. MULTIGRID RESEARCH

Some of the techniques described above are of course still subject to initial investigation, but in all cases the potential has already at least been demonstrated in some simple examples. In many types of problems we are indeed still in the stage of learning "how to do it right". Quite often our thinking habits, from our experience with one-grid methods, are misleading. It sometimes takes very simple problems, designed to tackle one difficulty at a time, to discern fully efficient multigrid approaches.

There are two different branches in multigrid research, with little interaction between them. One branch deals with rigorous mathematical analysis of the convergence of multigrid algebraic solvers. For a growing class of problems the basic multigrid assertion is rigorously proven, namely, that the algorithm will solve the algebraic system of n equations (n unknowns on the finest grid) to the level of truncation errors in less than Cn computer operations, where C is independent of n. This is clearly the best one can do in terms of the order of dependence on n, hence the result is very satisfying.

The other branch of research deals with the practical development of efficient multigrid algorithms like the ones surveyed above. It turns out that the rigorous mathematical analysis is not useful for that purpose. Usually, the size of the constant C obtained by that analysis is extremely unrealistic, and so is the dependence of C on problem parameters other than n, so that for practical sizes of n the obtained estimates are inferior to those derived for much slower methods. This is a crucial deficiency, since the whole purpose of multigrid methods is not mere convergence, but real practical efficiency. The unrealistic estimates historically led to several wrong practical conclusions (Brandt, 1982, Section 14). Even in those particularly simple cases (5-pint Poisson equations, essentially) where more reasonable values of C are obtained, the relative values of C in two competing multigrid approaches do not point to their relative efficiency in practice. Since one wants to understand the differences between several candidate multigrid algorithms, all of which may have "Cn" performance but some of which may still be orders of magnitude faster than some others, very different theoretical approaches, such as *local mode analyses*, are introduced, which are not necessarily rigorous. By neglecting some of the less work-consuming aspects (such as the details of treating smooth errors or the influence of boundaries on non-smooth errors), one gains a clear and precise picture of the more important processes. The predictions so obtained can be made accurate enough to serve in program optimization and debugging. They give us the precise figure of the ideal performance the algorithm *can* achieve, so that we do not ignorantly rest until that ideal is approached. Experience has taught us that careful incorporation of such theoretical studies is essential for producing reliable programs which fully realize the potential of the multigrid method.

A central line in the development of multigrid research can be viewed as a gradual "de-algebraization", or a gradual liberation from algebraic concepts, and the development of methods that increasingly exploit the underlying *differential* nature of problems (Brandt, 1982, Section 13). Another, reversed trend is the development of purely algebraic multi-level algorithms, where the geometrical meaning of the discrete equations is not used, hence the code is more generally applicable (Brandt, 1982, Section 13.1, and Brandt et al., 1982). Other works neglect the differential meaning of the equations but still use the geometry of the grids (Dendy, 1982).

For a further view of current multigrid research, see the papers in (Hackbusch and Trottenberg, 1982), and references therein.

3. ELEMENTARY ACQUAINTANCE WITH MULTIGRID

If you have no multigrid experience, the easiest way to acquire an elementary acquaintance with the method is to go through a simple example. Thus, reading Sections 2, 3.1, 4 and Appendix B in (Brandt, 1977a), just 12 pages, would acquaint you with the most basic concept, with elementary mode analysis, and with a simple algorithm together with its short Fortran code and output.

The general formulation of multigrid algorithms is also described in this reference, as well as in many other papers, such as (Brandt, 1977b, 1979, 1980b, 1982), (Brandt and Dinar, 1979) and (Stueben and Trottenberg, 1982). For a comprehensive treatment of *model problems* by a variety of multigrid algorithms, mode analyses and numerical experiments, for a model program and for a general introduction to the basic algorithms, see (Stueben and Trottenberg, 1982). These model programs and those of (Brandt, 1977a, App. B), and several other model programs and multigrid software are available in (MUGTAPE 82, 1982).

4. GUIDE TO FURTHER DEVELOPMENT

The FAS multigrid version can be learned from any one of several papers, such as those referenced in the previous section. The local refinement and grid adaptation techniques are described in the first three. Practical *finite element* multigrid formulations can for example be found in (Bank, 1981) and (Brandt 1980a).

Part I of (Brandt, 1982) organizes existing multigrid know-how in an order which corresponds to actual stages in developing fast multigrid solvers. For each stage the corresponding theoretical tools are explained through which one can predict the obtainable efficiency and thus debug his program. Part II of this reference summarizes more advanced multigrid techniques, and points out some basic insights important for further development.

Research not yet surveyed in here can be found in other papers in (Hackbusch and Trottenberg, 1982). For the state of the art in the rigorous analysis branch, see for example (Hackbusch, 1982), (Braess, 1982) and references therein. Additional material, and in fact summaries of all new multigrid papers, appear quarterly in the Multigrid Newsletter, obtainable in North and South America from its Editor (Steve McCormick, Department of Mathematics, Colorado State University, Fort Collins, CO 80523, USA) and in other countries from the Managing Editor (Kurt Brand, GMD/IMA, Postfach 1240, D-5205 St. Augustin 1, Federal Republic of Germany). Periodically, it publishes a complete list of all past papers.

REFERENCES

BANK, R.E. 1981 Multi-level Iterative Method for Nonlinear Elliptic Equations. In Elliptic Problem Solvers (M. Schultz, Ed.) Academic Press, New York, pp. 1-16.

BRAESS, D. 1982 The convergence rate of a multigrid method with Gauss-Seidel relaxation for the Poisson equation. In

Multigrid Methods (Conference Proceedings) (W. Hackbusch and I. Trottenberg, Eds.) Springer-Verlag.

BRANDT, A. 1977a Multi-level adaptive solutions to boundary-value problems, *Math. Comp.* **31**, pp. 333-390.

BRANDT, A. 1977b Multi-level adaptive techniques (MLAT) for partial differential equations: Ideas and Software. In Mathematical Software III (J.R. Rice, Ed.) Academic Press, New York pp. 273-314 (ICASE Report 77-20).

BRANDT, A. 1979 Multi-level adaptive solutions to singular-perturbation problems. In Numerical Analysis of Singular Perturbation Problems (P.W. Hemker and J.J.H. Miller, Eds.) Academic Press. pp. 53-142.

BRANDT, A. 1980a Multi-level adaptive finite-elements methods: I Variational problems. In Special Topics of Applied Mathematics (J. Frehse, D. Pallaschke and U. Trottenberg, Eds.) North Holland, pp. 91-128.

BRANDT, A. 1980b Multi-level adaptive computations in fluid dynamics, *AIAA J.* **18**, pp. 1165-1172.

BRANDT, A. 1981 Multigrid Solvers on Parallel Computers. In Elliptic Problem Solvers (M. Schultz, Ed.) Academic Press, New York, pp. 39-84.

BRANDT, A. 1982 Guide to Multigrid Development. In Multigrid Methods (Conference Proceedings) (W. Hackbusch and U. Trottenberg, Eds.) Springer-Verlag.

BRANDT, A. and DINAR, N. 1979 Multigrid solutions to elliptic flow problems. In Numerical Methods for Partial Differential Equations (S. Parter, Ed.) Academic Press, pp. 53-147.

BRANDT, A., McCORMICK, S. and RUGE, J. 1981 Multigrid algorithms for differential eigenproblems, submitted to *SIAM J. Sci. Stat. Comp.*

BRANDT, A., McCORMICK, S. and RUGE, J. 1982 Algebraic multigrid (AMG) for automatic algorithm design and problem solution, a preliminary report.

DENDY, J.E. (Jr) 1982 Black box multigrid, LA-UR-812337 Los Alamos National Laboratory, Los Alamos, New Mexico, *J. Comput. Phys.* (to appear).

HACKBUSCH, W. 1982 In Multigrid Methods (Conference Proceedings) (W. Hackbusch and U. Trottenberg, Eds.) Springer-Verlag.

HACKBUSCH, W. and TROTTENBERG, U. 1982 Multigrid Methods Proceedings of Conference (Köln-Porz, November 1981), Springer-Verlag.

MEINGUET, J. 1979 Multivariate interpolation at arbitrary points made simple, *J. Appl. Math. Phys. (ZAMP)* **30**, pp. 292-304.

MEISSLE, P. 1980 A priori prediction of roundoff error accumulation in the solution of a super-large geodetic normal equation system, NOAA Professional Paper 12, National Oceanic and Atmospheric Administration, Rockville, Maryland, USA.

MUGTAPE 82, 1982 A tape of multigrid software and programs, including GRIDPACK; MUGPACK; simple model programs (CYCLE C, FASCC, FMG1 and an eigenproblem solver); Stokes equations solver; SMORATE; BOXMG (In Dendy, 1982); MGOO and MG01 (In Stueben and Trottenberg, 1982). Available at the Department of Applied Mathematics, Weizmann Institute of Science, Rehovot, Israel, and at the GMD-IMA Postfach 1240, Schloss Birlinghoven, D-5205, West Germany.

NARAYANAN, K.A., O'LEARY, D.P. and ROSENFELD, A. 1981 Multiresolution relaxation, TR-1070 MCS-79-23422, University of Maryland, College Park, Maryland.

STUEBEN, K. and TROTTENBERG, U. 1982 Multigrid methods: fundamental algorithms, model problem anlaysis and applications. In Multigrid Methods, (Conference Proceedings) (W. Hackbusch and U. Trottenberg, Eds.) Springer-Verlag.

TANIMOTO, S.L. 1981 Template matching in pyramids, *Computer Graphics and Image Processing* **16**, pp. 356-369.

A SURVEY OF SEVERAL FINITE DIFFERENCE CODES USED IN TWO-DIMENSIONAL UNSTEADY FLOW PROBLEMS

N.E. Hoskin

(Atomic Weapons Research Establishment, Aldermaston)

and

W. Herrmann

(Sandia National Laboratories, Albuquerque, USA)

1. INTRODUCTION

Reactor safety analysts have been using large computer codes for many years to assist them in assessing the consequences of hypothetical core disruptive excursions. It is essential to demonstrate the codes' ability to simulate conditions within the reactor geometry and their accuracy in predicting loadings and strains on important reactor components. As part of this process of verification the US Energy Research and Development Administration (ERDA, now DOE) in 1975 proposed a technical interchange and comparison of computational techniques between various fast reactor safety analysis groups in the US, Europe and Japan.

In the first phase of the APRICOT project (Analysis of PRImary COntainment Transients) nine groups performed calculations of three agreed problems, using ten computer codes (US Dept. Energy, 1977). Two of these problems were simple test problems for which analytical solutions existed, namely an ideal gas shock tube and a suddenly pressurised spherical cavity in an elastic medium. The third problem was that of an explosion in a partially water-filled overstrong cylindrical containment vessel. This has, in turn, been a benchmark test in an inter-European programme of code validation experiments (COVA) and high quality experimental data were available (Rees et al., 1976).

In total the results of twelve calculations to the first problem, APRICOT 1 (shock tube) were submitted, eight of APRICOT 2 (spherical cavity) and eleven of APRICOT 3 (COVA benchmark test). The calculational results were assessed by independent experts and discussed by the participants. A critique of the results of this phase has been published (US Dept. Energy, 1977) and showed that essentially all the calculations produced solutions whose major features were sufficiently accurate for engineering purposes for these relatively simple problems when the material response was well specified.

This provided a basis of confidence in the state of the art and a desire by the participants to proceed to a second phase where problems of increasing complexity could be considered. For this phase four problems were agreed. First, the third problem of phase 1 was extended in time so that the codes' capability to calculate impact of water on the roof of the containment vessel could be examined. Then three model tests carried out by Stanford Research Institute (SRI) on behalf of Argonne National Laboratory (ANL) as part of ANL's code validation programme were included (Cagliostro and Romander, 1979).

They form a set of model tests of increasing degree of complexity and, referring to them as APRICOT 5, 6 and 7, APRICOT 5 was, like the COVA benchmark test, an explosion in a partially filled overstrong cylindrical containment vessel, although there were major differences in the COVA and SRI model tests in model geometry and charge behaviour. APRICOT 6 had a deformable containment vessel in place of the overstrong one of APRICOT 5, and in APRICOT 7 the simulated core barrel immediately outside the charge was allowed to deform whereas in APRICOT 5 and 6 it too had been overstrong. All three tests were well instrumented with pressure gauges on floor, roof and outer containment and with strain gauges on deformable components. APRICOT 6 and 7 thus provide valuable data on the codes' ability to model fluid-structure coupling.

Eight groups performed calculations of the four problems (not all groups did all the problems) up to July 1979 and in total submitted thirty-three calculations using eleven different computer codes. Of these there were seven calculations of APRICOT 4 (the COVA test to later times), eight of APRICOT 5, ten of APRICOT 6 and eight of APRICOT 7.

The review of the results of this second phase raised questions on the modelling of fluid-structure interaction and in 1980 a third phase of benchmark problems was agreed. This was a series of eight problems of increasing complexity examining fluid-structure modelling in detail. This phase is still continuing although a preliminary review has been made and discussed between the participants.

It is clear that the accumulated set of calculations form a significant data bank which can be used to assess the relative performances of the various techniques used in the codes. The codes (which all consider two-dimensional axisymmetric flow) are based on various finite difference algorithms (Lagrangian, Eulerian and coupled Lagrangian-Eulerian) for the calculation of the fluid flow, with the exception of one code in phase 1 which used a finite element formulation. The flexible vessel movement in problems 6 and 7 was calculated by either finite element theory, by finite difference methods using thin shell theory or by simulating the vessel by representing it by compressible Lagrangian zones. A definitive analysis of all the calculations submitted is obviously a major task and, while a fairly complete critique of phase 1 has been made, a full assessment of the results of phase 2 is far from complete and the review of phase 3 still continues.

Furthermore, these reviews have been primarily directed towards the reactor safety analyst's point of view. There are many aspects of numerical analysis which are illustrated within this data bank and it is the purpose of this report to discuss a few of these aspects, albeit in fairly general terms since details of the computational algorithms of some of the codes are not available. Thus, this is not a detailed survey of the form Sod produced for one-dimensional codes (Sod, 1979) although the first benchmark problem (a one-dimensional shock tube) is similar to that which he used as a test problem.

2. THE COMPUTER CODES

The computational techniques varied widely. Several of the codes used Lagrangian finite difference methods, one was a Lagrangian finite element method, there were two Eulerian implicit codes, one Eulerian explicit code, a coupled Eulerian-Lagrangian code and an arbitrary Lagrangian-Eulerian code. However, not all the codes were employed on each of the benchmark problems (see Table 1). Several of the Lagrangian codes used the "surface integral" spatial algorithm (Wilkins and Giroux, 1963) while at least two others used a Taylor expansion algorithm. As pointed out by Herrmann the two methods are equivalently second order accurate for regular meshes but for distorted meshes first order errors occur and these will be different for the two methods. He has shown, however, that the errors in spatial gradients are comparable for the two methods (Herrmann, 1964).

The stability criteria used varied widely in form and, while they have a minor effect on the quality of the solutions obtained, there was a significant effect on the computer time required for the calculations. Similarly there was a large range of artificial viscosity terms used. All the Lagrangian codes included a von Neumann-Richtmyer form (von Neumann and Richtmyer, 1950) but with a range of viscosity coefficients. Some used an additional term (Landshoff, 1955) linear in the dilatation rate (the VN form is quadratic) but the coefficients used varied by some orders of magnitude. There were also differences in the viscosity switch off - some linear viscosity terms were set to zero on expansion and some on compression. In the later two-dimensional problems several codes included other mesh stabilising viscosities related to the rate of mesh distortion, while several attempted to minimise the effect of such distortion by rezoning. While the rezoning algorithms were all reasonably accurate there were wide choices in the criteria used to decide when to rezone.

In the later problems, which involved aspects of fluid-structure interaction, the majority of the codes modelled the deformable vessels as thin shells with biaxial stress-strain behaviour, but one (PISCES 2DL) modelled them by Lagrangian meshes. Some codes unfortunately did not give details of the structural modelling.

Modelling details of the separate codes are given in Tables 1 and 2. References are quoted where it is known that they are available but in some cases no details have been given.

3. COMPUTER RESOURCES AND PROBLEM PARAMETERS

It must be remembered that the APRICOT exercise was carried on within the participants' normal research programmes and was conceived as a form of comparison and validation of independent calculations carried out according to local practices. Thus, there were large variations in problem parameters (artificial viscosity coefficients, stability criteria, number of rezones, etc.) and eight different types of computer were used. A crude comparison of computer time used by each calculation has been made by multiplying the reported CPU time by the following rough conversion factors:-

Computer	Factor	Computer	Factor
CDC 7600	5	CYBER 70/74	1
IBM 370/195	3	IBM 370/165	1
IBM 370/168	$1\frac{1}{4}$	CYBER 172	$\frac{1}{2}$
CDC 6600	1	SIEMENS 4004/150	$\frac{1}{3}$

With this conversion (which should not be accepted too rigidly) results are indicated in Tables 3 and 4 under the heading "Equiv CPU sec". For small word length machines the letter S is appended when the calculations are known to have been

TABLE 1

Participating Codes

Code Name	Spatial Algorithm		Ref.	Time Integral	Artificial Viscosities			Shell Model	Problems Calculated						
					Quad.	Linear	Mesh Stabilstn.		1	2	3	4	5	6	7
ALICE	Arbit. Lag.-Eul. ALE method		Hirt et al, 1974 Chu, 1979	Implicit	(3)	(3)	(3)	(3)						X	X
ARES	Lag.	Surface integral	Doerbecker, 1972	Explicit	C[5]	-	-	Thin Shell	X	X	X	X	X	X	X
ASTARTE	Lag.	Surface integral		Explicit	C	C	Yes	Thin Shell			X	X	X	X	X
CEFRA	Lag.	Mid point method		Explicit	C	E[6]	Yes	-	X				X		
CSQ	Eul.	Donor Cell	Thompson, 1979	Explicit	C	C	-	-	X		X	X			
EURDYN 1H[1]	Lag.	Constant pressure quad. elements		Explicit	Yes	Yes	-	-	X	X	X				
ICECO	Eul.	ICE method	Harlow and Amsden, 1971a Wang, 1975	Implicit	None	None	None	Thin Shell				X	X	X	X
PISCES 2DL	Lag.	Surface integral		Explicit	C	E	Yes	Shell nodes	X	X	X	X	X	X	
PISCES 2DELK	CE/L[2]	Surface integral	Cowler and Hancock, 1979	Explicit	C	E	Yes	Thin Shell	X[4]	X	X		X	X	X
REXCO	Lag.	Mid point method	Chang and Gvildys, 1975	Explicit	C	E	Yes	Thin shell	X	X	X	X	X	X	X
SEURBNUK	Eul.	ICE method	Cameron et al, 1977	Implicit	None	None	None	Thin shell	X				X	X	
SIRIUS	Lag.	Surface integral		Explicit	(3)	(3)	Yes	Thin shell							X
STEALTH	Lag.	Surface integral	EPRI, 1976	Explicit	C	E	-	-	X	X	X	X			
TOODY	Lag.	Surface integral		Explicit	C	E	Yes	-	X	X	X	X			

Notes:
(1) Finite element code: all others are finite difference
(2) Coupled Euler-Lagrange
(3) Not known
(4) Both Lagrangian and Eulerian calculations submitted
(5) C . artificial viscosity term active in compression only
(6) E . artificial viscosity term active in expansion only

TABLE 2

Structure Descriptions

	Nickel Vessel Wall				Aluminium Core Barrel			Lead Core Barrel		
Code Name	Description		Strain Hardening	Initial Yield, MPa	Description	Strain Hardening	Initial Yield, MPa	Description	Strain Hardening	Initial Yield, MPa
ASTARTE	90 Node Shell	5 Layers	4 Sublayers	128.8	Membrane	?	283	2 × 7 Solid Meshes	9 Element Isotropic	7.586
SEURBNUK	Shell	5 Layers	5 Sublayers	96.5	n.a.			n.a.		
ALICE (2)	Shell									
ICECO	Shell	5 Layers	4 Sublayers	100	Shell	4 Sublayers		2 × 8 Solid Meshes	Perfectly Plastic	10.0
REXCO	Shell	5 Layers	4 Sublayers	100	Shell	4 Sublayers		2 × 8 Solid Meshes	Perfectly Plastic	10.0
SIRIUS	Shell		Power Law	93.7	Shell	Power Law	150	4 × 8 Solid Meshes	Power Law	9.0
ARES	Shell	4 Layers	5 Element Isotropic	113.8	Shell	4 Element Isotropic	120	4 × 16 Solid Meshes	4 Element Isotropic	7.0
PISCES 2DELK	22 Node Shell	4 Layers	4 Element Isotropic	103.4	9 Node Shell	2 Element Isotropic	296	2 × 8 Solid Meshes	3 Element Isotropic	8.48
PISCES 2DL	2 × 23 Solid Meshes		1 Element Isotropic	158 upper (1) 131 lower	n.a.			n.a.		

(1) Different properties were used for the upper and lower vessel wall
(2) No information was furnished

performed in single precision, while a letter D denotes double precision calculations. Since, because of the wide variation in stability criteria, the number of cycles varied by large amounts in each problem, the equivalent CPU seconds have been divided by the number of cycles and the total number of elements or meshes to obtain the CPU time per mesh-cycle (in units of milliseconds/mesh-cycle). Variations within a factor 2 or so mean little but it was found that some calculations expended considerably larger computer resources to obtain results whose numerical quality did not necessarily correlate with the resources used.

The seven benchmark problems will be described briefly below, together with the results of the calculations - fuller specifications of the problems are given in (Hoskin, 1982).

4. PROBLEM 1 - ONE-DIMENSIONAL SHOCK TUBE

This was a simple shock tube with an initial pressure and density discontinuity; it tests the axial equations of motion, checks for any undesirable radial dependence (the problem is considered as a two-dimensional problem in cylindrical geometry) and requires some artificial viscosity formulation to capture the shocks. This needs to be chosen with some care if dispersion and numerical noise around the shock wave and rarefaction wave is to be minimised. Participants were asked to report the radial variation of the solution, the total energy error accumulating during the calculation and the computer resources used. These are shown in Table 3 and a visual indication of the quality of the results is given in the pressure-distance profiles at a common time (500 μs) plotted to the same scale in Fig. 1.

All calculations reproduce the major features of the exact solution to an acceptable degree with the strength of the shock predicted accurately, while its position and that of the mid-point of the rarefaction wave are within one mesh width of the exact position. Radial variations and total energy errors in each calculation were acceptably small. However, there are considerable variations in the dispersion and noise around these two waves.

We may examine the numerical prediction of the shock wave from Fig. 1. The overshoot, expressed as a percentage of wave amplitude, and the wave width, crudely estimated by assuming that the wave has the shape of a half sine wave, are given in Table 3. Also given in Table 3 are the magnitudes of the von Neumann-Richtmyer (quadratic) and Landshoff (linear) viscosity coefficients which were used (note that a C has been added when the viscosity term is active only when the material is

TABLE 3

Problem 1 Details

Code Name	Type	Viscosity		Shock Wave			Rarefaction Wave		Computer		No. of Cycles	Equiv. CPU sec	ECPU ms/ Mesh Cycle
		Quad.	Linear	Overshoot %	Width	Theor.	Overshoot %	Width	Type	CPU sec			
CEFRA	L F-D	1.2 C	0.0	10.0	3.5	3.44	10.0	4.0	SIEMENS 4004/150	702 S	223	234 S	2.1 S
ARES	L F-D	1.225 C	0.0	8.6	4.0	3.51	8.5	4.4	CYBER 172	400	167	200	1.2
PISCES 2DL	L F-D	2.0 C	0.001 E	6.5	5.0	5.75	10.0	5.0	CYBER 70/74	263	139	263	3.7
PISCES 2DL	L F-D	2.0 C	0.01 E	7.5	4.5	5.75	10.0	5.5	CDC 7600	37.2	136	186	2.7
PISCES 2DELK	L F-D	2.0 C	0.1 E	5.0	5.0	5.75	6.6	6.5	CDC 7600	14.0	106	70	1.3
TOODY 3	L F-D	2.0 C	0.1 C	2.5	6.0	5.75	8.3	4.5	CDC 6600	200	148	200	2.7
STEALTH	L F-D	2.0 C	0.2 E	5.0	4.5	5.75	5.0	7.5	CDC 7600	266	138	1330	20 (1)
REXCO	L F-D	3.0 C	0.05 C	< 1	8.0	8.60	10.0	4.0	IBM 370/195	129 D	750	387 D	1.0 D
EURDYN 1H	L F-E	1.6	0.0	5.0	4.0	4.59	7.5	4.5	IBM 370/165	33 D	75	33 D	0.88 D
SURBOUM (2)	E F-D	n.a.	n.a.	20.0	4.0	n.a.	< 1	7.5	IBM 370/165	118 S	150	118 S	1.6 S
PISCES 2DELK	E F-D	1.0 C	0.0	9.0	6.0	2.87	< 1	12.5	CDC 7600	20.4	89	102	2.3
CSQ	E F-D	2.0 C	0.1 C	< 1	7.0	5.74	< 1	8.0	CDC 6600	415	165	415	5.0 (3)

C Compression Only   S Single Precision

E Expansion Only   D Double Precision

Notes: (1) This time reflects use of a developmental version of STEALTH. A later calculation with the production version of the code reduced this value to 3.58.

(2) Early version of SEURBNUK.

(3) Time used in iterative loop in solution of the equation of state.

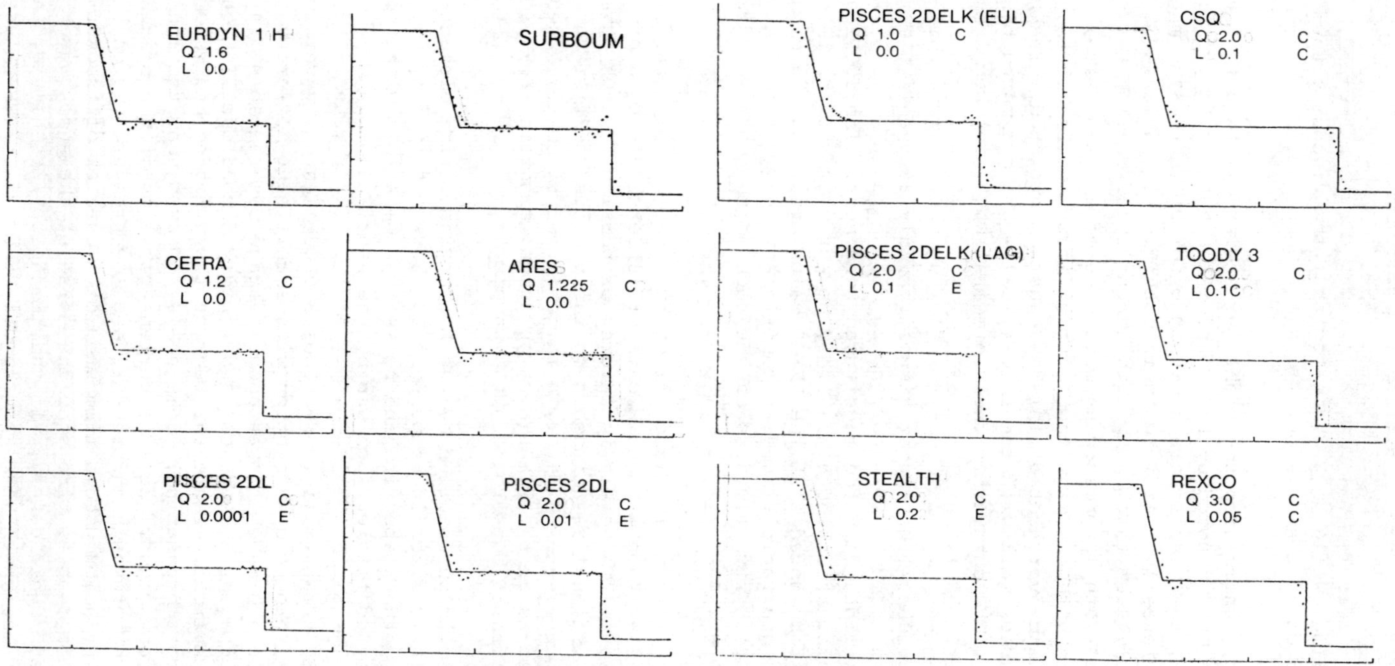

Fig. 1 Problem 1 results, pressure v distance at t = 500 μs.

compressing, while an E indicates that the viscosity term is included only on expansion).

First, considering the eight Lagrangian finite-difference calculations (denoted by L F-D under the heading "Type" in Table 3), the results are shown in order of increasing magnitude of the viscosity coefficients in both Fig. 1 and Table 3. It is clearly seen that, within the accuracy of reading the graphs, the wave width (expressed as multiples of the mesh size) increases with the magnitude of the viscous coefficients, while the maximum overshoot diminishes, just as expected. The wave width agrees well with the theoretical width estimated by von Neumann and Richtmyer in their original work (von Neumann and Richtmyer, 1950) (see Table 3).

At the shock the linear viscosity (only active if a C is indicated) damps oscillatory noise but has only a small effect on shock width with the magnitudes used here. The TOODY and REXCO calculations both give very acceptable shock profiles but it should be noted that the REXCO calculation had the highest value of quadratic viscosity coefficient and (perhaps in consequence) took five times as many cycles as the corresponding TOODY solution.

At the rarefaction wave all eight L F-D calculations showed the familiar "von Neumann-Richtmyer" undershoot and a broadening of the wave. These are in inverse proportion to each other but their functional relationship to the viscous coefficients is not readily apparent. However, a large linear viscosity, used on expansion as in the STEALTH calculation, significantly reduces the undershoot but broadens the wave.

The solution obtained by the Lagrangian finite element code EURDYN 1H is directly comparable to those obtained by the L F-D codes and the dispersion and noise are compatible with the viscosity coefficients chosen by the user.

The Eulerian implicit F-D code SURBOUM (an early version of SEURBNUK) produced a solution with significant differences. Shock width broadening was about average but the solution exhibited a large overshoot and rapidly damped oscillation at the shock front. At the rarefaction there was no undershoot but significant broadening. Some noise appeared near the location of the contact surface (the current position of the original interface between high pressure and low pressure regions). This calculation used a time step of a quarter of the CFL limit but with a doubled time step the noise at the shock wave increased significantly. If the time step were halved, the noise at the shock wave was slightly reduced. SURBOUM uses the implicit ICE algorithm (Harlow and Amsden, 1971) which should be unconditionally stable, but it appears that one cannot take advantage of this property without introducing unacceptable noise.

The other two Eulerian codes (PISCES 2DELK and CSQ) performed about as expected. The CSQ solution in the vicinity of the shock was smoother and broader than a L F-D solution with comparable viscosity (cf, TOODY 3). The PISCES 2DELK solution employed the smallest viscosity coefficients of all codes and shows an overshoot comparable to the L F-D solutions with low viscosities. Both the codes (as SURBOUM) predict very smooth rarefaction waves with no visible overshoot but considerable broadening.

There was a wide variation in the number of time cycles used in each calculation from 75 to 750 (if the CFL limit were used, only 37 cycles would have been required). The finite element code appears to have operated with the highest safety factor. The majority of codes either started with a conservatively small time step and increased it slowly towards their stability limit, or operated with a large stability margin throughout. The quality of the solution does not seem to have been affected by the choice of operation and factoring out the variation in number of cycles the variation in CPU ms/mesh cycle is most probably due to overheads (equation of state calculations, input-output, etc.) being significant in a very small calculation.

5. PROBLEM 2 - SUDDENLY PRESSURISED SPHERICAL CAVITY IN AN ELASTIC MEDIUM

This problem considered a pressurised hollow gas sphere in an infinite elastic medium, the whole system being initially at rest. It is a good test of a code's representation of narrow pressure pulses since the outward going pulse in the elastic region is a sharply peaked triangular wave. It also tests the ability of a large-strain code to handle small displacement elastic disturbances and to cope with the large impedance mismatch which occurs at the gas-solid interface. Calculators were asked to report the circumferential variation of the solution and the total energy error accumulated during the calculations - for the results submitted these were all within acceptable limits.

The problem has an exact analytic solution (Cooper, 1966), and Fig. 2 compares calculation and exact solution for velocity-distance profiles at a specified time. Eight calculations were submitted, using seven codes all of which were Lagrangian (this is an extremely severe test of an Eulerian code with small deformations within a fixed mesh and no calculations were submitted). In one PISCES 2DL calculation the gas region was

TABLE 4

Problem 2 Details

Code Name	Type	Viscosity			Computer		No. of Cycles	No. of Meshes	Equiv. CPU sec	ECPU ms Mesh Cycle
		Quad.	Linear	Other	Type	CPU sec				
ARES	L F-D	1.225	-	-	CYBER 172	420	105	915	210	2.2
EURDYN 1H	L F-E	0 (1)	0.03 (1)	-	IBM 370/165	106 D	88	1020	106 D	1.2 D
PISCES 2DELK (2)	L F-D	2.0 C	0.1 E	-	CDC 7600	24	102	901	120	1.3
PISCES 2DL (3)	L F-D	2.0 C	0.1 E	-	CDC 7600	45	105	900	225	2.4
PISCES 2DL	L F-D	2.0 C	0.01 D (4)	-	CYBER 70/74	294	102	915	294	3.1
REXCO	L F-D	1.2 C	0.05 (5)	Tensor, Rotational	IBM 370/195	251 D	433	960	753 D	1.8 D
STEALTH	L F-D	2.0 C	0.1 E	-	CDC 7600	198	106	915	990	10.2 (6)
TODDY 3	L F-D	2.0 C	0.1 C	-	CDC 6600	275	66	1275	275	3.3

C Compression Only

D Double Precision

E Expansion Only

Notes:
(1) In the gas quadratic 1.6 linear 0
(2) The gas was represented by a single Eulerian mesh
(3) The gas was not included, but represented by a pressure boundary
(4) In the gas quadratic 2.0 linear 0.0001
(5) Spherical portion of the "Navier-Stokes" viscosity
(6) This time reflects use of a developmental version of STEALTH. A later calculation using the production version of the code reduced the value to 3.5

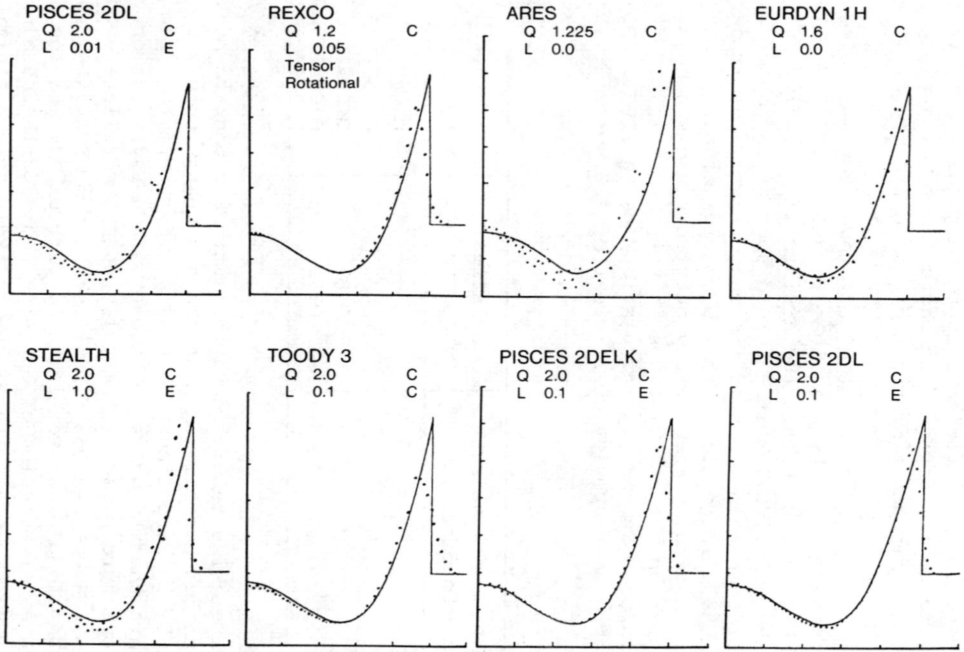

Fig. 2 Problem 2 results, velocity v distance at t = 8.66 μs.

replaced by a pressure condition at the solid boundary (see Table 4), while the other defined the gas region by Lagrangian meshes. In the Eulerian-Lagrangian code (PISCES 2DELK) calculation the gas bubble was represented by a single Eulerian mesh which was coupled to the Lagrangian meshes of the elastic medium. None of these variants appeared to have a markedly superior effect, indicating that those codes which explicitly calculated the impedance mismatch did so correctly.

Problem 2 is an excellent test of dispersion and noise introduced at a sharp pulse by numerical techniques. Generally, as seen from Fig. 2, the higher the artificial viscosity, the smoother the solution, but the more dispersed the shock wave and the greater the attenuation of its peak value. The linear viscosity has a large effect in the control of noise, particularly in the region immediately outside the gas bubble, but it is clear from the calculations that it has a major effect on the peak value. Subsidiary calculations using PISCES 2DL, all with a quadratic viscosity coefficient of 2.0, provided the following results:

Linear Viscosity Coefficient	Relative Peak Pressure	Relative Noise
0	0.95	0.442
0.01	0.92	0.060
0.05	0.85	0.047
0.10	0.79	0.016

The relative noise tabulated is the estimated peak-to-peak relative to the exact maximum at the shock and increases unacceptably at zero linear viscosity. Furthermore, from Fig. 2 it appears that those calculations with noisy solutions in the negative pressure phase of the pulse do not oscillate about the exact solution but have an offset. The REXCO solution, which used an additional mesh regularisation term based on the differential rates of rotation of opposite sides of the mesh, showed a very low noise level with a comparatively good definition of the peak value of the shock pressure.

It is clear that a quadratic viscosity alone is not sufficient to ensure an acceptable solution (too noisy) and that a linear viscosity must be included. However, an overlarge coefficient in this term will attenuate the peak stress too much so care is needed in the choice of viscosity coefficients.

Computer resources used are shown in Table 4. Again the number of cycles used to complete the calculation varied by a large factor, but the variation in CPU ms/mesh cycle was less than a factor of three which for this small problem is not thought to be significant.

6. PROBLEMS 3 AND 4 - ENERGY RELEASE IN AN OVERSTRONG WATER FILLED VESSEL

This problem is described fully in (US Dept. Energy, 1977). Briefly, a rigid cylindrical vessel is filled with water almost to the roof. A small high explosive charge is detonated on the axis somewhat below the centre of the vessel. Pressure histories had been measured experimentally at 20 distinct gauge locations on the floor, wall and roof of the vessel (Rees et al., 1976) and calculators were asked to provide pressure and impulse histories at these locations. For problem 3 participants were asked to take the calculation only as far as 1 ms which was just before impact of the main body of water on the roof. Thus, it is a test of the codes' ability to model several aspects of wave motion in a dense fluid, including shock wave reflection at the solid boundaries and at the free surface of the water. It also tests their ability to follow large fluid deformations as the water moves up towards the roof. Problem 4 was an extension to 2 ms thereby testing the codes' capability to calculate non-planar impact of the water on the solid roof, the shock formed because of this impact and the subsequent loading on the walls and floor as this shock moved downwards.

Considering problem 3, despite differences in viscosity coefficients, stability margins and rezones all the Lagrangian calculations produced results in fair agreement with each other and with experiment, at least on the floor and the walls - Fig. 3 shows two typical pressure histories on the floor one close to the axis and one further away. Both the Lagrangian (ASTARTE) and the Eulerian (SEURBNUK) codes predict the arrival times of the two waves accurately but the Lagrangian code overestimates the magnitude of the peak pressures, while the Eulerian code underpredicts them. Also note that the Lagrangian calculation has a lot of noise in the section of the pressure-time history between the two peaks (but, since the linear viscosity coefficient was zero, this is perhaps not unexpected). Wave arrival times were in good agreement with measurements for most of the calculations while total impulses (including roof reflections) were, with one or two exceptions, within 10 to 15% of the experimental values. The relative accuracy of peak pressures depended very much on the viscosity coefficients chosen and some calculations predicted waves between the major waves which were not present in the experimental records (some of these discrepancies

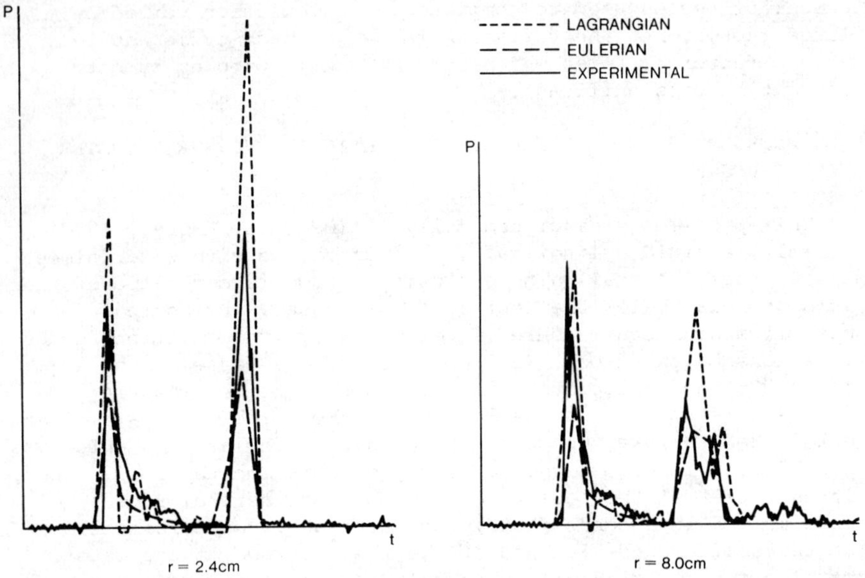

Fig. 3 Problem 3 pressure histories on the base .

were later shown to be due to elastic response of the walls (Morris and Bryant, 1978) which had been assumed initially to be perfectly rigid). However, even if the pressure history plots deviated in this way from experiment this had little effect on the total momentum delivered since the codes all conserve momentum very well.

The majority of the codes provided acceptable predictions of impulses on the floor and wall of the test vessel. The Eulerian code CSQ produced significant under-estimations even for a finely zoned calculation but this error arose in the early stages of the gas expansion due to the treatment of mixed cells. Subsequent calculations with an improved treatment of this region reduced the discrepancies significantly.

The data on computer resources furnished by the calculations of problem 3 have not been included here but there was a large variation in the number of cycles to take the problem to 1 ms. Considering the Lagrangian codes, except ARES, for which computer times were not furnished, there appeared to be an inverse relation between the number of cycles and the number of rezones used. The mesh configurations at 1 ms are shown in Fig. 4 and the distorted mesh occurring with zero or minimal rezoning shows

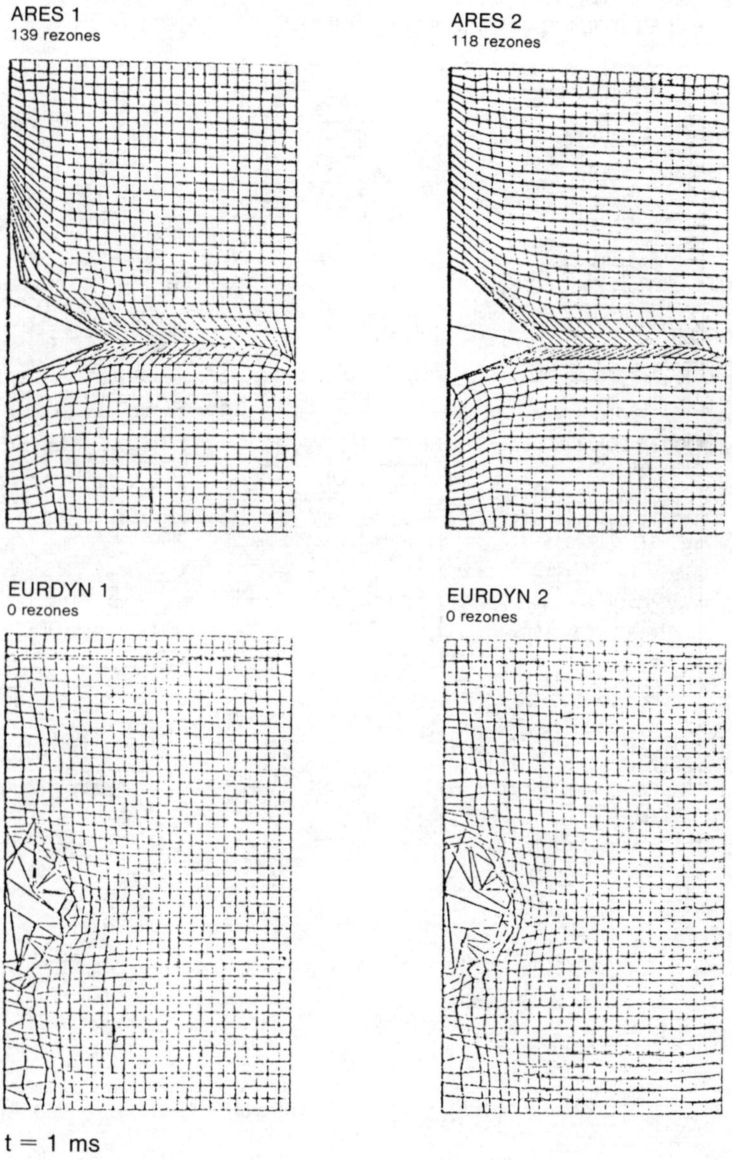

Fig. 4 Mesh configurations for Problem 3 at t = 1 ms.

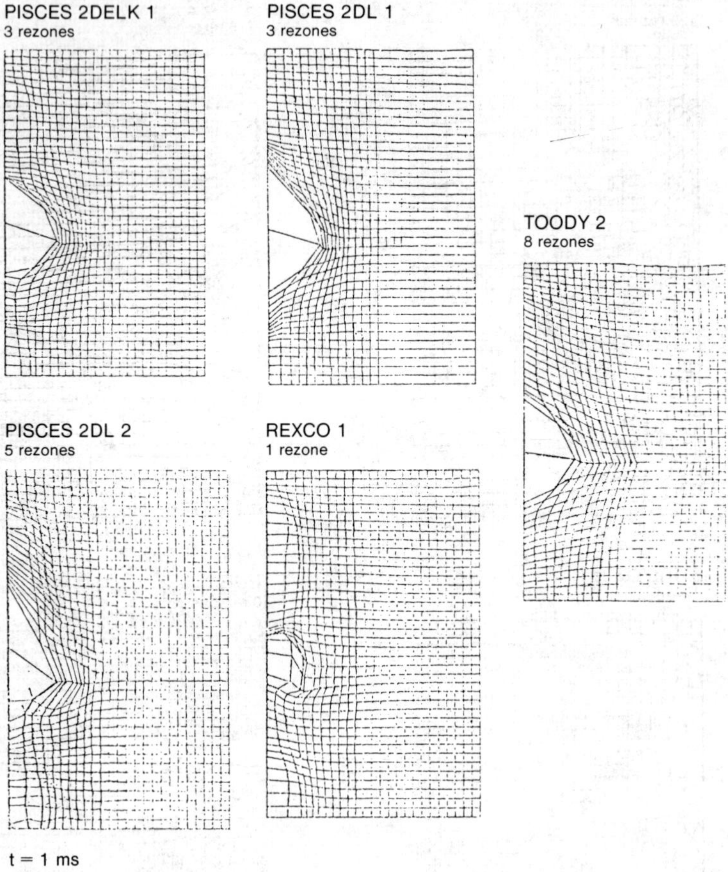
PISCES 2DELK 1
3 rezones
PISCES 2DL 1
3 rezones
TOODY 2
8 rezones
PISCES 2DL 2
5 rezones
REXCO 1
1 rezone
t = 1 ms

Fig. 4 (Cont.)

why time steps will be small and numbers of cycles large. However, comparison of the separate figures shows significant differences in the shape and position of the gas bubble. The two PISCES 2DL calculations were made by different participants, in different countries and presumably using different criteria since one calculation included 3 rezones and the other 5 rezones. The large variations in bubble detail must cast doubt on the accuracy of the solution in this region after multiple rezones. However, in the present problem the gas pressure, and hence energy, drops rapidly to small values so errors in predicting the gas bubble behaviour have little effect on the loadings on the outer boundaries. It is of interest to note that the CPU time required to rezone offset to some extent the gain due to the reduction of the number of cycles and the equivalent CPU time taken by the Lagrangian codes varied by little more than a factor two.

The two CSQ calculations took almost an order of magnitude more equivalent CPU time than the Lagrangian calculations, but most of this was due to the fact that the code used a complex iterative solution of a multiphase water equation of state instead of the simple explicit form used by the other codes. This relatively lengthy procedure was employed also in the calculation of problem 4 when the calculation was continued on to 2 ms and the same discrepancy seen. The equivalent CPU times for all the other codes, including the Eulerian codes ICECO and SEURBNUK, varied by a factor of less than four but it was difficult to correlate quality of predictions with equivalent CPU time used.

6.1 Extension into problem 4

The continuation of problem 3 past impact of the water surface on the roof was treated as a separate problem. Not all the calculators who submitted solutions to problem 3 chose to continue them on while, since problem 4 formed part of phase 2 of the APRICOT exercise, new participants submitted solutions to the problem up to 2 ms. However, these were received after the review of problem 3 and it has not proved practicable to retrieve all these solutions at 1 ms and add to the submissions for problem 3.

In problems 1 to 3 the codes' abilities to predict wave (both shocks and rarefactions) behaviour in gases, solids and liquids, and to cope with impedance mismatches and wave reflections, were tested extensively. Problem 4 became a test of codes' ability to predict the fluid acceleration caused by a shock reflection at a free surface and the impact of two materials (the accelerated fluid and the rigid roof). It also tested even further the codes' treatment of long-time material motion. Thus, all the Lagrangian codes but one used mesh stabilisation techniques or rezones to enable the calculation to be extended to 2 ms. The exception, PISCES 2DL, did not continue the solution past 1.3 ms.

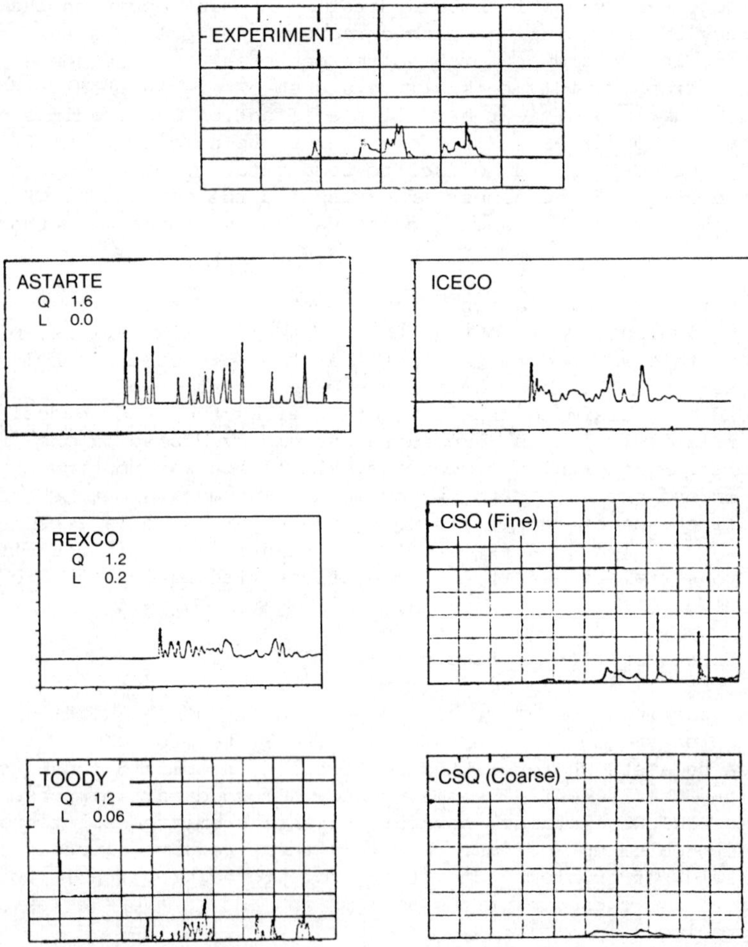

Fig. 5 Problem 4 predictions of pressure-time histories on the roof (near axis).

Fig. 5 illustrates predictions of a particular pressure-time history on the roof by most of the codes submitting problem 4 calculations. This was for a gauge near the centre of the roof and the experimental record shows a single pulse around 0.8 ms (which arose from the surface of the water spalling off), followed by the main impact pulse at about 1.1 ms, a second pulse at about 1.25 ms and a third at 1.5 ms. Examination of the pressure records shows very large differences in the results from the different codes. Use of the PMIN (minimum pressure) option for the equation of state of water allows the Lagrangian meshes just below the free surface to expand freely. The minimum value of pressure of -25 bars had been determined in auxiliary experiments to produce the correct free surface velocity. It is not surprising therefore that the codes predict the arrival time of the first pulse correctly. The three Lagrangian codes ASTARTE, REXCO and TOODY all have produced spiky pressure records and the severity of this spikiness correlates with the magnitude of the linear viscosity coefficient used. The Eulerian codes ICECO and CSQ have produced smoother pressure records, but the early pulse at 0.8 ms is too large in ICECO, is too small in the fine zoned CSQ calculation and is non-existent in the coarse zoned CSQ results. The later pulses are predicted in time and magnitude with varying degree of success.

As in problem 2, the effect of linear viscosity has been examined in a series of subsidiary calculations, this time with the code ASTARTE. Fig. 6 shows calculations of roof impact pressures for a somewhat similar geometry. The standard ASTARTE calculation (quadratic viscosity only) consists of a series of well separated spikes, but the addition of a linear viscosity produces a significantly different form of pressure record. The initial impact produces a pressure spike but this is followed by a well defined smooth pressure pulse. Modifying the way in which the linear viscosity is applied makes differences in detail only. The Eulerian code SEURBNUK prediction is very similar to the highly damped Lagrangian calculations.

Despite the inability of the codes to predict the correct pressure-time history the total impulse predictions, except from the CSQ code, were in reasonable agreement with measured values over the whole of the roof. It is clear that there is no unique way of reducing or eliminating numerical noise and that further investigation is necessary to determine the correct way of modelling the fluid-impact phase.

7. PROBLEMS 5 to 7 - ENERGY RELEASE IN SMALL-SCALE CONTAINMENT EXPERIMENTS

These form a sequence of experiments in which a gas-producer charge within a massive core barrel produces an expanding gas

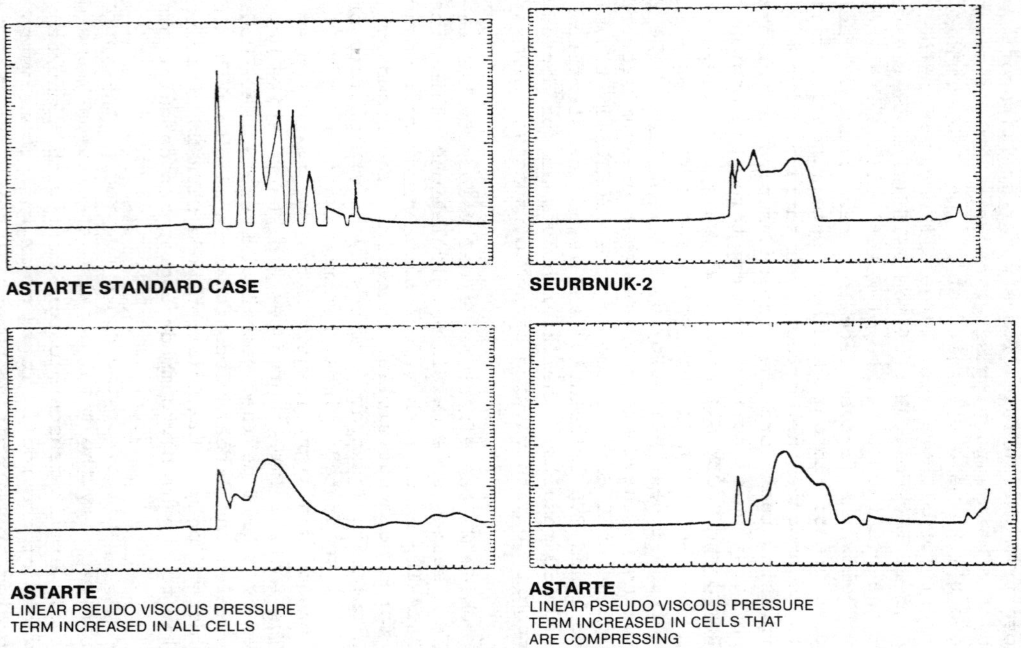

Fig. 6 Effect of pseudo-viscous term on pressure-time history on roof.

bubble in a water filled tank. In problem 5 both the core barrel and the water tank are rigid. In problem 6 the outer tank is deformable while the core barrel remains rigid, while in problem 7 both the outer tank and the core barrel deform. The deformable outer tank is thin relative to its radius so thin shell theory should be adequate to model its movement, but the core barrel is thick and a complete stress tensor should be used to calculate its deformation. Pressure and impulse histories were required at specified positions in all problems while strain histories were required at specified positions on the deformable components. These positions correspond to gauge positions in the experiments and measured data exist for comparison with the code predictions (Cagliostro and Romander, 1979). However, several calculations were received in which the plotted results were on scales which did not correspond to the scales specified in the instructions, so detailed comparison could not be made.

Detailed analysis of the code predictions can only be achieved by close study of all the pressure, impulse and strain histories throughout the complete geometry but, because of space considerations, only a typical comparison is included here. More detailed comparisons for a Lagrangian code (REXCO) and an Eulerian code (PISCES 2DELK) are given in (Hoskin, 1982) and all the comparisons are included in the review of phase 2 to be published shortly by the US Department of Energy. In total eight groups using ten different codes submitted twenty six individual calculations and for detailed comparisons the above mentioned review must be consulted - only general conclusions are given here.

Problem 5 provided a good test of the ability to compute fluid flow in a relatively complicated geometry (more so than problems 3/4). In contrast to problems 3/4 the initial mesh configuration was not specified and participants were encouraged to define the mesh as they thought most appropriate to the efficient utilisation of their particular code. While most codes employed a rectangular mesh (not necessarily of uniform mesh size throughout the total volume), the Lagrangian codes ASTARTE and SIRIUS used an initially non-orthogonal mesh, presumably because these were found to be most effective for the subsequent calculation. However, no arguments justifying the choice of mesh are available. Most calculators used the same mesh for problems 5, 6 and 7 but the PISCES 2DL calculations for problems 5 and 6 used different meshes.

All the codes did quite well in predicting the first pressure pulse on the vessel wall and floor, but overpredicted impulse on the roof by some 10 to 50%. The reflected wave from the roof travelling along the wall to the floor was also over-predicted, but otherwise the agreement between calculation and experiment was good. There were differences in detail due to discretisation, time step and choice of artificial viscosities. Larger values of artificial viscosities produced smoother solutions but attenuated transient pressure peaks, while calculations with the smaller values of artificial viscosities had more numerical noise. Eulerian codes all gave smoother calculations than Lagrangian codes.

Various modellings for the deformation of the flexible outer vessel (and deformable core barrel) were used and these are listed in Table 2. Most calculations used a thin shell representation with various strain hardening plasticity models. One calculation (PISCES 2DL for problem 6) used two columns of solid Lagrangian meshes to represent the flexible vessel.

Problem 6 taken in conjunction with problem 5 provided a test of fluid-structure interaction, albeit a very complicated one. The introduction of a deformable outer boundary wall altered the fluid motion so that wall loadings were, in general, reduced, but the initial pulses on the wall and floor were generally predicted well and the pressure loadings on the roof were generally in better agreement with experiment than in problem 5. However, when considering strains there appeared to be a tendency for the codes to under-predict deformations due to low level loading, while deformations due to high stress loadings were over-predicted. Most codes under-estimated the strains on the lower half of the vessel. These points are illustrated in Fig. 7, which shows comparisons between prediction from REXCO, a well developed Lagrangian code, and experiment for Problem 6. The pressure gauge P2 is on the core barrel adjacent to the energy source, P4 is on the floor of the vessel, P5 midway up the deformable vessel, P7 near the top of the vessel while P8 and P9 are on the roof. Of the strain gauges S1 measures hoop strain and S8 axial strain at the position of P7 while S4, S11 and P5 form a similar triplet. It can be seen that agreement between experiment and prediction for pressure or impulse histories does not necessarily ensure agreement for strain histories. These comparisons are typical although there were differences between the several codes accentuated by differences in yield point and strain hardening behaviour while, particularly in the low level loading regime, the effects of high frequency noise or excessive attenuation arising from too little or too much smoothing with artificial viscosity can become significant.

Problem 7 added the flexible core barrel which produced only small changes in the over-all response and the codes behaved in a similar fashion to that of problem 6, except perhaps for some improvement in comparison with experiment.

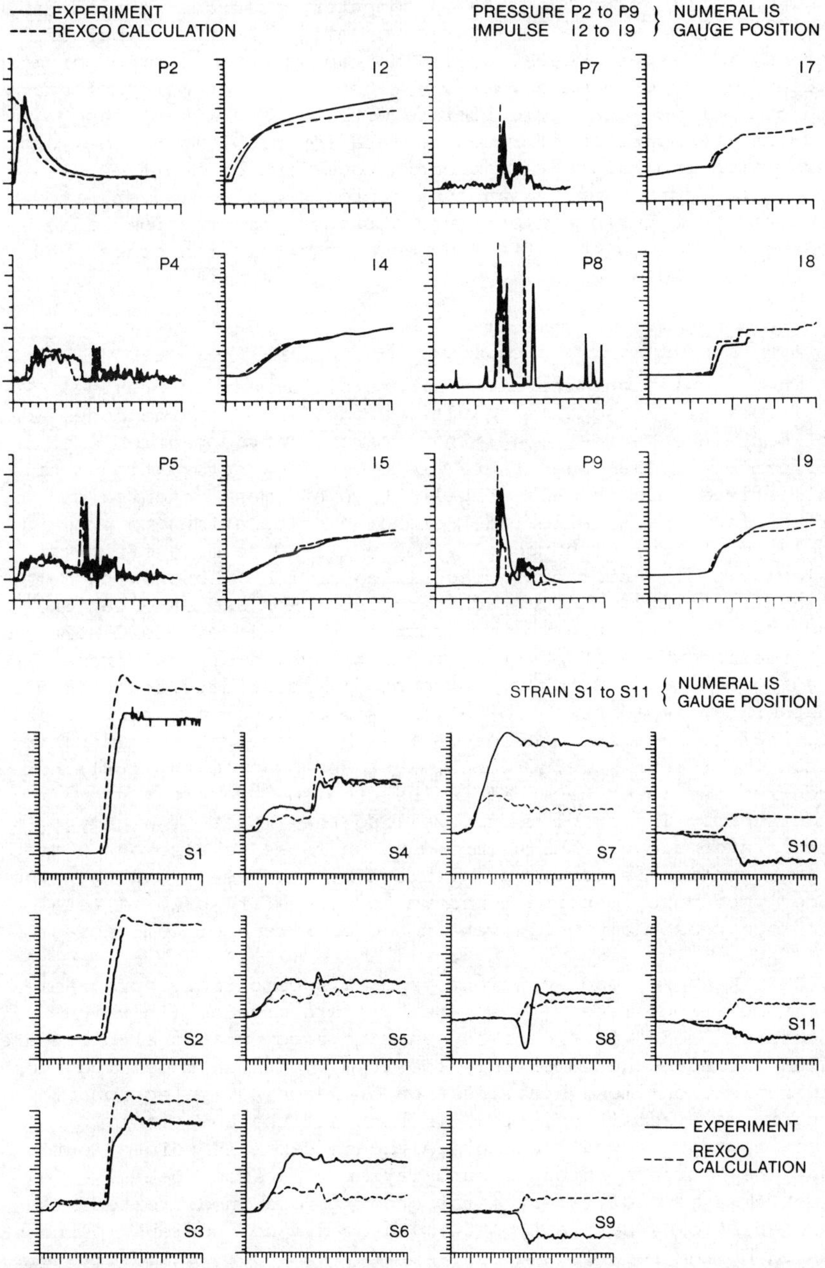

Fig. 7 Problem 6, pressure, impulse and strain histories. REXCO (Lagrangian) calculations.

When considering the data on computer resources furnished by the calculators there were large variations in the number of meshes, number of cycles, etc. The equivalent CPU time per mesh cycle for say problem 5 varied by about an order of magnitude and did not necessarily correlate with the quality of the results. Comparison of resource data for problems 6 and 7 was less revealing, since variation in modelling of the flexible vessel and core barrel cannot be factored out. However, it can be noted that inclusion of fluid-structure interaction had a greater effect on the efficiencies of the Eulerian codes than on the Lagrangian codes.

8. CONCLUSIONS

The calculations represented here all appear to provide qualitatively reasonable results, suggesting that the codes do not contain any gross modelling errors. They appear able to perform well, even when there are gross flow deformations while palliatives such as mesh stabilisation and mesh rezoning can prove effective in allowing Lagrangian calculations to proceed to longer times, although they may not lead to more efficient calculations. Artificial viscosities must be chosen with care, balancing noise reduction against dispersion and attenuation, and this is particularly important in problems of fluid-structure interaction where at any time the local value of the stress loading relative to the yield point of the structure determines whether the immediate deformation is elastic or plastic. This topic of fluid-structure interaction has received considerable attention recently (e.g. Jones, 1981 and Staniforth, 1981) but much remains to be done before the problems are clearly understood. The problem of modelling the impact of highly accelerated flows on high impedance surfaces is also one where more attention is required, although it has been shown that the codes used here conserve momentum and therefore predict total impulses reasonably well even in these extreme circumstances.

What has been made apparent from this comparative programme of calculations made with several different codes (in which we interpret "code" in the broad sense of a computational procedure which includes the numerical algorithm, material modelling, mesh discretization, mesh stabilisation or rezoning, selection of artificial viscosities, etc.) is that all the numerical algorithms are soundly based. Although Herrmann pointed out that the "surface integral" and Taylor expansion spatial algorithms have different first order errors in distorted mesh configurations (Herrmann, 1964), these are not obvious (perhaps masked by the rezonings used). Differences between results are much more dependent upon parameters under user control such as artificial viscosities, mesh housekeeping, etc., and these differences may be significant in problems involving fluid-structure interaction.

While numerical analysts have expended considerable effort in considering what is the optimum representation of the equations of motion, little work has been done into analysing the effects of the factors under user control in order to determine what is a "best" representation. If such an understanding could be obtained, it would clearly lead to a much more effective use of computer resources.

9. ACKNOWLEDGEMENT

We thank Science Applications Incorporated of Palo Alto, California, the coordinators of the APRICOT programme, for permission to use extracts from the draft report of phase 2 (prior to publication).

REFERENCES

CAGLIOSTRO, D.J. and ROMANDER, C.H. 1979 Experiments on the Response of Rigid and Flexible Reactor Vessel Models to a Simulated Hypothetical Core Disruptive Accident. Stanford Research Institute, 5th Interim Report to Contract 31-109-38-2655 (April 1979).

CAMERON, I.C., HANKIN, B.C., WARHAM, A.G.P., BENUZZI, A. and YERKESS, A. 1977 The Computer Code SEURBNUK-2 for Fast Reactor Containment Safety Studies. Paper B2/1, Fourth SMIRT Conference, San Francisco.

CHANG, Y.W. and GVILDYS, J. 1975 REXCO-HEP, a Two-Dimensional Computer Code for Calculating the Primary System Response in Fast Reactors, ANL/RAS-75-11.

CHU, H.Y. 1979 ALICE - An Arbitrary-Lagrangian-Eulerian Code for Analysis of FBR Containment Response to HCDA. Paper E1/4, Fifth SMIRT Conference, Berlin.

COOPER, H.F. 1966 Generation of an Elastic Wave by Quasi-Static Isentropic Expansion of a Gas in a Spherical Cavity; Comparison between Finite-Difference Predictions and the Exact Solution. AFWL-TR-66-63.

COWLER, M.S. and HANCOCK, S.L. 1979 Dynamic Fluid Structure Analysis of Shells Using the PISCES 2DELK Computer Code. Paper B1/6, Fifth SMIRT Conference, Berlin.

DOERBECKER, K. 1972 ARES - Ein Zweidimensionales Rechenprogram zur Beschreibung der Kurzzeitigen Auswickungen einer

Hypothetischen Uncontrollierten Nuklearen Exkursion auf Reaktortank, Drehdeckel und Tankeinbauten, Gezeigt am Beispiel des SNR-300. Reaktortagung des DAtF, Hamburg.

EPRI, 1976 STEALTH, A Lagrange Explicit Finite Difference Code for Solids Structural and Thermohydraulic Analysis. EPRI NP-1976 Technical Report 1.

HARLOW, F.H. and AMSDEN, A.A. 1971a A Numerical Fluid Dynamics Calculation Method for all Flow Speeds. *J.Comput.Phys.* **8**, 197.

HARLOW, F.H. and AMSDEN, A.A. 1971b Fluid Dynamics - a LASL monograph. LA4700.

HERRMANN, W. 1964 Comparison of Finite Difference Expressions Used in Lagrangian Fluid Flow Calculations. AFWL-TR-64-104.

HIRT, C.W., AMSDEN, A.A. and COOK, J.L. 1974 An Arbitrary Lagrangian-Eulerian Computing Method for All Flow Speeds. *J.Comput. Phys.* **14**, 227-253.

HOSKIN, N.E. 1982 A Survey of Several Finite Difference Codes used in Two-Dimensional Unsteady Flow Problems. AWRE/44/92/25.

JONES, A. 1981 Fluid-Structure Coupling in Lagrange-Lagrange and Euler-Lagrange Descriptions. Euratom Report EUR 7424 EN.

LANDSHOFF, R. 1955 A Numerical Method for Treating Fluid Flow in the Presence of Shocks. LA-1930.

MORRIS, E. and BRYANT, A.R. 1978 Response of Fluid Filled Vessels to Internal Explosions. Paper Presented at Annual Meeting of American Nuclear Society, San Diego.

REES, N.J.M., TATTERSAL, R.E. and VERZELETTI, G. 1976 Results of Repeat Firings of High Explosive Charges in Water Filled Vessels. AWRE/44/97/1, TRG Report 2909 (R/X), JRC Ispra EE/01/76.

SOD, G.A. 1979 A Survey of Several Finite Difference Methods for Systems of Non-Linear Hyperbolic Conservation Laws. *J. Comput. Phys.* **27**, 1-31.

STANIFORTH, R. 1981 Stability of Thin-Shell/Fluid Coupling in Lagrangian Codes. Paper B1/2, Sixth SMIRT Conference, Paris.

THOMPSON, S.L. 1979 CSQ II - An Eulerian Finite Difference Program for Two Dimensional Material Response - Part 1. Material Sections SAND 77-1339.

US DEPARTMENT OF ENERGY 1977 SAN-1112-1.

VON NEUMANN, J. and RICHTMYER, R.D. 1950 A Method for the Numerical Calculation of Hydrodynamic Shocks. *J.Appl.Phys.* **21**, 232.

WANG, C.Y. 1975 An Implicit Eulerian Method for Calculating Fluid Transients in Fast Reactor Containment. ANL-75-81.

WILKINS, M.J. and GIROUX, R. 1963 Calculation of Elastic-Plastic Flow. UCRL-7322.

CALCULATION OF SHOCKS IN OIL RESERVOIR MODELLING AND POROUS FLOW

P. Concus

(Lawrence Berkeley Laboratory, University of California, USA)

1. INTRODUCTION

In recent years the numerical modelling of fluid displacement through a porous medium has received increased attention, stimulated by the development of enhanced recovery methods for obtaining petroleum from underground reservoirs and the advent of larger, higher-speed computers. For many recovery methods of interest, propagating fronts arise that may be steep or discontinuous. One example is the water flooding of a petroleum reservoir, in which there is forced out residual oil that remains after outflow by decompression has declined.

Some of the work being carried out in the Mathematics Group of the Lawrence Berkeley Laboratory for developing high-resolution numerical methods to solve porous flow problems having propagating discontinuities is discussed here. Such discontinuities usually pose substantial difficulty for conventional discretization methods. Our investigations centre on some alternative numerical methods that incorporate analytical information concerning the discontinuities. Such methods have been effective in treating hyperbolic conservation laws arising in gas dynamics and can be adapted in many cases to the equations of porous flow.

One method of interest, the random choice method, can track solution discontinuities sharply and accurately for one space dimension. The first phase of our study adapted this method for solving the Buckley-Leverett equation for immiscible displacement in one space dimension. Extensions to more than one space dimension for the random choice method were carried out subsequently in our study by means of fractional splitting. Because inaccuracies could be introduced for some problems at discontinuity fronts propagating obliquely to the splitting directions, our efforts are currently being directed at investigating alternatives for multi-dimensional cases.

A front tracking method for multi-dimensional problems based on the SLIC scheme developed by Noh & Woodward (1976) was extended to the Burgers equation and porous flow cases. The method assigns to mesh cells a value representing the fraction of the cell lying behind the front. The cell fractions are then appropriately advanced at each time step. Away from the front, the random choice method, Godunov's method, or other methods may be used.

Recently a higher order version of Godunov's method, which utilizes piecewise linear rather than piecewise constant segments, as introduced in the MUSCL scheme of van Leer (1979), has been extended to gas dynamics in Eulerian coordinates. Preliminary results for the extension of this work to the porous flow equations for one space dimension have been promising, and an investigation of extension to higher dimensions is being carried out.

Specialized numerical approaches taken by others for porous flow problems with steep fronts can be found, for example, in papers SPE 10499 to SPE 10502 in the Proceedings of the Sixth SPE Reservoir Simulation Symposium (1982) and the references therein, in (Miller and Miller, 1981) and in (Glimm, Isaacson, Marchesin and McBryan, 1981).

2. EQUATIONS FOR IMMISCIBLE DISPLACEMENT

The equations for two-phase immiscible incompressible displacement in a porous medium in the absence of capillary pressure are (Peaceman, 1977)

$$\phi\frac{\partial s}{\partial t} + \underline{q}\cdot\nabla f(s) - \gamma\frac{\partial}{\partial z} g(s) = 0 \qquad (2.1)$$

$$\nabla\cdot\underline{q} = Q \qquad (2.2)$$

$$\underline{q} = -\lambda(s)[\nabla p - \gamma\tilde{g}(s)\underline{e}_k] \qquad (2.3)$$

The porous medium is taken to be homogeneous and isotropic and the interior of the domain free of sources or sinks (i.e., injection or production wells).

In the above equations $s(\underline{x},t)$, $0 \le s \le 1$, is the saturation of the wetting fluid (fraction of available pore volume occupied by the fluid). The saturation of the non-wetting fluid is then $1-s$. The independent variables $\underline{x}$ and t are the space and time, respectively, and $\underline{q}(\underline{x},t)$ is the total velocity (sum of the individual velocities of the two fluids). If gravity is present it is assumed to act in the negative z direction, with $\underline{e}_k$ the unit vector in the positive z direction. The quantity $p(\underline{x},t)$ is the excess over gravitational head of the reduced pressure; here the reduced pressure is the average of the individual phase pressures less the gravitational head. The quantity Q represents the sources and sinks of fluid on the boundary of the domain, and ϕ is the porosity, which will be assumed constant. The quantity γ, the coefficient of the gravitational term, is the product of the acceleration due to gravity times the density difference between the wetting and nonwetting phases.

Equation (2.1) is the Buckley-Leverett equation, which for a given $\underline{q}$ is hyperbolic. Equation (2.2) is the incompressibility condition, and (2.3) is Darcy's law. For a given s, (2.2), (2.3) is elliptic.

The functions of saturation $f(s), g(s), \lambda(s)$ and $\tilde{g}(s)$ can be expressed in terms of the empirically determined phase mobilities (ratios of permeability to viscosity) λ_n and λ_w of the non-wetting and wetting fluids. For immiscible displacement these are

$$f(s) = \lambda_w/\lambda, \quad g(s) = \lambda_n f$$

$$\tilde{g}(s) = \lambda_n/\lambda, \quad \lambda(s) = \lambda_n + \lambda_w.$$

The quantities f, g, and $\tilde{g}$ are non-negative, and λ is positive.

A distinguishing feature of the immiscible displacement equations is that f and g are non-convex: f typically has one inflection, as depicted for a model case in Fig. 1, and g has two, as depicted in Fig. 2. Thus weak solutions may have combinations of propagating shock and expansion waves in contact.

Attempts to solve (2.1), (2.2), (2.3), subject to appropriate boundary conditions, by standard discretization methods such as finite difference or finite element methods can give rise to substantial difficulty. Inaccuracies may arise near a moving front, or an incorrect weak solution may be obtained. To circumvent these difficulties, the first phase of our study initiated an attempt to adapt the random choice method to solving problems of fluid displacement in porous media.

3. RANDOM CHOICE METHOD

The random choice method, which was formulated originally for solving the equations of gas dynamics, is a numerical method that incorporates the accurate propagation of solution discontinuities. It is based on a mathematical construction of

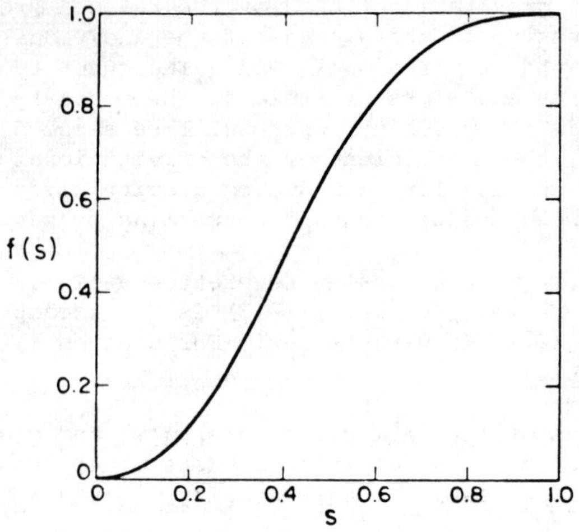

Fig. 1

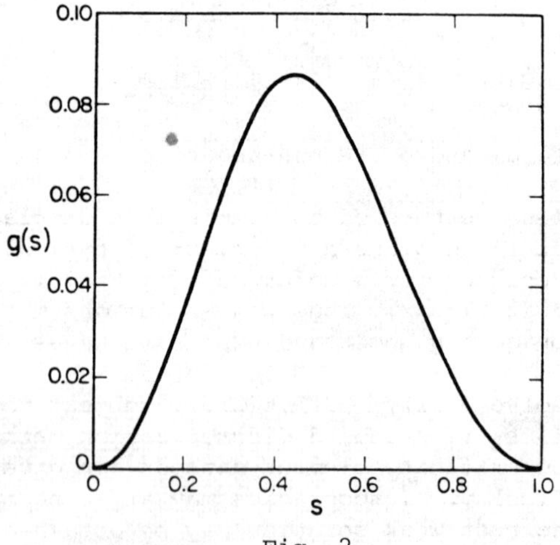

Fig. 2

Glimm (1965) that was developed into a practical and efficient computational algorithm by Chorin (1976, 1977). It was first adapted to porous flow problems in (Concus and Proskurowski, 1979), and subsequently in (Glimm, Marchesin and McBryan, 1979).

For a single nonlinear conservation law,

$$\frac{\partial s}{\partial t} + \frac{\partial}{\partial z}\psi(s) = 0, \tag{3.1}$$

to which (2.1) reduces in one space dimension, the random choice method advances a solution in time as follows. The solution $s(z,t_j)$ at time t_j is represented by a piecewise-constant function on a spatial grid of spacing Δz, where the function is equal to $s_i^j = s(z_i, t_j)$ in the interval $z_i - \frac{1}{2}\Delta z < z \leq z_i + \frac{1}{2}\Delta z$. Then the solution of (3.1) is constructed analytically by the method of characteristics for this piecewise-constant initial data by solving a sequence of Riemann problems (3.1) with initial data

$$s(z,t_j) = \begin{cases} s_i^j, & z \leq z_i + \frac{1}{2}\Delta z \\ s_{i+1}^j, & z > z_i + \frac{1}{2}\Delta z. \end{cases} \tag{3.2}$$

So long as the time increments Δt satisfy the Courant-Friedrichs-Lewy condition $(\Delta t/\Delta z)\max|\psi'(s)| < \frac{1}{2}$ (<1 for forms of the method using half time-steps on staggered grids), the waves propagating from the individual mesh-point discontinuities will not interact during a given time step. This permits the solution of (3.1) to be obtained during the step by joining together the separate Riemann problem solutions. The above procedure is common with other methods, such as Godunov's method. The distinguishing feature of the random choice method is that it obtains the new piecewise-constant representation of the solution at the new time by sampling the exact solution at a point within each spatial interval. In this way moving discontinuities remain perfectly sharp, since no intermediate values are introduced by the method, at the price of introducing a small amount of statistical uncertainty. The method is essentially first order and is observed to give good results for one-dimensional problems.

4. RIEMANN PROBLEMS

The practicality of the random choice method depends upon being able to solve the Riemann problems efficiently. For the immiscible displacement problem the function $\psi(s)$, which is a linear combination of $f(s)$ and $g(s)$, has either one or two inflections, depending upon the relative magnitudes of q and γ. If in (2.1) the gravity term $\gamma g(s)$ is small compared with the transport term $qf(s)$ then there is only one inflection, for which the Riemann problem solution is given in (Concus and Proskurowski, 1979). For the case of two inflections the solution is given in (Albright, Anderson and Concus, 1980), (Anderson and Concus, 1980), and for a special case in (Proskurowski, 1980).

A typical example for a case in which two inflections occur

is depicted in Fig. 3, which is the case given in Fig. 3 of (Anderson and Concus, 1980). The Riemann problem solution is obtained by applying the following general conditions, which must hold along any curve of discontinuity of $s(z,t)$.

Let

$$s_- = \lim_{z \to z_-} s(z,t)$$

and

$$s_+ = \lim_{z \to z_+} s(z,t)$$

be the limiting values from the left and right at a discontinuity. Then there must hold (Lax, 1973), (Oleinik, 1963),

(i) Rankine-Hugoniot jump condition: the curve of discontinuity is a straight line with slope

$$\frac{dz}{dt} = \frac{\psi(s_+) - \psi(s_-)}{s_+ - s_-}$$

(ii) Generalized entropy condition:

$$\frac{\psi(s_+) - \psi(s)}{s_+ - s} \leq \frac{\psi(s_+) - \psi(s_-)}{s_+ - s_-}$$

for any s between s_+ and s_-.

For the case of $s_i^j = 0$ and $s_{i+1}^j = 1$ one obtains the solution of (3.1),(3.2) depicted in Fig. 4. Fig. 3 depicts the corresponding concave hull of $\psi(s)$, whose points of tangency with $\psi(s)$ determine the shock propagation speeds. The two shocks shown in Fig. 4 propagate to the left and right, respectively, from the initial discontinuity. The characteristics from the left of the discontinuity intersect the leftward travelling shock, and those from the right intersect the rightward travelling shock. Between the two shocks is an expansion wave with its fan of characteristics emanating from the initial discontinuity.

The solution depicted in Fig. 4 is then sampled at a value of z at the later time to obtain the new value to be used for the interval in the piecewise constant representation of s. The sampling details and further discussion of the Riemann problem for these equations are given in (Albright, Anderson and Concus, 1980) and (Anderson and Concus, 1980).

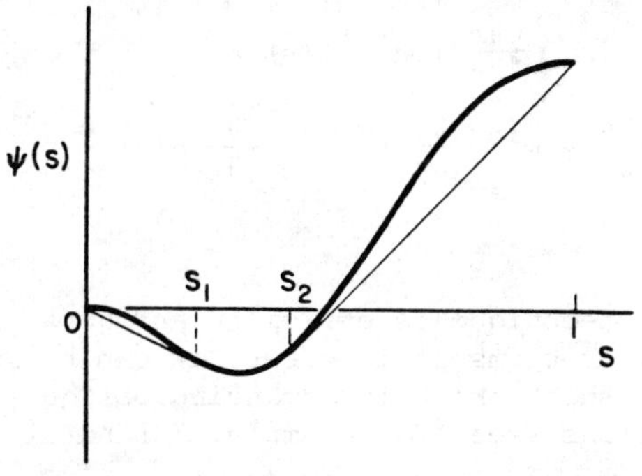

Fig. 3

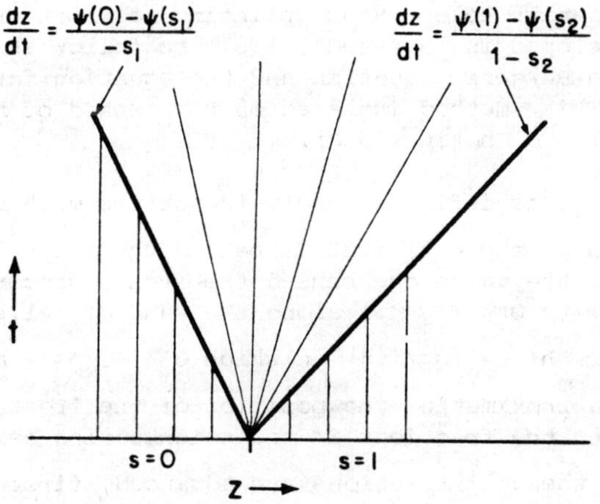

Fig. 4

5. FRACTIONAL SPLITTING FOR MULTI-DIMENSIONAL PROBLEMS

For a one-dimensional problem, (2.1), (2.2), (2.3) reduces to the single conservation law of the form (3.1), since q is constant in the interior for this case. For a multi-dimensional problem the standard technique for solving (2.1), (2.2), (2.3) is to solve successively (2.2), (2.3) for p (and q) taking s to be fixed at its approximate solution for the current time, and then to advance (2.1) one time step considering q fixed, to obtain an approximate solution for s at the new time.

Advancing (2.1) is carried out using the random choice method in (Albright and Concus, 1980), (Albright, Anderson and Concus, 1980), and (Anderson and Concus, 1980) by means of fractional splitting. Specifically, one solves successively the one-dimensional problems

$$\phi\frac{\partial s}{\partial t} + u^{(n)} f(s) = 0$$

$$\phi\frac{\partial s}{\partial t} + w^{(n)} \frac{\partial}{\partial z} f(s) - \gamma\frac{\partial}{\partial z} g(s) = 0,$$

where $\underline{q} = (u,w)$.

Although this technique is efficient and gives acceptable results for many problems of interest, it can be inaccurate for cases in which a shock front is advancing obliquely to the splitting directions (see, for example, (Crandall and Majda, 1980)).

6. FRONT TRACKING METHOD

As an alternative to operator splitting, a front tracking method was developed in (Lotstedt, 1981) to follow shock discontinuities in Burgers' equation and the equation for two-phase porous flow. This method is based on the method of Noh and Woodward (1976) and Chorin (1980).

A function F_{ij} is defined whose value at the mesh cell (i,j) is the fraction of the cell that is behind the front. Thus most cells have the value one behind the front, or zero ahead of the front, with $0 < F_{ij} < 1$ along the band of cells enclosing the front. For the cells (i,j) in which $0 < F_{ij} < 1$ a line segment is drawn approximating the position of the front, based on not only F_{ij} but the fractions in the neighbouring cells. At each time step the cell fractions are advanced, first in the x-direction and then in the y-direction.

In (Lotstedt, 1981) the details of the front tracking scheme are given. The line segments approximating the front within each cell are permitted to have oblique slope and need not be parallel to the mesh lines. The method is tested on the inviscid Burgers equation and porous flow equations in two space dimensions. The numerical solution is calculated for discontinuous initial data and is found to agree very well with known solutions. For the case of a physically unstable interface, which can occur for porous displacement when a more mobile fluid displaces a less mobile one, the method is able to resolve and follow the fingering of the surface as it develops.

Fig. 5 depicts successive positions of the discontinuity front as calculated by the method. This example, which is presented in (Lotstedt, 1981), is for two-phase miscible displacement with a source at (0,0) and a sink at (1.1). For this miscible displacement case, (2.1), (2.2), (2.3) hold with $f(s) = s$ and $\lambda(s) = (s+\tau(1-s))^4$, $\tau>0$. Initially the square is occupied entirely by the fluid to be displaced, and at the boundary the normal derivative of s is zero. A uniform spatial grid with spacing 1/40 was used for the calculation. For this problem the front is physically unstable, a property that Fig. 5 depicts as being captured by the method. Other, more stabilized, front-tracking methods, such as the one in (Glimm, Isaacson, Marchesin and McBryan, 1981), are reported to yield for this same problem a solution that does not have observable fingering.

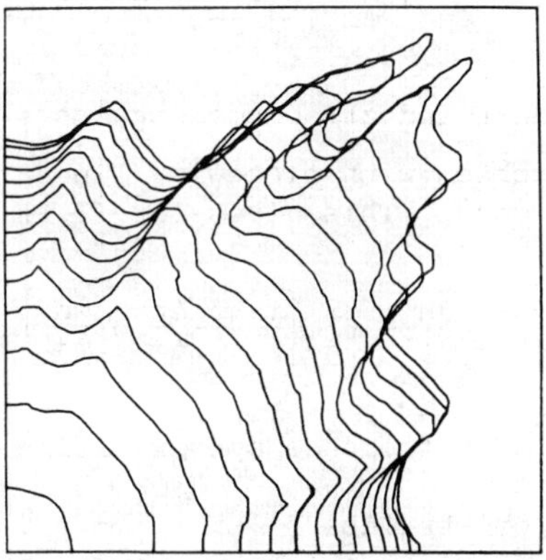

Fig. 5

7. HIGHER ORDER GODUNOV METHOD

Another alternative to the random choice method with splitting for multi-dimensional problems included in our study is that of a higher order Godunov method. Although first order Godunov methods generally are too diffusive to follow steep fronts sharply, the higher order methods introduced recently by van Leer (1979) for gas dynamics in Lagrangian coordinates appear to circumvent a large portion of this shortcoming. Godunov's method (first order) was introduced in (Godunov, 1959). It and the random choice method are similar in that they both require the solution of Riemann problems corresponding to piecewise constant initial data. Whereas the random choice method obtains the new approximation at the new time by sampling, Godunov's method obtains it by a conservative differencing scheme. Shocks are not kept sharp by it, as intermediate values are introduced. However, when Godunov's method is combined with operator splitting to solve multi-dimensional problems, shocks travelling obliquely to the mesh do not give rise to the same degree of difficulty as might be encountered in the random choice method. The second-order Godunov methods hold promise of giving good results for solving multi-dimensional porous flow problems.

For solving (3.1) with $\gamma=0$, the condition that $\psi'(s)>0$ simplifies Godunov's method to first order upwind differencing

$$s_{j+\frac{1}{2}}^{n+\frac{1}{2}} = s_j^n$$

$$s_j^{n+1} = s_j^n + \frac{\Delta t}{\Delta z}\left\{\psi(s_{j-\frac{1}{2}}^{n+\frac{1}{2}}) - \psi(s_{j+\frac{1}{2}}^{n+\frac{1}{2}})\right\}.$$

The second order (in t and z) scheme obtains $s_{j+\frac{1}{2}}^{n+\frac{1}{2}}$ by solving an initial value problem with piecewise linear rather than piecewise constant data. This gives for (3.1) with $\psi'(s)>0$

$$s_{j+\frac{1}{2}}^{n+\frac{1}{2}} = s_j^n + \delta_j\left\{\frac{1}{2} - \frac{\Delta t}{2\Delta z}\psi'(s_j^n)\right\},$$

where δ_j is a centred difference approximation to $\Delta z \left.\frac{\partial s}{\partial z}\right|_j$, subject to certain constraints.

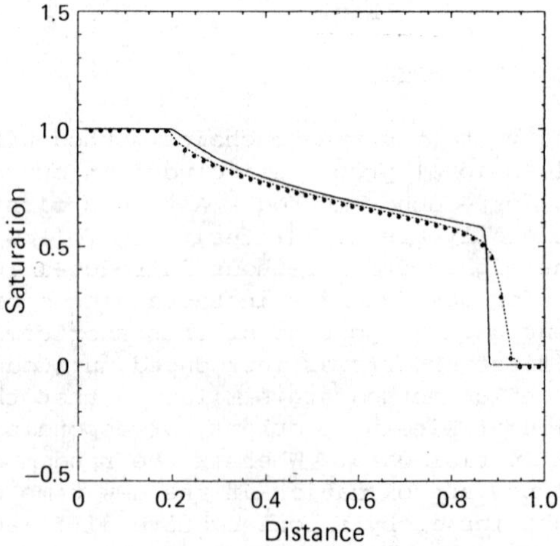

Fig. 6

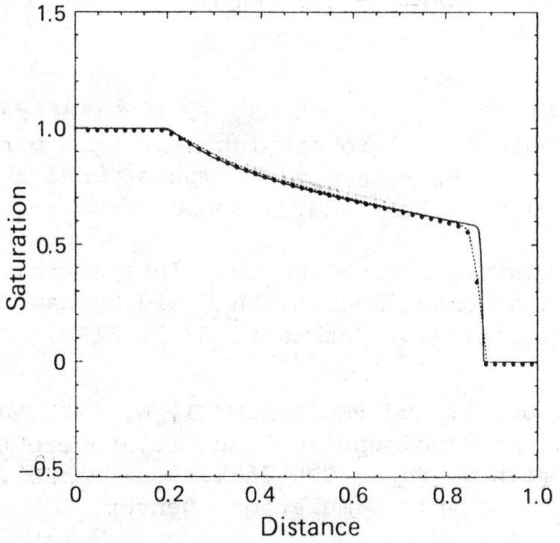

Fig. 7

The extensions to van Leer's approach developed in (Woodward and Colella, 1980) and (Colella, 1982) are being adapted to the porous flow equations with promising results (Bell, Colella, Concus and Glaz, 1982). A comparison between the first order and second order Godunov methods for a one dimensional case is depicted in Figs. 6 and 7. For these problems $s(z,t) = 1$ for $0 \le z \le 0.2$ and $s(z,0) = 0$ for $0.2 < z \le 1$. The dashed curves represent the solution of (3.1) with $\psi(s) = f(s) = s^2/(s^2+0.5(1-s)^2)$ and $\Delta z = 0.02$. The first order method is depicted in Fig. 6 and the second order one in Fig. 7. The plotting routine indicates the data points by placing circles below them, more or less tangent to the interpolating curve. The solid lines in the figures represent the solution for $\Delta z = 0.0025$ using the second order method, which is essentially the exact solution for this case. (Data points are not indicated.) The time step Δt was taken to be $0.1\ \Delta z$ for all cases, which corresponds to a CFL number of about 0.2. Precise computer running time comparisons for optimally designed programs are not yet available, but for the experimental programs the computer running time for the second-order method was substantially less than twice that for the corresponding first-order method. The improvement in the solution using the second-order over the first-order method is easily seen for this problem.

ACKNOWLEDGEMENTS

This work was supported in part by the Director, Office of Energy Research, Office of Basic Energy Sciences, Engineering,

Mathematical and Geosciences Division of the US Department of Energy under contract DE-AC03-76SF00098.

REFERENCES

ALBRIGHT, N., ANDERSON, C. and CONCUS, P. 1980 The random choice method for calculating fluid displacement in a porous medium. In Boundary and Interior Layers - Computational and Asymptotic Methods (J.J.H. Miller, Ed.), Boole Press.

ALBRIGHT, N. and CONCUS, P. 1980 On calculating flows with sharp fronts in a porous medium. In Fluid Mechanics in Energy Conservation (J.D. Buckmast, Ed.), SIAM, Philadelphia, pp. 172-184.

ALBRIGHT, N., CONCUS, P. and PROSKUROWSKI, W. 1979 Numerical solution of the multi-dimensional Buckley-Leverett equation by a sampling method, Paper SPE 7681, Soc. Petrol. Eng. Fifth Symp. on Reservoir Simulation, Denver, CO.

ANDERSON, C. and CONCUS, P. 1980 A stochastic method for modelling fluid displacement in petroleum reservoirs. In Analysis and Optimization of Systems (A. Bensoussan and J.L. Lions, Ed.), Springer-Verlag, *Lecture Notes in Control and Information Sciences*, **28**, pp. 827-841.

BELL, J.B., COLELLA, P., CONCUS, P. and GLAZ, H. 1982 Higher order Godunov methods for the Buckley-Leverett equation, presented at SIAM 30th Anniversary Meeting, Stanford, California, USA (to appear).

CHORIN, A.J. 1976 Random choice solution of hyperbolic systems, *J. Comput. Phys.*, **22**, pp. 517-533.

CHORIN, A.J. 1977 Random choice methods with applications to reacting gas flow, *J. Comput. Phys.*, **25**, pp. 253-272.

CHORIN, A.J. 1980 Flame advection and propagation algorithms, *J. Comput. Phys.*, **35**, pp. 1-11.

COLELLA, P. 1982 A direct Eulerian MUSCL scheme for gas dynamics, Lawrence Berkeley Laboratory Report LBL-14014.

CONCUS, P. and PROSKUROWSKI, W. 1979 Numerical solution of a nonlinear hyperbolic equation by the random choice method, *J. Comput. Phys.*, **30**, pp. 153-166.

CRANDALL, M. and MAJDA, A. 1980 The method of fractional steps for conservation laws, *Num. Math.* **34**, pp. 285-314.

GLIMM, J. 1965 Solutions in the large for nonlinear hyperbolic systems of equations, *Commun. Pure Appl. Math.* **18**, pp. 697-715.

GLIMM, J., ISAACSON, E., MARCHESIN, D. and McBRYAN, O. 1981 Front tracking for hyperbolic systems, *Advances Appl. Math.*, **2**, pp. 91-119.

GLIMM, J., MARCHESIN, D. and MCBRYAN, O. 1979 The Buckley-Leverett equation: Theory, computation and application, Proc. Third Meeting of the International Society for the Interaction of Mechanics and Mathematics, Edinburgh.

GODUNOV, S.K. 1959 Finite difference methods for numerical computation of discontinuous solutions of the equations of fluid dynamics, *Mat. Sbornik,* **47**, pp. 271-306.

LAX, P.D. 1973 Hyperbolic Systems of Conservation Laws and the Mathematical Theory of Shock Waves, SIAM Regional Conf. Series in Appl. Math.

LOTSTEDT, P. 1981 A front tracking method applied to Burgers' equation and two-phase porous flow, Lawrence Berkeley Laboratory Report LBL-13302 (to appear in *J. Comput. Phys.*).

MILLER, K. and MILLER, R. 1981 Moving finite elements I, *SIAM J. Num. Analysis,* **18**, pp. 1019-1032.

NOH, W.F. and WOODWARD, P. 1976 SLIC (Simple Line Interface Calculation). In Proc. 5th International Conf. on Numerical Methods in Fluid Dynamics (A.I. van de Vooren and P.J. Zandbergen, Eds.), Springer-Verlag, *Lecture Notes in Physics* **59**, pp. 330-340.

OLEINIK, O.A. 1963 Uniqueness and stability of the generalized solution of the Cauchy problem for a quasilinear equation, *Amer. Math. Soc. Translat. II Ser.,* **33**, pp. 285-290.

PEACEMAN, D.W. 1977 Fundamentals of Numerical Reservoir Simulation, Amsterdam,Oxford,New York: Elsevier.

PROSKUROWSKI, W. 1981 A note on solving the Buckley-Leverett equation in the presence of gravity, *J. Comput. Phys.,* **41**, pp. 136-141.

VAN LEER, B. 1979 Towards the ultimate conservative difference scheme V. A second-order sequel to Godunov's method, *J.Comput. Phys.,* **32**, pp. 101-136.

WOODWARD, P.R. and COLELLA, P. 1980 High resolution difference schemes for compressible gas dynamics. In Proc. 7th International Conf. on Numerical Methods in Fluid Dynamics (W.C. Reynolds and R.W. MacCormack, Eds.), Springer-Verlag, *Lecture Notes in Physics*, **141**, pp. 434-441.

SHOCK MODELLING IN AERONAUTICS*

S. Osher

(University of California, USA)

INTRODUCTION

Numerical simulation of problems in transonic flow, both steady and time dependent, are carried out around the world, as a very important guide to the design of aeroplanes flying near the speed of sound. Recently, a great deal of interest has also been generated in supersonic and hypersonic flow calculations.

There are three important models for inviscid compressible gas dynamics used to study these flow problems. They are: the transonic small disturbance equation (TSD), the transonic full potential equation (FP), and the full Euler equations for compressible gas dynamics (EU).

It is my purpose here to describe shock capturing methods recently developed for each model, and to emphasize the design principles behind the algorithms.

These principles can be stated as follows:

(-1) Consistency
(0) Conservation form
(1) Monotone and sharp steady discrete shocks
(2) No nonphysical limit solutions (e.g. expansion shocks)

The often used Murman algorithm for TSD (Murman and Cole, 1971) is well known to satisfy (-1), (0), (1), however it occasionally violates (2). (See e.g. (Jameson, 1976)). The algorithms I propose in this paper satisfy all of the above principles. This is always proved at the level of rigorous mathematics.

---

*Research partially supported by NSF Research Grant, No. MCS 78-0152

I have recently been involved in the following research projects.

Bjorn Engquist and I developed an algorithm for TSD, which we then extended to general scalar conservation laws, (Engquist and Osher, 1980a, 1980b, 1981).

Our TSD algorithm involves very simple modifications in codes using Murman's algorithm. It has proved to be much more robust and reliable. As a result, it has replaced the standard Murman scheme in production codes at NASA Langley, NASA Ames, and in various aircraft companies. See, e.g. (Goorjian and Van Buskirk, 1981); and (Edwards et al., 1982).

Moreover, the scalar conservation law approximation was shown to be very successful in non-linear singular perturbation calculations, when the convection dominates (Osher, 1981a, 1981b; Abrahamsson and Osher, 1982).

I later extended the Engquist-Osher scalar scheme to general hyperbolic systems in (Osher, 1981a). Then, with Fred Solomon, (Osher and Solomon, 1982) the details were worked out for EU (isentropic and nonisentropic) and for the Lagrangian formulation, all in cartesian coordinates. We also presented numerical results in one and two space dimensions. Sharp, skew, two-dimensional, discrete shocks were found.

Last year, Sukumar Chakravarthy and I (Chakravarthy and Osher, 1982; Osher and Chakravarthy, 1982), extended this algorithm in a very simple manner to the full nonisentropic EU, for general regions in several space dimensions, using body fitted coordinates. Using this algorithm, we also treated the boundary conditions in a new and natural manner. We obtained excellent numerical results for a variety of flows, including two-dimensional Mach 8 supersonic flow past a circular cylinder.

My algorithm for the space differencing is now the basis for an ongoing joint project with Chakravarthy. We intend to develop a shock capturing production EU code for supersonic flows.

The FP model has, until now, eluded this approach for the following reason: if FP is written as a hyperbolic system of conservation laws in several space dimensions, e.g. (Chipman and Jameson, 1979) it is a very bad hyperbolic system. It fails to be strictly hyperbolic, and is not even linearly, a well-posed problem. Thus, the general algorithm I devised in (Osher, 1981b) and applied to the one-dimensional FP with Engquist in (Engquist and Osher, 1981), seems not applicable to this model for transonic flows in several dimensions. (Some of the considerations involving characteristics were used to improve a very useful hypersonic FP model devised by Shankar and Osher, 1982.)

Recently, Mohamed Hafez suggested that I use the scalar potential formulation for FP, and try to devise a scalar algorithm, satisfying design principles (-1) to (2), directly.

Several enlightening conversations with him have helped me to construct a new and simple algorithm approximating FP, for which I can prove the validity of the four basic design principles. I shall present the algorithm in this paper and then state the main theorems.

In future work with Hafez, my new algorithms will be tested numerically, and proofs of the main theorems will be given.

The format of the paper is as follows. In Section 1, I shall discuss the approximations to TSD and then to general scalar conservation laws. A "before" and "after" algorithm for TSD is given here. In Section 2, I describe my algorithm for general hyperbolic systems. In Section 3, the detailed algorithm for EU, in general geometries, is presented, and in Section 4 the results of some calculations (Chakravarthy and Osher, 1982; Osher and Solomon, 1982) for EU, are presented. In Section 5, the new FP algorithm is given, and two new theorems are stated.

1. TSD AND SCALAR CONSERVATION LAWS

The differential equations for TSD, steady and low frequency time dependent, cases are

$$(K\phi_x - \tfrac{1}{2}(\gamma+1)\phi_x^2)_x + \phi_{yy} = 0 \qquad (1.1)$$

and

$$(K\phi_x - \tfrac{1}{2}(\gamma+1)\phi_x^2)_x + \phi_{yy} - 2\phi_{tx} = 0, \qquad (1.2)$$

respectively.

Here, $\phi(x,y,t)$ is the velocity potential, and K,γ are positive constants.

During the last decade, many numerical calculations using these equations have been performed. (See the bibliography in (Engquist and Osher, 1980a).) Although TSD contains some sacrifice in accuracy, it allows a great simplification of the equations and boundary conditions.

In a very important 1971 paper Murman and Cole (1971) suggested type dependent differencing for the steady problem. Separate difference formulae were suggested in the elliptic (subsonic, $\phi_x < K/\gamma+1 = \bar{u}$ = sonic point), and hyperbolic (supersonic, $\phi_x > \bar{u}$) regions, to account properly for the local domain of dependence of the differential equation.

In a succeeding paper, (Murman, 1974) devised his shock and sonic point operators, so as to ensure conservation form of the difference approximations. Thus, the method became a shock capturing algorithm, ensuring that design principles (-1) and (0) were valid.

It was noticed by Jameson (1976), that the Murman differencing allowed nonphysical expansion shocks as exact solutions to one-dimensional steady problems. Moreover, some of these nonphysical shocks were shown to be numerically stable. Various remedies were suggested to remove these expansion shocks, involving the addition of localized artificial viscosity, or changing the relaxation procedure.

Engquist and I began our study of this phenomenon in 1978, continuing some related work I had earlier done with Andrew Majda (Majda and Osher, 1978; 1979) which concerned the Lax-Wendroff scheme.

In the first E-O paper, (Engquist and Osher, 1980a), we proposed and tested a new type dependent switching scheme. We also proved theoretically that this new algorithm was monotone - hence nonlinearly stable, and that limit solutions were physically correct. We also verified, analytically and numerically, the existence of steady, monotone, sharp, numerical profiles, which have a two point transition region. Our algorithm involves a very small change in existing codes using the standard Murman switch, and no increase in computer storage.

While working under a NASA Ames - University Consortium agreement, we suggested to Peter Goorjian of NASA Ames, that he experiment with our scheme by modifying the existing TSD production code which used the Murman scheme. He did this with van Buskirk in (Goorjian and van Buskirk, 1981). The results were remarkable.

Goorjian and van Buskirk found the following. The resulting algorithms are more stable, hence calculations can be done more efficiently. For steady flows, the convergence is around 35% faster, and for unsteady flows, the allowable time step is around 30 times larger.

The steady profiles for both methods are very similar, except that nonlinear instabilities were triggered using Murman's scheme for a blunt airfoil, while ours converged with no problem.

For unsteady flows, the new method allowed time steps at least an order of magnitude larger, but, perhaps more importantly, one case illustrated the following phenomenon.

The Murman switching can trigger transient numerical instabilities; although these instabilities will not cause calculations to diverge, they will cause large errors in the pressure profiles. Many users of these codes, such as aeroelasticians, are particularly interested in integrated quantities, such as the unsteady aerodynamical loads. These users could be unaware of these large errors, unless they monitored, in addition, the calculation of the pressure coefficients.

Additional evidence of the dramatic possibilities of our scheme came in some recent work at NASA Langley Research Center. There, Edwards et al. (1982) modified their existing code by using our switch, and were able to calculate large amplitude motion and large angles of attack. Thus, transonic flutter solutions, which could, previously, not even be calculated using existing production codes, were obtained, and were found to be quite accurate.

I shall now give the details of the methods. Let u and v be the velocities in the x and y directions, respectively, with

$$\phi_x = u, \quad \phi_y = v. \tag{1.3}$$

Define the convex function

$$f(u) = - Ku + (\gamma+1)\, u^2/u. \tag{1.4}$$

Then, (1.1) can be rewritten

$$-(f(u))_x + v_y = 0, \tag{1.5}$$

and, as already mentioned, this equation is hyperbolic if $u > \bar{u}$, elliptic if $u < \bar{u}$, for $\bar{u}$ the sonic speed.

$$\bar{u} = K/(\gamma+1) \tag{1.6}$$

We shall solve a difference equation for a discrete potential function, ϕ_{jk}, approximating $\phi(j\Delta x, k\Delta y)$. Here, for simplicity only, we define a uniform grid $(x_j, y_k) = (j\Delta x, k\Delta y)$, for j, $k = 0, \pm 1, \pm 2, \ldots$

We also define discrete velocities:

$$D_-^x \phi_{jk} = v_{jk}, \quad D_-^y \phi_{jk} = v_{jk}. \tag{1.7}$$

Here, and in what follows, for any grid function f_{jk}:

$$D_\pm^x f_{jk} = \pm \frac{(f_{j\pm1,k} - f_{jk})}{\Delta x}, \quad \Delta_\pm^x f_{jk} = \pm (f_{j\pm1,k} - f_{jk}). \tag{1.8}$$

$D_\pm^y$ and $\Delta_\pm^y$ are defined analogously.

Murman's scheme may be written as:

$$-D_+^x f(u_{jk}) + \Delta_-^x (\theta_{jk} D_+^x f(u_{jk})) + D_+^y v_{jk} = 0, \tag{1.9}$$

where the switch, θ_{jk}, is defined by:

$$\theta_{jk} \equiv 1 \text{ if } \frac{u_{jk} + u_{j+1,k}}{2} \geq \bar{u}$$

$$\theta_{jk} \equiv 0 \text{ if } \frac{u_{jk} + u_{j+1,k}}{2} < \bar{u}.$$

Our difference approximation to (1.1) is

$$-D_-^x f(\max(u_{jk}, \bar{u})) - D_+^x f(\min(u_{jk}, \bar{u})) + D_+^y v_{jk} = 0 \tag{1.10}$$

Fig. 1 shows the computational grid for both schemes away from points for which $u = \bar{u}$.

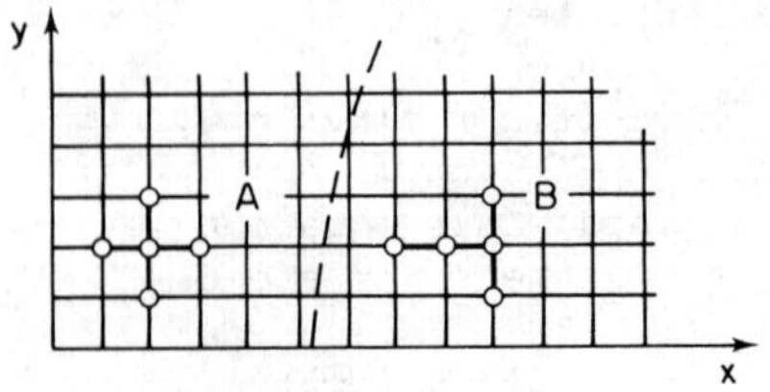

Fig. 1 Part of the computational grid for ϕ with a subsonic (A) and a supersonic (B) difference stencil. The dashes represent the sonic line.

The schemes are identical, except near such points and they both always use centred differencing for the ϕ_{yy} term. Thus, to explain their differences, we consider the unsteady, y independent case. We rescale time trivially, and rewrite (1.2) for this case as:

$$u_t + f(u)_x = 0. \qquad (1.11)$$

An explicit* time difference approximation to (1.11), using Murman's type dependent space differencing, would be (letting $u_j^n \approx u(j\Delta x, \Delta\eta t)$)

$$u_j^{n+1} = u_j^n - (\Delta t/\Delta x)\Delta_- \; h_M(u_{j+1}^n, u_j^n), \qquad (1.12)$$

where the numerical flux function is:

$$h_M(u_{j+1}, u_j) = (1-\theta_j)\; f(u_{j+1}) + \theta_j f(u_j). \qquad (1.13)$$

This numerical flux function has a very natural geometric and physical (or, in some cases, nonphysical) interpretation.

Let u_j^n be the constant value taken on at $t^n = n\Delta t$, for

$$x_{j-\frac{1}{2}} = x_j - \tfrac{1}{2}\Delta x \le x < x_j + \tfrac{1}{2}\Delta x = x_{j+\frac{1}{2}},$$

for each j. Then at $x = x_{j+\frac{1}{2}}$, $t = t^n$, construct a shock solution to (1.11) moving with the correct jump condition speed.

$$S_{j+\frac{1}{2}}^n = \frac{f(u_{j+1}^n) - f(u_j^n)}{u_{j+1}^n - u_j^n} = (\gamma+1)\left(\frac{u_j^n + u_{j+1}^n}{2} - \bar{u}\right) \qquad (1.14)$$

and advancing in time.

In <u>all</u> cases, Murman's flux function is thus defined to be:

$$\begin{aligned} h_M(u_{j+1}^n, u_j^n) &= f(u_j^n) \quad \text{if } S_{j+\frac{1}{2}}^n > 0 \\ &= f(u_{j+1}^n) \quad \text{if } S_{j+\frac{1}{2}}^n \le 0 \end{aligned}$$

*Explicit time differencing is too slow to be recommended computationally. We use it here as an illustrative device.

(see Fig. 2). This is the true flux, evaluated at the solution to the initial value problem, at $x = x_{j+\frac{1}{2}}$, when shocks only are used.

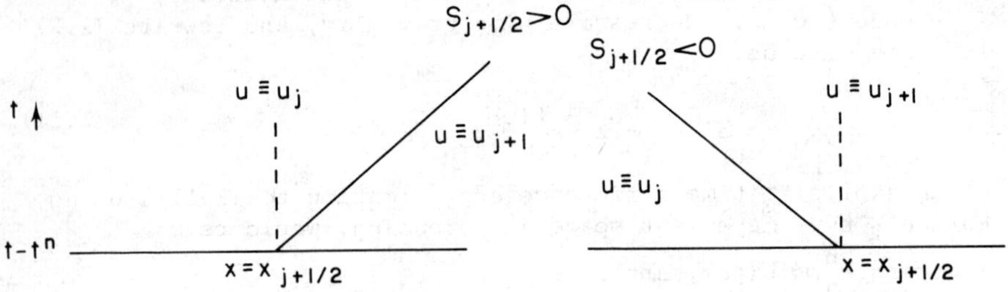

Fig. 2 Schematic of Murman's numerical flux .

The scheme is linearly stable if $|(\Delta t/\Delta x) f'| < 1$, i.e., if the CFL stability condition is valid.

It is now obvious that if $f(u^L) = f(u^R)$, with $u^L < \bar{u} < u^R$, then Murman's scheme will keep as steady non-physical shock solution:

$$u_j^n \equiv u^L,\ j \le 0$$
$$u_j^n \equiv u^R,\ j > 0. \qquad (1.16)$$

Numerical experiments have demonstrated the stability of these non-physical shocks. (Goorjian and van Buskirk, 1981) and (Jameson, 1976).

One remedy to this obviously non-physical construction is to impose the correct physical solution at $x = x_{j+\frac{1}{2}}$, to this Riemann initial value problem. The true solution uses rarefaction waves when $u_j < u_{j+1}$. This is the method of Godunov (1959). Godunov's method will thus give the same numerical flux function as Murman's, except in the case $u_j < \bar{u} < u_{j+1}$.

We have

$$h_G(u_{j+1}, u_j) = h_M(u_{j+1}, u_j),\ \text{unless } u_j < \bar{u} < u_{j+1}$$
$$h_G(u_{j+1}, u_j) = f(\bar{u}),\ \text{when } u_j < \bar{u} < u_{j+1}. \qquad (1.17)$$

We turn now to our method, which we believe to be superior to Godunov's, for reasons of simplicity and smoothness of flux function. Both h_G and h_M fail to be differentiable functions of their arguments when

$$u_j = u^L > \bar{u} > u_{j+1} = u^R,$$

with $f(u^L) = f(u^R)$, i.e., at a standing shock. This is significant for implicit methods, which are usually based on linearization of the flux functions.

Our numerical flux function is defined by

$$h_{EO}(u_{j+1}, u_j) = f(\min.\ (u_{j+1}, \bar{u})) + f(\max.\ (u_j, \bar{u})). \quad (1.18)$$

The resulting explicit difference scheme is

$$u_j^{n+1} = u_j^n - \frac{\Delta t}{\Delta x} \Delta_- h_{EO} \quad (u_{j+1}^n, u_j^n) = G_{EO}(u_{j+1}^n, u_j^n, u_{j-1}^n).$$

(1.19)

The function G_{EO} is a non-decreasing function of its arguments if $|(\Delta t/\Delta x)f'| < 1$, i.e., the scheme is then monotone. Thus the solution to (1.19) is known to converge to the correct physical solution with rate $O((\Delta t)^{\frac{1}{2}})$ in the L^1 norm (Kuznetsov, 1977) and Sanders, 1982). (The same fact is true for Godunov's scheme).

Theorems implying correct "entropy" production (= no non-physical limit solutions) for our full two-dimensional TSD algorithm, were obtained in (Engquist and Osher, 1980a). Existence, uniqueness, and stability of one dimensional, sharp, discrete, shocks was obtained there, and in (Engquist and Osher, 1980a; Engquist and Osher, 1981).

Bram van Leer (1982) has presented a beautiful and useful interpretation of our algorithm. Without any change in our scheme, we can subtract the constant, $f(\bar{u})$, from $h_{EO}(u_{j+1}, u_j)$. Then, just as Godunov's solves the true initial value Riemann problem, and Murman's uses shocks only, ours uses rarefactions, or overturned (multivalued), compression waves only. We solve the Riemann problem in the case $u_j > u_{j+1}$, by allowing the solution manifold to overturn following characteristics, and then by connecting the top and bottom parts of this manifold via a rarefaction fan. We then add up each contribution of $f(u)$ at $x=x_{j+\frac{1}{2}}$, according to the sign of the normal to the manifold. This is precisely h_{EO}.

This interpretation explains why our flux functions are

fairly smooth. They are C^1, with Lipschitz continuous derivations, in the TSD case.

The production codes I mentioned above always use implicit time, or artificial time, methods. We merely recommend using our method for the x space differencing, including, of course, any linearizations.

We now present the "before" and "after" implicit algorithms, used in (Goorjian and van Buskirk, 1981) and (Edwards et al, 1982), and originating in the AF idea devised in (Ballhaus & Steger, 1975).

L TRAN 2 SOLUTION ALGORITHM (Murman differencing)

$$2M^2\phi_{xt} = [1-M^2-M^2(\gamma+1)\phi_x]\phi_{xx} + \phi_{yy}$$

ADI SOLUTION METHOD

$$\frac{2M^2}{\Delta t} D_-^x(\bar{\phi}_{jk} - \phi^n_{jk}) = D_x f_{jk} + D_+^y D_-^y \phi^n_{jk} \qquad (x - \text{sweep})$$

$$\frac{2M^2}{\Delta t} D_-^x(\phi^{n+1}_{jk} - \bar{\phi}_{jk}) = \tfrac{1}{2} D_+^y D_-^y (\phi^{n+1}_{jk} - \phi^n_{jk}) \qquad (y - \text{sweep})$$

$$f_{jk} = \tfrac{1}{2}[(1-M^2)\phi^n_x + (1-M^2-M^2(\gamma+1)\phi^n_x)\bar{\phi}^n_x]$$

$$D_x f_{jk} = \frac{1}{\Delta x}[(1-\theta_j)(f_{j+\frac{1}{2},k}-f_{j-\frac{1}{2},k}) + \theta_{j-1}(f_{j-\frac{1}{2},k}-f_{j-3/2,k})]$$

(ENGQUIST-OSHER Differencing)

$$\frac{2M^2}{\Delta t} D_-^x(\bar{\phi}^n_{jk} - \phi^n_{jk}) = D_+^x \bar{f}_{j-\frac{1}{2},k} + D_-^x \hat{f}_{j-\frac{1}{2},k} + D_+^y D_-^y \phi^n_{jk} \qquad (x - \text{sweep})$$

(y - sweep is the same as above)

$$\bar{f}_{j-\frac{1}{2},k} = \tfrac{1}{2}[(1-M^2)\bar{u}_{j-\frac{1}{2},k} + (1-M^2-M^2(\gamma+1)\bar{u}_{j-\frac{1}{2},k}) D_-^x \bar{\phi}_{jk}]$$

$$\bar{u}_{j-\frac{1}{2},k} = \min\left(\frac{1-M^2}{M^2(\gamma+1)}, D_-^x \phi^n_{jk}\right)$$

$$\hat{f}_{j-\frac{1}{2},k} = \tfrac{1}{2}[(1-M^2)\hat{u}_{j-\frac{1}{2},k} + (1-M^2-M^2(\gamma+1)\hat{u}_{j-\frac{1}{2},k}) D_-^x \hat{\phi}_{jk}]$$

$$\hat{u}_{j-\frac{1}{2},k} = \max\left(\frac{1-M^2}{M^2(\gamma+1)},\ D_-^x \phi_{jk}^n\right)$$

For a general scalar conservation law of the form

$$u_t + f(u)_x = 0, \tag{1.20}$$

the E-O space differencing is done as follows:

$$\text{Define} \quad f_+(u) = \int_0^u \max(f'(s), 0)\, ds$$
$$f_-(u) = \int_0^u \min(f'(s), 0)\, ds, \tag{1.21}$$

i.e., the increasing and decreasing parts of the true flux function. Then, our space differencing is:

$$f(u)_x \sim D_- f_+(u_j) + D_+ f_-(u_j), \tag{1.22}$$

and the explicit version of our algorithm is:

$$u_j^{n+1} = u_j^n - (\Delta t/\Delta x)\ (D_- f_+(u_j^n) + D_+ f_-(u_j^n)),$$

which is monotone for arbitrary $f(u)$, subject only to the CFL condition:

$$(\Delta t/\Delta x)\ |f'(u)| < 1.$$

Results about sharp shock profiles, etc., are given in (Engquist and Osher, 1981). This scheme is a direct generalisation of the version of our scheme for strictly convex f, given in (1.18), (1.19). Bram van Leer's geometric interpretation of our scheme can sometimes be extended to the non-convex case. Godunov's method can be extremely complicated for non-convex $f(u)$.

2. GENERAL NONLINEAR HYPERBOLIC SYSTEMS OF CONSERVATION LAWS

Consider the hyperbolic system of conservation laws

$$q_t + f(q)_x = 0, \tag{2.1}$$

with $$q = (q_1, \ldots, q_d)^T, \quad f = (f_1, \ldots, f_d)^T.$$

The square $d \times d$ Jacobian matrix, $\partial f(q) = \{\partial f_i/\partial q_j\}$, is assumed to have real distinct eigenvalues.

To approximate $f(q)_x$, using ideas of the previous section, begin by rewriting the scalar E - O differencing of (1.23) as

$$f(u)_x \sim D_-\left[\frac{f(u_{j+1}) + f(u_j)}{2} - \tfrac{1}{2}\int_{u_j}^{u_{j+1}} |f'(u)|\,du\right] \qquad (2.2)$$

This is the form which is generalised to systems.

Replace the integral above by an integral in d dimensional phase space, with $|f'(u)|$ replaced by $|\partial f(q)|$. (The absolute value of a matrix with distinct real eigenvalues is a continuous function of its elements.) It still remains to choose the path in phase space connecting q_j to q_{j+1}.

Define the eigenvalues of $\partial f(q)$ to be $\lambda_1 < \lambda_2 \ldots < \lambda_d$, with associated right eigenvectors $r_k(q)$. Consider Γ_k, curves in phase space, which are integrals of the ordinary differential system:

$$\frac{dq}{ds} = r_k(q). \qquad (2.3)$$

Along such a curve, it follows that:

$$\begin{aligned}\int_{\Gamma_k} |\partial f(q)\,dq &= \int |\partial f(q)|\,r_k(q)\,ds \\ &= \int |\lambda_k(q)|\;r_k(q)\,ds.\end{aligned} \qquad (2.4)$$

The integral "decouples", and thus simplifies.

Moreover, for many hyperbolic systems of physics, including EU, the characteristic fields are either linearly degenerate, or genuinely nonlinear, as defined by Lax (1972).

A field is linearly degenerate if the equation

$$\nabla_q \lambda_k \cdot r_k \equiv 0 \qquad (2.5)$$

is valid. For such fields, λ_k is constant along Γ_k, and

$$\int_{\Gamma_k} |\partial f(q)|\,dq = (\mathrm{sgn}\lambda_k)\,[\,f(q^u) - f(q^L)]\,, \qquad (2.6)$$

where q^u, q^L are the upper and lower limits of integration.

A field is genuinely nonlinear if r_k can be normalised, so that

$$\nabla_q \lambda_k \cdot r_k \equiv 1. \tag{2.7}$$

For these fields, λ_k is strictly monotone along Γ_k. Thus, on Γ_k or its extension, there exists a unique sonic point $\bar{q}$, for which $\lambda_k(\bar{q}) = 0$. In this case one has:

$$\int_{\Gamma_k} |\partial f(q)|dq = 2f(q) \quad \text{at} \quad \begin{cases} q^u & \text{if } \lambda_k(q^u) > 0 \\ \bar{q} & \text{if } \lambda_k(q^u) \leq 0 \\ q^L & \text{if } \lambda_k(q^L) > 0 \\ \bar{q} & \text{if } \lambda_k(q^L) \leq 0 \end{cases} \tag{2.8}$$

$$-f(q) \quad \text{at} \quad \begin{cases} q^u \\ q^L \end{cases}$$

The full path Γ^j connecting q_j to q_{j+1} is obtained by constructing a path starting from q_j using the d wave, k = d in (2.4). Then one connects it via a d-1 wave, etc. finishing with a one wave passing through q_{j+1}. By the implicit function theorem, there is exactly one such path, for $|q_j - q_{j+1}|$ small. For EU this small variation hypothesis is unnecessary, as long as cavitation does not occur. Flows with very strong shocks have been computed this way, see (Chakravarthy and Osher, 1982), (Osher & Chakravarthy, 1982) and section 4 below.

The numerical flux is thus defined in terms of the true flux f(q), evaluated at various intermediate points. These points are easily computed for EU and for other physical systems, because along the curves Γ_k, there are d-1 independent, and well-known, functions $\psi_\nu^k(q)$, $\nu = 1, \ldots, d$, $\nu \neq k$, for which

$$\nabla_q \psi_\nu^k(q) \cdot r_k \equiv 0. \tag{2.9}$$

Hence the $\psi_\nu^k(q)$ are constant along Γ_k, and thus each Γ_k is defined by setting

$$\psi_{\nu}^{k}(q) \equiv C_{\nu}, \quad \nu = 1, \ldots, d \quad \nu \neq k \tag{2.10}$$

for constant C_{ν}.

These ψ_{ν}^{k} are the so-called k-Riemann invariants.

Thus, my scheme is relatively easy to construct and to program for EU and for some other systems (Chakravarthy and Osher, 1982), (Osher and Solomon, 1982).

The observation made by van Leer (1982) in the scalar case also generalizes here, (Osher & Chakravarthy, 1982). The Riemann problem is solved by replacing shocks with overturned compressions, and averaging the resulting multivalued solution.

Godunov's scheme, (Godunov, 1959) uses the true Riemann solver, and is quite complicated to program. Compressions, rarefactions, and contacts are much easier than Hugoniot curves to tabulate. I merely use (2.10) for each k. Hence, my scheme is obtained in a relatively simple closed form, and has C^1 (at least) flux functions. The "entropy" condition (no non-physical limit solutions (Lax, 1972)), is valid, and steady, sharp, monotone, discrete shocks exist and are unique (Osher and Solomon, 1982).

Boundaries can now be treated in a natural way. Suppose $x = 0$ is the physical, or computational, boundary. Assume at the boundary that there is a constant index for which

$$\lambda_k \leq 0 < \lambda_{k+1}.$$

(See (Osher and Chakravarthy, 1982) for some minor additional technical hypotheses, which are valid for EU.) The boundary operator may be written as $B(q) = 0$, where $B(q)$ is a d-k vector valued function. Instead of setting up an extra cell at the boundary and using reflection, or something of the sort, we solve the <u>initial-boundary</u> value problem exactly, in the overturned compression, rarefaction, and contact, context. This gives us a complete set of boundary values, and the numerical flux at the boundary is obtained in the usual fashion. The simple details are worked out in (Chakravarthy and Osher, 1982), (Osher and Chakravarthy, 1982).

My approximation for systems can now be written in either of the two equivalent ways:

$$f(q)_x \approx D_-\left[\frac{f(q_{j+1}) + f(q_j)}{2} - \tfrac{1}{2}\int_{q_j}^{q_{j+1}} |\partial f(q)|\,dq\right]$$

$$= \frac{1}{\Delta x}\left[\int_{q_{j-1}}^{q_j} (\partial f)^+ dq + \int_{q_j}^{q_{j+1}} (\partial f)^- dq\right]. \tag{2.11}$$

Here $(\partial f)^+$, $(\partial f)^-$ are the natural restrictions of ∂f onto its positive and negative eigenspaces.

3. EU IN GENERAL GEOMETRIES

Here I present the algorithm of the previous section applied to two-dimensional EU in general coordinate systems. The simplicity and elegance of the numerical fluxes is a result is a result of the coordinate invariance of EU itself - the Riemann invariants are truly invariant.

This is joint work with S. Chakravarthy, (Chakravarthy and Osher, 1982), (Osher and Chakravarthy, 1982). In those papers we also present our equally simple boundary treatment and a number of successful numerical calculations. Some of these are also presented in the next section.

We consider Euler's equations in two dimensions using computational coordinates $\tau = t$, $\xi = \xi(x,y,t)$, $\eta = \eta(x,y,t)$. The equations, when transformed from a cartesian base coordinate system, become

$$q_\tau + \xi_t q_\xi + \xi_x E_\xi + \xi_y F_\xi + \eta_t q_\eta + \eta_x E_\eta + \eta_y F_\eta = 0 \tag{3.1}$$

where

$$q = \begin{bmatrix} \rho \\ \rho u \\ \rho v \\ e = p/(\gamma-1) + (u^2+v^2)/2 \end{bmatrix}, \quad E = \begin{bmatrix} \rho u \\ \rho u^2 + p \\ \rho uv \\ (e+p)u \end{bmatrix}, \quad F = \begin{bmatrix} \rho v \\ \rho vu \\ \rho v^2 \\ (e+p)v \end{bmatrix} \tag{3.2}$$

Alternate formulations involving fully conservative forms of the equations with and without Jacobians of the mapping can be used, and are discussed in the papers (Chakravarthy and Osher, 1982), (Osher and Chakravarthy, 1982). We concentrate on this form for the moment. The Osher scheme is novel in its treatment of the space derivatives along ξ and η coordinates, which we now discuss.

We set up a grid in ξ,η space, with $\xi_j = j\Delta\xi$, $\eta_k = k\,\Delta\eta$ for j, k integers. Let $q_{j,k} = q(j\Delta\xi, k\Delta\eta)$ (ignoring momentarily the time dependence). Then we approximate (now dropping the k dependence for convenience).

$$\left. \xi_t q_\xi + \xi_x E_\xi + \xi_y F_\xi \right|_{\xi_j,\eta_j} \sim (1/\Delta\xi)\{ \int_{q_{j-1}}^{q_j} ((\xi_t)_j I + (\xi_x)_j A(q) + (\xi_y)_j B(q))^+ dq\} + (1/\Delta\xi)\ \{ \int_{q_j}^{q_{j+1}} ((\xi_t)_j I + (\xi_x)_j A(q) + (\xi_y)_j B(q))^- dq\}. \tag{3.3}$$

The matrices A and B are the Jacobian matrices $\partial E/\partial q$ and $\partial F/\partial q$, respectively, and the matrix I is the identity matrix.

These integrals must now be described. We begin by stressing that each integral will result in a sum of values of

$$((\xi_t)_j q_j + (\xi_x)_j E(q_{j\nu}) + (\xi_y)_j F(q_{j\nu}))$$

multiplied by ± 1 at various values $q_{j\nu}$ between the end states. This is done exactly, no approximate integral is used. The full approximate derivative always therefore uses sum of values of the true fluxes evaluated at various points and is easily programmed.

Define the contravariant velocities

$$\hat{U} = \frac{\xi_t + u\xi_x + v\xi_y}{(\xi_x^2 + \xi_y^2)^{1/2}}, \quad \hat{V} = \frac{v\xi_x - u\xi_y}{(\xi_x^2 + \xi_y^2)^{1/2}} \tag{3.4}$$

and $c = (\gamma\, p/\rho)^{1/2}$.

Between q_{j-1} and q_j, define the intermediate points $q_{j-2/3}$ and $q_{j-1/3}$ as follows. Let

$$\rho_{j-1/3}^{(\gamma-1)/2} = \frac{((\gamma-1)/2\ (\hat{U}_j-\hat{U}_{j-1}) + c_j+c_{j-1})}{(c_j(1+(p_{j-1}/p_j)^{1/(2\gamma)}(\rho_{j-1}/\rho_j)^{-1/2}))}\ \rho_j^{(\gamma-1)/2} \tag{3.5a}$$

$$\rho_{j-2/3}^{(\gamma-1)/2} = \frac{((\gamma-1)/2\ (\hat{U}_j-\hat{U}_{j-1}) + c_j+c_{j-1})}{(c_{j-1}(1+(p_j/p_{j-1})^{1/(2\gamma)}(\rho_j/\rho_{j-1})^{-1/2}))}\ \rho_j^{(\gamma-1)/2} \tag{3.5b}$$

$$p_{j-2/3} = p_{j-1/3} = p_{j-1}\ (\rho_{j-1}/\rho_{j-2/3})^{-\gamma} \tag{3.5c}$$

$$\hat{U}_{j-2/3} = \hat{U}_{j-1/3} = \hat{U}_{j-1} - 2/(\gamma-1)(c_{j-1}-(\gamma p_{j-2/3}/\rho_{j-2/3})^{1/2}) \tag{3.5d}$$

$$\hat{V}_{j-2/3} = \hat{V}_{j-1} \tag{3.5e}$$

$$\hat{V}_{j-1/3} = \hat{V}_j\,. \tag{3.5f}$$

We also define sonic points $\underline{q}_{j-1/3}$ and $\underline{q}_{j-2/3}$ as follows:

$$\underline{\rho}_{j-1/3}^{(\gamma-1)/2} = (\gamma-1)/(\gamma+1)\ (\hat{U}_j+2/(\gamma-1)c_j)/c_j(\rho_j^{(\gamma-1)/2}) \tag{3.6a}$$

$$\underline{p}_{j-1/3} = p_j(\underline{\rho}_{j-1/3}/\rho_j)^{\gamma} \tag{3.6b}$$

$$\underline{\hat{U}}_{j-1/3} = \hat{U}_j + 2/(\gamma-1)c_j\ (1-(\underline{\rho}_{j-1/3}/\rho_j)^{(\gamma-1)/2}) \tag{3.6c}$$

$$\underline{\hat{V}}_{j-1/3} = \hat{V}_j \tag{3.6d}$$

$$\underline{\rho}_{j-2/3}^{(\gamma-1)/2} = (\gamma-1)/(\gamma+1)(2/(\gamma-1)c_{j-1} - \hat{U}_{j-1})/c_{j-1}(\rho_{j-1}^{(\gamma-1)/2}) \tag{3.6e}$$

$$\underline{p}_{j-2/3} = p_{j-1}(\underline{\rho}_{j-2/3}/\rho_{j-1})^{\gamma} \tag{3.6f}$$

$$\underline{\hat{U}}_{j-2/3} = \hat{U}_{j-1} - 2/(\gamma-1)c_{j-1}(1-(\underline{\rho}_{j-2/3}/\rho_{j-1})^{(\gamma-1)/2}) \tag{3.6g}$$

$$\underline{\hat{V}}_{j-2/3} = \hat{V}_{j-1}. \tag{3.6h}$$

In the above, it is straightforward to decode for u, v from $\hat{U}$, $\hat{V}$ from the definition of the contravariant velocities.

Then we have

$$\begin{aligned} \int_{q_{j-1}}^{q_j} (\xi_t I + \xi_x A(q) + \xi_y B(q))^+ dq &= \underbrace{\int_{q_{j-1}}^{q_{j-2/3}} (\xi_t I + \xi_x A(q) + \xi_y B(q))^+ dq}_{\text{subintegral 1}} \\ &+ \underbrace{\int_{q_{j-2/3}}^{q_{j-1/3}} (\xi_t I + \xi_x A(q) + \xi_y B(q))^+ dq}_{\text{subintegral 2}} \\ &+ \underbrace{\int_{q_{j-1/3}}^{q_j} (\xi_t I + \xi_x A(q) + \xi_y B(q))^+ dq}_{\text{subintegral 3}} . \end{aligned} \tag{3.7}$$

Now

$$\text{subintegral 1} = \xi_t q + \xi_x E(q) + \xi_y F(q) \left| \begin{array}{l} \underline{\text{upper limit}} \\ q_{j-2/3} \quad \text{if } (\hat{U} + c)_{j-2/3} > 0 \\ \underline{q}_{j-2/3} \quad \text{if } (\hat{U} + c)_{j-2/3} \le 0 \\ \hline \underline{\text{lower limit}} \\ q_{j-1} \quad \text{if } (\hat{U} + c)_{j-1} > 0 \\ \underline{q}_{j-2/3} \quad \text{if } (\hat{U} + c)_{j-1} \le 0 \end{array} \right. \tag{3.8a}$$

When $(\hat{U} + c) \le 0$ at both j-2/3 and j-1, this subintegral is identically zero and sonic points need not be evaluated.

subintegral 2

$$= \xi_t q + \xi_x E(q) + \xi_y F(q) \Big|_{q_{j-2/3}}^{q_{j-1/3}} \quad \text{if } \hat{U}_{j-2/3} > 0$$

$$= 0 \quad \dots \quad \dots \quad \dots \quad \dots \quad \dots \quad \text{if } \hat{U}_{j-2/3} \le 0 \tag{3.8b}$$

subintegral 3

$$= \xi_t q + \xi_x E(q) + \xi_y F(q) \Bigg| \begin{array}{l} \underline{\text{upper limit}} \\ q_j \quad \text{if } (\hat{U} - c)_j > 0 \\ \underline{q}_{j-1/3} \quad \text{if } (\hat{U} - c)_j \le 0 \\ \hline \underline{\text{lower limit}} \\ q_{j-1/3} \quad \text{if } (\hat{U} - c)_{j-1/3} > 0 \\ \underline{q}_{j-1/3} \quad \text{if } (\hat{U} - c)_{j-1/3} \le 0 \end{array} \tag{3.8c}$$

When $(\hat{U} - c) \le 0$ at both j-1/3 and j, this subintegral is identically zero and sonic points need not be evaluated. Then we compute simply:

$$\int_{q_j}^{q_{j+1}} (\xi_t I + \xi_x A(q) + \xi_y B(q))^- dq$$

$$= \xi_t q + \xi_x E(q) + \xi_y F(q) \Big|_{q_j}^{q_{j+1}} - \int_{q_j}^{q_{j+1}} (\xi_t I + \xi_x A(q) + \xi_y B(q))^+ dq \tag{3.9}$$

and define the last integral from (3.7) and (3.8) with j replaced by j+1.

The next few paragraphs in conjunction with Fig. 3 explain the algorithm stated above. The explanation of these results comes from the following definitions. The positive part of a matrix A^+ is defined to be the projection on to its positive eigenspace, and $A^- = A - A^+$. The path of integration Γ_j between q_{j-1} and q_j is obtained using subpaths as follows. Let $\lambda_1 = \hat{U}-c$, $\lambda_2 = \hat{U}$ and $\lambda_3 = \hat{U}+c$ be the eigenvalue of $\xi_t I+\xi_x A(q)+\xi_y B(q)$

(with λ_2 having multiplicity two). Associated with each λ_i are the Riemann invariants

$$\begin{aligned} \lambda_1 &: p/\rho^\gamma, \quad \hat{U} + 2/(\gamma-1)c, \quad \hat{V} \\ \lambda_2 &: p \quad , \quad \hat{U} \\ \lambda_3 &: p/\rho^\gamma, \quad \hat{U} - 2/(\gamma-1)c, \quad \hat{V} \end{aligned} \tag{3.10}$$

We take a curve Γ_j^1 passing through q_j for which the 1 Riemann invariants are constant and connect q_j to a point $q_{j-1/3}$ determined thus:

$$\begin{aligned} (p/\rho^\gamma)_{j-1/3} &= (p/\rho^\gamma)_j \\ (\hat{U} + 2/(\gamma-1)c)_{j-1/3} &= (\hat{U} + 2/(\gamma-1)c)_j \\ \hat{V}_{j-1/3} &= \hat{V}_j . \end{aligned} \tag{3.11}$$

Then connect $q_{j-1/3}$ to a point $q_{j-2/3}$ using any curve Γ_j^2 lying in the plane for which the 2 Riemann invariants are constant:

$$\begin{aligned} \hat{U}_{j-1/3} &= \hat{U}_{j-2/3} \\ p_{j-1/3} &= p_{j-2/3} . \end{aligned} \tag{3.12}$$

Finally, we connect $q_{j-2/3}$ to q_{j-1} using Γ_j^3 for which the three Riemann invariants are constant:

$$\begin{aligned} (p/\rho^\gamma)_{j-2/3} &= (p/\rho^\gamma)_{j-1} \\ (\hat{U} - 2/(\gamma-1)c)_{j-2/3} &= (\hat{U} - 2/(\gamma-1)c)_{j-1} \\ \hat{V}_{j-2/3} &= \hat{V}_{j-1} . \end{aligned} \tag{3.13}$$

We now have eight equations in eight unknowns for $q_{j-1/3}$ and $q_{j-2/3}$ which can be solved exactly. The solution was given in (3.5). The sonic points were calculated to be the points on Γ_j^1, Γ_j^3 for which $\hat{U} = c$ or $\hat{U} = -c$, respectively, which means that the associated eigenvalues vanish there.

The paths of integration in (3.8a), (3.8b) and (3.8c) are now well defined, and a simple calculation gives us the results in those equations.

Fig. 3 is a schematic of the scheme, and Fig. 4 a schematic of the boundary treatment.

● Grid points
○ Points in state space
c = Speed of sound
σ = Eigen value
$\hat{U} = (\xi_t + u\xi_x + v\xi_y)/(\xi_x^2 + \xi_y^2)^{1/2}$
P = Pressure
ρ = Density

Riemann invariants

Paths ①,④ : $\hat{U} - \dfrac{2}{\lambda-1}c$, P/ρ^λ , $v\xi_x - u\xi_y$

Paths ②,⑤ : $\hat{U}$,P

Paths ③,⑥ : $\hat{U} + \dfrac{2}{\lambda-1}c$, P/ρ^λ , $v\xi_x - u\xi_y$

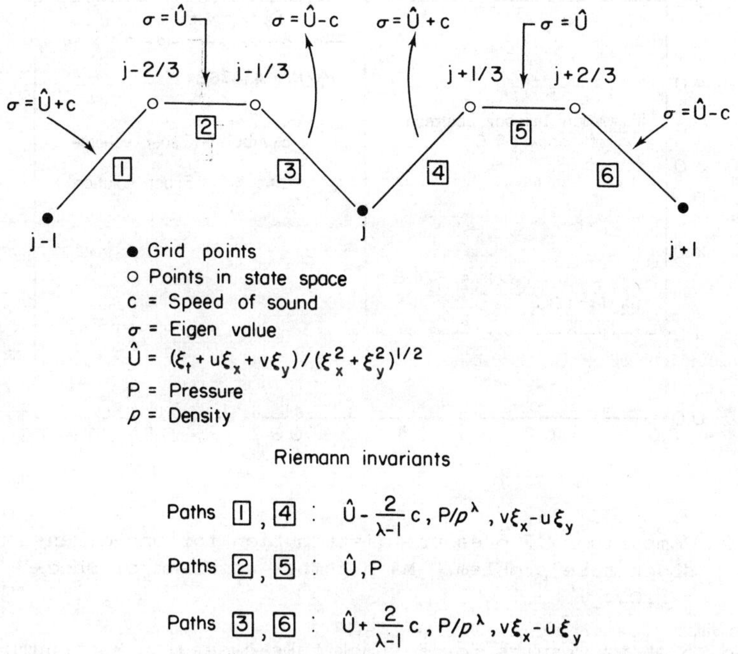

Fig. 3   Schematic representation of Osher scheme.

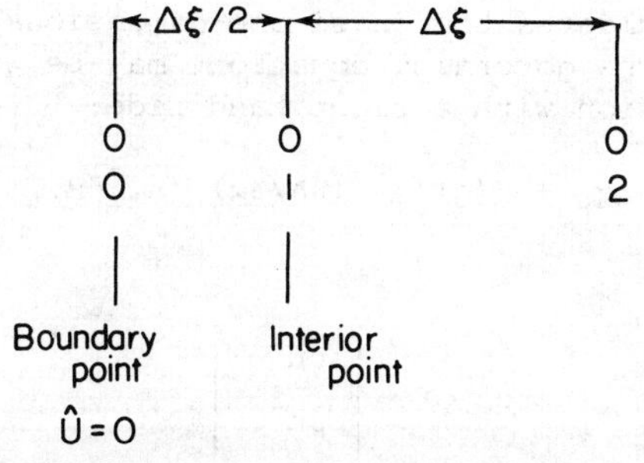

Fig. 4   Boundary point discretization.

4. RESULTS OF EU CALCULATIONS

The first results come from the work with Chakravarthy, (Osher and Chakravarthy, 1982), and use our algorithm presented in the previous section.

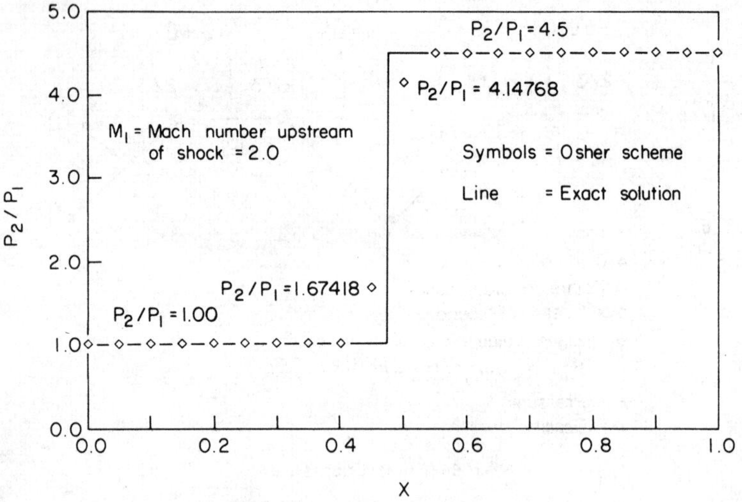

Fig. 5 Comparison of pressure distribution for one-dimensional shock tube problem. Mach number upstream of shock = 2.00.

Fig. 5 shows results for a shock tube case with Mach number 2.0 upstream of the shock. Fig. 6 shows the same kind of problem with Mach number of 1,000.0. In both cases, but for two transition points, the numerical and exact solutions match up to computer precision.

Next, we considered the quasi-one-dimensional flow through a Laval nozzle. The governing equations may be written as EU in one space dimension with a right hand side:

$$q_t + f(q)_x = (\partial A/\partial x)\ (A)^{-1} H. \tag{4.1}$$

with

$$H = \begin{pmatrix} \rho u \\ \rho u^2 \\ (e+\rho) u \end{pmatrix} \tag{4.2}$$

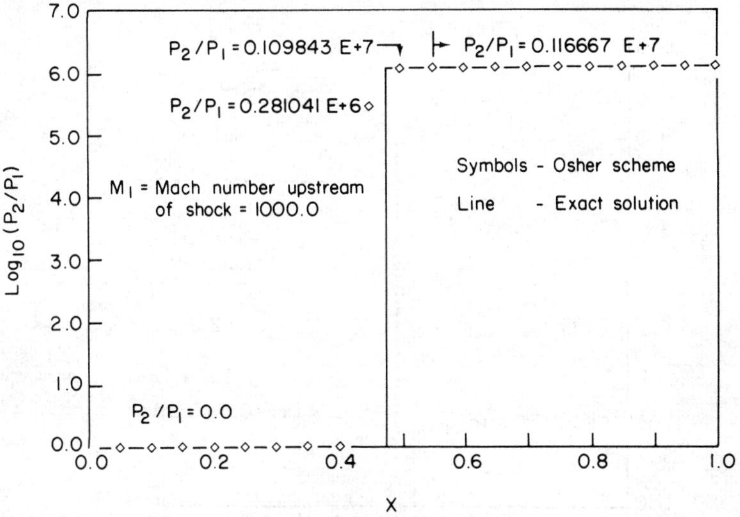

Fig. 6 Comparison of pressure distribution for one-dimensional shock tube problem. Mach number upstream of shock = 1000.0.

while for our problem

$$A(x) = 0.5 + 0.25x^2 . \tag{4.3}$$

For an inflow Mach number of 2.0, we compare the pressure distribution we obtained, with the exact solution in Fig. 7. Excellent agreement and sharp shock transitions are obvious.

The third example we considered is supersonic flow past a wedge. The wedge surface is aligned with the x axis, and the free stream flow is inclined to the surface. For negative angle of incidence α, a straight oblique shock, attached to the leading edge results, while for positive α, a Prandtl-Meyer expansion at the leading edge results. We solved this by changing coordinates to $\xi = x$, $\eta = y/x$, and ignoring ξ derivatives. The results presented are for $M_\infty = 2.0$.

In Fig. 8, we compare numerical and exact solutions for $\alpha = -10.0^\circ$. In Fig. 9, we consider $\alpha = 10.0^\circ$. The agreement is excellent, as is the absence of smearing and oscillations.

Finally, in (Osher and Chakravarthy, 1982), we turned our attention to two-dimensional flow past a circular cylinder, with free stream Mach number of 8.0. The flow was begun impulsively by prescribing free stream quantities at all points, and then imposing the boundary conditions. Convergence from such a start

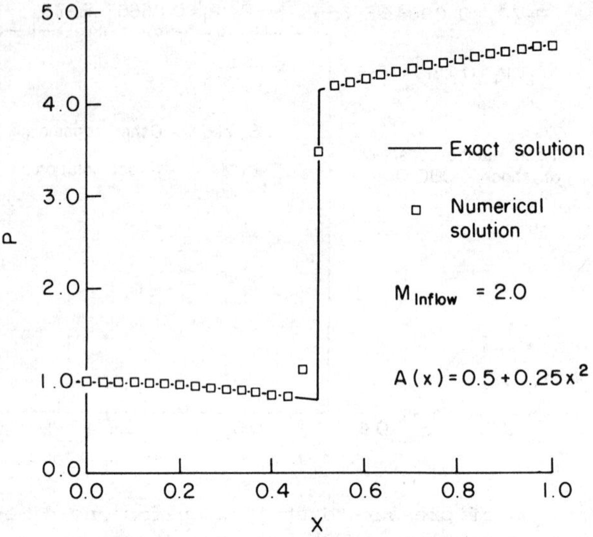

Fig. 7 Flow through quasi-one-dimensional Laval nozzle.

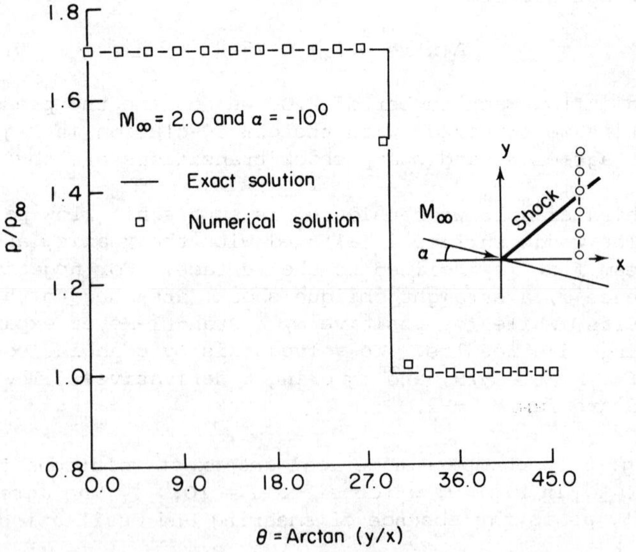

Fig. 8 Wedge in supersonic flow of $M_\infty = 2.0$ and $\alpha = -10^\circ$.

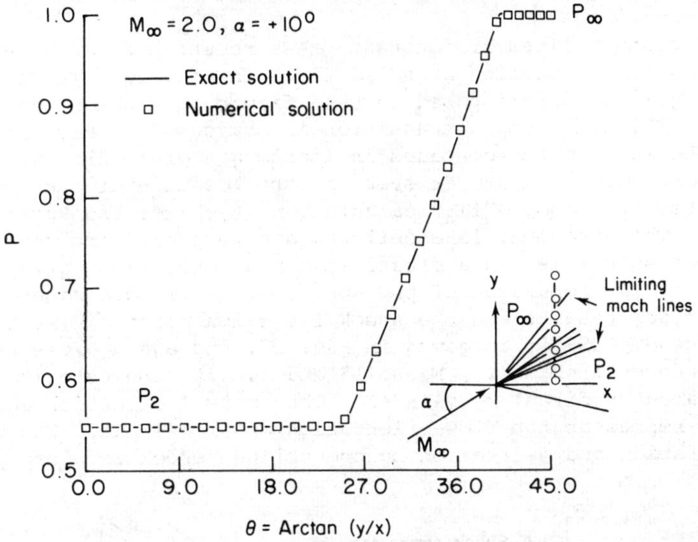

Fig. 9 Wedge in supersonic flow with M_∞ = 2.0, α = +10°.

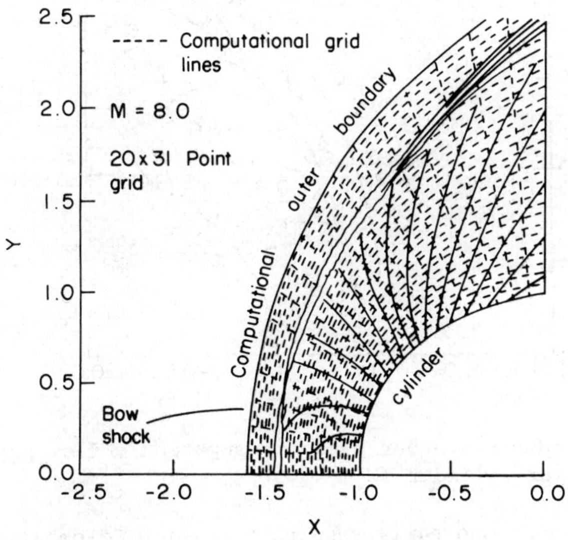

Fig. 10 Pressure contours for supersonic flow past a circular cylinder, M_∞ = 8.0.

clearly demonstrated the robustness of the method.

The contour lines of constant pressure which we obtained numerically are plotted as solid lines in Fig. 10. The computational grid is superimposed in this figure as a set of dashed lines. The crisp shock transition is obvious. In Fig. 11 we show the Mach number contours in increments of (0.2). Here we superpose, with triangular symbols, the location of the shock as predicted by a shock "fit" calculation (Lyubimov and Rusanov, 1973). The agreement is excellent, and we remind the reader that the scheme is, in a classical sense, only first order accurate. A comparison of the surface pressure distribution between our results and the shock fit calculation of (Lyubimov and Rusanov, 1973), is given in Fig. 12, and again gives excellent agreement. Finally, Figs. 13 and 14 give the velocity field about a circular cylinder. The arrows' direction and length represent the flow orientation and magnitude. The sharp deceleration and deflection by the oblique shock is clearly seen.

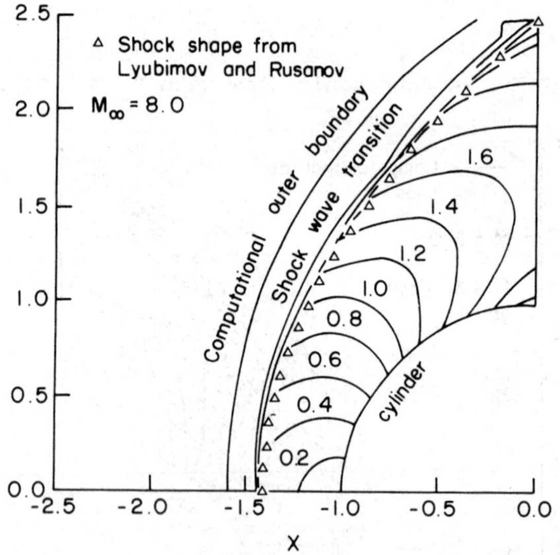

Fig. 11 Mach number contour plot for supersonic flow past circular cylinder ($M_\infty = 8.0$).

Next, I present some early results done with Solomon, (Osher and Solomon, 1982). Here we considered the isentropic EU in two space dimensions. The simple cartesian coordinate algorithm is presented there. This was before the boundary treatment of Chakravarthy and Osher (1982), Osher and Chakravarthy (1982), was arrived at, so in (Osher and Solomon, 1982) we imposed overdetermined boundary data, corresponding to steady

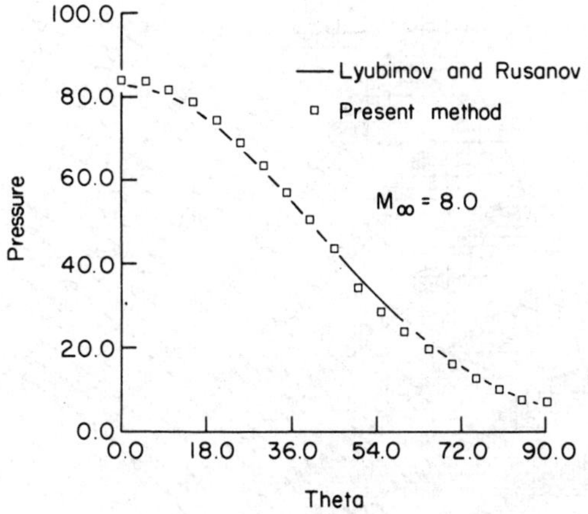

Fig. 12 Surface pressure distribution for supersonic flow past circular cylinder (M_∞ = 8.0).

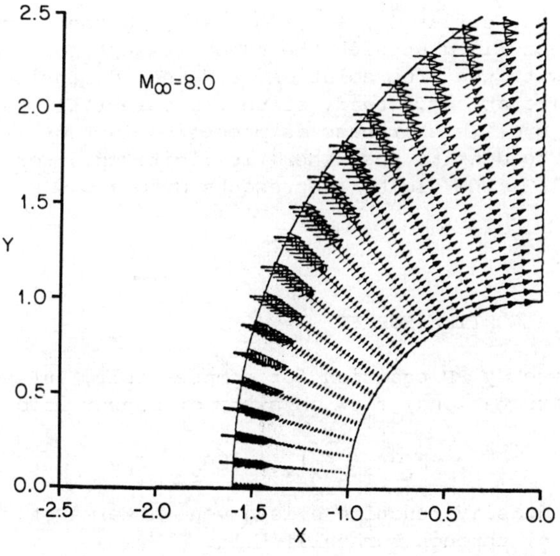

Fig. 13 Velocity vector plot for M_∞ = 8.0 supersonic flow past circular cylinder.

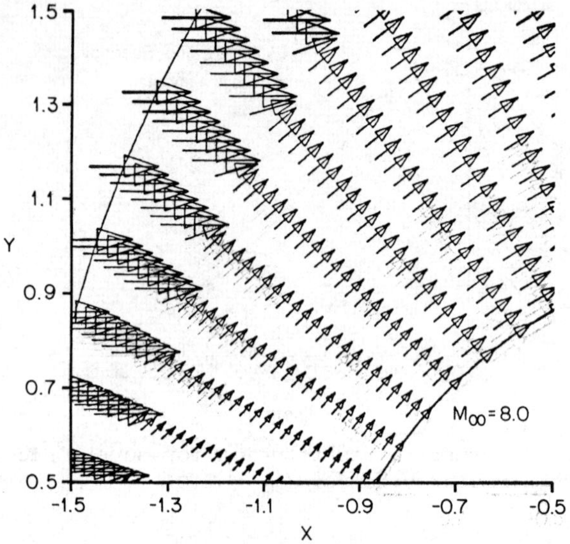

Fig. 14 Close up of velocity vector plot.

shocks oriented at 0°, 15°, 30°, 45°, to the x axis, and the grid was not lined up with the shock (except for 0°, of course). We began with the true solution, and used dimensional splitting. After 20 iterations, steady state was essentially achieved. The degree of profile sharpness is presented in Figs. 15-18. The value of the density seen there is digitised. For ρ between 0.50 and 1.0, the digit k represents $0.50 + 0.05k \le \rho < 0.55 + 0.05k$.

5. A NEW ALGORITHM FOR FP

The unsteady FP equation for compressible, inviscid, isentropic, and irrotational flow consists of conservation of mass

$$\rho_t + (\rho\phi_x)_x + (\rho\phi_y)_y = 0, \tag{5.1}$$

where the density function ρ is given in terms of velocity potential ϕ, through Bernoulli's law

$$\frac{A\gamma}{\gamma-1}\rho^{\gamma-1} + \phi_t + \frac{\phi_x^2 + \phi_y^2}{2} = H \tag{5.2}$$

for γ, A, H, positive constants, $\gamma \simeq 1.4$, and $A\gamma = M_\infty^{-2}$ is the standard aerodynamics notation.

```
0 0 0 0 0 0 0 0 0 0 9 9 9 9 9 9 9 9 9 9 9        0 0 0 0 0 0 0 0 0 0 9 9 9 9 9 9 9 9 9 9 9
0 0 0 0 0 0 0 0 0 0 9 9 9 9 9 9 9 9 9 9 9        0 0 0 0 0 0 0 0 0 1 9 9 9 9 9 9 9 9 9 9 9
0 0 0 0 0 0 0 0 0 0 9 9 9 9 9 9 9 9 9 9 9        0 0 0 0 0 0 0 0 0 1 9 9 9 9 9 9 9 9 9 9 9
0 0 0 0 0 0 0 0 0 0 9 9 9 9 9 9 9 9 9 9 9        0 0 0 0 0 0 0 0 0 1 9 9 9 9 9 9 9 9 9 9 9
0 0 0 0 0 0 0 0 0 0 9 9 9 9 9 9 9 9 9 9 9        0 0 0 0 0 0 0 0 0 1 9 9 9 9 9 9 9 9 9 9 9
0 0 0 0 0 0 0 0 0 0 9 9 9 9 9 9 9 9 9 9 9        0 0 0 0 0 0 0 0 0 1 9 9 9 9 9 9 9 9 9 9 9
0 0 0 0 0 0 0 0 0 0 9 9 9 9 9 9 9 9 9 9 9        0 0 0 0 0 0 0 0 0 1 9 9 9 9 9 9 9 9 9 9 9
0 0 0 0 0 0 0 0 0 0 9 9 9 9 9 9 9 9 9 9 9        0 0 0 0 0 0 0 0 0 1 9 9 9 9 9 9 9 9 9 9 9
0 0 0 0 0 0 0 0 0 0 9 9 9 9 9 9 9 9 9 9 9        0 0 0 0 0 0 0 0 0 1 9 9 9 9 9 9 9 9 9 9 9
0 0 0 0 0 0 0 0 0 0 9 9 9 9 9 9 9 9 9 9 9        0 0 0 0 0 0 0 0 0 1 9 9 9 9 9 9 9 9 9 9 9
0 0 0 0 0 0 0 0 0 0 9 9 9 9 9 9 9 9 9 9 9        0 0 0 0 0 0 0 0 0 1 9 9 9 9 9 9 9 9 9 9 9
0 0 0 0 0 0 0 0 0 0 9 9 9 9 9 9 9 9 9 9 9        0 0 0 0 0 0 0 0 0 1 9 9 9 9 9 9 9 9 9 9 9
0 0 0 0 0 0 0 0 0 0 9 9 9 9 9 9 9 9 9 9 9        0 0 0 0 0 0 0 0 0 1 9 9 9 9 9 9 9 9 9 9 9
0 0 0 0 0 0 0 0 0 0 9 9 9 9 9 9 9 9 9 9 9        0 0 0 0 0 0 0 0 0 1 9 9 9 9 9 9 9 9 9 9 9
0 0 0 0 0 0 0 0 0 0 9 9 9 9 9 9 9 9 9 9 9        0 0 0 0 0 0 0 0 0 1 9 9 9 9 9 9 9 9 9 9 9
0 0 0 0 0 0 0 0 0 0 9 9 9 9 9 9 9 9 9 9 9        0 0 0 0 0 0 0 0 0 1 9 9 9 9 9 9 9 9 9 9 9
0 0 0 0 0 0 0 0 0 0 9 9 9 9 9 9 9 9 9 9 9        0 0 0 0 0 0 0 0 0 1 9 9 9 9 9 9 9 9 9 9 9
0 0 0 0 0 0 0 0 0 0 9 9 9 9 9 9 9 9 9 9 9        0 0 0 0 0 0 0 0 0 1 9 9 9 9 9 9 9 9 9 9 9
0 0 0 0 0 0 0 0 0 0 9 9 9 9 9 9 9 9 9 9 9        0 0 0 0 0 0 0 0 0 1 9 9 9 9 9 9 9 9 9 9 9
0 0 0 0 0 0 0 0 0 0 9 9 9 9 9 9 9 9 9 9 9        0 0 0 0 0 0 0 0 0 1 9 9 9 9 9 9 9 9 9 9 9
0 0 0 0 0 0 0 0 0 0 9 9 9 9 9 9 9 9 9 9 9        0 0 0 0 0 0 0 0 0 0 9 9 9 9 9 9 9 9 9 9 9

                  t = 0                                  t = .1 (20 iterations)
```

Fig. 15 Euler Equations 2-D Isentropic Shock Orientation Angle - 0°.
Density Plot.

```
O O O O O O O O O O 9 9 9 9 9 9 9 9 9 9 9 | O O O O O O O O O O 9 9 9 9 9 9 9 9 9 9 9
O O O O O O O O O 9 9 9 9 9 9 9 9 9 9 9 9 | O O O O O O O O O 5 9 9 9 9 9 9 9 9 9 9 9
O O O O O O O O O 9 9 9 9 9 9 9 9 9 9 9 9 | O O O O O O O O 1 7 9 9 9 9 9 9 9 9 9 9 9
O O O O O O O O O 9 9 9 9 9 9 9 9 9 9 9 9 | O O O O O O O O 2 9 9 9 9 9 9 9 9 9 9 9 9
O O O O O O O O O 9 9 9 9 9 9 9 9 9 9 9 9 | O O O O O O O O 4 9 9 9 9 9 9 9 9 9 9 9 9
O O O O O O O O 9 9 9 9 9 9 9 9 9 9 9 9 9 | O O O O O O O O 6 9 9 9 9 9 9 9 9 9 9 9 9
O O O O O O O O 9 9 9 9 9 9 9 9 9 9 9 9 9 | O O O O O O O 2 8 9 9 9 9 9 9 9 9 9 9 9 9
O O O O O O O O 9 9 9 9 9 9 9 9 9 9 9 9 9 | O O O O O O O 3 9 9 9 9 9 9 9 9 9 9 9 9 9
O O O O O O O 9 9 9 9 9 9 9 9 9 9 9 9 9 9 | O O O O O O O 6 9 9 9 9 9 9 9 9 9 9 9 9 9
O O O O O O O 9 9 9 9 9 9 9 9 9 9 9 9 9 9 | O O O O O O 1 7 9 9 9 9 9 9 9 9 9 9 9 9 9
O O O O O O O 9 9 9 9 9 9 9 9 9 9 9 9 9 9 | O O O O O O 2 9 9 9 9 9 9 9 9 9 9 9 9 9 9
O O O O O O O 9 9 9 9 9 9 9 9 9 9 9 9 9 9 | O O O O O O 4 9 9 9 9 9 9 9 9 9 9 9 9 9 9
O O O O O O 9 9 9 9 9 9 9 9 9 9 9 9 9 9 9 | O O O O O O 6 9 9 9 9 9 9 9 9 9 9 9 9 9 9
O O O O O O 9 9 9 9 9 9 9 9 9 9 9 9 9 9 9 | O O O O O 1 7 9 9 9 9 9 9 9 9 9 9 9 9 9 9
O O O O O O 9 9 9 9 9 9 9 9 9 9 9 9 9 9 9 | O O O O O 2 9 9 9 9 9 9 9 9 9 9 9 9 9 9 9
O O O O O O 9 9 9 9 9 9 9 9 9 9 9 9 9 9 9 | O O O O O 4 9 9 9 9 9 9 9 9 9 9 9 9 9 9 9
O O O O O 9 9 9 9 9 9 9 9 9 9 9 9 9 9 9 9 | O O O O O 6 9 9 9 9 9 9 9 9 9 9 9 9 9 9 9
O O O O O 9 9 9 9 9 9 9 9 9 9 9 9 9 9 9 9 | O O O O 2 8 9 9 9 9 9 9 9 9 9 9 9 9 9 9 9
O O O O O 9 9 9 9 9 9 9 9 9 9 9 9 9 9 9 9 | O O O O 4 9 9 9 9 9 9 9 9 9 9 9 9 9 9 9 9
O O O O 9 9 9 9 9 9 9 9 9 9 9 9 9 9 9 9 9 | O O O O 7 9 9 9 9 9 9 9 9 9 9 9 9 9 9 9 9
O O O O 9 9 9 9 9 9 9 9 9 9 9 9 9 9 9 9 9 | O O O O 9 9 9 9 9 9 9 9 9 9 9 9 9 9 9 9 9

          t = O                                    t = .1 (20 iterations)
```

Fig. 16 Euler Equations 2-D Isentropic Shock Orientation Angle - 15°.
Density plot.

```
O O O O O O O O O O 9 9 9 9 9 9 9 9 9 9 9 | O O O O O O O O O O 9 9 9 9 9 9 9 9 9 9 9
O O O O O O O O O O 9 9 9 9 9 9 9 9 9 9 9 | O O O O O O O O O 3 9 9 9 9 9 9 9 9 9 9 9
O O O O O O O O O 9 9 9 9 9 9 9 9 9 9 9 9 | O O O O O O O O 1 7 9 9 9 9 9 9 9 9 9 9 9
O O O O O O O O O 9 9 9 9 9 9 9 9 9 9 9 9 | O O O O O O O O 4 8 9 9 9 9 9 9 9 9 9 9 9
O O O O O O O O 9 9 9 9 9 9 9 9 9 9 9 9 9 | O O O O O O O 1 7 9 9 9 9 9 9 9 9 9 9 9 9
O O O O O O O O 9 9 9 9 9 9 9 9 9 9 9 9 9 | O O O O O O O 4 9 9 9 9 9 9 9 9 9 9 9 9 9
O O O O O O O 9 9 9 9 9 9 9 9 9 9 9 9 9 9 | O O O O O O 3 7 9 9 9 9 9 9 9 9 9 9 9 9 9
O O O O O O 9 9 9 9 9 9 9 9 9 9 9 9 9 9 9 | O O O O O 1 6 8 9 9 9 9 9 9 9 9 9 9 9 9 9
O O O O O O 9 9 9 9 9 9 9 9 9 9 9 9 9 9 9 | O O O O O 4 8 9 9 9 9 9 9 9 9 9 9 9 9 9 9
O O O O O 9 9 9 9 9 9 9 9 9 9 9 9 9 9 9 9 | O O O O 1 7 9 9 9 9 9 9 9 9 9 9 9 9 9 9 9
O O O O O 9 9 9 9 9 9 9 9 9 9 9 9 9 9 9 9 | O O O O 4 8 9 9 9 9 9 9 9 9 9 9 9 9 9 9 9
O O O O 9 9 9 9 9 9 9 9 9 9 9 9 9 9 9 9 9 | O O O 1 7 9 9 9 9 9 9 9 9 9 9 9 9 9 9 9 9
O O O O 9 9 9 9 9 9 9 9 9 9 9 9 9 9 9 9 9 | O O O 4 9 9 9 9 9 9 9 9 9 9 9 9 9 9 9 9 9
O O O 9 9 9 9 9 9 9 9 9 9 9 9 9 9 9 9 9 9 | O O 3 7 9 9 9 9 9 9 9 9 9 9 9 9 9 9 9 9 9
O O 9 9 9 9 9 9 9 9 9 9 9 9 9 9 9 9 9 9 9 | O 1 6 8 9 9 9 9 9 9 9 9 9 9 9 9 9 9 9 9 9
O O 9 9 9 9 9 9 9 9 9 9 9 9 9 9 9 9 9 9 9 | O 3 8 9 9 9 9 9 9 9 9 9 9 9 9 9 9 9 9 9 9
O 9 9 9 9 9 9 9 9 9 9 9 9 9 9 9 9 9 9 9 9 | O 6 9 9 9 9 9 9 9 9 9 9 9 9 9 9 9 9 9 9 9
O 9 9 9 9 9 9 9 9 9 9 9 9 9 9 9 9 9 9 9 9 | O 8 9 9 9 9 9 9 9 9 9 9 9 9 9 9 9 9 9 9 9
9 9 9 9 9 9 9 9 9 9 9 9 9 9 9 9 9 9 9 9 9 | 9 9 9 9 9 9 9 9 9 9 9 9 9 9 9 9 9 9 9 9 9
9 9 9 9 9 9 9 9 9 9 9 9 9 9 9 9 9 9 9 9 9 | 9 9 9 9 9 9 9 9 9 9 9 9 9 9 9 9 9 9 9 9 9
9 9 9 9 9 9 9 9 9 9 9 9 9 9 9 9 9 9 9 9 9 | 9 9 9 9 9 9 9 9 9 9 9 9 9 9 9 9 9 9 9 9 9

              t = O                       |         t = .1 (2O iterations)
```

Fig. 17 Euler Equations 2-D Isentropic Shock Orientation Angle - 30°. Density plot.

```
0 0 0 0 0 0 0 0 0 0 0 0 0 9 9 9 9 9 9 9 9
0 0 0 0 0 0 0 0 0 0 0 0 9 9 9 9 9 9 9 9 9
0 0 0 0 0 0 0 0 0 0 0 9 9 9 9 9 9 9 9 9 9
0 0 0 0 0 0 0 0 0 0 9 9 9 9 9 9 9 9 9 9 9
0 0 0 0 0 0 0 0 0 9 9 9 9 9 9 9 9 9 9 9 9
0 0 0 0 0 0 0 0 9 9 9 9 9 9 9 9 9 9 9 9 9
0 0 0 0 0 0 0 9 9 9 9 9 9 9 9 9 9 9 9 9 9
0 0 0 0 0 0 9 9 9 9 9 9 9 9 9 9 9 9 9 9 9
0 0 0 0 0 9 9 9 9 9 9 9 9 9 9 9 9 9 9 9 9
0 0 0 0 9 9 9 9 9 9 9 9 9 9 9 9 9 9 9 9 9
0 0 0 9 9 9 9 9 9 9 9 9 9 9 9 9 9 9 9 9 9
0 0 9 9 9 9 9 9 9 9 9 9 9 9 9 9 9 9 9 9 9
0 9 9 9 9 9 9 9 9 9 9 9 9 9 9 9 9 9 9 9 9
9 9 9 9 9 9 9 9 9 9 9 9 9 9 9 9 9 9 9 9 9
9 9 9 9 9 9 9 9 9 9 9 9 9 9 9 9 9 9 9 9 9
9 9 9 9 9 9 9 9 9 9 9 9 9 9 9 9 9 9 9 9 9
9 9 9 9 9 9 9 9 9 9 9 9 9 9 9 9 9 9 9 9 9
9 9 9 9 9 9 9 9 9 9 9 9 9 9 9 9 9 9 9 9 9
9 9 9 9 9 9 9 9 9 9 9 9 9 9 9 9 9 9 9 9 9
9 9 9 9 9 9 9 9 9 9 9 9 9 9 9 9 9 9 9 9 9
9 9 9 9 9 9 9 9 9 9 9 9 9 9 9 9 9 9 9 9 9
```

t = 0

```
0 0 0 0 0 0 0 0 0 0 0 0 0 9 9 9 9 9 9 9 9
0 0 0 0 0 0 0 0 0 0 0 3 7 9 9 9 9 9 9 9 9
0 0 0 0 0 0 0 0 0 0 3 7 9 9 9 9 9 9 9 9 9
0 0 0 0 0 0 0 0 0 3 7 9 9 9 9 9 9 9 9 9 9
0 0 0 0 0 0 0 0 3 7 9 9 9 9 9 9 9 9 9 9 9
0 0 0 0 0 0 0 3 7 9 9 9 9 9 9 9 9 9 9 9 9
0 0 0 0 0 0 3 7 9 9 9 9 9 9 9 9 9 9 9 9 9
0 0 0 0 0 3 7 9 9 9 9 9 9 9 9 9 9 9 9 9 9
0 0 0 0 3 7 9 9 9 9 9 9 9 9 9 9 9 9 9 9 9
0 0 0 3 7 9 9 9 9 9 9 9 9 9 9 9 9 9 9 9 9
0 0 3 7 9 9 9 9 9 9 9 9 9 9 9 9 9 9 9 9 9
0 3 7 9 9 9 9 9 9 9 9 9 9 9 9 9 9 9 9 9 9
0 7 9 9 9 9 9 9 9 9 9 9 9 9 9 9 9 9 9 9 9
9 9 9 9 9 9 9 9 9 9 9 9 9 9 9 9 9 9 9 9 9
9 9 9 9 9 9 9 9 9 9 9 9 9 9 9 9 9 9 9 9 9
9 9 9 9 9 9 9 9 9 9 9 9 9 9 9 9 9 9 9 9 9
9 9 9 9 9 9 9 9 9 9 9 9 9 9 9 9 9 9 9 9 9
9 9 9 9 9 9 9 9 9 9 9 9 9 9 9 9 9 9 9 9 9
9 9 9 9 9 9 9 9 9 9 9 9 9 9 9 9 9 9 9 9 9
9 9 9 9 9 9 9 9 9 9 9 9 9 9 9 9 9 9 9 9 9
9 9 9 9 9 9 9 9 9 9 9 9 9 9 9 9 9 9 9 9 9
```

t = .1 (20 iterations)

Fig. 18 Euler Equations 2-D Isentropic Shock Orientation Angle - 45°.
Density plot.

The sound speed c is defined through

$$c^2 = A\gamma\rho^{\gamma-1} \tag{5.3}$$

and the absolute speed by

$$q = \sqrt{(\phi_x^2 + \phi_y^2)}. \tag{5.4}$$

We also define, for future use, the angle of the flow θ, for $-\pi \le \theta \le \pi$, with

$$\cos\theta = \phi_x/q, \quad \sin\theta = \phi_y/q. \tag{5.5}$$

An attempt was made by Chipman and Jameson (1979) to solve this by viewing it as a first order hyperbolic system. One does this, by taking the space gradient of (5.2), and defining velocities

$$u = \phi_x, \quad v = \phi_y. \tag{5.6}$$

This leads to the system of conservation laws:

$$\begin{pmatrix} \rho \\ u \\ v \end{pmatrix}_k + \begin{pmatrix} \rho u \\ \frac{1}{2}q^2 + \dfrac{c^2}{\gamma-1} \\ 0 \end{pmatrix}_x + \begin{pmatrix} \rho u \\ 0 \\ \frac{1}{2}q^2 + \dfrac{c^2}{\gamma-1} \end{pmatrix}_y \tag{5.7}$$

The quasi-linear system becomes

$$q_t + Aq_x + Bq_y = 0 \tag{5.8}$$

with $q = (\rho, u, v)^T$.

$$A = \begin{pmatrix} u & \rho & 0 \\ \dfrac{c^2}{\rho} & u & v \\ 0 & 0 & 0 \end{pmatrix}, \quad B = \begin{pmatrix} v & 0 & \rho \\ 0 & 0 & 0 \\ \dfrac{c^2}{\rho} & u & v \end{pmatrix}$$

Unfortunately, this system is not strictly hyperbolic. The eigenvalues of $A\xi + B\eta$, for ξ, η real, $\xi^2 + \eta^2 = 1$, are $\hat{u} \pm c, 0$, where $\hat{u}$ is the contravariant velocity $\hat{u} = u\xi + v\eta$. When $\hat{u} = +c$ or $-c$, the resulting matrix is not diagonalizable. This means, computationally, that any numerical vorticity which develops will necessarily generate a numerical instability.

Of course, in spite of this fact, numerical methods for the steady version of (5.1), (5.2), mostly based on upwind differencing, have been carried out successfully, e.g. (Jameson, 1976), (Holst, 1979). See (Goorjian, 1980) for an unsteady algorithm which also works fairly well. Thus, I shall concentrate on the scalar potential formulation, steady state first. The unsteady algorithm will appear in the near future.

I suggest the following approach for the steady version of (5.1), which is rewritten for convenience as:

$$\begin{aligned} &\text{(a)} \quad (\rho q \cos\theta)_x + (\rho q \sin\theta)_y = 0 \\ &\text{(b)} \quad c^2/(\gamma-1) + q^2/2 = H. \end{aligned} \tag{5.9}$$

The second equation gives us ρ as a function of q, satisfying the mass flux relationship:

$$\frac{d}{dq}\,\rho q = \rho\left(1 - \frac{q^2}{c^2}\right). \tag{5.10}$$

The sonic point, at which $q = c$, is thus the minimum of the convex function ρq.

This unique sonic value is c^*, the associated sonic density is ρ^*, and the sonic mass flux is $\rho^* c^*$. These values are

$$c^* = \left(2H \frac{(\gamma-1)}{(\gamma+1)}\right)^{\frac{1}{2}} = q^* \tag{5.11a}$$

$$\rho^* = \left(\frac{(c^*)^2}{A\gamma}\right)^{1/\gamma-1} \tag{5.11b}$$

$$\rho^* c^* = (A\gamma)^{-1/\gamma-1} (c^*)^{(\gamma+1)/\gamma-1}. \tag{5.11c}$$

In the spirit of past flux splitting, define:

$$(\rho q)_+ = \int_0^q \rho(s)\left(\max\left(1 - \frac{s^2}{c^2(s)}\right), 0\right) ds. \tag{5.12}$$

So

$$(\rho q)_+ = \rho q \quad \text{if } q \le q^* \iff q \le c$$
$$(\rho q)_+ = \rho^* q^* \quad \text{if } q > q^* \iff q > c. \tag{5.13}$$

Similarly:

$$(\rho q)_- = \int_0^q \rho(s)\ (\min\ (1 - \frac{s^2}{c^2(s)}),\ 0)\ ds \tag{5.14}$$

and thus:

$$(\rho q)_- \equiv 0 \text{ if } q \le q^* \iff q \le c$$
$$= \rho q - \rho^* q^* \text{ if } q > q^* \iff q > c. \tag{5.15}$$

Let $\Phi = [\Phi_{jk}]$ be a discrete potential function, and define the operations

$$(A)^+ = \max\ (A,0)$$
$$(A)^- = \min\ (A,0)\ . \tag{5.16}$$

Then on the grid, let the absolute speed be defined through

$$q = \sqrt{((D_-^x\Phi)^+)^2 + ((D_+^x\Phi)^-)^2 + ((D_-^y\Phi)^+)^2 + ((D_+^y\Phi)^-)^2}. \tag{5.17}$$

This defines p and c uniquely.

My proposed approximation to (5.9), using the potential formulation, is:

$$D_+^x\left[(\rho(D_-^x\Phi)^+) - \Delta x D_-^x((\rho q)_- \frac{(D_-^x\Phi)^+}{q})\right]$$
$$+ D_-^x\left[(\rho(D_+^x\Phi)^-) + \Delta x D_+^x((\rho q)_- \frac{(D_+^x\Phi)^-}{q})\right]$$
$$+ D_+^y\left[(\rho(D_-^y\Phi)^+) - \Delta y D_-^y((\rho q)_- \frac{(D_-^y\Phi)^+}{q})\right]$$
$$+ D_-^y\left[(\rho(D_+^y\Phi)^-) + \Delta y D_+^y((\rho q)_- \frac{(D_+^y\Phi)^-}{q})\right] = 0\ . \tag{5.18}$$

This method yields traditional second order accurate central differencing in subsonic regions. However, since it uses momentum rather than density biasing, it differs from the usual upwinding in supersonic regions, and at sonic and shock points.

It is clear that principles (-1) and (0) are valid for (5.18). A nonphysical shock solution to (5.9) is a weak solution to (5.7), which is time independent, and for which characteristics flow out of the surface of discontinuity, i.e., an expansion shock. I have proved the following.

Theorem 1

Let $\{\Phi_{jk}\}$ be a solution to (5.18) which converges boundedly almost everywhere, together with $\{D_-^x\Phi_{jk}\}$, $\{D_-^y\Phi_{jk}\}$ as $\Delta x, \Delta y \to 0$ to Φ, u, and v. Then Φ, u, and v yield a steady weak solution to (5.1), (5.2) which does not have weak expansion shocks.

Theorem 1 is valid also for strong shocks in the one-dimensional case. A steady shock solution is a piecewise constant solution

$$u \equiv u^L \quad x < 0$$

$$u \equiv u^R \quad x > 0$$

with $\rho(q^L)u^L = \rho(q^R)u^R$.

There are two possible families of such non-expansive shock solutions. These are

One shocks, for which $u^L - c^L > 0 > u^R - c^R$,

and two shocks, for which $u^L + c^L > 0 > u^R + c^R$.

The following is true (letting $D_-^x\Phi_j = u_j$):

Theorem 2

For any such one shock there exists a discrete solution to (5.18) of the form

$$u_j \equiv u^L \quad j < 0$$

$$u_j \equiv u^R \quad j > 0$$

$$u^L \geq u_0 > \bar{u} \geq u_1 > u^R.$$

The two intermediate points are related through the above inequality and:

$$\rho(u_1)\, u_1 + \rho(u_0)\, u_0 = \rho(u^L)\, u^L + \rho^* q^*$$

Analogous statements hold for two shocks

These are the only possible (up to translation) discrete shock solutions. Thus, as in (Engquist and Osher, 1981), we have a unique one parameter family of discrete, monotone, sharp, steady shock profiles, which allows for any shock location intermediate on the grid.

REFERENCES

ABRAHAMSSON, L. and OSHER, S. 1982 Monotone difference schemes for singular perturbation problems, SINUM, to appear.

BALLHAUS, W.F. and STEGER, J.L. 1975 Implicit approximate factorization schemes for the low frequency transonic equation, NASA TM X-73, 082.

CHAKRAVARTHY, S. and OSHER, S. 1982 Numerical experiments with the Osher upwind scheme for the Euler equations, AIAA Paper 82-0975, St. Louis, Missouri.

CHIPMAN, R. and JAMESON, A. 1979 Fully conservative numerical solutions for unsteady irrotational flow about airfoils, AIAA Paper 79-1555.

EDWARDS, J.W., BENNETT, R.M., WHITLOW, W. Jnr. and SEIDEL, D.A. 1982 Time marching transonic flutter solutions including angle-of-attach effects, AIAA Paper 82-0685, New Orleans, Louisiana.

ENGQUIST, B. and OSHER, S. 1980a Stable and entropy satisfying approximations for transonic flow calculations, *Maths. Comput.* **34**, pp. 45-75.

ENGQUIST, B. and OSHER, S. 1980b One-sided difference schemes and transonic flow, Proceedings of National Academy of Sciences, USA **77**, pp. 3071-3074.

ENGQUIST, B. and OSHER, S. 1981a One-sided difference equations for non-linear conservation laws, *Maths. Comput.* **36**, pp. 321-352.

ENGQUIST, B. and OSHER, S. 1981b Upwind schemes for systems of conversation laws - potential flow equations, MRC Technical Summary Report No. 2186, University of Wisconsin.

GODUNOV, S.K. 1959 A finite difference method for the numerical computation of discontinuous solutions of the equations of fluid dynamics, *Math. Sb.*, **47**, pp. 271-290.

GOORJIAN, P.M. 1980 Implicit computations of unsteady transonic flow governed by the full potential equation in conservation form, AIAA Paper 80-0150, Pasadena, California.

GOORJIAN, P.M. and VAN BUSKIRK, R. 1981 Implicit calculations of transonic flow using monotone methods, AIAA Paper 81-331, St. Louis, Missouri.

HOLST, T.L. 1979 A fast, conservative algorithm for solving the transonic-potential equation, AIAA Paper 79-1456.

JAMESON, A. 1976 Numerical solutions of nonlinear partial differential equations of mixed type. In Numerical solutions of Partial Differential Equations III, Academic Press, New York, pp. 275-320.

KUZNETSOV, N.N. 1977 On stable methods for solving non-linear first order partial differential equations in the class of discontinuous functions. In Topics in Num. Anal. II, Proc. Roy. Irish Acad. Conference on Num. Anal., (J.J.H. Miller, Ed), Academic Press, pp. 183-197.

LAX, P.D. 1972 Hyperbolic systems of conservation laws and the mathematical theory of shock waves, SIAM Regional Conference Series Lectures in Applied Mathematics **11**.

LAX, P.D. and WENDROFF, B. 1960 Systems of conservation laws, *Comm. Pure Appl. Math.* **13**, pp. 217-237.

LYUBIMOV, A.N. and RUSANOV, V.V. 1973 Gas flows past blunt bodies, NASA-TT-F 715.

MAJDA, A. and OSHER, S. 1978 A systematic method for correcting nonlinear instabilities: the Lax Wendroff scheme for scalar conservation laws, *Num. Math.* **30**, pp. 429-452.

MAJDA, A. and OSHER, S. 1979 Numerical viscosity and the entropy condition, *Comm. Pure Appl. Maths.* **32**, pp. 797-838.

MURMAN, E. 1974 Analysis of embedded shockwaves calculated by relaxation methods, *AIAA J.* **12**, pp. 626-633.

MURMAN, E.M. and COLE, J.D. 1971 Calculations of plane steady transonic flow, *AIAA J.* **9**, pp. 114-121.

OSHER, S. 1981a Numerical solution of singular perturbation problems and hyperbolic systems of conservation laws, North Holland Mathematics Studies No. 47 (O. Axelsson, L.S. Frank and A. van der Sluis, Eds), pp. 179-205.

OSHER, S. 1981b Nonlinear singular perturbation problems and one-sided difference schemes, *SINUM* **18**, pp. 129-144.

OSHER, S. and CHAKRAVARTHY, S. 1982 Upwind schemes and boundary conditions with applications to Euler's equations in general geometries, submitted to *J. Comput. Phys*.

OSHER, S. and SOLOMON, F. 1982 Upwind schemes for hyperbolic systems of conservation laws, *Maths. Comput.*, **38**, pp. 339-374.

SANDERS, R. 1982 On convergence of monotone finite difference schemes with variable spatial differencing, *Maths. Comput.* to appear.

SHANKAR, V. and OSHER, S. 1982 An efficient full potential implied method based on characteristics for analysis of supersonic flows, AIAA Paper 82-0974, St. Louis, Missouri.

VAN LEER, B. 1982 On the relation between the upwind-differencing schemes of Godunov, Engquist-Osher, and Roe, *SIAM J. Scientific and Stat. Comp.*, to appear.

WHITLOW, W. Jnr. and BENNETT, R.M. to appear.

FLUCTUATIONS AND SIGNALS - A FRAMEWORK FOR NUMERICAL EVOLUTION PROBLEMS

P.L. Roe

(Royal Aircraft Establishment, Bedford)

1. INTRODUCTION

It has been said that when a problem is well-posed it has no more interest for a physicist. Behind this epigram is the idea that when a physicist understands a problem well enough to express it as a set of equations with enough auxiliary constraints to ensure existence and uniqueness, then his job is done. He could hand the equations over to a mathematician who need not understand the physics at all. The mathematician has only to avoid technical errors, and any solutions which he may find to those equations are guaranteed to be true statements about the real world.

As a working relationship this might not be very satisfying, but it would be quite possible. I mention it, however, simply to introduce a related question which I find very interesting. Suppose the problem turns out to be too difficult for the mathematician, who is replaced by a computer programmer, equally competent and just as ignorant of physics. How should the physicist now communicate his knowledge? The equations by themselves are not always enough, even if supplemented by a "numerical method", which would converge to a correct solution if the computer became infinitely large and accurate. On the other hand, one feels that the physicist ought not to have to specify every little detail of the algorithm.

The problem is particularly acute in fluid mechanics, where a continuous problem has to be replaced by a discrete model. It is perhaps at the level of this model that communication should take place. The physicist says "think of it as if ....", and goes on to explain things in terms perhaps of particles or other discrete objects more in scale with the programmers finite resources. The description need not be "true", but at the chosen level it must be complete, like the book of rules for playing some game.

To devise such rules is, however, no easy task, even for situations where the true physics is extremely simple. For example, we could hardly have a simpler physical situation than passive advection of a contaminant, borne along with uniform velocity a by a fluid in a parallel pipe. If the concentration of the contaminant is $u(x,t)$, and if we know the initial distribution $U(x,0)$, then

$$u(x,t) = U(x - at) \tag{1.1}$$

which is a perfectly complete mathematical model. Equally complete is the partial differential equation

$$u_t + au_x = 0 \tag{1.2}$$

which has (1.1) as its exact solution.

A simple numerical model of the situation is to introduce u_i^n as an approximation to $u(i\Delta x, n\Delta t)$ and then to proceed by the marching process,

$$u_i^{n+1} = \sum_k c_k \, u_{i-k}^n \tag{1.3}$$

All that needs to be done is to choose the coefficients c_k. From a formal mathematical point of view this is also quite easy. The c_k can be chosen so that polynomial solutions of (1.1), for example, are exactly recovered up to some order. However, the resulting algorithms may not be entirely satisfactory. Are there conceptual models which would help us to choose the c_k more wisely? A likely-looking start is to observe that unless the Courant number $(a\Delta t/\Delta x)$ is an integer the exact velocity cannot be represented on the chosen grid. Suppose therefore we think of the problem as if it involved particles moving with "quantised" velocities $a_k = k\Delta x/\Delta t$, $k \in Z$ (that is, with integral Courant numbers). Any of the following three models would be consistent with Eqn (1.3).

(i) The particles are of equal mass. A proportion c_k of them move with velocity a_k

(ii) Every velocity a_k is equally likely, but each particle moving with velocity a_k has a mass equal to c_k.

(iii) All particles are identical, but their motion is random. During each time interval there is a probability c_k that a given particle will move with velocity a_k.

We could attempt to choose the c_k by thinking about the behaviour of any of these models, but in fact all three run up against the same snag. It is easily shown that if we attempt to make the truncation error of the numerical solution less than the scale of the mesh, then at least some of the c_k must be negative. Our models therefore force us to contemplate either negative distributions, or negative masses, or negative probabilities. I do not know any simple model of advection which is both highly accurate and highly realistic for non-integer Courant numbers. Modelling on the scale of the mesh is easy, but to model events on any finer scale appears to require either that additional information is carried, or that rules are introduced which do not have a direct physical meaning. Nevertheless it may happen that these artificial rules will evolve over a period of time so as to offer a satisfying blend of economy and effectiveness.

In my own attempts to discover such rules I have found it useful to think of numerical evolution problems in terms of two basic concepts. A <u>fluctuation</u> is something detected in the data, indicating that it has not yet reached equilibrium, and a <u>signal</u> is an action performed on the data so as to bring it closer to equilibrium. I do not claim that these concepts automatically generate good models. Indeed, they are sufficiently vague that most methods, good or bad, can be cast in these terms. What I do believe is that they create a very suitable language for discussing numerical methods, which lays about equal stress on the real physics, and on the rules by which we hope to simulate the physics. Nor do I claim these concepts as an original invention; indeed it is encouraging that others have also found them to be stimulating and fruitful. In this paper I hope to demonstrate the utility of these concepts by using them to discuss various topics of current interest. In some cases the aim will be to shed new light on old methods; sometimes new results will be found with less effort, and in other cases quite new methods will emerge.

2. FLUCTUATIONS AND SINGULAR BEHAVIOUR

We discuss now the scalar, but possibly non-linear, conservation law

$$u_t + f(u)_x = 0 \tag{2.1}$$

for which again we seek solutions on a uniform grid $x_i = i\Delta x$, $t^n = n\Delta t$, and where u_i^n approximates $u(x_i, t^n)$. Define a numerical fluctuation to be, for $1 \leq i \leq imax - 1$,

$$\phi_{i+\frac{1}{2}}^n = f(u_{i+1}^n) - f(u_i^n) \tag{2.2}$$

We note that this quantity indicates a lack of local equilibrium, or a propensity for change, but that for non-linear f(u), such as $f = \frac{1}{2}u^2$, a zero value of $\phi_{i+\frac{1}{2}}$ does not always imply $u_{i+1}^n = u_i^n$. We now imagine that each fluctuation generates a signal whose magnitude is $(\Delta t/\Delta x)\ \phi_{i+\frac{1}{2}}^n$. We next choose a target, which is either x_i or x_{i+1}, and subtract the signal from the value of u at that target. We choose our target so that a positive signal is used to reduce the greater of (u_i^n, u_{i+1}^n), or a negative signal to increase the lesser. It is easily shown that the target selected by this rule is $x_{i+\frac{1}{2}+\frac{1}{2}\sigma}$ where

$$\sigma = \text{sign}\ \frac{f_{i+1}^n - f_i^n}{u_{i+1}^n - u_i^n} \tag{2.3}$$

If we wish to be pedantic we can replace this expression by sign $f'(u_i^n)$ whenever $u_{i+1}^n = u_i^n$, but in that case there is no action to be taken anyway. The quantity σ also tells us whether the average wavespeed in the interval $[x_i, x_{i+1}]$ is positive or negative at time t^n, thereby establishing a clear link between the local stability of a method, and the direction in which signals are propagated.

Consider the case where σ has the same sign everywhere, say positive. All signals are passed to the right (Fig. 1(a)) and we have merely reinvented upwind differencing. We may impose one boundary condition, in a realistic manner, by specifying u_1^n as a function of n. If u_1^n is held constant, all fluctuations will eventually decay to zero, for n sufficiently large.

Now consider the case where σ changes sign from positive on the left to negative on the right (Fig. 1(b)). Correct physics

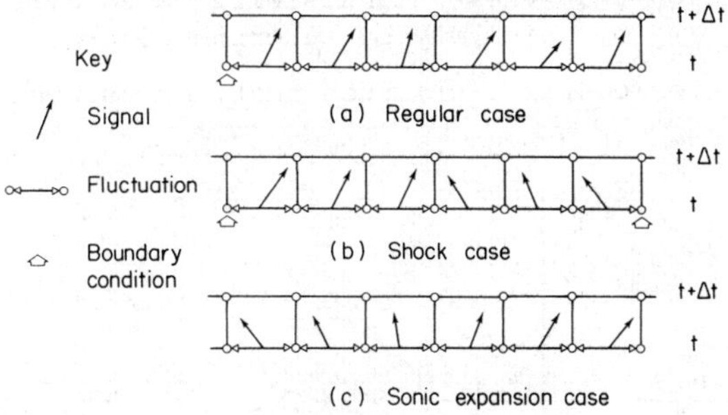

Fig. 1 Flow of information for a simple upwind algorithm applied to a scalar conservation law.

requires us to specify a boundary condition at each end, by prescribing u_1^n and u_{imax}^n. There is a point x_s in the middle, which we shall call the shock point, which receives two signals. Notice that information is lost at this point; in its vicinity we cannot reconstruct $\{u^n\}$ from $\{u^{n+1}\}$. Notice also that not all the fluctuations can decay to zero, even if we specify u_1^n, u_{imax}^n to be constants (U_L, U_R) such that $f(U_L) = f(U_R)$. Although the exact solution for large time is then a shockwave separating two constant states, what happens numerically is this. The constant states propagate inward from the boundaries until only one point in the grid has a value different from both U_L and U_R. Let that point be x_s and let the value there be u_s then

$$\phi_{s-\frac{1}{2}} = f_s - f_L \tag{2.4a}$$

$$\phi_{s+\frac{1}{2}} = f_R - f_s \tag{2.4b}$$

and <u>if</u> both these intervals send their signal to x_s, then that point will receive a total increment

$$\frac{-\Delta t}{\Delta x}(\phi_{s-\frac{1}{2}} + \phi_{s+\frac{1}{2}})$$

$$= \frac{\Delta t}{\Delta x}(f_L - f_R)$$

$$= 0$$

and the numerical solution becomes steady.

The condition that these signals will be so directed is that $\sigma_{s+\frac{1}{2}} < 0$, $\sigma_{s-\frac{1}{2}} > 0$, which may be interpreted graphically as in sketch (i). For convex f(u), such that, as assumed, $f'(U_R)<0$, $f'(U_L)>0$, the point M can lie anywhere between R and L.

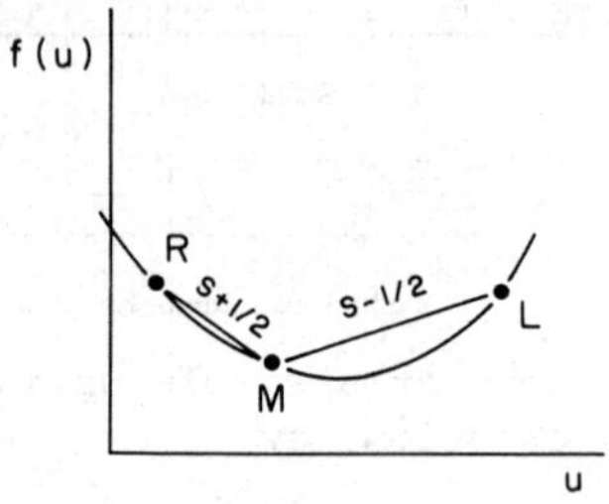

Sketch (i)

I would like to describe such a numerical model of a shockwave by a stronger term than "shock capturing". Perhaps we can say that the shock is "trapped". It generates signals, but they do not escape. Later we shall look at the internal structure of trapped shockwaves for systems of equations.

So far there is a very satisfying match between properties of the model, and the true situation. This match fails however in the remaining case, where σ changes from negative on the left to positive on the right (Fig. 1(c)), so that we attempt to simulate a rarefaction wave. The physics does not permit any boundary conditions to be imposed, and there is now a point which receives no signals. Again let that singular point be x_s. Numerical experiments now reveal the following typical behaviour. The value of u_s does not change (since it receives no signal). One of its neighbours ($u_{s-\sigma}$, where σ = sign $f'(u_s)$) approaches a value for which $f_{s-\sigma} = f_s$, and a steady state is reached consisting of two uniform regions, separated by an "expansion shock". That is, a discontinuity on either side of which the characteristics diverge rather than converging as they should.

Such expansion shocks are prevented if we include "artifical viscosity", i.e., if we have some smoothing operator which effectively links the two halves of the solution and brings them back into communication. However, such a remedy is not attractive if one of our objectives is to study real but small viscous effects. A better prospect, in my opinion, is to observe that expansion shocks cannot be represented with one point in the middle like the true shockwave can. The analogue of Sketch (i) is Sketch (ii), but now, for any M between L and R we have

$\sigma_{s-\frac{1}{2}} = -1$, $\sigma_{s+\frac{1}{2}} = +1$, and signals are generated which spread outward. One remedy is then to insert, inside that interval across which the wave speed changes sign, an extra grid point where the value of U is $U_M \varepsilon [U_L, U_R]$. Notice that it does not at first matter what value we give to U_M, or where we put it, because it will receive no signals itself anyway! On the other hand, an expansion shock will eventually reform with adjacent states M, M', such that $f(U_{M'}) = f(U_M)$, so actually the best value for U_M is the sonic value u_s. If we do this, then for expansion waves only, we will be using Osher's method which was described in the previous paper. Such a combination is easy to implement for scalar problems, but I do not yet see a really clean way to incorporate it into the rules of the game for systems.

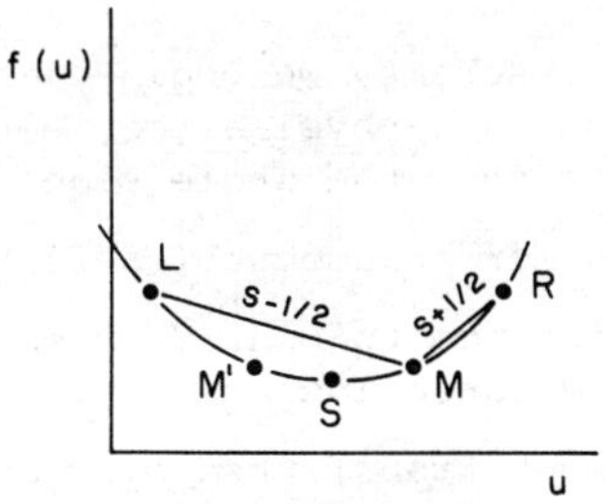

Sketch (ii)

3. HIGHER-ORDER SCHEMES AND MONOTONICITY

The scheme described in Section 2 was merely the simplest first-order version of a fluctuation-signal algorithm. A second-order version requires the incorporation of curvature effects, which I regard as the simplest case of sub-grid modelling, liable (as we saw in the introduction) to require either the carriage of additional information or else the introduction of artificial modelling rules. Here I choose the latter alternative, and use the fluctuation $\phi_{i+\frac{1}{2}}$ to send signals to both x_i and x_{i+1}. Only one of these is a realistic target. We will choose weights α, β such that $\alpha + \beta = 1$, and define the signals to be subtracted from u_i, u_{i+1} as

$$\delta u_i = (\alpha\Delta t/\Delta x)\phi_{i+\frac{1}{2}} \quad \delta u_{i+1} = (\beta\Delta t/\Delta x)\phi_{i+\frac{1}{2}}. \tag{3.1}$$

For the constant-coefficient case (F = au) it is easy to analyse the accuracy and stability of such a scheme by Taylor expansions and Fourier analysis. The results are summarised in Fig. 2, where the coefficient β of the "correct" signal is plotted against the Courant number, $\nu = a\Delta t/\Delta x$.

In that figure the values of β which gives rise to disguised versions of various well-known algorithms are plotted. I have also shown the region for which the scheme is stable, and the region for which the scheme is monotone. This second region follows from a lemma of Sweby and Baines (1981). The particular value of β which makes the scheme second-order accurate is

$$\beta = \tfrac{1}{2}(1 + \nu) \tag{3.2}$$

which reproduces the Lax-Wendroff scheme. This is of course not monotone, and leads to spurious oscillations near discontinuities.

To obtain a monotone scheme, we begin by thinking of the above process (3.1) as a non-physical correction to the first-order scheme. We first send the whole signal $(\Delta t/\Delta x)\phi_{i+\frac{1}{2}}$ to $x_{i+\frac{1}{2}+\frac{1}{2}\sigma}$, and then transfer an amount $(\frac{1}{2}\Delta t/\Delta x)(1 - |\nu|)\phi_{i+\frac{1}{2}}$ to the other target. Reference to Fig. 2 shows that to transfer any larger amount would cause instability. We could, if we wished, reduce the amount transferred by $O(\Delta x^2)$, without losing accuracy, because this would only effect the outcome at a given point by $O(\Delta x^3)$.

In fact this is exactly what we propose to do. A motivation for our rule is to remark that we are trying to explain events inside a region of (x,t) space given by $[x_i, x_{i+1}] \times [t^n, t^{n+1}]$, and at the moment are doing so purely by reference to the fluctuation $\phi_{i+\frac{1}{2}}$ which, so to speak, occupies that region at $t = t^n$. We could equally explain those events by relating them to the fluctuation $\phi_{i+\frac{1}{2}-\sigma}$ which is entering the region (Roe and Baines, 1981). We replace the Lax-Wendroff correction term

$$t_{LW} = \tfrac{1}{2}(\Delta t/\Delta x)(1 - |\nu|)\ \phi_{i+\frac{1}{2}} \tag{3.3}$$

by the modified correction term

$$t_m = \tfrac{1}{2}(\Delta t/\Delta x)(1 - |\nu|)\ B(\phi_{i+\frac{1}{2}}, \phi_{i+\frac{1}{2}-\sigma}) \tag{3.4}$$

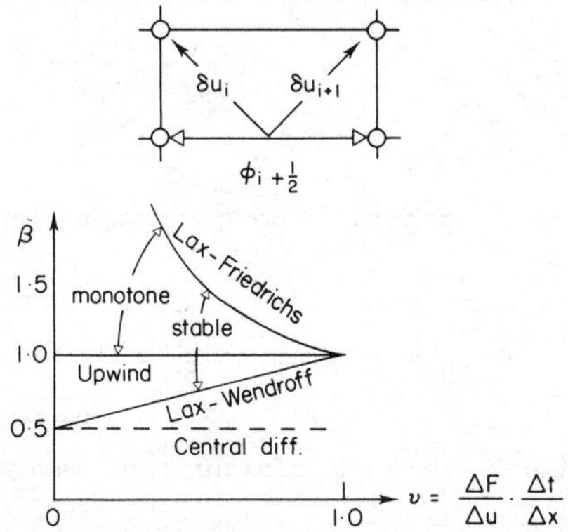

Fig. 2 Some properties of two-target algorithms.

where the function B is an averaging function which is biased toward the smaller of its arguments. For shock-like problems a good choice is to select as the value of B whichever argument has the smallest absolute value, but many other choices are possible. Evidently the modified scheme retains second-order accuracy in smooth flows provided

$$B(b_1,b_2) = \tfrac{1}{2}(b_1 + b_2) + O(b_1 - b_2). \qquad (3.5)$$

It can be shown that the scheme is monotone in the case F = au under the conditions

$$\sup[B/b_2] - \inf\,[B/b_1] < 2/|\nu| \qquad (3.6)$$

$$\sup[B/b_1] - \inf\,[B/b_2] < 2/(1-|\nu|) \qquad (3.7)$$

and this seems to be a good guide to the choice of B in more general situations.

Papers are in preparation which develop this strategy in more detail. The concept is broadly similar to the Flux Corrected Transport approach developed by Boris and co-workers (see Zalesak (1979)), but it is somewhat more compact in that the information used to improve the solution is made a little more selective, by starting with an upwinded algorithm. A conventional FCT algorithm would, in effect, compare $\phi_{i+\frac{1}{2}}$ with the fluctuations in both the neighbouring cells. There is, also, a very close relationship between the present work and that of van Leer (1979).

4. EXTENSION TO ONE-DIMENSIONAL SYSTEMS

We now consider the case where the functions $\underline{u}$ and $\underline{F}$ in

$$\underline{u}_t + \underline{F}(\underline{u})_x = 0 \tag{4.1}$$

are each vectors of length m. The corresponding numerical fluctuations

$$\underline{\phi}_{i+\frac{1}{2}} = \underline{F}(\underline{u}_{i+1}) - \underline{F}(\underline{u}_i) \tag{4.2}$$

are then also vectors, in an m-dimensional phase space.

The key to applying the ideas of the previous section to the vector case is to construct expansions of the form

$$\underline{\phi}_{i+\frac{1}{2}} = \sum_{k=1}^{k=m} \underline{\phi}^{(k)}_{i+\frac{1}{2}} \tag{4.3}$$

where the kth component describes that part of the fluctuation due to the kth wave system. An expansion of the form (4.3) may perhaps be called an explanation of a fluctuation. If all the waves happen to be travelling in the same direction (i.e., supersonic flow) we do not need to discriminate between them to create a first-order upwind scheme. The whole of the signal, $(\Delta t/\Delta x)\underline{\phi}_{i+\frac{1}{2}}$, is sent to the same target, which is either x_i or x_{i+1}. However, if some waves are travelling in each direction, then the signals due to each component must be separated out and sent to the appropriate target.

For one example of such an explanation, we resort for a moment to the linear system

$$\underline{u}_t + A\,\underline{u}_x = 0 \tag{4.4}$$

where A is a constant mxm matrix. We project $\underline{\phi}_{i+\frac{1}{2}}$ on to the eigenvectors of A. Let these eigenvectors be $\underline{e}_1, \underline{e}_2, \ldots \underline{e}_m$, and the corresponding eigenvalues $\lambda_1, \lambda_2, \ldots, \lambda_m$. The component of $\underline{\phi}_{i+\frac{1}{2}}$ due to the kth wave family is the projection $\underline{\phi}^{(k)}_{i+\frac{1}{2}}$ of $\underline{\phi}_{i+\frac{1}{2}}$ onto $\underline{e}_k$, and the speed of that wave is λ_k. The signals sent to left and right are, respectively,

$$\delta \underline{u}_i = \left(\frac{\Delta t}{\Delta x}\right) \sum_{\lambda_k < 0} \underline{\phi}^{(k)}_{i+\frac{1}{2}} \qquad \delta \underline{u}_{i+1} = \left(\frac{\Delta t}{\Delta x}\right) \sum_{\lambda_k > 0} \underline{\phi}^{(k)}_{i+\frac{1}{2}}.$$

Note that any component for which $\lambda_k = 0$ makes no contribution.

The question is whether such an explanation can be devised in the non-linear case also. In fact, many of the "physically motivated" finite-difference schemes, which involve solving Riemann problems, can be thought of as based upon explanations of this kind. The classical Godunov (1959) method is reformulated in this way below.

When two adjacent states $\underline{u}_i$, $\underline{u}_{i+1}$ are observed to be dissimilar, Godunov's method begins by making the rather crude assumption that the left half of the interval $[x_i, x_{i+1}]$ is wholly occupied by fluid in the state $\underline{u}_i$, and the right half by fluid in the state $\underline{u}_{i+1}$, so that a discontinuity exists at $x_{i+\frac{1}{2}} = \frac{1}{2}(x_i + x_{i+1})$. We then solve exactly the Riemann problem for this discontinuity and observe that the solution depends only on one similarity variable. It could be stated as $\underline{u}(s)$, where $s = (x - x_{i+\frac{1}{2}})/t$. Let $\underline{F}_{i+\frac{1}{2}} = \underline{F}(\underline{u}(0))$. Then the signals sent to left and right are

$$\delta \underline{u}_i = \left(\frac{\Delta t}{\Delta x}\right) (\underline{F}_{i+\frac{1}{2}} - \underline{F}_i) \qquad \delta \underline{u}_{i+1} = \left(\frac{\Delta t}{\Delta x}\right) (\underline{F}_{i+1} - \underline{F}_{i+\frac{1}{2}}). \qquad (4.5)$$

Now although this scheme is interesting, ingenious, and historically important it violates the golden rule (waste not, want not) advocated elsewhere at this conference by Brandt. The exact Riemann solution is an expensive and elaborate calculation; it is an extravagance to combine it with the crude assumption that the data are flattened into homogenised slabs.

However, there is no need to make the assumption of piecewise constant data. It is better to keep a completely open mind about the behaviour of the fluid between x_i and x_{i+1}. The fact that $\underline{u}_{i+1}$ is not equal to $\underline{u}_i$ is all we know, and all we need to know. As an explanation of what may be going on, (4.5) is as good as any, and better than most. For example, if $\underline{u}_i$, $\underline{u}_{i+1}$ straddle a shock, or lie within a simple wave, the explanation (4.5) will tell us so. As a first step in creating a high-order upwind scheme, (4.5) is a very adequate beginning.

The question may now suggest itself, are there any other explanations of $\underline{\phi}_{i+\frac{1}{2}}$, as powerful or almost as powerful as this but less expensive or more convenient to compute? Indeed there are many, and Osher's method, described in the previous paper, can also be regarded in this way (see Roe, 1982). My own preference in this matter is based on the discovery of matrices which represent a kind of large-scale Jacobian in the following sense.

Let (4.1) be written as

$$\underline{u}_t + A(\underline{u})\,\underline{u}_x = 0 \tag{4.6}$$

where $A(u) = \partial \underline{F}/\partial \underline{u}$ is the usual Jacobian matrix. In the non-linear case at least some of the elements $a_{ij} = \partial F_i/\partial u_j$ depend on $\underline{u}$. The eigenvalues and eigenvectors of $A(u)$ also depend on $\underline{u}$, but in analysing the small scale behaviour of smooth solutions to (4.6) they play the same role as their constant counterparts in the linear case. Now we make a further generalisation, introducing a matrix $\tilde{A}_{i+\frac{1}{2}}(u_i, u_{i+1})$ such that, for arbitrary u_i, u_{i+1},

$$\tilde{A}_{i+\frac{1}{2}}(\underline{u}_{i+1} - \underline{u}_i) = \underline{F}_{i+1} - \underline{F}_i \tag{4.7}$$

In addition we require that $\tilde{A}(u,u) = A(u)$, and that $\tilde{A}$ has a full set of real, linearly independent eigenvectors. Such matrices were demonstrated for the steady and unsteady polytropic Euler equations by Roe (1981b), and shown to exist for any set of conservation laws having an entropy function by Harten and Lax (Harten, 1981).

In general they are not at all unique, except for the special case $m = 1$, when $\tilde{A}_{i+\frac{1}{2}}$ reduces to the scalar quantity $(f_{i+1} - f_i)/(u_{i+1} - u_i)$ (cf. Section 2). Computational speed and convenience were the motivation for those matrices which I first put forward. Subsequently Colella (1982) has found an alternative version which is not limited to a polytropic equation of state.

It follows immediately from the definition of such a matrix that whenever the states u_i, u_{i+1} can be connected by a single shockwave, that is to say $(\underline{F}_{i+1} - \underline{F}_i) = S(\underline{u}_{i+1} - \underline{u}_i)$ for some scalar S equal to the shock speed, then $\underline{F}_{i+1} - \underline{F}_i$ is an eigenvector of $\tilde{A}_{i+\frac{1}{2}}$. The projection of $\underline{\phi}_{i+\frac{1}{2}}$ will be solely onto that eigenvector, and the "explanation" offered by the analysis will be correct. It also follows that if u_i, u_{i+1} are part of a simple wave, then the error involved in the explanation is at most $O(\Delta u^3)$, because the shock paths and wave paths through a given point in phase space differ only by such an amount (Courant and Friedrichs, 1948, p. 156).

A very great advantage of using such matrices to analyse the fluctuations is that they carry over into the non-linear case the highly developed language and technology of linear algebra, without in fact requiring that any expensive matrix arithmetic be performed. At least for the Euler equations, very simple closed form expressions are available for the projected fluctuations $\underline{\phi}^{(k)}_{i+\frac{1}{2}}$ and for the associated wavespeeds $a^{(k)} = \lambda_k$ (see Roe, 1981b)). These quantities can then be used in exactly the same way as their scalar counterparts to build up high-order monotonicity preserving schemes, provided only that due recognition is given to the direction $\sigma^{(k)} = \text{sign}\ (\lambda_k)$ with which each wave propagates inside a given interval. For an example based on a predictor-corrector scheme, see (Sells, 1980), or, for a condensed description of the general strategy, (Roe, 1981a).

One possible second-order scheme is to analyse each $\underline{\phi}_{i+\frac{1}{2}}$ into components $\underline{\phi}^{(k)}_{i+\frac{1}{2}}$, and then to send signals $(\Delta t/\Delta x)\underline{\phi}^{(k)}_{i+\frac{1}{2}}$ to $x_{i+\frac{1}{2}+\frac{1}{2}\sigma^{(k)}}$. Now we make the bounded backward transfer

$$\frac{1}{2}\frac{\Delta t}{\Delta x}\left(1 - \left|\nu^{(k)}_{i+\frac{1}{2}}\right|\right)\ B\left(\phi^{(k)}_{i+\frac{1}{2}}\ ,\ \phi^{(k)}_{i+\frac{1}{2}-\sigma^{(k)}}\right) \qquad (4.8)$$

where $\nu^{(k)}_{i+\frac{1}{2}} = (\Delta t/\Delta x)\ a^{(k)}_{i+\frac{1}{2}}$.

We choose $B(b_1, b_2) = 0$ if $b_1 b_2 < 0$, and whichever of b_1, b_2 has the smaller absolute value otherwise. In this form, the method produces the results shown in Fig. 3, for the standard shocktube problem studied by Sod (1978). CPU time on a DEC KL-10 computer was about half a microsecond per data point per time step.

There would still seem to be considerable latitude in the precise rules of the game. For example the $(1-|\nu|)$ factor in

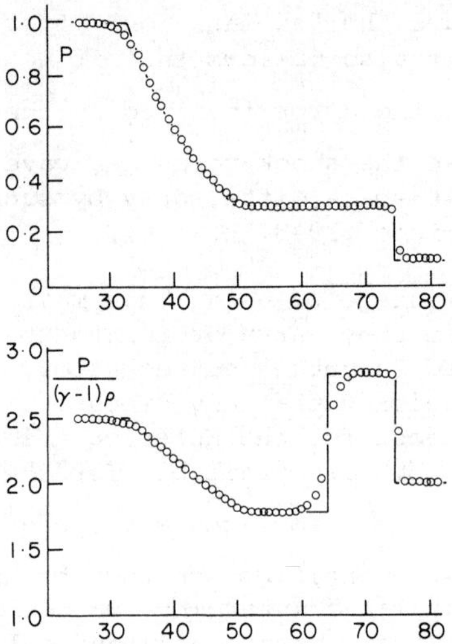

Fig. 3 Pressure and internal energy calculated by a second-order monotone algorithm for Sod's shock-tube problem.

(4.8) can be variously defined, or it can be removed whilst attaching two similar terms to the arguments of B. All these possibilities represent legitimate generalisations of the linear case. We might choose those alternatives which are easiest to program, or we might hope to find choices for which convergence of the method can be proved under the widest circumstances. With luck these will not turn out to be too different.

5. STRUCTURE OF TRAPPED SHOCKWAVES

In this section I would like to consider an interesting and important point of detail that seems to be dealt with rather naturally by the fluctuation-signal approach. In Section 2, and particularly with reference to Fig. 1b, we showed the existence of stationary "trapped" shockwaves, which created a pair of non-zero, but self-cancelling fluctuations (Eqn. 2.4). We now ask whether such shock trapping behaviour is possible for systems of equations, and the answer is affirmative. Osher's scheme, for example, traps shockwaves inside three intervals, see (Osher and Solomon, 1982). Glimm's scheme, as described in these proceedings by Concus, apparently traps shockwaves inside one interval, but because of the random element in its construction never reaches a truly steady state. One could say that the shock remains inside a trap, but that the trap itself undertakes a random walk over the grid (which after a sufficiently long time will take it indefinitely far from the proper place).

Actually, it should not be possible to trap shocks inside one interval, for the following reason. Consider compressible flow through a one-dimensional nozzle with exactly specified initial data and properly posed boundary conditions. Then at large time the exact solution will reveal a unique position for the shockwave (if it exists) which will not generally be at a point represented on the numerical grid. To indicate a shockwave displaced in this way we need an intermediate point (see Sketch (iii)). The variation of λ between 0 and 1 would show the shock location shifting between mid-points of consecutive intervals.

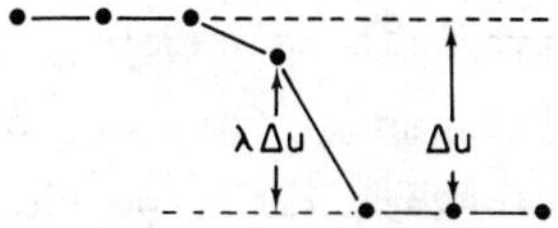

Sketch (iii)

So we see that a trap consisting of two intervals is actually the most compact possible, and it turns out that the existence of such traps is ensured by the matrices $\tilde{A}$ $(\underline{u}_i, \underline{u}_{i+1})$ described in the last section. A full proof and discussion will be published elswhere, but we give a short account below.

As in the scalar case, we can consider a problem in which the initial data are arbitrary, but on the left boundary we prescribe $\underline{u}_1^n = \underline{u}_L$, and on the right boundary $\underline{u}_{imax}^n = \underline{u}_R$, for all n. The governing conservation laws are again

$$\underline{u}_t + \underline{F}(\underline{u})_x = 0 \tag{5.1}$$

and if a steady shock solution exists then

$$\underline{F}(\underline{u}_R) = \underline{F}(\underline{u}_L). \tag{5.2}$$

If, in addition, the shock is to be a physically admissible shock, the boundary data must satisfy the evolutionary condition. This states that for precisely one integer k between 1 and m, the wave speed $a^{(k)}$ changes from positive to negative across the shockwave, i.e., $a^{(k)}(\underline{u}_L)>0$, $a^{(k)}(\underline{u}_R)<0$.

This condition is thought to select the physically admissible solutions because, as shown by Jeffrey (1976), it is the necessary and sufficient condition to make the solution of the differential equations stable against small disturbances. Now we hypothesise that a numerical solution exists in the form shown in Fig. 4, where $\underline{u}_M$ is a member of a one-parameter family of states whose end-points are $\underline{u}_L$ and $\underline{u}_R$. The situation we would like to have with regard to the average wavespeeds inside the intervals $(s \pm \frac{1}{2})$ is as shown. Inside $(s - \frac{1}{2})$ we would like $a^{(k)}_{s-\frac{1}{2}}$ to be positive. It is certainly positive when $\underline{u}_M = \underline{u}_L$, and because the matrix $\tilde{A}$ correctly recognises fluctuations due to shockwaves, it will be zero when $\underline{u}_M = \underline{u}_R$, so we will assume for the moment that it is never negative. Similarly, inside $(s + \frac{1}{2})$ we would like $a^{(k)}_{s+\frac{1}{2}}$ to be negative. It is negative when $\underline{u}_M = \underline{u}_R$ and zero when $\underline{u}_M = \underline{u}_L$, so we will assume that it is never positive. If these assumptions are in fact true, we have only to ensure that all those signals which might be directed away from the shock are actually zero. There are $(k - 1)$ of those in $(s - \frac{1}{2})$, and $(m - k)$ of them in $(s + \frac{1}{2})$, making a total of $(m - 1)$ constraints to be applied to the m-vector $\underline{u}_M$. So $\underline{u}_M$ appears to have just one degree of freedom, which is what we want. The proof can be made rigorous "in the small", i.e., if $\underline{u}_L$, $\underline{u}_R$ are sufficiently close. However, it can also be made rigorous "in the large" in the important special case that the shock is a 1-shock or an m-shock and that the difference scheme correctly recognises fluctuations due to single shockwaves. The situation then is that we can actually exhibit the set of states $\underline{u}_M$, and that the conjecture about the behaviour of $a^{(k)}$ can be verified by a simple direct calculation.

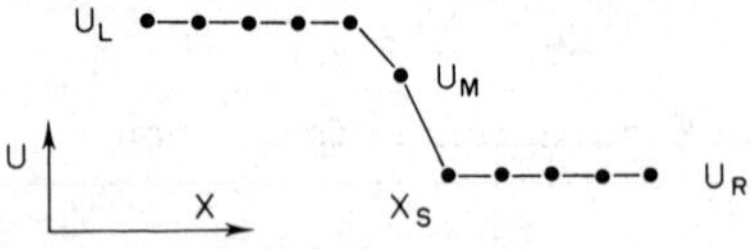

(a) Hypothetical steady solution

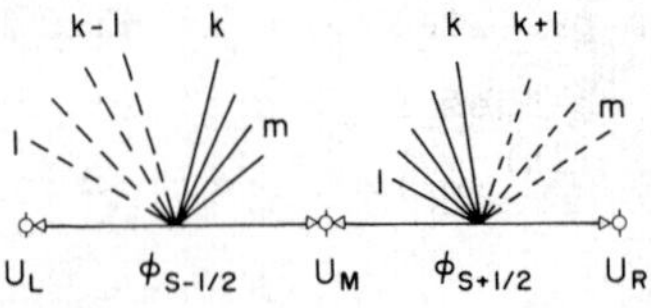

(b) Pattern of signals which must vanish near a trapped shock

Fig. 4 Behaviour of signals due to a k-shock.

Without loss of generality we consider a 1-shock, because the result for an m-shock follows by symmetry. In this case, all the waves in $(s + \frac{1}{2})$ except for the 1-wave must vanish. Therefore $\underline{\phi}_{s+\frac{1}{2}}$ must project wholly on to the first eigenvector $\tilde{A}_{s+\frac{1}{2}}$, that is

$$\underline{u}_M - \underline{u}_R = S(\underline{F}_M - \underline{F}_R) \tag{5.3}$$

which implies that $\underline{u}_M$ lies on a Hugoniot curve through $\underline{u}_R$. This Hugoniot curve represents the set of states from which it is possible to arrive at $\underline{u}_R$ by means of a 1-shock. This set certainly includes $\underline{u}_L$, and so (5.3) is an equation for $\underline{u}_M$ representing a path from $\underline{u}_L$ to $\underline{u}_R$. The sole remaining conjecture that $a^{(1)}_{s-\frac{1}{2}}$ is always positive can now be verified numerically.

A subtlety in the proof is that there are two 1-shock Hugoniot curves connecting $\underline{u}_L$ to $\underline{u}_R$. The other one represents the set of states that can be reached by setting out from $\underline{u}_L$, but it does not serve our purpose, because it generates other waves in $(s + \frac{1}{2})$.

Using these results we can now investigate the structure of trapped shockwaves for, say, the unsteady Euler equations. We show in Fig. 5 some properties of the state $\underline{u}_M$ for the case of a polytropic gas $(\gamma = 1.4)$ when the onset Mach number is 2.0. The pressure ratio across the shock is then $p_R/p_L = 4.5$, and we have chosen the pressure in the middle (p_M/p_L) to be the parameter which characterises the state $\underline{u}_M$. Fig. 5a shows the value of the mass flow ratio, and shows that it is very significantly different from its "real" value of 1.0. The trapped shockwave behaves like a doublet, with mass being created and destroyed by equal amounts inside $(s - \frac{1}{2})$, $(s + \frac{1}{2})$. Similar "doublet" behaviour is found for the other fluxes, $(p + \rho u^2)$ and $u(e + p)$. Fig. 5b shows a measure of the entropy (p/ρ^γ), which increases monotonically with shock strength. To make this latter point, we could simply have appealed to known properties of the Hugoniot curves. Not only entropy, but also pressure, density, and temperature are known to vary monotonically along them (Courant and Friedrichs, 1948, p. 148). However, it has been found in practice that when a scheme with these properties is used to compute two-dimensional flows, by the techniques discussed in the next section, spikes of temperature or entropy may occur inside the shock. There seems to be much more to two-dimensional shock-capturing than is revealed by one-dimensional theory.

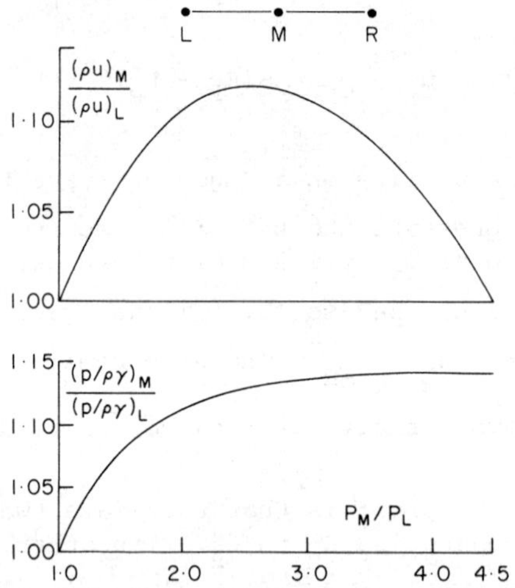

Fig. 5 Values of mass flux and entropy at the central point of a trapped 1-shock. Onset Mach number = 2.0.

6. MULTIDIMENSIONAL PROBLEMS

It would be generally agreed that problems which evolve in more than one dimension are very much harder to solve than those which evolve in one dimension only. And yet the great majority of real problems are in fact multidimensional. The many different methods which are employed in practice may be broadly categorised into the ones which are "genuinely multidimensional", and the ones which are to a greater or lesser extent based on "dimensional splitting". We need to keep a careful distinction between methods which employ one-dimensional operations and those which are based on one-dimensional thinking. Any numerical problem which involves manipulating a multidimensional array is obviously dealt with most conveniently row by row. This is especially true if we wish to make efficient use of vector processing computers. And yet this need not imply that we have a one-dimensional conceptual model of the problem. For example, if we solve Laplace's equation with the usual five-point finite-difference approximation, line relaxation will not give us a significantly different final answer from point relaxation.

Where there is a one-dimensional conceptual model, it is usually based on some such idea as this. Suppose we have a two-dimensional conservation law, written as

$$\underline{u}_t + \underline{F}(\underline{u})_x + \underline{G}(\underline{u})_y = 0. \tag{6.1}$$

We now imagine time divided into equal intervals of duration Δt, with these intervals numbered and then labelled as odd or even. Consider a new set of unknowns $\underline{v}(x, y, t)$, which during the odd intervals obey

$$\underline{v}_t + 2\underline{F}(\underline{v})_x = 0 \tag{6.2}$$

and during the even intervals obey

$$\underline{v}_t + 2\underline{G}(\underline{v})_y = 0. \tag{6.3}$$

It is shown by Yanenko (1971) that if $\underline{u}$, $\underline{v}$ have the same initial data, then the solution $\underline{v}$ of (6.2) and (6.3) tends almost everywhere to the solution $\underline{u}$ of (6.1) as Δt tends to zero, the error $\underline{v} - \underline{u}$ being $O(\Delta t)$. A lot of the enthusiasm that has been shown for operator splitting of this kind can be traced back to early theoretical studies which indicate that the efficiency of split methods was considerably higher than that of non-split methods, in terms of time advanced per unit of computational work (Strang, 1968; MacCormack and Paullay, 1972). Secondly, when attempting to construct schemes with superior shock-capturing properties, the most reliable one-dimensional algorithms have exploited the one-dimensional theory of characteristics, which is very tidy and fully developed. However, the multidimensional characteristic theory does not have nearly so clear an interpretation. There is, therefore, a great temptation to put together a good set of one-dimensional operators, and to hope that they will somehow provide a good representation of the multidimensional physics. If the operators involved are merely first-order accurate, then this hope does seem to be realised (see, e.g., the previous paper by Osher) but the attempt to generate higher-order algorithms is fraught with many difficulties.

It would be beyond the scope of this paper to discuss at greater length the general merits of split or non-split methods. We merely treat the much more modest question of whether the fluctuation-signal concept has anything relevant to offer. Certainly the problem can be formulated in this way, but to arrive at a complete strategy we need to solve three sub-problems, each of them more difficult than its one-dimensional version:

(i) How do we identify a fluctuation?

(ii) How do we explain it as a set of physical or quasi-physical events?

(iii) What signals do we allow it to generate?

With regard to (i), at least two alternatives have been tried, which we illustrate in Fig. 6. Both are inspired by the finite volume approach, which makes it rather easy to define and generate conservative algorithms. In Fig. 6(a) values of $\underline{u}$ have been associated with the centres of quadrilateral cells whose indices are (i, j), and (i+1, j). We will associate the concept of fluctuation with the interfaces between cells, in this case

$$\underline{\phi}_{i+\frac{1}{2},j} = (\underline{F}_{i+1,j} - \underline{F}_{i,j})\,(y_P - y_Q) - (\underline{G}_{i+1,j} - \underline{G}_{i,j})\,(x_P - x_Q) \tag{6.4}$$

which represents the exchange of conserved quantities between the adjacent cells, per unit time.

This definition is rather close to operator splitting. Even if the flow is close to equilibrium, the fluctuations, so defined, may not be small because they must be balanced against exchanges which occur in other directions; this sounds like an undesirable property. On the other hand, it is easy to set up an "explanation" for (6.4) in terms of wave motions normal to PQ. The matrix which maps $(\underline{u}_{i+1,j} - \underline{u}_{i,j})$ onto $\underline{\phi}_{i+\frac{1}{2},j}$ can be constructed and used exactly as in the one-dimensional case (see Sells (1980) for details). Signals obtained in the form $\Delta t\underline{\phi}_{i+\frac{1}{2},j}$ have the physical dimensions of the conserved quantities and can be thought of as incrementing the mass, momentum, or energy in the cells to which they are sent. When we add up the total changes of mass (say) over the whole flow field, the cell (i,j) contributes to four such changes, and its total contribution, from (6.4) is

$$-\underline{F}_{i,j}\,[(y_P-y_Q) + (y_Q-y_R) + (y_R-y_S) + (y_S-y_P)] + \underline{G}_{i,j}\,[(x_P-x_Q) + (x_Q-x_R) + (x_R-x_S) + (x_S-x_P)] \tag{6.5}$$

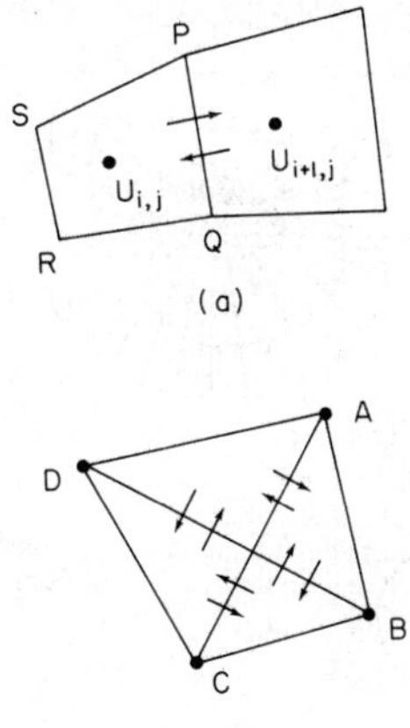

Fig. 6 Illustrating two possible definitions of fluctuation for a two-dimensional unsteady flow.

which is zero, however those changes are distributed. Once a signal is assigned to a particular cell, however, it should be converted back into a density change by dividing by the cell area.

Note that this strategy has bypassed several awkward details. Although we are dealing with a quite general mesh, there has been no need to reformulate the equations, or to get involved with the metric coefficients of a coordinate transformation. Also we have avoided a defect of several early finite-volume codes, that they failed to preserve a steady uniform flow if it were given as data on a non-uniform grid. This problem was perhaps first publicised by Thomas and Lombard (1978), but the solution proposed there would only work for rather simple differencing schemes. With the fluctuation-signal approach, the fluctuations are, by definition, zero in a uniform flow, and so the problem cannot arise.

Sells (1980) has written a two-dimensional aerofoil program based upon the definition of a fluctuation in (6.4). This program has been used within the Aerodynamics Department of RAE to calibrate computations based upon less accurate models of transonic flow. Its main deficiencies are that it is slow to converge, and that shockwaves which lie obliquely across the mesh are not always captured very compactly. Part of the computing grid for a NACA 0012 aerofoil is shown in Fig. 7a, and the pressure contours are presented in Fig. 7(b) and (c) for a subsonic and supersonic free stream. Note that the upper surface shock in Fig. 7(b), and the trailing edge shock in Fig. 7(c) are

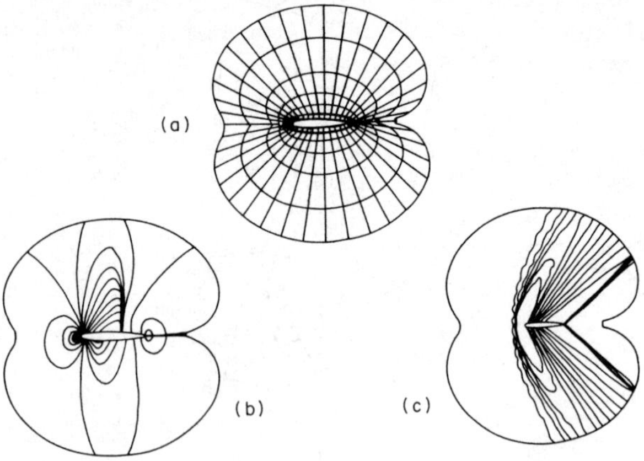

Fig. 7 Computations of transonic flow past a NACA 0012 aerofoil by the Sells program. (a) Part of computing grid. (b) Subsonic free stream; pressure contours at $M_\infty = 0.80$, $\alpha = 1.25^\circ$. (c) Supersonic free stream, density contours at $M_\infty = 1.30$, $\alpha = 0$.

each very cleanly captured (in fact nearly as well as the optimally trapped one-dimensional shocks discussed in the previous section). However, the bow shock in Fig. 7(c) becomes rather diffuse at large distances, and it is felt that this may result from the rather one-dimensional nature of the definition in (6.4), which makes it inappropriate when the shock is actually oblique to the grid.

To obtain a more "genuinely two-dimensional" definition we return to Fig. 6(b), wherein values of $\underline{u}$ have been associated with the vertices of a quadrilateral, although we could equally well associate them with the mid-points of four adjacent cells. Then we define the fluctuation within that quadrilateral to be

$$\underline{\phi}_{i+\frac{1}{2},j+\frac{1}{2}} = \tfrac{1}{2}\Big[(\underline{F}_B - \underline{F}_D)(y_A - y_C) - (\underline{F}_A - \underline{F}_C)(y_B - y_D) - (\underline{G}_B - \underline{G}_D)(x_A - x_C) + (\underline{G}_A - \underline{G}_C)(x_B - x_D)\Big]. \tag{6.6}$$

Observe that this is the sum of two terms such as (6.4), one from each diagonal of the quadrilateral. The quadrilateral will be in equilibrium if these diagonal terms cancel each other. To complete the definition of an algorithm, we specify four weights $\alpha, \beta, \gamma, \delta$, such that $\alpha+\beta+\gamma+\delta = 1$, and then increase the conserved quantities (not their densities) residing at point A by $\alpha\Delta t\phi$, at B by $\beta\Delta t\phi$, and so on. The question now arises how to choose $\alpha,\beta,\gamma,\delta$.

Two authors who base their methods on (6.6) have taken rather different approaches. Ni (1981) chooses weights which ensure second-order accuracy in smooth regions of the flow, and which effectively reproduce the two-dimensional Lax-Wendroff scheme, just as (3.2) reproduces the one-dimensional Lax-Wendroff scheme. His scheme therefore makes little attempt to model the wave behaviour, and seems to rely on heavy numerical damping to stablise the flow near shockwaves. On the other hand, he claims that the fluctuation-signal formulation leads very naturally to a rather novel type of multi-grid algorithm, for which he quotes very impressive convergence rates. However, several practical details are omitted from the account given.

Denton (1982) also takes (6.6) as a starting point. He then declares that his only interest will be in obtaining steady-state solutions to (6.1) and that to achieve this with second-order accuracy he wishes merely to make (6.6) small almost everywhere (he recognises that this cannot be done close to shockwaves). The weights $\alpha,\beta,\gamma,\delta$ are then thought of as parameters available to stablise a kind of local relaxation process. They are determined as the calculation proceeds by a dynamic feedback technique, established for the Euler equations by numerical experiment. In its present form, Denton's scheme is fast and has been widely used for turbomachinery calculations, but appears to lose accuracy in regions where flow direction changes rapidly.

Baines (1980, 1981) has studied the two- and three-dimensional versions of the fluctuation-signal schemes from a mathematical point of view, chiefly applied to the linear scalar case $F = au$, $G = bu$. He finds that if he starts with (6.6), then the condition that the whole algorithm be second-order accurate in both space and time places only three constraints on the four weights $\alpha,\beta,\gamma,\delta$. The degree of freedom which remains can be used to influence the stability of the algorithm, and to impart a directional bias to it. However, in order to achieve a property analogous to monotonicity, Baines finds it necessary to modify the definition (6.6). For more details, the reader is referred to the quoted papers, or to Roe and Baines (1982).

Clearly, the fluctuation-signal approach does not yet solve all the problems associated with multidimensional flows, but then neither does any other method. A lot of work remains to be done at a very fundamental level.

7. EVOLUTION ON AN IRREGULAR GRID

Several contributors to these proceedings have mentioned the difficulties created by irregular computing grids. Such grids arise in practice for two different reasons. First they arise from a wish to pack the grid more densely into eventful regions of the flow, such as shockwaves, boundary layers, or geometrical singularities. In such cases there will be an attempt to match the grid in some suitable way to anticipated features of the solution. However, it is seldom easy to insert extra grid lines only into the regions where they are needed. They tend to "overflow" into other parts of the grid, where they are a liability rather than an asset. If they are cut off, then that also creates difficulties. Mostly the problem is studied by trying to identify what constitutes a "good" grid, and then trusting to the ingenuity of the program user to provide a good grid for whatever practical problem is to be solved. However, it would in the long run be much more convenient if algorithms were available which gave good solutions even on poor grids.

I am now going to give a preliminary account of a method which has just this property. It has been developed to the point where it can be applied to one-dimensional systems, and it seems that useful, simple, lessons can be learnt from it. It could probably be formulated without using the fluctuation-signal concept, but there do seem to be at least two reasons why that approach is helpful. The first has already been discussed in the section on multidimensional problems; it is the fact that fluctuations arise from changes of densities, but signals produce a redistribution of mass. This makes it easy to ensure that the conservation principle and the principle of a null response to uniform flow are both built into the method. The second is that the ability to take account of flow direction is helpful in keeping the algorithm compact. This compactness property seems particularly important on irregular grids. I am indebted to Dr. D.E. Davies of RAE for the observation that the future effect of a wave should not depend strongly, in any sensible scheme, on the geometry of the region through which it has just passed.

The problem was formulated and studied by Sweby (1981), and subsequently refined by Pike (RAE, to be published) to whom the numerical results shown later are also due. The few details set out below are intended only to illustrate the general approach, and are given for a linear scalar conservation law

$$u_t + au_x = 0 \qquad (a>0). \tag{7.1}$$

The generalisation to non-linear systems follows the lines of Section 4.

Consider the real line divided into finite non-overlapping segments by grid points x_i. Let the length of a typical segment be

$$\Delta x_{i+\frac{1}{2}} = x_{i+1} - x_i. \tag{7.2}$$

Let a fluctuation $\phi_{i+\frac{1}{2}} = a(u_{i+1} - u_i)$ generate three signals $(\alpha_{i+\frac{1}{2}}, \beta_{i+\frac{1}{2}}, \gamma_{i+\frac{1}{2}})(\Delta t\, \phi_{i+\frac{1}{2}})$, which increment the conserved quantity u at x_i, x_{i+1}, x_{i+2}, respectively. The reason why we need as many as three signals will become apparent; meanwhile, the targets have been selected so as to bias the scheme towards transferring information in the proper direction (we have assumed $a>0$).

The statement of the conservation principle is extremely simple; it is

$$\alpha_{i+\frac{1}{2}} + \beta_{i+\frac{1}{2}} + \gamma_{i+\frac{1}{2}} = 1 \tag{7.3}$$

for all i. Now we have to define accuracy. If u has the dimensions of density, then the total mass increment at the point i is

$$\Delta t[\alpha_{i+\frac{1}{2}}\, \phi_{i+\frac{1}{2}} + \beta_{i-\frac{1}{2}}\, \phi_{i-\frac{1}{2}} + \gamma_{i-3/2}\, \phi_{i-3/2}]. \tag{7.4}$$

We will <u>define</u> the increase in density at the point i to be (7.4) divided by $\frac{1}{2}(\Delta x_{i-\frac{1}{2}} + \Delta x_{i+\frac{1}{2}})$, and we will <u>define</u> the method to be first-order accurate if we recover from this process the exact increase in density whenever the data values of density are linear in x (NB <u>not</u> when they are linear in i). The condition for the method to be first-order accurate in this sense comes out as

$$\alpha_{i+\frac{1}{2}}\, \Delta x_{i+\frac{1}{2}} + \beta_{i-\frac{1}{2}}\, \Delta x_{i-\frac{1}{2}} + \gamma_{i-3/2}\, \Delta x_{i-3/2} = \tfrac{1}{2}[\Delta x_{i+\frac{1}{2}} + \Delta x_{i-\frac{1}{2}}]. \tag{7.5}$$

Now we get to the real point of the analysis. If all the Δx are equal, Eqn. (7.5) reduces to

$$\alpha_{i+\frac{1}{2}} + \beta_{i-\frac{1}{2}} + \gamma_{i-3/2} = 1 \tag{7.6}$$

and if we further choose (as we may) to make α,β,γ independent of i, then (7.3) and (7.6) are identical. However, if the Δx are very different, there seem to be no sensible definitions of accuracy which make such an identification possible. To find the weights α,β,γ we must solve (7.3) and (7.5). If we wish, we can get another condition by insisting that (7.4) produces the correct answer whenever the data values are quadratic in x, and we then have a sufficient system of equations to solve for the α,β and γ. There are still a number of details to be settled, largely to do with keeping the algorithm stable, and restoring monotonicity to the second-order method, but these have all been overcome, and it is hoped to publish a complete account of them before long.

Some numerical results for the same shocktube problem treated in Section 4 are given in Fig. 8. Consecutive segments of the computing mesh were taken as having lengths (h,h,3h,3h,h,h,3h,3h,...) as compared with (2h,2h,2h,2h,...) in Fig. 3. In Fig. 8(a) the only equation which has been satisfied is (7.3) (by taking $(\alpha,\beta,\gamma) = (0,1,0)$ everywhere) so that the algorithm is conservative, acknowledges wave directions, and leaves uniform flow untouched, but otherwise has no constraints placed upon its accuracy. In Fig. 8(b), both (7.3) and (7.6) have been satisfied, taking $\gamma_{i+\frac{1}{2}} \equiv 0$. In Fig. 8(c), the condition for second-order accuracy has also been satisfied. At this stage, as with a regular grid, the algorithm fails to preserve monotonicity, and so a strategy similar to that of Section 4 is followed. The basic algorithm is modified, however, only if the flow is varying rapidly, not if it is just the grid that is varying rapidly. In addition to these results, Pike has studied the same test problem, and others, on various near-pathological grids. He has included the canonical case of piecewise uniform grids, and also grids in which the interval lengths are chosen at random. From these studies which confirm and extend the earlier results of Sweby (1981) the following conclusions can be drawn.

(i) None of the methods (zeroth, first, or second-order) suffers significantly from problems with wave reflection at grid irregularities.

(ii) The resolution of contact discontinuities is set by the coarsest part of the grid through which the contact has passed.

(iii) The resolution of shockwaves is set by the local grid size at the present or very recent position of the shock. What happens is that shocks diffuse on passing through a coarse mesh, but reform if the grid is subsequently refined. The over-all spread is typically two or three local mesh widths, for strong, but moving, shocks. Very weak shocks behave more like contacts.

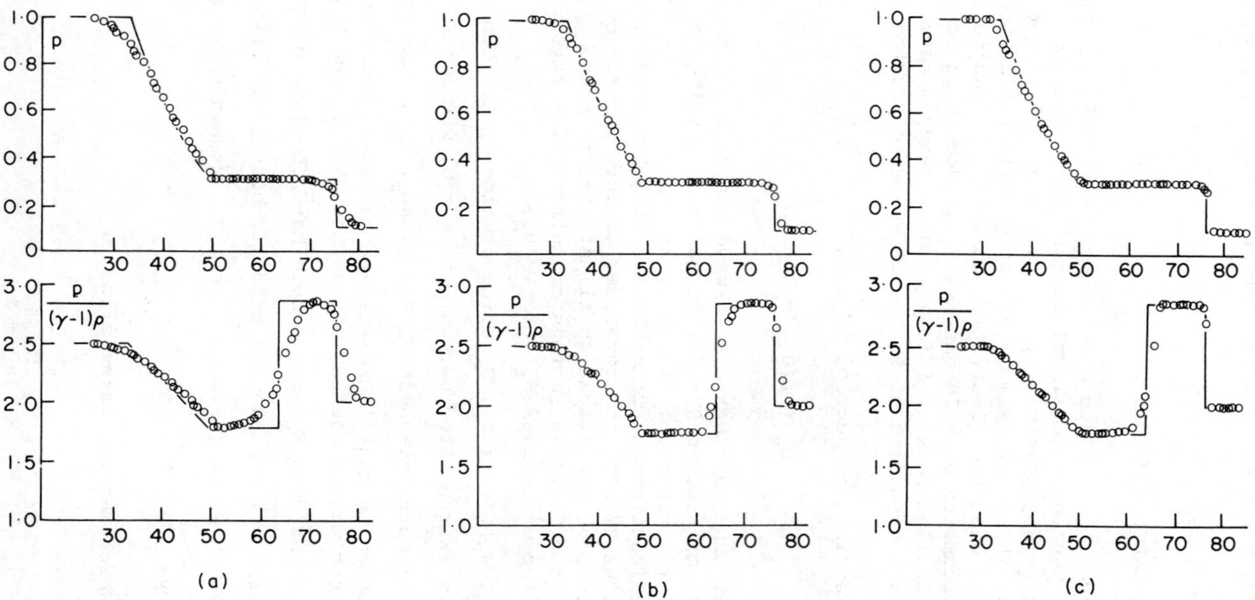

Fig. 8 Pressure and internal energy calculated by Pike for the same problem as in Fig. 3, but on an irregular grid. (a) Zeroth order algorithm; (b) First-order algorithm; (c) Second-order algorithm.

(iv) Apart from (ii) above, the results that can be obtained by transmitting three signals on a highly irregular grid are very nearly as good as those that are obtained by transmitting two signals on a regular grid.

8. LEVEQUE'S LARGE TIME STEPS

In this Section I want to discuss some work which at present is limited to the one-dimensional scalar case. I am not sure whether there is any possibility of its being applied more generally, but it is both simple and spectacular, and a remarkable example of what the fluctuation concept can lead to under favourable circumstances. In particular it demonstrates the carefully controlled loss of information, already observed in Sections 2 and 5 as a part of successful shock capturing.

Our equation in this section is then

$$u_t + f(u)_x = 0.$$

Leveque (1982) observes that a shockwave can be described by an ordered triple of numbers $S = (x, \Delta u, a)$ representing, respectively, its position, strength, and speed. In general all three will be functions of time; I have slightly amended his notation to conform with the remainder of this paper. He then goes on to consider the situation illustrated in Sketch (iv), where at time $t = 0$, three regions of uniform flow exist, separated by two shockwaves $S_A = (x_A, \Delta u_A, a_A)$ $S_B = (X_B, \Delta u_B, a_B)$. If S_A, S_B merge at time t_M to form a shock S_M, then S_M is extrapolated backward to $t = 0$, where its representation would have been $S_M = (x_M, \Delta u_M, a_M)$. The point of this operation is that if we wish to calculate the flow for $t>t_M$ it does not matter whether we use the data S_A and S_B, or whether we use the data S_M, since both have the same solution. In fact it turns out that we are actually better off using S_M, for reasons which we will discuss shortly. Meanwhile we state

LEVEQUE'S LEMMA

The time at which merging takes place is

$$t_M = \frac{x_B - x_A}{a_A - a_B} \tag{8.1}$$

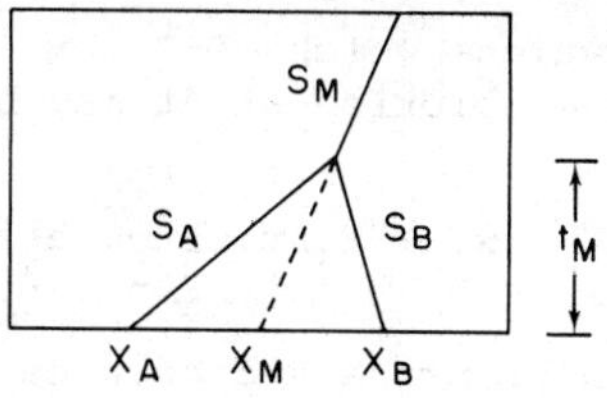

Sketch (iv)

and the rules for obtaining S_M from S_A, S_B are

$$\Delta u_M = \Delta u_A + \Delta u_B \tag{8.2a}$$

$$x_M = \frac{\Delta u_A x_A + \Delta u_B x_B}{\Delta u_A + \Delta u_B} \tag{8.2b}$$

$$a_M = \frac{\Delta u_A a_A + \Delta u_B a_B}{\Delta u_A + \Delta u_B} \tag{8.2c}$$

Although the proof follows at once from conservation of $\int u dx$, the simplicity of these results is very striking, as is their apparent lack of dependence on the flux function f(u). In fact f(u) does appear in disguised form, since $a = \Delta f/\Delta u$, but the form involving a is more elegant. Another striking observation is that the rules obey a mechanical analogy. Let S be represented by a particle of mass Δu placed in a cartesian plane coordinate system at the point (x,a). The rule for merging two shocks is exactly the same as the rule for combining two masses by moving them to their centre of gravity. By now it has become obvious that the outcome of merging several shocks will not depend on the order in which the mergers are calculated.

Based on this lemma, Leveque devises the following algorithm which allows the solution to be advanced by apparently unlimited Δt. All fluctuations present in the initial data are represented in the "ordered triple" format. Those fluctuations which correspond to expansions are broken up into a number of smaller, coincident fluctuations. Then the fluctuations are examined a pair at a time to determine which of them are destined to merge within the time step Δt. The outcome is to replace the original data set by a reduced set

$$S_j^O = (x_j, \Delta u_j, a_j) \quad (j = 1,2,\ldots, j_{max}) \tag{8.3}$$

consisting of fluctuations which will not interact before $t = \Delta t$. Therefore the solution at Δt may be denoted by

$$S_j^1 = (x_j + a_j \Delta t,\ \Delta u_j,\ a_j) \tag{8.4}$$

and this solution can be projected back on to a conventional representation.

The nominal accuracy of the method is only first-order, but in fact the results are amazingly good, and get better as Δt increases. Fig. 9a shows some results for Burger's equation $f = \frac{1}{2}u^2$, and Fig. 9b shows results for the non-convex Buckley-Leverett equation $f = u^2/(u^2 + \frac{1}{2}(1 - u)^2)$.

When such a simple method produces such surprising results, it is time for us to revise our intuitions, and by now I have thought about Leveque's results for long enough that I no longer find them surprising. In a conventional method, if we advance to $t = k(\delta t)$ by k steps of (δt) we have to perform a set of interpolation operations on k separate occasions, each of which involves a loss of information. This happens in a rather unstructured way which has as much to do with the grid on to which the solution is being interpolated, as with the actual behaviour of the solution. Leveque's merging operations, which allow him to advance through the same total time in just one step, also lose information, but they do so in a way which models very closely the way that information is lost in the exact solution, and makes the bare minimum of reference to the underlying grid. From this viewpoint the method and the results come to seem quite natural.

9. AN ADAPTIVE LEAPFROG METHOD

The last idea I want to discuss is also rather novel, and has not yet been tried out on any practical problem. However, like the previous idea, it has some fascinating properties, and may serve at least to extend our feeling for what is possible in numerical schemes. It will be described in the context of a non-linear scalar conservation law.

$$u_t + f(u)_x = 0. \tag{9.1}$$

Consider the (x, t) space tessellated as in Fig. 10; the central unit of the scheme is the typical quadrilateral shown in Fig. 10, with its vertices labelled to indicate left, right, past and future states. We will rewrite (9.1) in the integral form

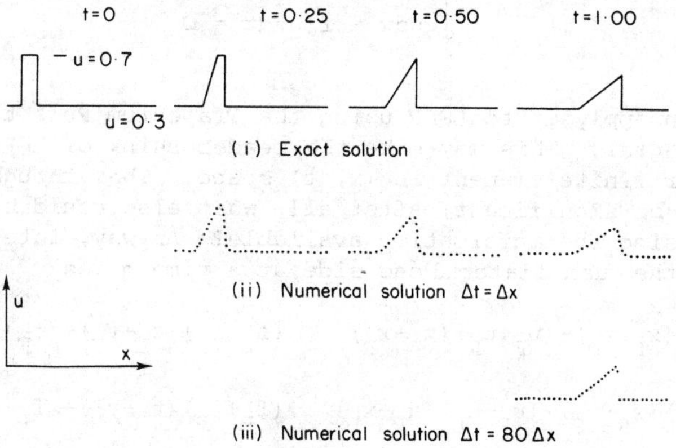

(a)

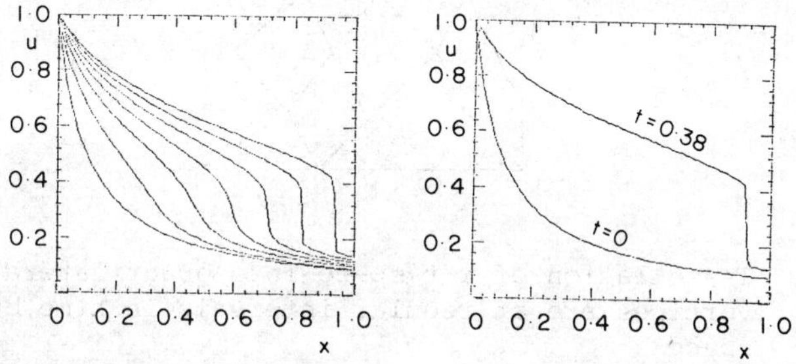

(b)

Fig. 9 Solutions to scalar test problems calculated by Leveque.
(a) Burger's equation at Courant numbers of 0.7 and 56.
(b) Buckley-Leverett equation at Courant numbers of 6.3 and 37.8.

$$\oint udx - f(u)dt = 0 \tag{9.2}$$

and then apply it to LFRP using the Trapezium Rule to evaluate the integral. This may make the reader think of LFRP as a bilinear finite element in (x, t) space. That thought may or may not be significant; after all, what else could be done with (9.2) using the information available? Anyway, integrating around the quadrilateral one side at a time gives

$$\begin{aligned}&(u_L+u_F)(x_L-x_F)+(u_F+u_R)(x_F-x_R) \qquad (f_L+f_F)(t_L-t_F)+(f_F+f_R)(t_F-t_R)\\ &\qquad\qquad\qquad = \\ &+(u_R+u_P)(x_R-x_P)+(u_P+u_L)(x_P-x_L) \qquad (f_R+f_P)(t_R-t_P)+(f_P+f_L)(t_P-t_L).\end{aligned} \tag{9.3}$$

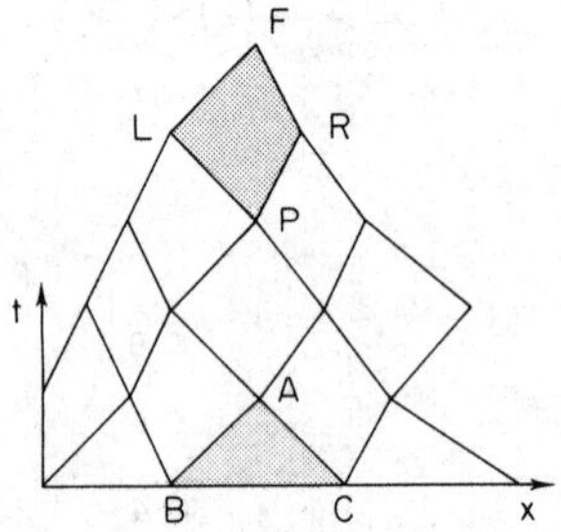

Fig. 10 Tessellation of x-t space into quadrilaterals. The vertices are at regular intervals in time but not in space.

From this form it is obvious that if (9.3) holds for all elementary quadrilaterals in some domain, then an analogous formula ($\Sigma\bar{u}\Delta x - \bar{f}\Delta t = 0$) holds around any region which is a union of such quadrilaterals. But to investigate a given quadrilateral, we can collapse (9.3) so that it reads

$$(u_F-u_P)(x_L-x_R) - (u_L-u_R)(x_F-x_P) = -2\Delta t\,(f_L-f_R). \tag{9.4}$$

If we rearrange (9.4) as

$$u_F-u_P = \frac{(u_L-u_R)(x_F-x_P)}{(x_L-x_R)} - \frac{2\Delta t(f_L-f_R)}{(x_L-x_R)} \tag{9.5}$$

we have a slight generalisation of the conventional leapfrog method. The grid is fixed, the data (u_L, u_R, u_P) are given on two levels, and the unknown quantity is u_F. The method is well known to be very accurate for smooth flows, neutrally stable against small disturbances, but highly vulnerable to non-linear instabilities which develop as the flow becomes less smooth.

Alternatively, we can rearrange (9.4) as

$$x_F - x_P = \frac{2\Delta t(f_L - f_R)}{(u_L - u_R)} + \frac{(u_F - u_P)(x_L - x_R)}{(u_L - u_R)} . \tag{9.6}$$

Here, we have inverted the problem, the grid and the data are still specified at two levels, but at the new time level we regard position x rather than state u as the unknown quantity, and suppose that u_F can be prescribed. The second term on the RHS of (9.6) is numerically dangerous, however; it could easily become singular. We can avoid this by prescribing $u_F = u_P$, so that (9.6) becomes

$$x_F - x_P = \frac{2\Delta t\ (f_L - f_R)}{(u_L - u_R)} . \tag{9.7}$$

The RHS of 9.7 is merely the average characteristic velocity multiplied by $2\Delta t$. It is never singular, and for rational f(u) the denominator will cancel. Thus, for $f(u) = \frac{1}{2}u^2$

$$x_F - x_P = \Delta t\ (u_L + u_R) \tag{9.8}$$

and for $f(u) = au$

$$x_F - x_P = 2a\Delta t. \tag{9.9}$$

Colloquially, (9.7) could be expressed as "don't send a signal, move the grid". The amount by which the grid is to be moved depends on the fluctuation between L and R. There are some striking properties. All information present in the data is preserved exactly (and in the case of (9.9), located exactly). Also the method is time reversible. Not only can the data be reconstructed from the solution but we can actually use the same algorithm to do so. Perhaps even more striking, though, is that these properties break down exactly when they should, that is, whenever a shockwave forms. Fig. 11 shows a very simple calculation for $f(u) = \frac{1}{2}u^2$ on a coarse grid ($\Delta x = 0.1$). The solution was started by integrating around triangles such as ABC, and assuming $u_A = \frac{1}{2}(u_B + u_C)$. This is a typical non-linear problem, with a shockwave forming out of a smooth initial compression. In the exact solution the shock would form at time $t = (2/\pi) = 0.636$. In the numerical solution, (9.8), what happens between $t = 0.6$ and $t = 0.7$ is that the grid points pass over each other and the numerical solution becomes multi-valued, just as the analytical solution does.

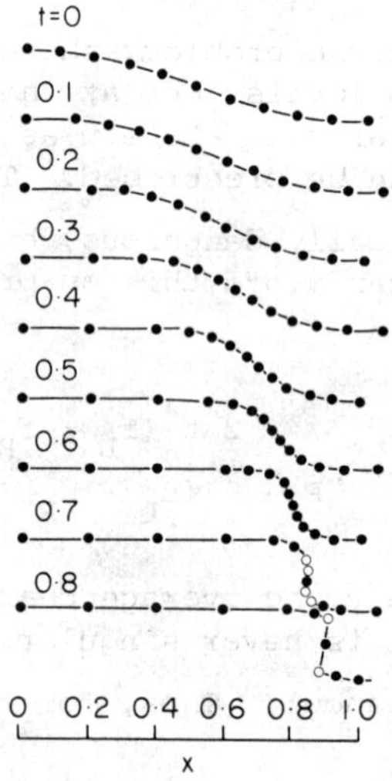

Fig. 11 Moving-grid solution for Burger's equation. Data is $u(x, 0) = \frac{1}{2}(1 + \cos \pi x)$.

The remedy is to cut out the tangled portion of the grid, and replace it with a new portion. Sketch (v)(a) shows two points p and q which have become crossed (in other words x_i is no longer a monotone function of i), and Sketch (v)(b) shows a new section of grid mesh with (p,q) replaced by a single point n. We need something similar to Leveque's Lemma, which will merge the information carried by (p,q). The only constraint which seems to be forced upon us is that the solution on the new mesh is conservative in a similar sense to the old solution.

This reduces to a requirement that $\Sigma\bar{u}\Delta x - \bar{f}\Delta t$ is the same along the path anc as it was along aqbpc. After some slight rearrangement we find

$$u_a(x_a-x_q) + \underline{u_q(x_a-x_b)} + u_b(x_q - x_p) + \underline{u_p(x_b-x_c)} + u_c(x_p-x_c) = u_a(x_a-x_n) + \underline{u_n(x_a-x_c)} + u_c(x_n-x_c) \qquad (9.10)$$

which is a linear relationship between u_n and x_n. It would seem that any sensible solution of (9.10) could be used, fixed perhaps by $x_n = \frac{1}{2}(x_p + x_q)$, but for what it may be worth we record an elegant solution obtained by setting the underlined and non-underlined terms in (9.10) separately equal. Then

$$u_n = \frac{u_p\,(x_c - x_b) + u_q\,(x_b - x_a)}{(x_c - x_a)} \qquad (9.11a)$$

$$x_n = \frac{x_p\,(u_c - u_b) + x_q\,(u_b - u_a)}{(u_c - u_a)}. \qquad (9.11b)$$

Since x_a, x_b, x_c were in ascending order, Eqn. (9.11a) ensures that u_n lies between u_p and u_q. Also, Eqn. (9.11b) has come out identical to (8.2b) of Leveque's Lemma.

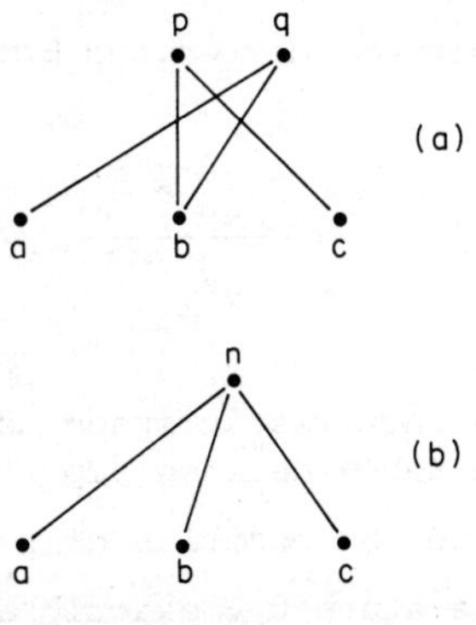

Sketch (v)

The evolving grid which corresponds to Fig. 11 is plotted in Fig. 12. When we recall that the diagonals of the quadrilateral elements are approximations to the characteristic lines, then the process of shock formation becomes clear. Incidentally, we can easily prove that these diagonals are collinear. Since the slope of the diagonal PF depends only on the states (L,R), and since the states (L,R) are repeated every other time step, then so are the diagonal slopes. So the approximate characteristics remain straight lines, just like the exact characteristics, until they disappear into a shockwave.

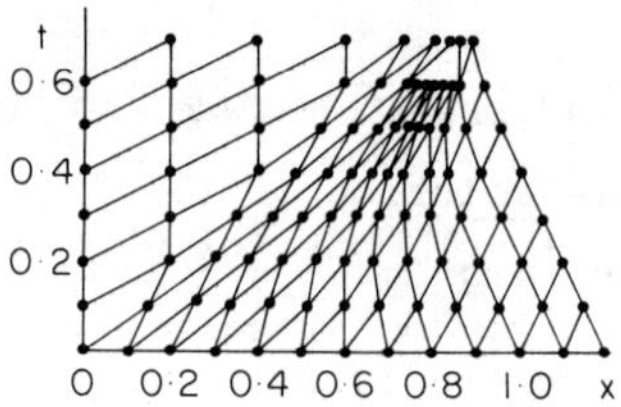

Fig. 12 Evolution of the grid for the solution shown in Fig. 11.

The question which now has to be asked is this. If this very elegant structure is progressively burdened with real-life complications, at what stage will it break down? For some time I thought that it broke down immediately if one attempted to apply it to a system of equations, so that (9.4) becomes a vector equation which is not easy to interpret, but some new results indicate that this may not be so. Returning to (9.7), we see that the displacement $x_F - x_P$ depends on the mean characteristic velocity associated with the fluctuation $\phi_{LR} = (f_R - f_L)$. Can we still define such a characteristic velocity even if ϕ_{LR} is a vector? In the context of a quite different adaptive mesh scheme, Harten and Hyman (1982) propose the following:

$$\hat{a} = \frac{\Sigma \alpha_k^{\,2}\, a^{(k)}}{\Sigma a_k^{\,2}} \tag{9.12}$$

where $a^{(k)}$ is the characteristic speed associated with the kth wave, and α_k is a suitably defined amplitude for that wave. If (9.12) is used to find the position of x_F, then (9.5) is available to find the state $\underline{u}_F$, exactly as in the scalar case. Note that if we have a simple k-wave, then the only non-zero

amplitude is α_k, so $\hat{a} = a^{(k)}$ and in fact the whole scheme goes through exactly as in the scalar case, since
$f_L - f_R = \hat{a}\ (u_L - u_R)$ and Eqn (9.4) reverts to a scalar equation.

10. CONCLUDING REMARKS

It seems to me that one of the main requirements in computational fluid dynamics is for a set of fundamental concepts sufficiently stable to serve as a background against which to view the welter of current research. For the computation of evolution problems, I believe that the attempt to exercise deliberate control over the flow of information provides such a concept, and I have tried to show how several areas of current research can be seen in this way. However, to make something more precise out of this rather general statement, it seems necessary to probe more deeply the nature of the information being processed.

For scalar problems, there is only one kind of information to consider, and this allows us to isolate the importance of directional bias, or "upwinding", in the numerical solution. When systems of equations are involved, it is important to identify the different kinds of information which yield the most physically meaningful description. Both the characteristic form and the conservation form of the governing equations are equally relevant, and in Section 4 I have tried to make use of both forms in order to construct "explanations" of fluctuating data. It should be said, however, that we are as yet very far from knowing just how detailed an explanation we should seek to obtain. The dilemma is summed up in the maxim that what is simple is always false, but anything which is not simple is useless. This dilemma is even more acute in higher dimensions.

When the problems are non-linear and shockwaves arise, it seems that the deliberate loss of information must form part of a realistic numerical model. The lesson which I extract from Sections 2, 5, 8 and 9 is that the more carefully we choose which information to retain and which to discard the better the resolution of our numerical models. I think this aspect of the problem will repay further study.

REFERENCES

BAINES, M.J. 1980 A numerical algorithm for the solution of conservation laws in two dimensions. Numerical Analysis Report 2/80, Dept. of Maths. University of Reading.

BAINES, M.J. 1981 Numerical algorithms for the solution of conservation laws in two and three dimensions. Numerical Analysis Report 4/81 Dept. of Maths. University of Reading.

COLELLA, P. 1982 Private Communication. Lawrence Livermore National Laboratory, California, USA.

COURANT, R. and FRIEDRICHS, K.O. 1948 Supersonic flow and shockwaves. Interscience.

DENTON, J.D. 1982 An improved time marching method for turbomachinery flow calculation. In Numerical Methods for Aeronautical Fluid Dyanmics (P.L. Roe, Ed.), Academic Press.

GODUNOV, S.K. 1959 A difference method for the numerical calculation of discontinuous solutions of hydrodynamic equations. *Mat. Sbornik* **47**, 3, p. 271. English translation by US Dept. of Commerce JPRS 7225, 1960.

HARTEN, A. 1981 On the symmetric form of systems of conservation laws with entropy. ICASE Report 81-34.

HARTEN, A. and HYMAN, M.J. 1981 A self-adjusting grid for the computation of weak solutions of hyperbolic conservation laws. I. One-dimensional problems. *J. Comput. Phys.* to appear.

JEFFREY, A. 1976 Quasilinear hyperbolic systems and waves. Pitman.

LEVEQUE, R.J. 1982 Large time step shock-capturing techniques for scalar conservation laws. *SIAM J. Num. Analysis* to appear.

McCORMACK, R.W. and PAULLAY, A.J. 1972 Computational efficiency achieved by time-splitting of finite-difference operators. AIAA Paper 72-154.

NI, R.H. 1981 A multiple grid scheme for solving the Euler equations. AIAA Paper 81-1025.

OSHER, S. and SOLOMON, F. 1982 Upwind difference schemes for hyperbolic systems of conservation laws. *Maths Comput.* to appear.

ROE, P.L. 1981(a) The use of the Riemann problem in finite difference schemes. In Seventh International Conference on Numerical Methods in Fluid Dynamics. (W.C. Reynolds and R.W. MacCormack, Eds) Springer.

ROE, P.L. 1981(b) Approximate Riemann solvers, parameter vectors, and difference schemes. *J. Comput. Phys.* **43**, No. 2, p. 357.

ROE, P.L. 1982 Numerical modelling of shockwaves and other discontinuities. In Numerical Methods for Aeronautical Fluid Dynamics (P.L. Roe, Ed.) Academic Press.

ROE, P.L. and BAINES, M.J. 1982 Algorithms for advection and shock problems. In Fourth GAMM - Conference on Numerical Methods in Fluid Mechanics (H. Viviand, Ed.) Vieweg.

SELLS, C.C.L. 1980 Solution of the Euler equations for transonic flow past a lifting aerofoil. RAE TR 80065.

SOD, G.A. 1978 A survey of finite difference methods for systems of nonlinear hyperbolic conservation laws. *J. Comput. Phys.* **27** p.1.

SWEBY, P.K. 1981 A high order monotonicity preserving algorithm on an irregular grid for non-linear conservation laws. Numerical Analysis Report 1/81. Dept. of Maths. University of Reading.

SWEBY, P.K. and BAINES, M.J. 1981 Convergence of Roe's scheme for the non-linear scalar wave equation. Numerical Analysis Report 6/81, Dept of Maths. University of Reading.

STRANG, G. 1968 On the construction and comparison of difference schemes. *SIAM J. Num. Analysis* **5** No. 3, p. 506.

THOMAS, P.D. and LOMBARD, C.K. 1978 Geometric conservation law: a link between finite-difference and finite-volume methods of flow computation on moving grids. AIAA Paper 78-1208.

VAN LEER, B. 1979 Towards the ultimate conservative difference scheme. V. A second-order sequel to Godunov's method. *J. Comput. Phys.* **32** 1, p. 101.

YANENKO, N.N. 1970 The method of fractional steps. English translation by M. Holt, Springer.

ZALESAK, S. 1979 Fully multidimensional flux-corrected transport algorithms for fluids. *J. Comput. Phys.* **31** p. 335.

THE USE OF IRREGULAR FINITE DIFFERENCE GRIDS FOR COASTAL SEA PROBLEMS

P.P.G. Dyke and G.D. Phelps

(Dept. of Offshore Engineering, Heriot-Watt University)

1. INTRODUCTION

In recent years there has been a rapid increase in the number of numerical techniques available to the mathematical modeller. One of the most intriguing of these is the use of an irregular finite difference grid whereby the standard rectangular network of points is abandoned in favour of a non-uniform distribution of points linked by not necessarily straight lines. The motivation for using such a grid is twofold. First, it has the advantage that grid points can be concentrated in areas of high interest without having to compute values at each point of a fine mesh over the whole domain. Secondly, the existence of irregularly spaced boundary data points can be easily dealt with. In short, the technique seems to offer all the advantages of local refinement inherent in the finite element method without the severe disadvantage of requiring large amounts of computer time to invert matrices at each time step.

Having rejected the rectangular mesh, the question arises as to how to decide where the points of the irregular grid will be. One could use a random number generator, but we prefer the more physical approach in that grid size and shape is dictated by boundary point spacing. In future, we will also locally refine the grid in areas of interest, but this has not been done here. A general discussion of mesh generation techniques may be found in Thacker (1980).

Another question postponed until later is the extension into three dimensions. It is expected that we will use the modal separation techniques of Davies (1980, 1982), but here we concentrate on the two-dimensional model that forms the basis on which this extension will be built.

We will also focus on what appears to be an added bonus. Because the variables are all evaluated at the same point, the equations for stability analysis can be solved exactly.

Particular attention is paid to the time at which the Coriolis terms are evaluated, and it is shown that when a scheme that evaluates these terms half at the earlier and half at the later time is adopted, a stable hence convergent finite difference formulation results. Finally, this stability analysis is explained diagramatically in a simple and lucid fashion.

2. BASIC EQUATIONS AND TIME DISCRETISATION

We shall consider motion in a shallow sea governed by the following equations:

$$\frac{\partial U}{\partial t} = -gD\frac{\partial H}{\partial x} + fV \qquad (2.1)$$

$$\frac{\partial V}{\partial t} = -gD\frac{\partial H}{\partial y} - fU \qquad (2.2)$$

$$\frac{\partial H}{\partial t} = -\left(\frac{\partial U}{\partial x} + \frac{\partial V}{\partial y}\right) \qquad (2.3)$$

The notation used is as follows: x and y are Cartesian coordinates, U and V are the x and y components of volume transport, D is the still water depth, H the surface elevation, g is gravitational acceleration and f the constant vertical component of the Coriolis acceleration. This is of course the two-dimensional, depth integrated, picture. The intention being to use the techniques of Davies (1980, 1982) to represent vertical structure. In order to use finite differences, one has to decide on a scheme by which to represent both the time and space derivatives. Here we use an entirely conventional centred time step for $\partial/\partial t$ as done by Thacker (1978). The Coriolis terms have to be treated carefully, since over-all stability is critically dependent upon the time level(s) at which they are evaluated (see section 4).

Using a superscript to denote the time step and a single subscript as an indicator of the position of each point in x, y space, the following finite difference equations result:

$$U_j^{n+\frac{1}{2}} = U_j^{n-\frac{1}{2}} - gD_j\tau\left(\frac{\partial H}{\partial x}\right)_j^n + \tfrac{1}{2}f\left(V_j^{n+\frac{1}{2}} + V_j^{n-\frac{1}{2}}\right) \qquad (2.4)$$

$$V_j^{n+\frac{1}{2}} = V_j^{n-\frac{1}{2}} - gD_j\tau\left(\frac{\partial H}{\partial y}\right)_j^n - \tfrac{1}{2}f\left(U_j^{n+\frac{1}{2}} + U_j^{n-\frac{1}{2}}\right) \qquad (2.5)$$

$$H_j^{n+1} = H_j^{n} - \tau[(\frac{\partial U}{\partial x})_j^{n+\frac{1}{2}} + (\frac{\partial V}{\partial y})_j^{n+\frac{1}{2}}] \quad (2.6)$$

The integer j spans all the grid points, D_j is the spot depth at the point labelled j and τ is the time step. This scheme is unusual for the numerical modelling of seas in that all variables are evaluated, or at least centred, at the same point in space. The standard practice is to use a staggered Hansen grid (see for example (Heaps, 1969)) but as will become clearer in the next section, the irregularity of the grid prevents this. In any case evaluating all variables at the same point turns out to have more advantages than disadvantages.

3. SPACE DISCRETISATION

Previous research using an irregular grid technique for sea modelling has been undertaken by Thacker (1977, 1978). Thacker used approximations for space derivatives that are only first order accurate. We use an entirely different technique based on Taylor's theorem which can be taken to any desired order of accuracy. This technique is similar to that used by Perrone and Kao (1975) and will now be described in some detail.

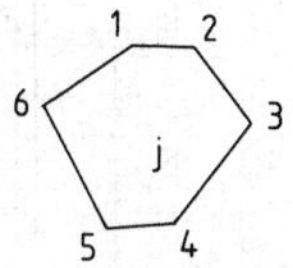

Fig. 1 The node j and its neighbours.

Suppose the point labelled j is surrounded by six others. For notational convenience, label these with integers 1 to 6 and put j = 0. Now we assume that the solutions for H, U and V are well enough behaved for two dimensional Taylor expansions to be used. Therefore, for any of these functions, F(x,y), its value at α (α = 1(1)6) is given by

$$F_{(\alpha)} = F_{(0)} + hF_{x(0)} + kF_{y(0)} + \tfrac{1}{2}h^2F_{xx(0)} + hkF_{xy(0)} + \tfrac{1}{2}k^2F_{yy(0)} + \tfrac{1}{2}h^2kF_{xxy(0)} + \ldots\ldots\ldots\ldots \quad \alpha = 1(1)6 \quad (3.1)$$

where $h = x_{(\alpha)} - x_{(O)}$, $k = y_{(\alpha)} - y_{(O)}$ and the suffix derivative notation has been adopted. The suffices in parenthesis denote the point at which the quantity is evaluated. Since we have six points surrounding our interior point labelled O, we have six equations with which to evaluate the six unknown derivatives F_x, F_y, F_{xx}, F_{xy}, F_{xx}, F_{xxy} (all evaluated at O). Approximations to the derivatives in terms of F_1, F_2, F_3, F_4, F_5 and F_6 would then be valid to second order. Two points should be made clear here. The fact that six points surround our interior point means that six terms of the Taylor expansion (putting F_O on the left hand side) need to be used. This in turn means the inclusion of one third order derivative. Now, the choice of which third order derivative to use is arbitrary, we choose F_{xxy} in order to involve both x and y, but the choice will not significantly influence the approximations to the derivatives since each choice is third order by definition and we are only second order accurate. The second point to emphasise is that by involving more surrounding points, and hence more terms in the Taylor expansion, we can work to any desired order of accuracy (truncation error). The six equations can be conveniently written in matrix form:

$$\begin{bmatrix} h_1 & k_1 & \frac{1}{2}h_1^2 & h_1k_1 & \frac{1}{2}k_1^2 & \frac{1}{2}h_1^2k_1 \\ h_2 & k_2 & \frac{1}{2}h_2^2 & h_2k_2 & \frac{1}{2}k_2^2 & \frac{1}{2}h_2^2k_2 \\ h_3 & k_3 & \frac{1}{2}h_3^2 & h_3k_3 & \frac{1}{2}k_3^2 & \frac{1}{2}h_3^2k_3 \\ h_4 & k_4 & \frac{1}{2}h_4^2 & h_4k_4 & \frac{1}{2}k_4^2 & \frac{1}{2}h_4^2k_4 \\ h_5 & k_5 & \frac{1}{2}h_5^2 & h_5k_5 & \frac{1}{2}k_5^2 & \frac{1}{2}h_5^2k_5 \\ h_6 & k_6 & \frac{1}{2}h_6^2 & h_6k_6 & \frac{1}{2}k_6^2 & \frac{1}{2}h_6^2k_6 \end{bmatrix} \begin{bmatrix} F_{x\,(O)} \\ F_{y\,(O)} \\ F_{xx(O)} \\ F_{xy(O)} \\ F_{yy(O)} \\ F_{xxy(O)} \end{bmatrix} = \begin{bmatrix} F_1 - F_O \\ F_2 - F_O \\ F_3 - F_O \\ F_4 - F_O \\ F_5 - F_O \\ F_6 - F_O \end{bmatrix} \qquad (3.2)$$

The 6x6 matrix is only singular when two of the surrounding points are on the same side of point O and all three lie on the same straight line. The mesh generation technique outlined below will not permit this.

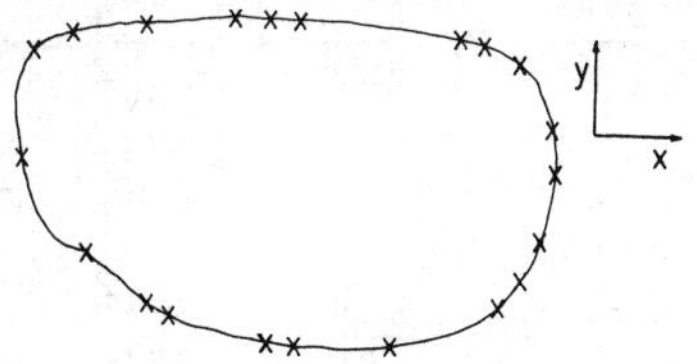

Fig. 2 A schematic sea area.

In order to evaluate the approximations to the derivatives at each point, 6x6 matrices of the type (3.2) will have to be inverted at each grid point. This is computationally straightforward and need only be done once.

We will now discuss how to generate the irregular mesh. Fig. 2 shows a typical sea area that one may wish to model. The boundary conditions are not arranged in a regular manner, and we shall use this irregularity to dictate the grid in the interior. The (x,y) coordinates are now transformed on to a rectangle in (ξ,η) space. We demand that

$$\frac{\partial^2 x}{\partial \xi^2} + \frac{\partial^2 x}{\partial \eta^2} = 0 \tag{3.3}$$

$$\frac{\partial^2 y}{\partial \xi^2} + \frac{\partial^2 y}{\partial \eta^2} = 0 \tag{3.4}$$

must hold throughout the rectangle and that on the rectangle itself, boundary point values are simply the (x,y) coordinates of the specified boundary conditions in (x,y) space. By solving equations (3.3) and (3.4) using SOR (say) we obtain the interior grid points which can be drawn on Fig. 2. Note that we have used the location of the boundary condition points, not the conditions themselves which will be no flow through solid boundaries and a prescribed velocity or elevation at an open boundary. If the actual net in (x,y) space is required, we shall have to solve the inverse Laplace problem

$$\alpha \frac{\partial^2 \xi}{\partial x^2} - 2\beta \frac{\partial^2 \xi}{\partial x \partial y} + \gamma \frac{\partial^2 \xi}{\partial y^2} = 0 \tag{3.5}$$

$$\alpha \frac{\partial^2 \eta}{\partial x^2} - 2\beta \frac{\partial^2 \eta}{\partial x \partial y} + \gamma \frac{\partial^2 \eta}{\partial y^2} = 0 \tag{3.6}$$

where

$$\alpha = \left(\frac{\partial x}{\partial \xi}\right)^2 + \left(\frac{\partial y}{\partial \xi}\right)^2$$

$$\beta = \frac{\partial x}{\partial \xi}\frac{\partial x}{\partial \eta} + \frac{\partial y}{\partial \xi}\frac{\partial y}{\partial \eta}$$

$$\gamma = \left(\frac{\partial x}{\partial \eta}\right)^2 + \left(\frac{\partial y}{\partial \eta}\right)^2$$

(see (Courant, 1936)).

This exercise is not necessary in general since only the points are required not the lines joining them. Having given the general procedure, let us consider a simple example for stability computation.

4. A SIMPLE EXAMPLE: STABILITY

In practice, if the number of points is large and the irregularities in boundary point spacing not too pronounced, then the interior points will be surrounded by six points in a manner close to the idealised picture of Fig. 3. In this particular case, stability analysis can take place. For the point O

$$(h_1,k_1) = (0,h) \quad (h_4,k_4) = (0,-h)$$

$$(h_2,k_2) = (h,0) \quad (h_5,k_5) = (-h,0)$$

$$(h_3,k_3) = (h,-h) \quad (h_6,k_6) = (-h,h)$$

giving the matrix (3.2) as

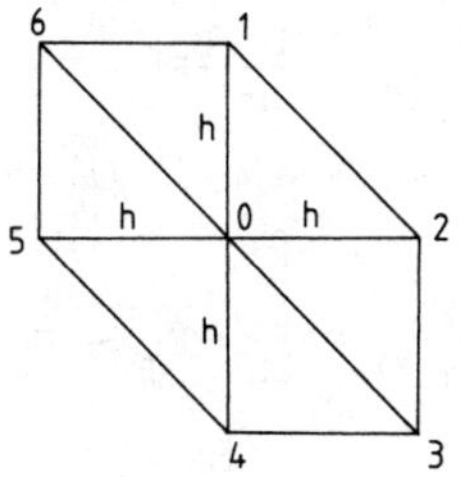

Fig. 3 An idealised mesh.

$$\begin{bmatrix} 0 & h & 0 & 0 & \frac{1}{2}h^2 & 0 \\ h & 0 & \frac{1}{2}h^2 & 0 & 0 & 0 \\ h & -h & \frac{1}{2}h^2 & -h^2 & \frac{1}{2}h^2 & \frac{1}{2}h^3 \\ 0 & -h & 0 & 0 & \frac{1}{2}h^2 & 0 \\ -h & 0 & \frac{1}{2}h^2 & 0 & 0 & 0 \\ -h & h & \frac{1}{2}h^2 & -h^2 & \frac{1}{2}h^2 & \frac{1}{2}h^3 \end{bmatrix} \tag{4.1}$$

with its inverse

$$\begin{bmatrix} 0 & 1/2h & 0 & 0 & -1/2h & 0 \\ 1/2h & 0 & 0 & -1/2h & 0 & 0 \\ 0 & 1/h^2 & 0 & 0 & 1/h^2 & 0 \\ 1/2h^2 & 1/2h^2 & -1/2h^2 & 1/2h^2 & 1/2h^2 & -1/2h^2 \\ 1/h^2 & 0 & 0 & 1/h^2 & 0 & 0 \\ -1/h^3 & 1/h^3 & -1/h^3 & 1/h^3 & -1/h^3 & 1/h^3 \end{bmatrix} \tag{4.2}$$

For the purpose of solving equations in the absence of friction, only first order space derivatives are required, hence only the first two rows of matrix (4.2) are used. The consequence of this is that to second order accuracy

$$\left.\frac{\partial F}{\partial x}\right|_O \simeq \frac{1}{2h}(F_2 - F_5) \tag{4.3}$$

and

$$\left.\frac{\partial F}{\partial y}\right|_O \simeq \frac{1}{2h}(F_1 - F_4). \tag{4.4}$$

Thus the standard centred difference formulae for first order x and y derivatives will hold to second order on this particular grid. This is not a surprising result, glancing at Fig. 3, but still a gratifying one.

For stability analysis, we write

$$(U,V,H) = (U_O,V_O,H_O)\exp i(\omega\tau n + \sigma_1 kh + \sigma_2 \ell h) \tag{4.5}$$

where h = space "step" (Fig. 3)

τ = time step

σ_1, σ_2 = wave numbers

k, ℓ, n = integers.

If we let $\lambda = \exp(\frac{1}{2}i\omega\tau)$, inserting the waves (4.5) into the difference equations (2.4), (2.5) and (2.6) utilising (4.3) and (4.4) leads to the following determinant

$$\begin{vmatrix} \lambda^2 - 1 & Q & \frac{2ig\tau D}{h}\sin\sigma_1 h \\ -Q & \lambda^2 - 1 & \frac{2ig\tau D}{h}\sin\sigma_2 h \\ \frac{2i\tau}{h}\sin\sigma_1 h & \frac{2i\tau}{h}\sin\sigma_2 h & \lambda^2 - 1 \end{vmatrix} = 0 \tag{4.6}$$

where D is the depth assumed constant over the seven points and

$$Q = \begin{cases} \lambda^2 f\tau & \text{if the Coriolis terms are evaluated at time step } n+\frac{1}{2} \\ \frac{1}{2}(\lambda^2+1) f\tau & \text{if the Coriolis terms are evaluated as in (2.4) and (2.5)} \\ f\tau & \text{if the Coriolis terms are evaluated at time step } n-\frac{1}{2} \end{cases}$$

Determinant (4.6) is satisfied if $\lambda^2 = 1$ or

$$(\lambda^2 - 1)^2 + Q^2 + \lambda^2\tau^2 a^2 = 0 \tag{4.7}$$

where

$$h^2 a^2 = gD(\sin^2\sigma_1 h + \sin^2\sigma_2 h).$$

For stability, it is sufficient that $|\lambda| < 1$. Although for some problems a growth is possible, we cannot allow this in our case (Richtmyer and Morton, 1967 p. 70). The results for the three values of Q given above are as follows.

Case 1: Coriolis term at the later time step $n+\frac{1}{2}$

Here the equation for λ^2 (4.7) becomes

$$\lambda^4(1 + f^2\tau^2) + \lambda^2(\tau^2 a^2 - 2) + 1 = 0 \tag{4.8}$$

from which a precise inequality for τ is difficult. However,

$$a^4\tau^2 \leqslant 4(a^2 + f^2) \tag{4.9}$$

is certainly sufficient to ensure $|\lambda| \leqslant 1$. In fact $|\lambda| < 1$ since equality occurs in (4.9) when

$$|\lambda| = (1 + 2f^2/a^2)^{-\frac{1}{2}} < 1. \tag{4.10}$$

Case 2: The "Crank-Nicolson" case; Coriolis terms evaluated half at the earlier and half at the later time steps.

Equation (4.7) is now

$$(1 + \frac{1}{4}f^2\tau^2)\lambda^4 + (\tau^2 a^2 - 2 + \frac{1}{2}f^2\tau^2)\lambda^2 + (1 + \frac{1}{4}f^2\tau^2) = 0 \tag{4.11}$$

which implies $|\lambda| \leq 1$ if

$$a^2\tau^2 \leq 4. \tag{4.12}$$

This condition is independent of f and corresponds to the Courant, Friederichs, Lewy condition (see Ramming and Kowalik, 1980). This case seems in some sense optimal since $|\lambda| = 1$ if (4.12) holds and there is neither growth nor damping of our solutions to this order of accuracy.

Case 3: Coriolis term evaluated at the earlier time step $n-\frac{1}{2}$

For this last case, equation (4.7) becomes

$$\lambda^4 + (\tau^2 a^2 - 2)\lambda^2 + (1 + f^2\tau^2) = 0 \tag{4.13}$$

from which

$$|\lambda|^2 = (1 + f^2\tau^2)^{\frac{1}{2}} > 1. \tag{4.14}$$

Hence the scheme is always unstable. Further discussion of these results is deferred until Section 6.

5. BOUNDARY CONDITION STABILITY

Before discussing the above results, we shall use a method outlined by Trapp and Ramshaw (1976) to test the stability of the schemes when we have a solid boundary. Using coordinates $\hat{\underline{n}}$ perpendicular to the boundary and $\hat{\underline{s}}$ tangential to it in such a fashion that with an upwards coordinate a right handed system results, no flow perpendicular to the boundary gives

$$\frac{\partial^2 H}{\partial t \partial n} + f\frac{\partial H}{\partial s} = 0. \tag{5.1}$$

The finite difference form of (5.1) is

$$(\lambda^2 - 1) + if\frac{\tau P}{h}\sin\sigma h = 0 \tag{5.2}$$

where λ retains its former definition, h and σ are the representative step length and wave number, respectively, and the factor P is given by

$$P = \begin{cases} \lambda^2 & \text{case 1} \\ \frac{1}{2}(\lambda^2 + 1) & \text{case 2} \\ 1 & \text{case 3} \end{cases} \tag{5.3}$$

for the three cases referred to in the previous section. For brevity, some of the routine details of solving (5.2) in the three cases are omitted.

The results are as follows:

$$|\lambda|^2 = \begin{cases} [1 + \frac{f^2\tau^2}{h^2}\sin^2\sigma h]^{-\frac{1}{2}} & \text{for case 1} \\ 1 & \text{for case 2} \\ [1 + \frac{f^2\tau^2}{h^2}\sin^2\sigma h]^{\frac{1}{2}} & \text{for case 3} \end{cases}$$

The conclusions are therefore that case 1 is stable but we have damping ($|\lambda| < 1$), case 2 is stable and case 3 is unstable.

6. CONCLUSIONS

The principal conclusion is that an optimally stable scheme results if the Coriolis terms are evaluated as in equations (2.4) and (2.5), that is half at the later time and half at the earlier time. This is consistent with the comments of research workers who have had to model storm surges, e.g. (Heaps, 1969) and (Sielecki, 1968).

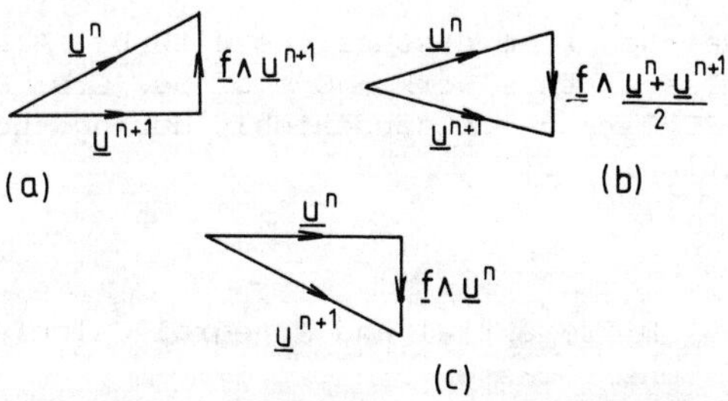

Fig. 4 A simple explanation of stability.
(a) Case 1 (b) Case 2 (c) Case 3.

With the benefit of hindsight, one can now draw the simple diagrams of Fig. 4. To draw these, we have used the fact that the Coriolis term is at right angles to the velocity term at the same time step. For Case 1, Fig. 4(a), we see that $|\underline{u}^n| > |\underline{u}^{n+1}|$ therefore kinetic energy is being lost and the velocity field is being numerically damped. For Case 2 $|\underline{u}^n| = |\underline{u}^{n+1}|$ and kinetic energy is conserved. For Case 3 $|\underline{u}^n| < |\underline{u}^{n+1}|$ and the kinetic energy is increasing leading to instability.

The somewhat loose phraseology used above can be given mathematical respectability by Banach space arguments. Since our problem is linear, and the finite difference problem is well posed, at each time step the solution points are points in a Banach space, and the linear transformation which is the result of solving equations (2.4) to (2.6) (and the Case 1 and Case 3 equivalents) is a Banach operator (Richtmyer and Morton, 1967, section 3.2). Following these arguments, we see that Cases 1, 2 and 3 above correspond to the modulus of the consistency operator operating on the initial values (that is the norm of the Banach space) being zero, one and unbounded, respectively. By the Lax Equivalence Theorem, we shall obtain convergent solutions in Cases 1 and 2 (for Case 1 eventually the zero solution).

We have used the scheme stated in equations (2.4) to (2.6) to analyse motion in a closed circular sea and have duplicated the results of Thacker (1978). We are at present engaged in using the scheme for a sea with an open boundary. It seems to us that irregular grids should have wider applications, in particular in solving the change of scale problems encountered with boundary layers mentioned elsewhere in this volume (Jones, 1982).

7. ACKNOWLEDGEMENTS

We have benefited from discussions with Dr. A.D. Jenkins during the course of this work. One of us (G.D. Phelps) was in receipt of an SERC research studentship during the course of this investigation.

REFERENCES

COURANT, R. 1936 Differential and Integral Calculus, Volume 2. Blackie and Sons, Glasgow.

DAVIES, A.M. 1980 On formulating a three-dimensional hydrodynamic sea model with an arbitrary variation of vertical eddy-viscosity, *Comp. Meth. Appl. Mech. and Eng.* **22**, 187-211.

DAVIES, A.M. 1982 Application of a Galerkin eigenfunction method to computing currents in homogeneous and stratified seas. In Numerical Methods for Fluid Dynamics (K.W. Morton and M.J. Baines, Eds.), (this volume) Academic Press.

HEAPS, N.S. 1969 A two-dimensional numerical sea model. *Phil. Trans. R. Soc.* **A265**, 93-137.

JONES, I.P. 1982 The use of strongly streatched grids in viscous flow problems. In Numerical Methods for Fluid Dynamics (K.W. (K.W. Morton and M.J. Baines, Eds.), (this volume) Academic Press.

PERRONE, N. and KAO, R. 1975 A general finite difference method for arbitrary meshes. *Computer Structures* **5**, 45-58.

RAMMING, H.G. and KOWALIK, Z. 1980 Numerical modelling of marine hydrodynamics: applications to dynamic physical processes. 368 pp. Elsevier Oceanography Series No. 26.

RICHTMYER, R.D. and MORTON, K.W. 1967 Difference methods for initial-value problems. 405pp. Interscience Publishers.

SIELECKI, A. 1968 An energy-conserving difference scheme for the storm-surge equations. *Mon. Weath. Rev.* **96**, 150-156.

THACKER, W.C. 1977 Irregular grid finite difference techniques: simulations of oscillations in shallow circular basins. *J. Phys. Oceanogr.* **7**, 284-292.

THACKER, W.C. 1978 Comparison of finite-element and finite-difference schemes. Part II: Two-dimensional gravity wave motion. *J. Phys. Oceanogr.* **8**, 680-689.

THACKER, W.C. 1980 A brief review of techniques for generating irregular computational grids. *Int. J. Num. Meth. Eng.* **15**, 1335-1341.

TRAPP, J.A. and RAMSHAW, J.D. 1976 A simple heuristic method for analysing the effect of boundary conditions on numerical stability. *J. Comput. Phys.* **20**, 238-242.

TIME-DEPENDENT MULTI-MATERIAL FLOW WITH LARGE FLUID DISTORTION

D.L. Youngs

(Atomic Weapons Research Establishment, Aldermaston)

1. INTRODUCTION

Fluid flow problems in which interfaces between different materials are present are most easily modelled by using a Lagrangian mesh. However, in many 2-D applications distortion of the Lagrangian mesh makes such a method impractical. This paper describes a 2-D, time-dependent, compressible, finite difference Eulerian method which has been developed to simulate such problems.

The change from a Lagrangian to an Eulerian technique introduces two problems (i) tracking interfaces and (ii) advection of fluid properties across cell boundaries. A simple method of representing interfaces is used, in which the portion of each interface in a cell is approximated by a straight line. This technique has been successfully applied to many complex problems in which several interfaces were present. The method is even capable of following thin layers of fluid a fraction of a cell thick. It has been found that the upwind advection method often gives rise to an unacceptable amount of numerical diffusion. The monotonic advection method of Van Leer (1977) has been used to reduce this false diffusion without introducing such defects as spurious oscillations or negative densities.

2. THE LAGRANGIAN AND ADVECTION PHASES

The explicit finite difference (or finite volume) method uses a staggered grid. Pressure (p), density (ρ) and specific internal energy (ε) are defined at the centre of each rectangular cell. Components of velocity, $\underline{u}$, are defined at cell corners. As in the Los Alamos code FLIC, (Gentry et al, 1966), the Livermore code BBC, (Sutcliffe, 1973), and a number of other codes, the calculation for each time step is divided into two phases.

The first phase may be thought of as a Lagrangian step in which the mesh moves with the fluid. In this phase the changes in velocity and internal energy due to the pressure terms are calculated. The equations integrated are

$$\rho \frac{\partial \underline{u}}{\partial t} = - \nabla p \text{ and } \rho \frac{\partial \varepsilon}{\partial t} = -p \text{ div } \underline{u}.$$

In the Lagrangian phase both spatial directions X and Y are calculated simultaneously.

In the second advection phase, transport of mass, internal energy and momentum across cell boundaries is computed. This may be thought of as remapping the displaced mesh at the end of phase 1 back to its original position. X and Y advection are calculated in separate steps. The order of calculation is X first, Y second and Y first, X second for alternate time steps. XY splitting greatly simplifies the interface tracking procedure and enables the 1-D advection method of Van Leer (1977) to be used.

For low Mach number flows the Lagrangian phase is divided into several, say L, steps each of which is explicit and second order accurate in time. This significantly improves the efficiency of the method as fewer of the more complex advection steps then need to be performed. The over-all time step, Δt must meet the following restrictions

$$\Delta t < L \frac{\min(\Delta x, \Delta y)}{c}, \frac{\Delta x}{|u|}, \frac{\Delta y}{|v|}$$

where c is the sound speed and u,v are the velocity components.

3. TREATMENT OF MATERIAL INTERFACES

A possible way of tracking interfaces would be to define each interface by a set of Lagrangian marker particles. However, this method becomes logically complicated if interfaces become highly distorted or if the geometry is complex. The author prefers to use a simpler method in which the material layout is described solely by the volume fractions of each fluid in each cell. The first method of this type to be widely used was the Donor-Acceptor Cell technique, variants of which are described by Ramshaw and Trapp (1976) and Hirt and Nichols (1981). Interfaces are not precisely located within each cell, but transport across cell sides is calculated by a method which inhibits different materials diffusing into one another. In a similar method, the SLIC technique of Noh and Woodward (1976), for the purpose of calculating transport across cell sides, interfaces are drawn parallel to either the X or Y axes.

The method used here represents interfaces more accurately. The portion of an interface in a cell is approximated by a straight line. The method of locating this straight line for the case of a cell containing two materials A and B is shown in Fig. 1(a) (b) (c) and (d). Volume fractions of material A for the cell under consideration and its eight surrounding cells are used to determine the slope of the interface. The position of the interface is then adjusted so that it divides the cell into two areas which match the two volume fractions. During this calculation the cell is treated as a square of size 1 unit × 1 unit.

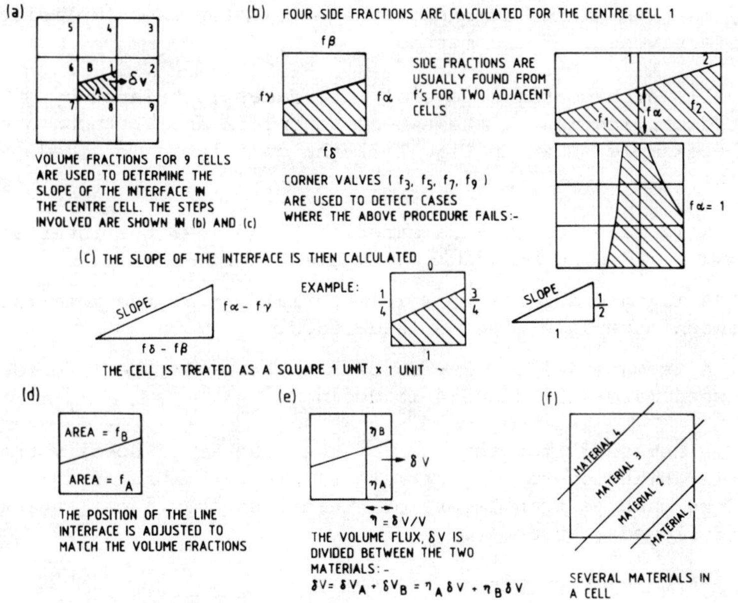

Fig. 1 The interface treatment.

As X-advection and Y-advection are calculated in separate steps it is sufficient to consider the flow across one side only. The interface position is used to divide the volume of fluid, δV, flowing across a cell side, between the two fluids, as indicated in Fig. 1(e),

$$\delta V = \delta V_A + \delta V_B .$$

The corresponding mass and internal energy fluxes are then

$$\delta m_A = \rho_A \delta V_A, \quad \delta m_B = \rho_B \delta V_B, \quad \delta E_A = \varepsilon_A \delta m_A, \quad \delta E_B = \varepsilon_B \delta m_B$$

where ρ_A, ρ_B, ε_A and ε_B are suitably defined density and internal energy values. During the X and Y advection steps the total mass and total internal energy of each material are conserved.

The portions of an interface in each cell are located independently and are not forced to join up. However, as illustrated by the results in section 5, near continuous interfaces are obtained unless structures of order the mesh size develop. The ability to treat each mixed material cell independently in this manner is one of the reasons for the simplicity and effectiveness of the method.

A cell which contains several interfaces is treated by repeated application of the two-component (A and B) technique. For the example shown in Fig. 1(f) the calculation proceeds as follows:

(a) A is material 1, B is material 2 + 3 + 4: the interface between materials 1 and 2 is defined.

(b) A is material 1 + 2, B is material 3 + 4: the interface between materials 2 and 3 is defined.

(c) A is material 1 + 2 + 3, B is material 4: the interface between materials 3 and 4 is defined.

It is necessary for the user of the program to specify the order in which layers of material are present. Use of this technique enables thin layers of fluid less than a cell thick to be followed accurately.

4. THE ADVECTION METHOD

The application of the method used in the hydrocode to the simple one dimensional problem,

$$\frac{\partial \rho}{\partial t} + u \frac{\partial \rho}{\partial x} = 0 \qquad \text{where } u = \text{a positive constant,}$$

is described first. As the advection phase is split into two steps X and Y, the extension to two dimensions is not difficult. The difference approximation used for the above equation has the following conservation form:

$$\frac{\rho^{n+1}_{j+\frac{1}{2}} - \rho^{n}_{j+\frac{1}{2}}}{\Delta t} + \frac{\bar{\rho}_{j+1} - \bar{\rho}_{j}}{\Delta x} = 0 \; .$$

The mass flowing across cell boundary j during the time step is $\bar{\rho}_j \, u\Delta t$, where the density value is given by

$$\bar{\rho}_j = \rho^n_{j-\frac{1}{2}} + \frac{1}{2}(1-\eta)\Delta x \, D_j \;, \text{ with } \; \eta = \frac{u\Delta t}{\Delta x} \; .$$

$\rho^n_{j-\frac{1}{2}}$ is the density value in the upwind cell at the beginning of the time step. The term containing $D_j \simeq \partial\rho/\partial x$ allows for the density gradient within the upwind cell. Possible choices for D_j are

First order: $D_j = 0$ (the upwind method)

Second order: $D_j = \dfrac{\rho^n_{j+\frac{1}{2}} - \rho^n_{j-\frac{1}{2}}}{\Delta x}$

Third order: $D_j = \dfrac{2-\eta}{3}\,\dfrac{\rho^n_{j+\frac{1}{2}} - \rho^n_{j-\frac{1}{2}}}{\Delta x} + \dfrac{1+\eta}{3}\,\dfrac{\rho^n_{j-\frac{1}{2}} - \rho^n_{j-\frac{3}{2}}}{\Delta x}$.

The numerical diffusion present in the first order method often gives poor results. This diffusion is reduced in the second or third order methods but non-physical oscillations and negative densities may then occur. The third order method, which has less numerical dispersion, gives rather better results than the second order method. The non-physical behaviour is eliminated if the monotonic advection methods of Van Leer (1977) are used. In these methods, non-linear cut-offs applied to D_j ensure that the new density value $\rho^{n+1}_{j+\frac{1}{2}}$ satisfies the following monotonicity conditions:

$$\min(\rho^n_{j-\frac{1}{2}}, \rho^n_{j+\frac{1}{2}}) \leq \rho^{n+1}_{j+\frac{1}{2}} \leq \max(\rho^n_{j-\frac{1}{2}}, \rho^n_{j+\frac{1}{2}}) \; .$$

The formula used here for D_j is

$$D_j\Delta x = S \min\{\bar{D}_j\Delta x, \; 2|\rho^n_{j+\frac{1}{2}} - \rho^n_{j-\frac{1}{2}}| \;, \; 2|\rho^n_{j-\frac{1}{2}} - \rho^n_{j-\frac{3}{2}}|\}$$

$$\text{where } S = \begin{cases} 0 & \text{if } \operatorname{sign}(\rho^n_{j+\frac{1}{2}} - \rho^n_{j-\frac{1}{2}}) \neq \operatorname{sign}(\rho^n_{j-\frac{1}{2}} - \rho^n_{j-\frac{3}{2}}) \\ \operatorname{sign}(\rho^n_{j+\frac{1}{2}} - \rho^n_{j-\frac{1}{2}}) & \text{otherwise.} \end{cases}$$

The standard advection method in the hydrocode uses the above third order approximation for $\bar{D}_j$. Alternative methods for calculating D_j are described in the recent paper (Van Albada et al. 1982).

Results for a test problem, the advection of a square pulse through 300 meshes, are shown in Fig. 2. The calculations used 100 meshes in the x-direction with periodic boundary conditions. η was equal to 0.1.

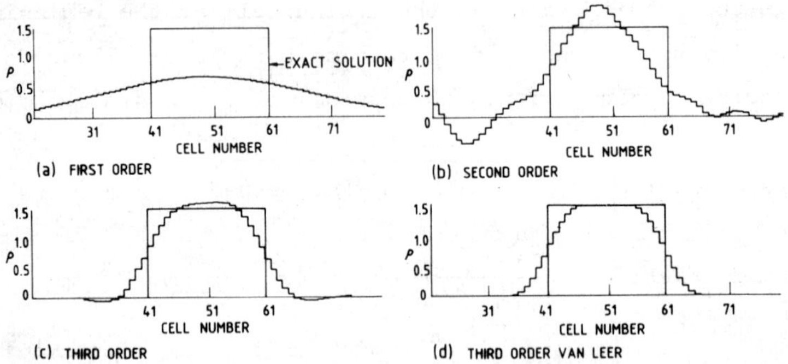

Fig. 2 Advection of a square pulse with uniform velocity.

The Van Leer advection method is used in the hydrocode at the end of the Lagrangian phase. The values of density, internal energy and velocity components at this stage are denoted by

$$\tilde{\rho}^n_{j+\frac{1}{2}\ k+\frac{1}{2}} = \rho^n_{j+\frac{1}{2}\ k+\frac{1}{2}}\ V^n_{j+\frac{1}{2}\ k+\frac{1}{2}}/\tilde{V}^n_{j+\frac{1}{2}\ k+\frac{1}{2}}\ ,\ \tilde{\varepsilon}^n_{j+\frac{1}{2}\ k+\frac{1}{2}}\ ,\ \tilde{u}^n_{jk}\ \text{and}\ \tilde{v}^n_{jk}.$$

The density value allows for the change in cell volume, $V^n_{j+\frac{1}{2}\ k+\frac{1}{2}}$ to $\tilde{V}^n_{j+\frac{1}{2}\ k+\frac{1}{2}}$, during the Lagrange phase. For the case when X-advection is calculated first and Y-advection is calculated second, $\tilde{\rho}^n_{j+\frac{1}{2}\ k+\frac{1}{2}}$ is updated to $\rho^*_{j+\frac{1}{2}\ k+\frac{1}{2}}$ during the X-step and $\rho^*_{j+\frac{1}{2}\ k+\frac{1}{2}}$ is updated to $\rho^{n+1}_{j+\frac{1}{2}\ k+\frac{1}{2}}$ during the Y-step. The Van Leer method uses three values of $\tilde{\rho}^n$ to calculate the flux across a cell side in the X-step and three values of ρ^* for a Y-step flux. If u and v are positive, the updated densities satisfy the following monotonicity conditions.

For the X-step

$$\min(\tilde{\rho}^{n}_{j-\frac{1}{2}\ k+\frac{1}{2}}, \tilde{\rho}^{n}_{j+\frac{1}{2}\ k+\frac{1}{2}}) \leq \rho^{*}_{j+\frac{1}{2}\ k+\frac{1}{2}} \leq \max(\tilde{\rho}^{n}_{j-\frac{1}{2}\ k+\frac{1}{2}}, \tilde{\rho}^{n}_{j+\frac{1}{2}\ k+\frac{1}{2}})$$

and for the Y-step

$$\min(\rho^{*}_{j+\frac{1}{2}\ k-\frac{1}{2}}, \rho^{*}_{j+\frac{1}{2}\ k+\frac{1}{2}}) \leq \rho^{n+1}_{j+\frac{1}{2}\ k+\frac{1}{2}} \leq \max(\rho^{*}_{j+\frac{1}{2}\ k-\frac{1}{2}}, \rho^{*}_{j+\frac{1}{2}\ k+\frac{1}{2}}).$$

Then for the advection phase as a whole $\rho^{n+1}_{j+\frac{1}{2}\ k+\frac{1}{2}}$ lies between the maximum and minimum of $\tilde{\rho}^{n}_{j-\frac{1}{2}\ k-\frac{1}{2}}$, $\tilde{\rho}^{n}_{j-\frac{1}{2}\ k+\frac{1}{2}}$, $\tilde{\rho}^{n}_{j+\frac{1}{2}\ k-\frac{1}{2}}$, $\tilde{\rho}^{n}_{j+\frac{1}{2}\ k+\frac{1}{2}}$. Similar results hold for different flow directions and for the variables $\varepsilon^{n+1}_{j+\frac{1}{2}\ k+\frac{1}{2}}$, u^{n+1}_{jk} and v^{n+1}_{jk}.

Experience in using the hydrocode for a wide range of applications has led to the following conclusions concerning the choice of advection method.

For advection of mass and internal energy the upwind method is usually adequate if variations of ρ and ε within each material are gradual. The interface treatment preserves the discontinuities in ρ and ε at material boundaries. If steep gradients of ρ or ε are present within a material then the Van Leer method is the best choice.

If the upwind method is used for momentum advection the numerical diffusion introduced dissipates kinetic energy and as a result distortion of material interfaces may be inhibited. An improved momentum advection method has often been found necessary. The Van Leer method greatly reduces the dissipation without producing an unstable solution.

Many of these points are illustrated in the following applications.

5. APPLICATIONS

5.1 Rayleigh-Taylor instability

In the starting configuration for this problem a fluid with $\rho = 2$ rests on top of a fluid with $\rho = 1$. Gravity acts downwards. The interface between the two fluids is initially horizontal but a velocity perturbation is present. As a result a spike of heavy fluid falls and bubbles of light fluid rise. Material properties have been chosen to give a low Mach number flow. The behaviour is then similar to that calculated for incompressible Rayleigh-Taylor instability, see (Daly, 1967).

Calculations have been performed in planar geometry using 40 zones in the horizontal direction and 100 zones in the vertical direction. For this application benefit is gained by using several, 4 in this case, Lagrange steps per advection step. An over-all time step of $\Delta t = 3\Delta x/c$ was used without loss of stability (shown by the pressure contours in Fig. 3(c)).

Figs. 3(a) and (b) show the effect of changing the momentum advection method. The vortices which are formed as the spike of denser fluid falls are partially damped out and the roll-up of the interface is reduced if the upwind method is used.

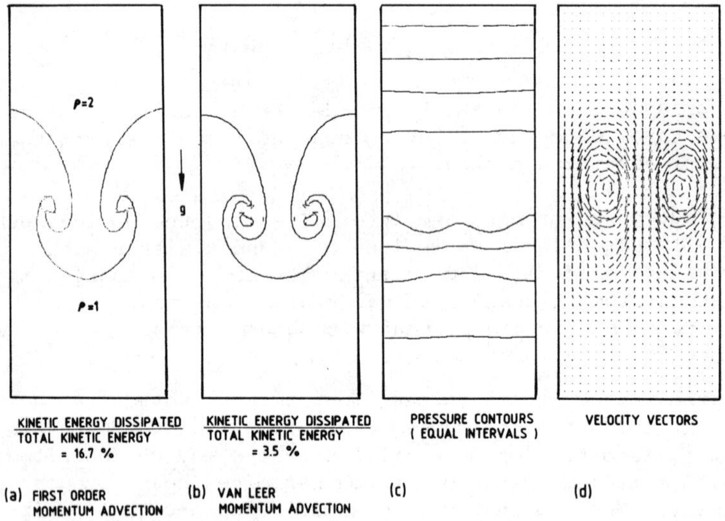

Fig. 3 Rayleigh-Taylor instability.

5.2 A shock tube problem

Calculations have been performed on the one-dimensional shock tube problem studied by Sod (1977). In Figs. 4(a) and 4(b) density profiles are shown for single material calculations. There is considerable smearing of the contact discontinuity when the upwind method is used for mass advection. The Van Leer advection method gives a large improvement. Alternatively the contact discontinuity may be treated as a material interface as illustrated in Fig. 4(c). Smearing of the contact discontinuity is then eliminated completely. As the methods of Van Leer (1979)

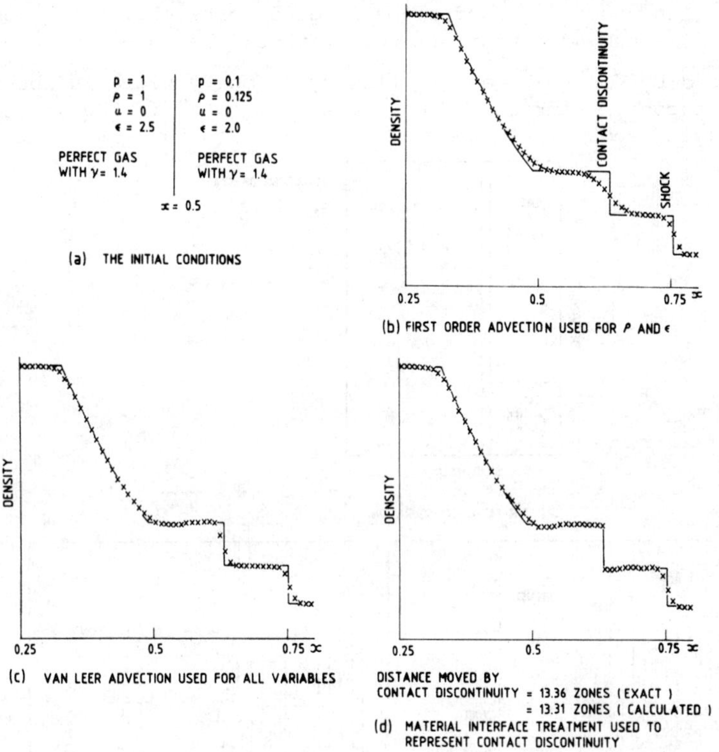

Fig. 4 Results for the shock tube problem of Sod (1978).

have not been used in the Lagrangian phase smearing of the shock front is still present.

5.3 An expanding pipe

This example demonstrates the ability of the interface tracking method to follow thin layers of fluid. A cylindrical pipe of liquid with $\rho = 10$ is filled with gas at $\rho = \frac{1}{2}$. Initially the gas at one end is pressurised. During the course of the calculation a shock wave travels through the gas and the pipe expands.

Interface and velocity vector plots are shown in Fig. 5. At the time illustrated the expanded end of the pipe is less than half a cell thick. The position of the expanded pipe calculated by the Eulerian code was checked against results from a

Lagrangian code and found to be in good agreement. If second order momentum advection is used, without the cut-offs on D_j proposed by Van Leer, an oscillatory velocity field is obtained. These supposed spurious effects are eliminated if the Van Leer method is used.

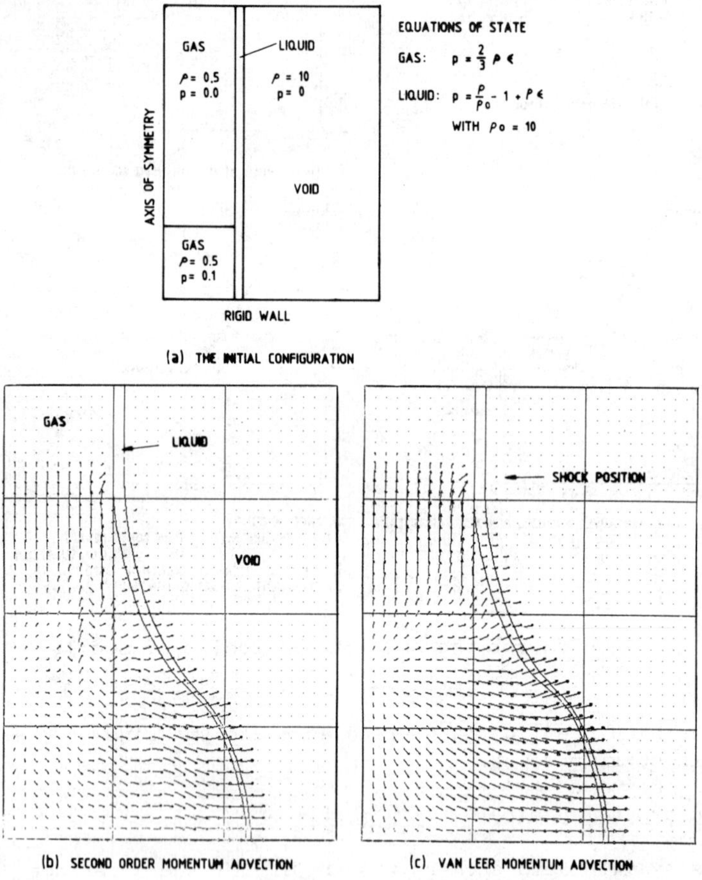

Fig. 5 Propagation of a shock wave inside an expanding pipe.

5.4 Explosively driven flows

Results are given here for application of the multi-material hydrocode to two experiments which were performed at AWRE.

In the first experiment a small explosive charge was detonated inside a brass cylinder, as illustrated in Fig. 6(a). The expansion of the outer surface of the cylinder in the experiment was obtained by high-speed photography. The calculation used

72 × 88 meshes to cover one quadrant of the flow. At the earlier time shown in Fig. 6 the observed interface position agrees well with the computer prediction. At the later time, the experiment lags slightly behind the calculation. This may be because the metal cylinder started to fracture by this time.

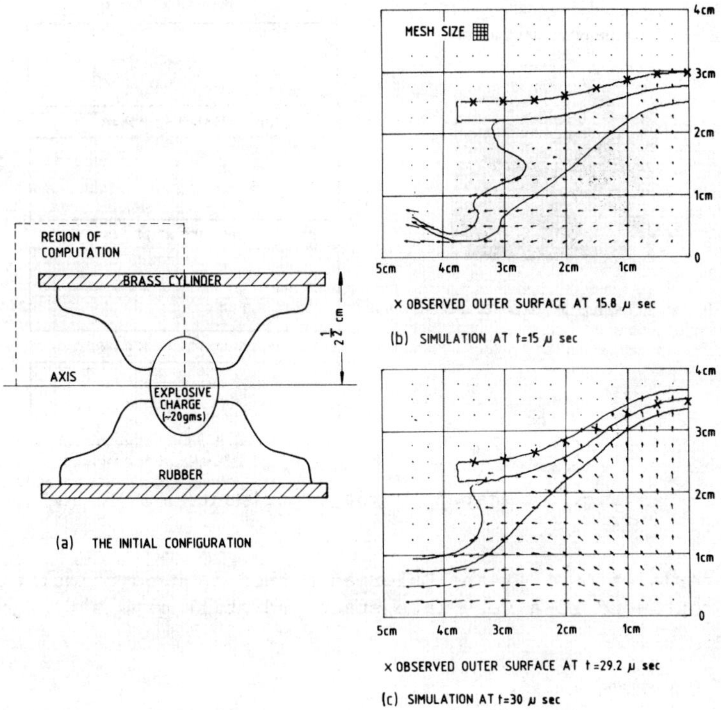

Fig. 6 Brass cylinder experiment.

In the second application the hydrocode is used to calculate a shaped charge experiment. High explosive drive is applied to a conical liner. As the liner collapses a high-velocity jet forms on the axis of symmetry. The computational region used 40 × 200 meshes as indicated in Fig. 7 and the calculation took approximately 10 minutes on a CRAY-1 computer. The interface plots in Fig. 7 show the formation of the jet and the late-time behaviour of the collapsed liner. Comparison with an X-ray of the experiment shows that the computed and observed liner shapes agree reasonably well at a late time.

These two applications show the advantages of the present Eulerian method over the conventional Lagrangian approach to

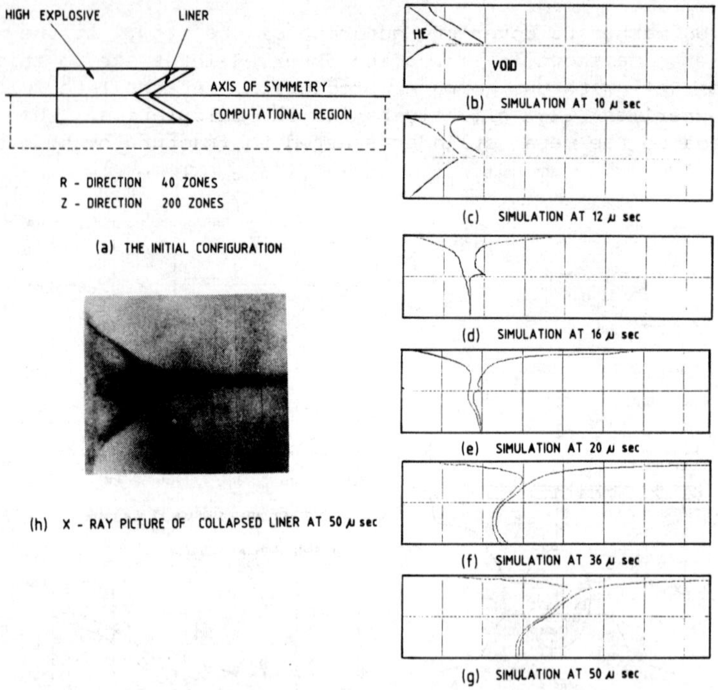

Fig. 7 Shaped charge experiment.

multimaterial flow. If the Eulerian method is used calculations may be continued to a very late stage and still give satisfactory results.

ACKNOWLEDGEMENTS

Experimental and computational results for the explosively driven flows described in section 5.4 were provided by Mr. E. Wade, Mr. R.V. Wells and Dr. R.E. Winter of AWRE.

REFERENCES

DALY, B.J. 1967 Numerical Study of Two Fluid Rayleigh-Taylor Instability. *Phys. Fluids* **10**, 297-307.

GENTRY, R.A., MARTIN, R.E. and DALY, B.J. 1966 An Eulerian Differencing Method for Unsteady Compressible Flow Problems. *J. Comput. Phys.* **1**, 87-118.

HIRT, C.W. and NICHOLS, B.D. 1981 Volume of Fluid (VOF) Method for the Dynamics of Free Boundaries. *J.Comput. Phys.* **39**, 201-225.

NOH, W.F. and WOODWARD, P. 1976 SLIC (Simple Line Interface Calculation), Lecture Notes in Physics **59**: Springer Verlag.

RAMSHAW, J.D. and TRAPP, J.A. 1976 A Numerical Technique for Low-Speed Homogeneous Two-Phase Flows with Sharp Interfaces. *J.Comput. Phys.* **21** 438-453.

SOD, G.A. 1978 A Survey of Several Finite Difference Methods for Systems of Non-linear Hyperbolic Conservation Laws. *J.Comput. Phys.* **27**, 1-31.

SUTCLIFFE, W.G. 1973 BBC Hydrodynamics. Report UCIR-760, Lawrence Livermore National Laboratory.

VAN ALBADA, G.D., VAN LEER, B. and ROBERTS, W.W. 1982 A comparative study of computational methods in cosmic gas dynamics. *Astron. Astrophys.* **108**, 76-84.

VAN LEER, B. 1977 Towards the Ultimate Conservative Difference Scheme. IV. A New Approach to Numerical Convection. *J.Comput. Phys.* **23**, 276-299.

VAN LEER, B. 1979 Towards the Ultimate Conservative Difference Scheme.V. A Second-Order Sequel to Godunov's Method. *J.Comput. Phys.* **32**, 101-136.

APPLICATION OF A GALERKIN-EIGENFUNCTION METHOD TO COMPUTING CURRENTS IN HOMOGENEOUS AND STRATIFIED SEAS

A.M. Davies

(Institute of Oceanographic Sciences, Birkenhead)

1. INTRODUCTION

Until recently accurate measurements of currents within the sea's boundary layers, namely at the sea surface and sea bed, could not be made due to difficulties of mooring instruments. However with the advent of remote sensing methods such as H.F. radar, it is now possible to measure currents in the surface layer. Also advances in current meter technology and moorings have enabled currents in the bottom boundary layer to be measured accurately.

These boundary layers are important physically since they are transition regions; at the sea surface, between air and sea, and at the sea bed, between water and sand/rock. Currents within these boundary layers may change very rapidly through the vertical, giving rise to regions of high shear. In this paper a method is developed to accurately model currents through the water column, particularly in these high shear layers.

A continuous representation of current from sea surface to sea bed is derived by expanding the two horizontal components of current in terms of time and horizontally dependent coefficients multiplying functions of the vertical co-ordinate (the basis functions).

In previous papers (Davies, 1977) a basis set of piecewise functions, namely B-splines, was employed although in subsequent papers trignometric functions (Davies, 1981a) and polynomial functions (Davies and Owen, 1979) were used.

In the eigenfunction method developed by Heaps (1972, 1981), eigenfunctions computed by solving an eigenvalue problem involving the vertical eddy viscosity were used as a basis set. This eigenvalue problem was solved analytically and consequently the vertical form of eddy viscosity was restricted to one for which an analytical solution was available (Heaps, 1972, 1981).

By solving the viscous eigenvalue problem numerically in terms of an expansion of functions, Davies (1982a, b) developed a method which could accommodate arbitrary vertical variations of eddy viscosity and still yield continuous functions through the vertical.

Both Heaps (1972, 1981) and Davies (1982a, b) solved the eigenvalue problem subject to a zero vertical derivative condition for each eigenfunction at the sea surface. The influence of this condition upon the accuracy of wind induced surface currents computed using these eigenfunctions as a basis set was examined by Davies (1982a). He showed that the effect of this condition was to significantly reduce the rate of convergence of the expansion at the sea surface.

In this paper a method is developed whereby the eigenvalue problem can still be solved but the vertical derivative of the eigenfunction at the sea surface can be specified. The effect of changes in the surface derivative of the eigenfunction upon the convergence of the expansion is examined for a number of eddy viscosity distributions.

The extension of the method to stratified seas is briefly discussed and eigenfunctions for homogeneous and stratified seas are compared.

2. HYDRODYNAMIC EQUATIONS DESCRIBING FLOW IN A HOMOGENEOUS SEA

Using Cartesian co-ordinates the linear three dimensional hydrodynamic equations are given by

$$\frac{\partial \zeta}{\partial t} = - \frac{\partial}{\partial x} \int_0^h U dz - \frac{\partial}{\partial y} \int_0^h V dz \tag{2.1}$$

$$\frac{\partial U}{\partial t} = \gamma V - g \frac{\partial \zeta}{\partial x} + \frac{\partial}{\partial z} \left(\mu \frac{\partial U}{\partial z} \right) \tag{2.2}$$

$$\frac{\partial V}{\partial t} = - \gamma U - g \frac{\partial \zeta}{\partial y} + \frac{\partial}{\partial z} \left(\mu \frac{\partial V}{\partial z} \right) , \tag{2.3}$$

where t denotes time, x, y, z a left handed set of co-ordinates, h the undisturbed depth of water, ζ free surface elevation, U, V the x and y components of current, γ geostrophic coefficients, g acceleration due to gravity and μ vertical eddy viscosity.

Sea surface boundary conditions evaluated at z = 0 are

$$-\rho\left(\mu\frac{\partial U}{\partial z}\right)_0 = F_S, \quad -\rho\left(\mu\frac{\partial V}{\partial z}\right)_0 = G_S \tag{2.4}$$

where F_S, G_S denote components of wind stress on the water surface, and ρ is the density of sea water.

Similarly using a stress condition at the sea bed gives

$$-\rho\left(\mu\frac{\partial U}{\partial z}\right)_h = F_B, \quad -\rho\left(\mu\frac{\partial V}{\partial z}\right)_h = G_B \tag{2.5}$$

Assuming a linear slip condition at the sea bed, with bottom stress in terms of bottom current, gives

$$F_B = k\rho U_h, \quad G_B = k\rho V_h \tag{2.6}$$

where k is the coefficient of bottom friction.

An alternative to applying a linear slip law (2.6) at the sea bed, is to use a no-slip condition, namely

$$U_h = V_h = 0. \tag{2.7}$$

Before solving (2.1), (2.2) and (2.3) using the Galerkin method it is convenient to transform these equations into depth-following sigma co-ordinates, using

$$\sigma = z/h. \tag{2.8}$$

Transforming (2.1), (2.2), (2.3) using (2.8) gives

$$\frac{\partial\zeta}{\partial t} = -\frac{\partial}{\partial x}\left\{h\int_0^1 U d\sigma\right\} - \frac{\partial}{\partial y}\left\{h\int_0^1 V d\sigma\right\} \tag{2.9}$$

$$\frac{\partial U}{\partial t} = \gamma V - g\frac{\partial\zeta}{\partial x} + \frac{1}{h^2}\frac{\partial}{\partial\sigma}\left(\mu\frac{\partial U}{\partial\sigma}\right) \tag{2.10}$$

$$\frac{\partial V}{\partial t} = -\gamma U - g\frac{\partial\zeta}{\partial y} + \frac{1}{h^2}\frac{\partial}{\partial\sigma}\left(\mu\frac{\partial V}{\partial\sigma}\right). \tag{2.11}$$

Considering the solution of (2.9), (2.10) and (2.11) using the Galerkin method, we expand the U and V components of velocity in terms of m depth-dependent functions $f_r(\sigma)$ and coefficients $A_r(x,y,t)$ and $B_r(x,y,t)$ thus:

$$U = \sum_{r=1}^{m} A_r f_r(\sigma) \tag{2.12}$$

$$V = \sum_{r=1}^{m} B_r f_r(\sigma). \tag{2.13}$$

Substituting (2.12) and (2.13) into the continuity equation (2.9) gives

$$\frac{\partial \zeta}{\partial t} = - \sum_{r=1}^{m} \left[\left\{ \frac{\partial}{\partial x} (A_r h) + \frac{\partial}{\partial y} (B_r h) \right\} . \int_0^1 f_r d\sigma \right] \tag{2.14}$$

Considering equation (2.10) and applying the Galerkin method, we multiply it by f_k and integrate with respect to σ. By integrating the term involving μ surface and bottom boundary conditions can be included. Here we consider the no-slip bottom boundary condition (2.7) an essential boundary condition, consequently

$$f_r(1) = 0 \text{ for all } r. \tag{2.15}$$

Satisfying condition (2.15), we obtain for the Galerkin form of (2.10)

$$\sum_{r=1}^{m} \frac{\partial A_r}{\partial t} \int_0^1 f_r f_k d\sigma = \gamma \sum_{r=1}^{m} B_r \int_0^1 f_r f_k d\sigma - g \frac{\partial \zeta}{\partial x} \int_0^1 f_k d\sigma \tag{2.16}$$

$$+ \frac{F_s}{\rho h} f_k(0) - \frac{1}{h^2} \sum_{r=1}^{m} A_r \int_0^1 \mu \frac{df_r}{d\sigma} \frac{df_k}{d\sigma} d\sigma \text{ where } k=1,2,\ldots,m.$$

From (2.11) the Galerkin form of the V-equation is obtained in a similar manner.

In general μ will vary in an arbitrary manner and this case has been considered by Davies (1981a), Davies and Owen (1979).

In many cases however the vertical variation of μ is fixed although its magnitude varies with horizontal position and time (see Davies, 1981b): when the vertical variation of μ is fixed we have

$$\mu = \psi(x,y,t)\Phi(\sigma), \tag{2.17}$$

where Φ is a fixed function representing the vertical variation of μ. In this paper we examine the case in which μ is given by (2.17) and consider the problem in which the basis functions used in (2.12) and (2.13) are solutions of an eigenvalue equation of the form

$$\frac{d}{d\sigma}\left[\Phi \frac{df}{d\sigma}\right] = - \varepsilon f. \tag{2.18}$$

In general Φ is arbitrary and hence it is necessary to compute numerically the set of eigenfunctions f and eigenvalues ε in (2.18).

Equation (2.18) can be readily solved using the Galerkin method. Following Davies (1982a, b), the rth eigenfunction f_r is expanded in terms of a set of $\bar{m}$ coefficients d_{ir} and B-splines M_i, giving

$$f_r = \sum_{i=1}^{\bar{m}} d_{ir} M_i(\sigma). \tag{2.19}$$

The B-splines in (2.19) are generated using the stable recurrence algorithm due to Cox (1972). With this method splines of an arbitrary order can be generated. However in this paper we shall consider only fourth order B-splines.

The piecewise nature of the B-splines enables essential boundary conditions such as (2.15) to be readily incorporated by linearly combining the splines (see Davies (1982a) for details).

Applying the Galerkin method to (2.18), the rth eigen-equation is multiplied by f_k and integrated over the region 0 to 1. Integrating the term involving Φ by parts gives

$$\Phi \frac{df_r}{d\sigma} f_k\bigg|_1 - \Phi \frac{df_r}{d\sigma} f_k\bigg|_0 - \int_0^1 \Phi \frac{df_r}{d\sigma}\frac{df_k}{d\sigma}\, d\sigma = - \varepsilon_r \int_0^1 f_r f_k d\sigma, \tag{2.20}$$

where $k=1,2,\ldots,m$.

The first term on the left hand side of (2.20) is zero because of the essential boundary condition (2.15). If we take

$$\Phi \left.\frac{df_r}{d\sigma}\right|_0 = \beta \left. f_r \right|_0 \tag{2.21}$$

with β a constant, then (2.20) gives

$$-\beta \left. f_r f_k \right|_0 - \int_0^1 \Phi \frac{df_r}{d\sigma}\frac{df_k}{d\sigma}\, d\sigma = -\varepsilon_r \int_0^1 f_r f_k d\sigma . \tag{2.22}$$

The use of condition (2.21) ensures that the eigenvalues of (2.22) are real. By normalizing the eigenfunctions such that $f_r(0) = 1$, then condition (2.21) becomes

$$\left.\frac{df_r}{d\sigma}\right|_0 = \beta/\Phi(0). \tag{2.23}$$

Since β in (2.23) is an arbitrary constant, then (2.23) enables the vertical derivative of the eigenfunction at the sea surface to be specified in an arbitrary manner. For the case in whicn $\beta = 0$, using (2.17) and (2.22), we obtain from (2.16)

$$\frac{\partial A_k}{\partial t}\int_0^1 f_k f_k d\sigma = \gamma B_k \int_0^1 f_k f_k d\sigma - g \frac{\partial \zeta}{\partial x}\int_0^1 f_k d\sigma$$

$$+ \frac{F_s}{\rho h} f_k(0) - \psi \frac{\varepsilon_k A_k}{h^2}\int_0^1 f_k f_k d\sigma , \tag{2.24}$$

where $k=1,2,\ldots,m$.

In deriving (2.24) advantage has been taken of the orthogonality property of eigenfunctions, namely

$$\int_0^1 f_r f_k d\sigma = 0 \qquad r \neq k. \tag{2.25}$$

A consequence of the orthogonality property (2.25) is that (2.24) is of the form of k partial differential equations, which are uncoupled and can be readily integrated forward through time.

For β non-zero (2.24) involves the additional terms $\beta f_r f_k \Big|_0$ which come from (2.22).

3. WIND INDUCED MOTION IN A SEA BASIN

3.1 A homogeneous sea

In order to determine the accuracy of current profiles computed using eigenfunctions with various values of β and to compare these current profiles with profiles computed previously using other basis functions (Davies, 1982a, Davies and Owen, 1979), wind induced motion in a simple rectangular sea area was examined.

The closed sea basin has dimensions and rotation representative of the North Sea and has been described previously (Davies and Owen, 1979). Motion was started from a state of rest by the sudden imposition of a uniform northerly wind stress of -15 dyn/cm^2. A conventional staggered finite difference grid scheme was used in the horizontal (see Davies and Owen, 1979). Parameters used in the calculation were: grid spacings, $\Delta x = 400/9$ km, $\Delta y = 800/17$ km, with depth $h = 65$m, $\gamma = 0.44\ h^{-1}$, $\rho = 1.025$ gm/cm^3, $g = 981$ cm^2/s.

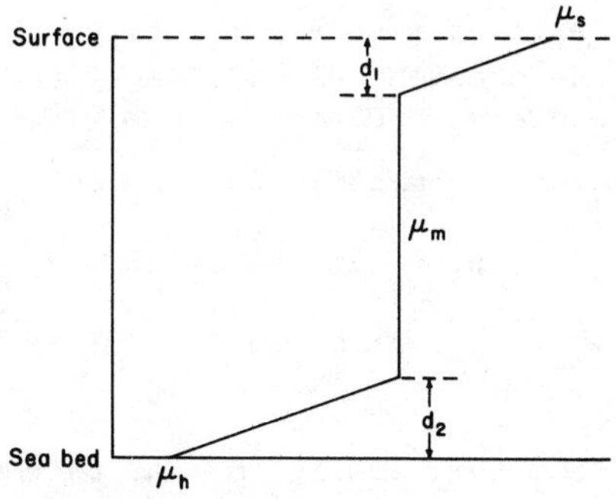

Fig. 1 Schematic diagram showing the depth distribution of eddy viscosity used in the homogeneous model.

The vertical variation of eddy viscosity used at each point in the basin is shown schematically in Fig. 1. In an initial series of calculations we have taken $\mu_s = 130$ cm^2/s, $\mu_m = 650$ cm^2/s, $\mu_h = 130$ cm^2/s, with d_1 and d_2 fixed at 11 m. The first five eigenfunctions of (2.18) using this distribution of vertical eddy viscosity, computed with twenty five equally spaced knots in the vertical, for β = 0.0, -0.20 and +0.20 are shown in Fig. 2.

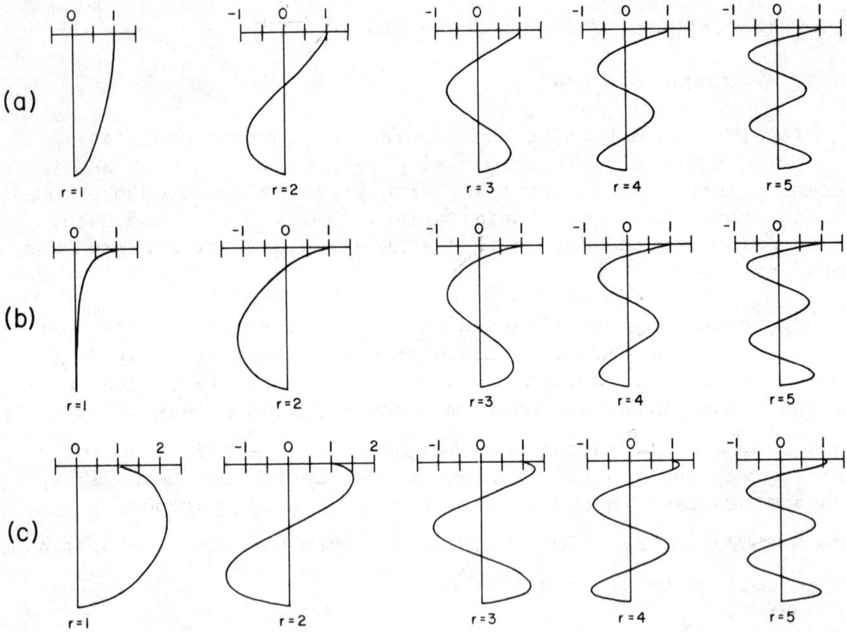

Fig. 2 Vertical variation of the first five eigenfunctions used in the homogeneous model computed with a no slip bottom boundary condition:

(a) $\beta = 0.0$ $\mu_s = 130$; $\mu_m = 650$, $\mu_h = 130\ cm^2/s$,

(b) $\beta = -0.2$ $\mu_s = 130$; $\mu_m = 650$, $\mu_h = 130\ cm^2/s$,

(c) $\beta = +0.2$ $\mu_s = 130$; $\mu_m = 650$; $\mu_h = 130\ cm^2/s$.

It is evident from this figure that as β changes from 0.0 to -0.20, the shear region at the sea surface increases, particularly in the lower modes. However, a positive value of β produces a vertical derivative at the sea surface in the opposite sense to that below it, particularly in the higher modes.

Sea surface and mid depth U and V components of current at the centre of the basin, fifteen hours after the onset of the wind field, are shown in Table 1. It is evident from this Table that changing β from 0.0 to -0.10, significantly increases the V component of surface current (the component in the direction of the wind stress). The rate of convergence of the V component

TABLE 1

U and V components of current, 15 hours after the onset of the wind, computed with an eddy viscosity distribution in which $\mu_s = 130\text{cm}^2/\text{s}$, $\mu_m = 650\text{cm}^2/\text{s}$, $\mu_h = 130\text{cm}^2/\text{s}$ and a range of β values.

β Value	Component	Depth	Number of Eigenfunctions (m)			10 Chebyshev polynomials (Davies and Owen, 1979)
			5	10	15	
0.0	U	0.0	-11.98	-12.22	-12.13	-11.40
	V	0.0	-27.04	-34.25	-36.66	-41.20
	U	0.5	3.42	3.46	3.46	4.01
	V	0.5	19.54	20.69	20.88	21.39
-0.10	U	0.0	-14.21	-13.11	-12.78	
	V	0.0	-39.72	-40.81	-41.12	
	U	0.5	3.73	3.48	3.45	
	V	0.5	19.74	20.57	20.69	
-0.15	U	0.0	-13.58	-12.84	-12.60	
	V	0.0	-45.94	-44.01	-43.31	
	U	0.5	3.49	3.25	3.32	
	V	0.5	17.78	19.30	19.77	
-0.20	U	0.0	-11.84	-12.06	-12.10	
	V	0.0	-49.47	-46.41	-45.11	
	U	0.5	3.05	3.05	3.13	
	V	0.5	15.26	17.55	18.44	
+0.20	U	0.0	- 4.82	- 7.07	- 8.30	
	V	0.0	-14.11	-24.27	-29.12	
	U	0.5	1.87	2.49	2.79	
	V	0.5	10.08	14.04	15.99	

of surface current as the number of basis functions increases is also greater for $\beta = -0.10$ than for $\beta = 0.0$. A further increase of β to -0.15 again increases the value of the V component of surface current. It is apparent that in this case using an expansion of five functions (i.e. $m = 5$) the V component of surface current is too high and more functions are required to accurately compute it. It appears that a value of $\beta = -0.1$ is near optimum in this case, and gives currents in good agreement with those computed by Davies and Owen (1979) using Chebyshev polynomials (see Table 1).

It is apparent from the Table that a positive value of β significantly reduces the magnitude of the surface current, and the convergence of the expansion in this case is particularly slow. The reason for this is that in this case the derivative of the eigenfunction at the surface is in the opposite sense to that found in the exact solution of this problem (see Davies and Owen, 1979) and this has the effect of reducing the accuracy of the solution.

To examine if the optimum value of β is affected by changes in the vertical variation of eddy viscosity or of its surface value, the previous calculation was repeated with $\mu_s = 2600\ cm^2/s$, $\mu_m = 650\ cm^2/s$, $\mu_h = 5\ cm^2/s$, values used previously (Davies, 1982a).

The first five eigenfunctions of this vertical eddy viscosity distribution computed with $\beta = 0.0$ are shown in Fig. 3. Comparing Fig. 3 and distribution (a) in Fig. 2, it is evident that reducing the vertical eddy viscosity near the sea bed increases the shear region in this area (see Davies, 1982a, for a full discussion).

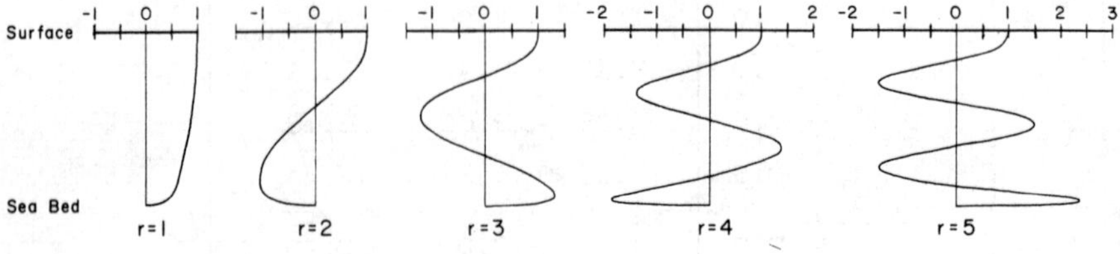

Fig. 3 Vertical variation of the first five eigenfunctions computed with

$\beta = 0.0$, $\mu_s = 2600$; $\mu_m = 650$; $\mu_h = 5\ cm^2/s$.

TABLE 2

U and V components of current, 15 hours after the onset of the wind, computed with an eddy viscosity distribution in which $\mu_s = 2600\text{cm}^2/\text{s}$, $\mu_m = 650\text{cm}^2/\text{s}$, $\mu_h = 5\text{cm}^2/\text{s}$ and a range of β values.

β Value	Component	Depth	Number of Eigenfunctions (m) 5	10	15	28 B-splines (Davies, 1982a)
0.0	U	0.0	-11.22	-11.24	-11.24	-11.23
	V	0.0	- 7.12	- 8.67	- 9.50	-10.21
	U	0.5	0.76	0.78	0.78	0.78
	V	0.5	20.74	20.62	20.61	20.61
-0.10	U	0.0	-11.87	-11.59	-11.43	
	V	0.0	- 8.37	- 9.36	- 9.88	
	U	0.5	0.73	0.73	0.72	
	V	0.5	20.79	20.61	20.56	
-0.15	U	0.0	-12.09	-11.71	-11.50	
	V	0.0	- 9.43	- 9.92	-10.17	
	U	0.5	0.72	0.71	0.69	
	V	0.5	20.43	20.40	20.45	
-0.20	U	0.0	-12.22	-11.79	-11.55	
	V	0.0	-10.73	-10.61	-10.51	
	U	0.5	0.72	0.69	0.66	
	V	0.5	19.84	20.08	20.27	

From Table 2, it is evident that as β increases the V component of surface current increases, with a β value of -0.2 giving the fastest rate of convergence.

It is apparent from these two examples that the value of β which gives the fastest rate of convergence in the expansion of eigenfunctions depends upon the vertical eddy viscosity. These two distributions of viscosity probably represent the extremes that will occur in wind induced circulation (Davies, 1982a) and it therefore appears appropriate to recommend a value of $\beta = -0.15$ with $m = 10$ in any practical calculation. Although such a value of β is not optimal, it is evident from Tables 1 and 2 that the choice of this value gives significantly more accurate results than using $\beta = 0.0$ with $m = 10$.

3.2 A stratified sea

In a stratified sea, density changes in the vertical, often going through a region of rapid density change (the pycnocline, see Fig. 4). Within the pycnocline turbulence is reduced and there is an associated decrease in eddy viscosity (Fig. 4). Fig. 5 shows the first four eigenfunctions computed using an expansion of twenty five B-splines with the eddy viscosity distribution shown in Fig. 4 and $\mu_1 = 500\ cm^2/s$, $\mu_2 = 200\ cm^2/s$, $\mu_3 = 10\ cm^2/s$ and $\mu_4 = 100\ cm^2/s$. In this particular case a slip bottom boundary condition (equation 2.6) was used and $\beta = 0.0$.

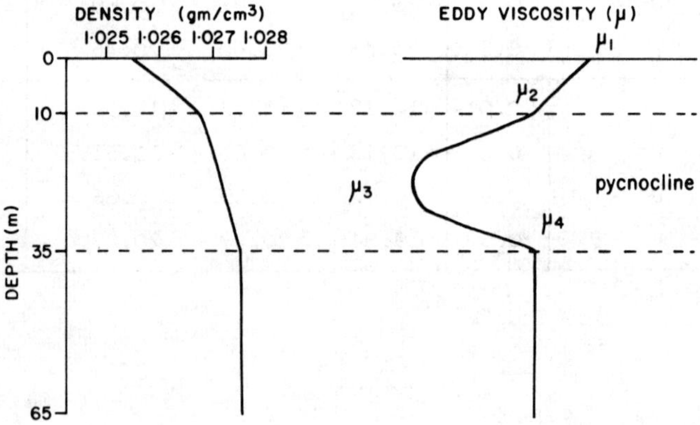

Fig. 4 Schematic diagram showing the depth distributions of density and eddy viscosity used in the stratified model.

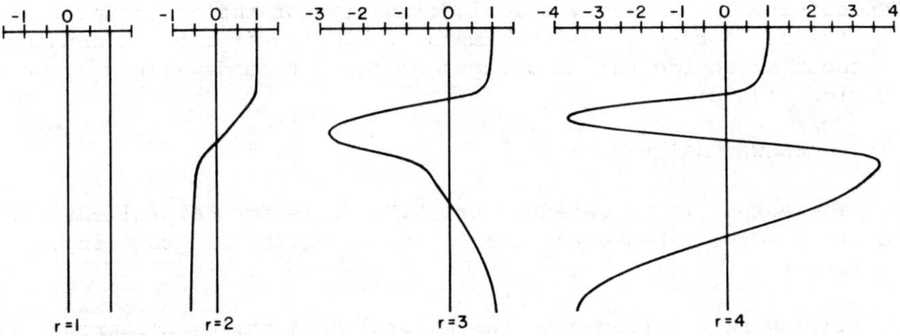

Fig. 5 Vertical variation of the first four eigenfunctions used in the stratified model, computed with $\beta = 0.0$, and $\mu_1 = 500$, $\mu_2 = 200$, $\mu_3 = 10$, $\mu_4 = 100\ cm^2/s$, using a slip bottom boundary condition.

Comparing Figs. 2, 3 and 5, it is evident that with a slip condition the rapid decrease of the eigenfunction to zero at the sea bed does not occur. Also, the effect of stratification is to produce a rapid vertical variation in the eigenfunctions in the region of the pycnocline.

The solution of the equations describing flow in a stratified sea have been solved using these modes as basis functions and used to examine inertial currents (Davies, 1982b). In a stratified sea the internal modes can be separated from the external mode and by integrating the modes using a time splitting algorithm (Davies, 1982b) a computationally very economic method is obtained.

4. CONCLUDING REMARKS

By solving the viscous eigenvalue problem in terms of an expansion of B-splines, eigenfunctions and eigenvalues for an arbitrary vertical variation of eddy viscosity can be readily computed. These eigenfunctions are subsequently used as a basis set in the solution of the three dimensional hydrodynamic equations.

Davies (1982a) has shown that this method has several advantages over using the B-splines directly as basis functions. The original method of Davies (1982a), however, suffered from the problem of poor convergence of the expansion when used to model wind induced surface currents.

The method developed in this paper overcomes this problem by specifying a non zero vertical derivative of the eigenfunctions at the sea surface. By this means rapid convergence is obtained without affecting the advantages of the method developed by Davies (1982a).

5. ACKNOWLEDGEMENTS

The author is indebted to Dr. N.S. Heaps for helpful comments on this work. Thanks are due to Mr. R. Smith for preparing diagrams.

This work was funded by the Natural Environment Research Council and the Department of Energy.

6. REFERENCES

COX, M. 1972 The numerical evaluation of B-splines, *J. Inst. Maths. Applics.* **10**, 134-149.

DAVIES, A.M. 1977 The numerical solution of the three-dimensional hydrodynamic equations using a B-spline representation of the vertical current profile. In Bottom Turbulence, Proceedings of the 8th Liège Colloquium on Ocean Hydrodynamics (J.C.J. Nihoul, Ed.), (Elsevier), 1-25.

DAVIES, A.M. 1981a Three-dimensional hydrodynamic numerical models. Part 1. A homogeneous ocean-shelf model. Part 2. A stratified model of the northern North sea, pp. 370-426. In Proceedings of the Norwegian Coastal Current Symposium, Geilo, Norway. Published by Bergen University.

DAVIES, A.M. 1981b Three dimensional modelling of surges, pp. 45-74. In Floods due to High Winds and Tides, (D.H. Peregrine, Ed.), Academic Press.

DAVIES, A.M. 1982a Formulation of a linear three-dimensional hydrodynamic sea model using a Galerkin-Eigenfunction method. (in press) *Int. J. Num. Meth. Fluids.*

DAVIES, A.M. 1982b Formulating a three-dimensional hydrodynamic sea model using a mixed Galerkin-Finite difference method. (in press). In Proceedings of the 4th International Conference on Finite Elements in Water Resources, Hannover June 1982.

DAVIES, A.M. and OWEN, A. 1979 Three-dimensional numerical sea model using the Galerkin method with a polynomial basis set, *Appl. Math. Modelling*, **3**, 421-428.

HEAPS, N.S. 1972 On the numerical solution of the three-dimensional hydrodynamic equations for tides and storm surges, *Mem. Soc. R. Sci. Liège Ser. 6*, **2**, 143-180.

HEAPS, N.S. 1981 Three-dimensional model for tides and surges with vertical eddy viscosity prescribed in two layers - I. Mathematical formulation. *Geophys. J.R. astr. Soc.* **44**, 291-302.

CHOICE OF TIME-INTEGRATION SCHEME IN PSEUDOSPECTRAL ALGORITHM FOR ADVECTION EQUATIONS

Z. Zlatev, R. Berkowicz and L.P. Prahm

(National Agency of Environmental Protection, Risø National Laboratory, Roskilde, Denmark)

1. STATEMENT OF THE PROBLEM

Partial differential equations (PDE's) of the form: $\partial c/\partial t = -u(x,y,t)\,\partial c/\partial x - v(x,y,t)\,\partial c/\partial y + Q(x,y,t)$ appear in the study of many advection phenomena in meteorological environments. This is especially true for the models describing long-range transport of air pollution in the atmosphere. Assume that trigonometric interpolants, which are truncated Fourier series, are applied in the discretization of the space derivatives $\partial c/\partial x$ and $\partial c/\partial y$. By this procedure, which is well-known as the pseudo-spectral (Fourier) method, the above PDE is transformed into a linear system of ordinary differential equations (ODE's). The use of two classes of time-integration schemes (explicit linear multistep methods and predictor-corrector methods) in the numerical solution of the system of ODE's is discussed. Several special time-integration schemes that exploit the salient features of the system of ODE's, obtained after the pseudospectral discretization, are proposed. It is demonstrated that in some special situations one of the new time-integration schemes performs better than the well-known and commonly used leap-frog (mid-point) rule. Two time-integration schemes of Adams type are also used in the experiments. Finally, some short remarks about the application of variable stepsize variable formula methods in the solution of the system of ODE's are made.

Consider

$$\frac{\partial c}{\partial t} = -u(x,y,t)\,\frac{\partial c}{\partial x} - v(x,y,t)\,\frac{\partial c}{\partial y} + Q(x,y,t) \qquad (1.1)$$

where $x\in[0,2\pi]$, $y\in[0,2\pi]$, $t\in[0,T]$ and

$$c(x,y,0) = f(x,y) \qquad \text{(f being a given function)} \qquad (1.2)$$

$$c(0,y,t) = c(2\pi,y,t), \qquad (1.3)$$

$$c(x,0,t) = c(x,2\pi,t). \qquad (1.4)$$

Let M and N be positive integers and let

$$X_M = \{x_m | x_0 = 0,\ x_{m+1} = x_m + \Delta x,\quad \Delta x = 2\pi/(2M+1),\ m = 0(1)2M,\ x_{2M+1} = 2\pi\}, \tag{1.5}$$

$$Y_N = \{y_n | y_0 = 0,\ y_{n+1} = y_n + \Delta y,\quad \Delta y = 2\pi/(2N+1),\ n = 0(1)2N,\ y_{2N+1} = 2\pi\}, \tag{1.6}$$

$$G(t) = \begin{bmatrix} c(x_0,y_0,t) & c(x_1,y_0,t) & \dots & c(x_{2M},y_0,t) \\ c(x_0,y_1,t) & c(x_1,y_1,t) & \dots & c(x_{2M},y_1,t) \\ \dots & \dots & \dots & \dots \\ c(x_0,y_{2N},t) & c(x_1,y_{2N},t) & \dots & c(x_{2M},y_{2N},t) \end{bmatrix} \tag{1.7}$$

where t is some fixed value in the interval $[0,T]$. Define the trigonometric polynomials

$$T_{M,n}(x,t) = A_n(t) + \sum_{k=1}^{M} [a_{k,n}(t)\cos kx + b_{k,n}(t)\sin kx],\quad n = 0(1)2N, \tag{1.8}$$

$$T^*_{N,m}(y,t) = A^*_m(t) + \sum_{k=1}^{N} [a^*_{k,m}(t)\cos ky + b^*_{k,m}(t)\sin ky],\quad m = 0(1)2M,$$

where

$$A_n(t) = \frac{1}{2M+1} \sum_{m=0}^{2M} c(x_m,y_n,t),\quad n = 1(1)2N, \tag{1.10}$$

$$a_{k,n}(t) = \frac{2}{2M+1} \sum_{m=0}^{2M} c(x_m,y_n,t)\cos kx_m,\quad n = 1(1)2N, \tag{1.11}$$

$$b_{k,n}(t) = \frac{2}{2M+1} \sum_{m=0}^{2M} c(x_m,y_n,t)\sin kx_m,\quad n = 1(1)2N, \tag{1.12}$$

$$A^*_m(t) = \frac{1}{2N+1} \sum_{n=0}^{2N} c(x_m, y_n, t), \quad m = 0(1)2M, \tag{1.13}$$

$$a^*_{k,m}(t) = \frac{2}{2N+1} \sum_{n=0}^{2N} c(x_m, y_n, t) \cos ky_n, \quad m = 1(1)2M, \tag{1.14}$$

$$b^*_{k,m}(t) = \frac{2}{2N+1} \sum_{n=0}^{2N} c(x_m, y_n, t) \sin ky_n, \quad m = 1(1)2M. \tag{1.15}$$

Let $L = (2M+1)(2N+1)$ and $g \in \mathbb{R}^L$ be the vector whose components are the elements of matrix $G(t)$ taken by rows (i.e. first the elements of the first row, then the elements of the second row and so on). By the use of vector g (1.1) can be approximated on the points of grid $X_M \times Y_N$ by a linear system of ordinary differential equations (ODE's) of the form (U, V and Q^* being functions of t):

$$\frac{dg}{dt} = (US + VPS^*P)g + Q^* \tag{1.16}$$

where (i) S is an $L \times L$ quasidiagonal matrix containing $2N+1$ diagonal blocks which are $(2M+1) \times (2M+1)$ matrices, (ii) S^* is an $L \times L$ quasidiagonal matrix containing $2M+1$ diagonal blocks which are $(2N+1) \times (2N+1)$ matrices, (iii) U and V are $L \times L$ diagonal matrices whose diagonal elements are values of functions u and v at the grid-points, (iv) Q^* is a vector with L components which are values of function Q at the grid-points, (v) P is a permutation matrix such that $P^{-1} = P$.

The choice of a time-integration scheme for solving (1.16) will be discussed in the next section. However, before the start of this discussion it is necessary to emphasize that:

a) the choice of the interval $[0,2\pi]$ is convenient in the representation of the trigonometric polynomials (1.8) and (1.9), however this is not a restriction (any finite interval can easily be transformed into $[0,2\pi]$),
b) the use of even numbers of grid-points in (1.5) - (1.6) is not a restriction either (all formulae can be modified for odd numbers of grid-points).

Both intervals different from $[0,2\pi]$ and odd numbers of the points in the grids will be used in the next sections.

2. TIME - INTEGRATION SCHEMES

Numerical time-integration algorithms of explicit type are normally used in the solution of (1.16). Consider the grid

$$T_K = \{t_k \mid t_0 = 0,\ t_{k+1} = t_k + \Delta t,\ \Delta t = T/K,\ k = 0(1)K-1,\ K \in \mathbb{N}\}. \tag{2.1}$$

Assume that all approximations g_j (to the exact solution $g(t_j)$ of the system of ODE's) up to some approximation g_k $(k=1(1)K)$ have been computed. Then an explicit linear multi-step (LM) formula is given by

$$g_k = \sum_{i=1}^{s} \alpha_i g_{k-i} + \Delta t \sum_{i=1}^{s} \beta_i f_{k-i} \tag{2.2}$$

where $f_j = (US + VPS*P)g_j + Q*$ for $j = 0(1)K$ (U, V and Q* depend on t_j).

A PECE (prediction-evaluation-correction-evaluation) scheme is defined by the following formulae:

$$g_k^* = \sum_{i=1}^{s} \alpha_i^* g_{k-i} + \Delta t \sum_{i=1}^{s} \beta_i^* f_{k-i}, \tag{2.3}$$

$$f_k^* = (US + VPS*P)g_k^* + Q* \quad \text{(U, V and Q* depend on } t_k), \tag{2.4}$$

$$g_k = \sum_{i=1}^{s} \alpha_i g_{k-i} + \Delta t \beta_0 f_k^* + \Delta t \sum_{i=1}^{s} \beta_i f_{k-i}, \tag{2.5}$$

$$f_k = (US + VPS*P)g_k + Q* \quad \text{(U, V and Q* depend on } t_k). \tag{2.6}$$

Consider the special system of ODE's:

$$\frac{dg}{dt} = Sg \quad \text{(S being a constant matrix)}. \tag{2.7}$$

Let λ_i $(i=1(1)2M+1)$ be the eigenvalues of S. Let λ be the eigenvalue of S for which

$$|\lambda| = \max(|\lambda_j|); \quad j = 1(1)2M+1, \quad Re(\lambda_j) \le 0, \tag{2.8}$$

is satisfied. It is well-known (Dahlquist (1963), Gear (1971),

Lambert (1973), Stetter (1973)) that an absolute stability region $\bar{S}$ can be associated with any LM formula or PECE scheme so that if

$$\lambda\Delta t \in \bar{S} \wedge \mathrm{Re}(\lambda_j) \leqq 0 \quad \text{for} \quad \forall j \in \{1,2,\ldots,2M+1\}, \tag{2.9}$$

then the computations are stable.

One might think that (2.7) is obtained after the pseudospectral (Fourier) discretization of $\partial c/\partial x$ in $\partial c/\partial t = \partial c/\partial x$ (which can formally be found from (1.1) by setting $u(x,y,t) \equiv -1$, $v(x,y,t) \equiv Q(x,y,t) \equiv 0$). If this is so, then S is antisymmetric and its eigenvalues are $\lambda_j = (-M+j-1)i$ ($j=1(1)2M+1$, $i^2=-1$); see (Kreiss and Oliger, 1972) or (Fornberg, 1975). This shows that in this special case we are not interested in the whole region $\bar{S}$ (as in the general theory of ODE's). The absolute stability interval $[0,h_{imag}]$ on the positive part of the imaginary axis is quite sufficient. Therefore the computations will be stable if

$$\Delta t \leqq \frac{h_{imag}}{M} . \tag{2.10}$$

In the case where $x \in [a,b]$ (instead of $x \in [0,2\pi]$) and the number of grid-points is 2M+1 (instead of 2M+2 as in (1.5)), then

$$\Delta t \leqq h_{imag} \left[\frac{M}{\pi(M-1)}\right] \Delta x \tag{2.11}$$

is necessary for stable computations in the solution of (2.7).

Let us now turn back to the general equation (1.16). Assume that the numbers of grid-points in (1.5) and (1.6) are odd (2M+1 and 2N+1 respectively). Let $x \in [a_1,b_1]$, $y \in [a_2,b_2]$, $\Delta x = (b_1-a_1)/2M$, $\Delta y = (b_2-a_2)/2N$ and denote

$$u^* = \max_{\substack{x \in [a_1,b_1] \\ y \in [a_2,b_2] \\ t \in [0,T]}} (|u(x,y,t)|), \tag{2.12}$$

$$v^* = \max_{\substack{x\in[a_1,b_1] \\ x\in[a_2,b_2] \\ t\in[0,T]}} (|v(x,y,y)|). \tag{2.13}$$

Then it should be expected that the computations in the solution of (1.16) will be stable if

$$\Delta t \leqq \frac{h_{imag}}{\pi}\left[\frac{u^*(M-1)}{M\Delta x}+\frac{v^*(N-1)}{N\Delta y}\right]^{-1}. \tag{2.14}$$

If $M = N \wedge \Delta x = \Delta y$, then (2.14) can be rewritten as

$$\Delta t \leqq \frac{h_{imag}}{\pi}\left[\frac{M}{(u^*+v^*)(M-1)}\right]\Delta x. \tag{2.15}$$

It is seen from (2.14) and (2.15) that the development of time-integration schemes with a large h_{imag} is desirable. Some such schemes will be discussed in the next section.

3. CONSTRUCTION OF SOME SPECIAL TIME - INTEGRATION SCHEMES

The following two results are deduced from a more general theorem proved by Jeltsch and Nevanlinna (1981).

Theorem 3.1

If F is any explicit LM formula, then $0 \leqq h_{imag} \leqq 1$.

Theorem 3.2

If F is any PECE scheme, then $0 \leqq h_{imag} \leqq 2$.

The above theorems show that the simple leap-frog (mid-point) rule is at least one of the best explicit LM formulae with regard to the stability properties. However, the order of this formula is only 2. In some situations the accuracy required cannot be achieved by the use of this rule when Δt is close to the bounds given in Section 2. This will be illustrated numerically in Section 5. In such situations (where the problem solved requires more accurate time-integration schemes) LM formulae or PECE schemes of higher order will be more efficient. Two-parameters families of LM formulae and PECE schemes have been studied by Thomsen and Zlatev (1979). These families, with parameters α^* and α, have been used to construct time-integration schemes with large stability regions by Zlatev (1978, 1981), Zlatev and Thomsen (1979, 1980). More recently the two parameters α^* and α have been used in order to construct time-integration schemes with large h_{imag} and of high order. Four such schemes are proposed in (Zlatev et al., 1981c). Here we shall use only one of them: the P_3EC_4E scheme (the indices show the orders of the predictor and of the corrector) obtained with $\alpha^* = -0.85$ and $\alpha = 1.80$. In the experiments this scheme will be compared with the well-known leap-frog (mid-point) rule, the third order Adams-Bashforth formula and the Adams P_3EC_4E scheme. The values of h_{imag} for these four time-integration schemes are given in Table 3.1. It should be pointed out that the time-integration algorithms developed in (Zlatev et al., 1981c) are suitable only for the special system (1.16) obtained after the discretization of the space derivatives in (1.1). Their absolute stability regions are very close to the imaginary axis (a typical example is shown in Fig. 3.1). Therefore they should not be used in the solution of general systems of ODE's.

Time-integration algorithm	h_{imag}
Leap frog (mid-point) rule	1.00
3rd order Adams-Bashforth	0.72
Adams P_3EC_4E scheme	1.18
(-0.85, 1.80) P_3EC_4E scheme	1.70

Table 3.1

The lengths of the stability intervals on the positive part of the imaginary axis (i.e. the values of h_{imag}) for the algorithms used in the numerical experiments in Section 5.

4. TEST - EXAMPLES

In the numerical illustrations, which will be presented in the next section, two test-examples are used.

TEST A (proposed by Molenkamp (1968) and Crowley (1968); used for example in (Anderson and Fattahi, 1974), (Burstein and Mirin, 1970), (Christensen and Prahm, 1976), (Fornberg, 1974), (Orszag, 1971)):

$$\frac{\partial c}{\partial t} = (1-y)\frac{\partial c}{\partial x} + (x-1)\frac{\partial c}{\partial y}\ , \quad x\in[0,2]\ , \quad y\in[0,2]\ , \quad t\in[0,T]\ , \tag{4.1}$$

$$x_0 = 0.5,\ y_0 = 1.0,\ r = 0.25,\ \bar{x} = \sqrt{(x-x_0)^2+(y-y_0)^2}, \tag{4.2}$$

$$c(x,y,0) = \begin{cases} 100(1 - \frac{\bar{x}}{r}) & \text{for } \bar{x} < r \\ 0 & \text{for } \bar{x} \geqq r \end{cases}. \tag{4.3}$$

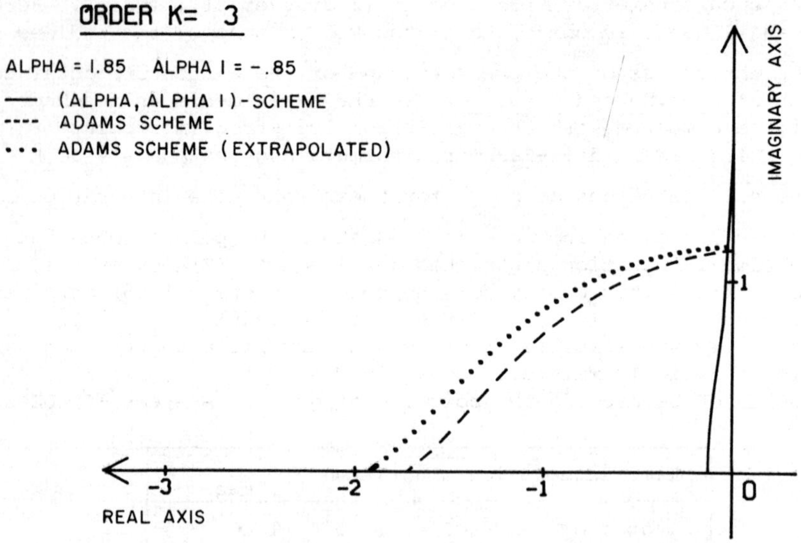

Fig. 3.1 The absolute stability regions of three predictor corrector schemes. The region of the (-0.85, 1.80) P_3EC_4E scheme is given by a continuous line. The region of the Adams P_3EC_4E scheme is drawn by dots. Dashes are used for the region of the Adams P_3EC_3E scheme (the Adams P_3EC_3E scheme is not used in the experiments in this paper).

TEST B This test-example (see (Long and Pepper, 1981) or (Pepper et al., 1979)) can be found from the above by replacing (4.3) with

$$c(x,y,0) = \begin{cases} 50(1 + \cos\frac{\pi\bar{x}}{r}) & \text{for } \bar{x} < r \\ 0 & \text{for } \bar{x} \geqq r \end{cases}. \tag{4.4}$$

The solution of both test-examples satisfies:

$$c(x,y,2k\pi) = c(x,y,0) \quad (k = 1,2,\ldots). \tag{4.5}$$

Some acceptability requirements (though sometimes in a very vague form) are always stated and the approximate solution is acceptable if the user has good reasons to expect that these requirements are satisfied. The acceptability criterion applied in our experiments is defined as follows. Let

$$\varepsilon = \frac{\| c_{ijk} - c(x_i, y_j, 2k\pi) \|_\infty}{\| c(x_i, y_j, 2k\pi) \|_\infty} \qquad (k = 1,2,\ldots), \qquad (4.6)$$

where c_{ijk} and $c(x_i, y_j, 2k\pi)$ are respectively the computed solution and the exact solution at the grid-point (x_i, y_j) for $t=2k\pi$. Then the results are accepted when $\varepsilon \leq 0.1$.

5. NUMERICAL RESULTS

Some results obtained in the numerical solution of the test-examples described in Section 4 will be presented in this section. All experiments were carried out on the IBM 3033 computer at NEUCC (Northern Europe University Computing Centre, Lyngby, Denmark). The FORTH compiler with OPT = 02 has been used. The computing times are always given in seconds. The package SURRENDER from the standard library at NEUCC has been applied to draw the three-dimensional plots.

In Table 5.1 the numerical results obtained in the solution of TEST A when $\Delta x = \Delta y = 0.0625$ and $T = 2\pi$ are given. The maximal Δt for which the results are both stable and sufficiently accurate (in the sense of the acceptability criterion formulated at the end of Section 4) is used. The stability requirements restrict Δt for all algorithms except the leap-frog rule. The latter will be stable if Δt is considerably larger (such that 420 time-steps are used); however, the results will not be acceptable. Nevertheless, the leap-frog is the best algorithm with regard to the computing time used.

The numerical results obtained in the solution of TEST A when $\Delta x = \Delta y = 0.0625$ and $T = 20\pi$ (i.e. after 10 revolutions) are given in Table 5.2. It is seen that the leap-frog rule performs worst (with regard to the computing time used) in this case. This method is inaccurate (its order is only two) and a small Δt has to be used in order to reach the acceptability requirements. The other formulae perform much better than the leap-frog rule in this situation.

Stable results can be obtained by the leap-frog rule if $\Delta t = \pi/220$; however, these results are not sufficiently accurate (the acceptability requirement is far from reached). The results obtained with $\Delta t = \pi/220$ are given in Table 5.3.

ALGORITHM	Number of steps	Computing time	Largest error	Largest value of the solution at $t=2\pi$	Smallest value of the solution at $t=2\pi$
Leap-frog (mid-point) rule	440	37.25	10%	93	-5
3rd order Adams-Bashforth	560	48.16	4%	96	-1
Adams P_3EC_4E scheme	348	58.01	2%	98	0
(-0.85, 1.80) P_3EC_4E scheme	272	45.04	4%	96	0

Table 5.1

Numerical results obtained in the solution of TEST A on a 32×32 space grid at $t=2\pi$ (the negative values of the calculated solution, which are larger than -0.5 are given as 0 in this table).

ALGORITHM	Number of steps	Computing time	Largest error	Largest value of the solution at $t=20\pi$	Smallest value of the solution at $t=20\pi$
Leap-frog (mid-point) rule	14000	658.74	10%	93	-5
3rd order Adams-Bashforth	5600	275.25	10%	90	-2
Adams P_3EC_4E scheme	3480	347.13	7%	93	-3
(-0.85, 1.80) P_3EC_4E scheme	2720	275.64	10%	90	-2

Table 5.2

Numerical results obtained in the solution of TEST A on a 32×32 space grid at $t=20\pi$ (this means that the numerical solution is given after 10 revolutions).

NUMBER OF TIME-STEPS	4400
Computing time	206.70
Largest error	17%
Largest value of the solution at $t=20\pi$	86*
Smallest value of the solution at $t=20\pi$	-12

Table 5.3

Numerical results obtained in the solution of TEST A by the leap-frog rule on a 32×32 grid at $t=20\pi$ (after 10 revolutions) when the maximal time-stepsize which ensures stable results is used (*the maximal value of the computed solution is at the point (0.5, $1.0+\Delta y$), while the maximal value of the exact solution is at the point (0.5, 1.0)).

The initial condition (or, in other words, the exact solution at the end of the integration interval; see (4.5)) and the calculated solutions (at $t=20\pi$) for the different time-integration algorithms are graphically presented in Fig. 5.1 - Fig. 5.6.

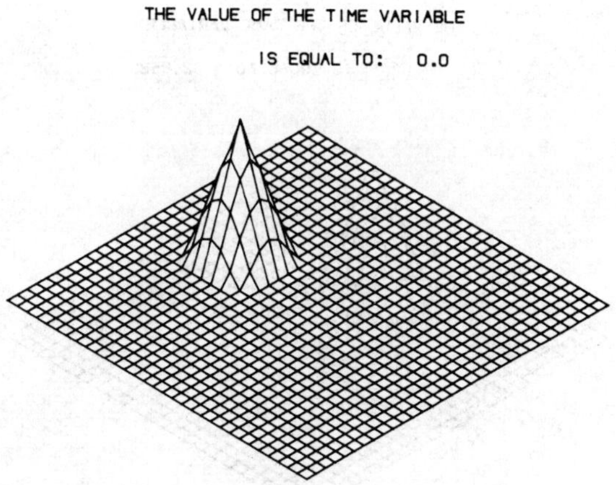
THE VALUE OF THE TIME VARIABLE
IS EQUAL TO: 0.0

Fig. 5.1 The initial values of the solution of TEST A on a 32×32 space grid.

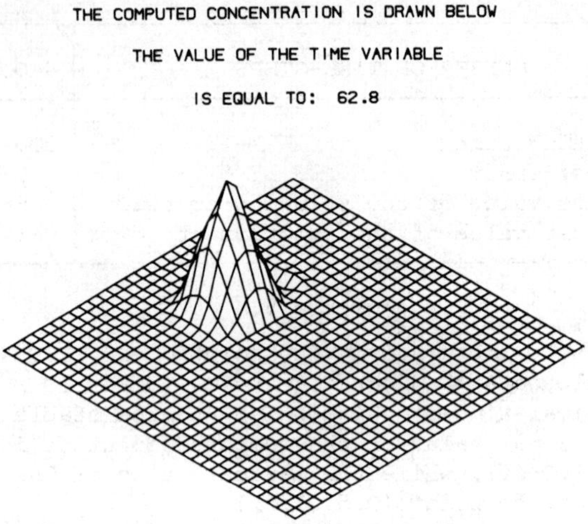

THE COMPUTED CONCENTRATION IS DRAWN BELOW
THE VALUE OF THE TIME VARIABLE
IS EQUAL TO: 62.8

Fig. 5.2 The solution of TEST A calculated by the leap-frog rule on a 32×32 space grid at $t=20\pi$ (14000 steps are carried out in this experiment; see Table 5.2).

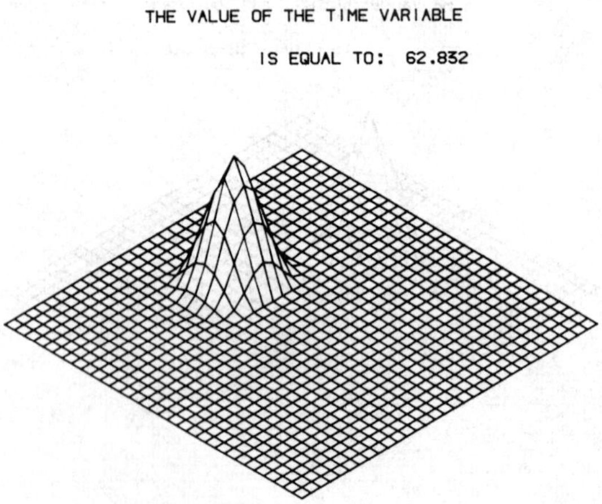

THE VALUE OF THE TIME VARIABLE
IS EQUAL TO: 62.832

Fig. 5.3 The solution of TEST A calculated by the third order Adams-Basforth formula on a 32×32 space grid at $t=20\pi$ (5600 steps are carried out in this experiment; see Table 5.2)

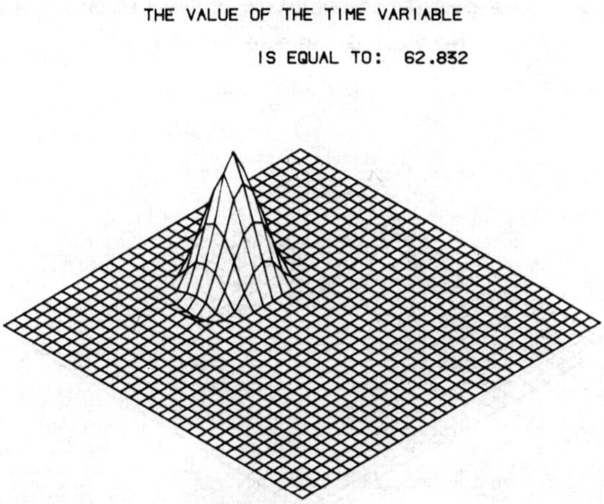
THE VALUE OF THE TIME VARIABLE
IS EQUAL TO: 62.832

Fig. 5.4 The solution of TEST A calculated by the Adams P_3EC_4E scheme on a 32×32 space grid at t=20π (3480 steps are carried out in this experiment; see Table 5.2).

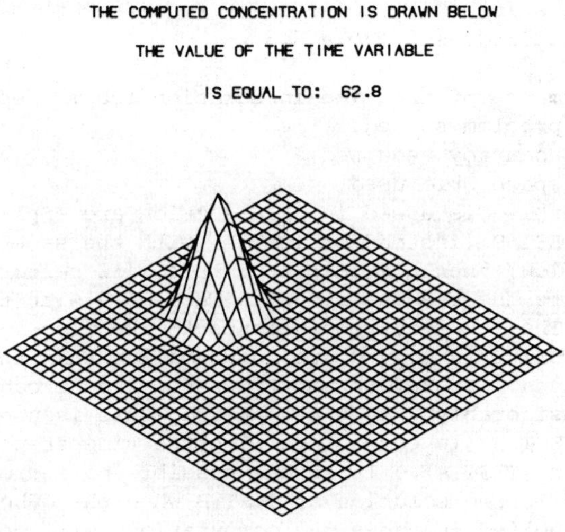
THE COMPUTED CONCENTRATION IS DRAWN BELOW
THE VALUE OF THE TIME VARIABLE
IS EQUAL TO: 62.8

Fig. 5.5 The solution of TEST A calculated by the (-0.85, 1.80) P_3EC_4E scheme on a 32×32 space grid at t=20π (2720 steps are carried out in this experiment; see Table 5.2).

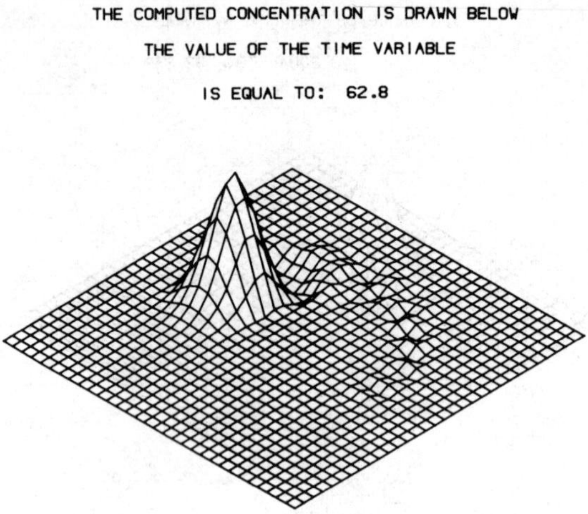
THE COMPUTED CONCENTRATION IS DRAWN BELOW
THE VALUE OF THE TIME VARIABLE
IS EQUAL TO: 62.8

Fig. 5.6 The solution of TEST A calculated by the leap-frog rule on a 32×32 space grid at $t = 20\pi$ (4400 steps are carried out in this experiment; the computations are stable but the accuracy requirement is not satisfied, see Table 5.3).

The performance of the time-integration scheme depends:

(i) on the problem solved,
(ii) on the accuracy required,
(iii) on the space grid used.

If the algorithms discussed in this section are applied in the solution of TEST B (instead of TEST A) with the same acceptability criterion, then the leap-frog rule will perform much better (compare the results given in Table 5.4 with those in Table 5.2). The accuracy requirements to the time-integration scheme when TEST B is solved are much smaller than those when TEST A is solved. Therefore the time-stepsize Δt can be increased considerably when the leap-frog rule is used in the solution of TEST B (in comparison with the time stepsize used in the solution of TEST A). It is not possible to increase the time-stepsize in the solution of TEST B when the other three schemes are applied (because the computations will become unstable). Therefore the computing times used in the solution of TEST B for these schemes are approximately the same as the corresponding computing times used in the solution of TEST A (see Table 5.2 and Table 5.4). However, note that these computing times are still better than the computing time used in the solution of TEST B by the leap-frog rule (see Table 5.4). Moreover, the accuracy achieved by these three schemes is much greater both than the accuracy achieved by the same schemes in the solution of TEST A and than the accuracy obtained by the

ALGORITHM	Number of steps	Computing time	Largest error	Largest value of the solution at $t=20\pi$	Smallest value of the solution at $t=20\pi$
Leap-frog (mid-point) rule	7800	366.54	10%	100	-6
3rd order Adams-Bashforth	5600	273.34	3%	97	-1
Adams P_3EC_4E scheme	3480	342.34	1%	99	-1
(-0.85, 1.80) P_3EC_4E scheme	2720	273.74	3%	97	-2

Table 5.4

Numerical results obtained in the solution of TEST B on a 32×32 space grid at $t=20\pi$ (this means that the numerical solution is given after 10 revolutions).

ALGORITHM	Number of steps	Computing time	Largest error	Largest value of the solution at $t=20\pi$	Smallest value of the solution at $t=20\pi$
Leap-frog (mid-point) rule	7720	87.81	10%	96	-5
3rd order Adams-Bashforth	3760	44.50	10%	90	-3
Adams P_3EC_4E scheme	1600	38.44	10%	92	-3
(-0.85, 1.80) P_3EC_4E scheme	1760	43.31	10%	90	-3

Table 5.5

Numerical results obtained in the solution of TEST A on a 16×16 space grid at $t=20\pi$ (this means that the numerical solution is given after 10 revolutions).

leap-frog rule in the solution of TEST B. In this situation the superiority of the time-integration schemes of high order will become more apparent when more stringent acceptability criteria are specified. For example, the time-stepsizes used with all schemes except the Adams P_3EC_4E scheme has to be decreased if the acceptability criterion of 10% relative accuracy is replaced by a requirement of 1% relative accuracy in the solution of TEST B.

It may be more profitable to use time-integration schemes of high order also in the case where the number of grid-points used in the space discretization is smaller. This is illustrated in Table 5.5, where some results obtained in the solution of TEST A on a 16×16 space grid are shown.

The results presented in this section indicate that in different situations different time-integration schemes have to be used if one wants to reduce the computing time needed to obtain acceptable results. Therefore, it is desirable to implement several time-integration schemes in a software for solving equations of type (1.1). In the software, which is in process of development at the Air Pollution Laboratory, eight different time-integration schemes are implemented. Each of these schemes can be chosen according to the properties of the problem solved, to the accuracy requirements and to the space discretization used. Some details about this software can be found in (Zlatev et al., 1981a,b,c; 1982a,b).

6. ON THE USE OF VARIABLE STEPSIZE VARIABLE FORMULA METHODS DURING THE TIME DISCRETIZATION PROCESS

It is difficult to use the bounds for Δt given in Section 2 when the time-interval is very long. This will normally lead to very inefficient computations. Indeed, in our interpretation u and v are wind velocities. This means that one or two storm days in a year will force us to carry out the whole computational process with a very small stepsize in order to ensure stable results. Therefore we are interested in replacing u^* and v^* from (2.12) - (2.13) by

$$\bar{u}_k = \max_{\substack{x\in[a_1,b_1]\\ y\in[a_2,b_2]\\ t\in T_k^*}} (|u(x,y,t)|), \quad \bar{v}_k = \max_{\substack{x\in[a_1,b_1]\\ y\in[a_2,b_2]\\ t\in T_k^*}} (|v(x,y,t)|), \qquad (6.1)$$

where T_k^* is some neighbourhood of t_k (the current integration point). The attempt to apply (6.1) in the determination of Δt implies the use of a variable stepsize. The time-integration algorithm can also be changed from one step to another. The implementation of these ideas leads to variable stepsize variable formula methods (VSVFM's); see e.g. Zlatev (1978, 1981). We plan to include in our package an option, whereby a VSVFM can be specified.

REFERENCES

ANDERSON, D. and FATTAHI, B. 1974 A comparison of numerical solutions of the advective equation, *J. Atmos. Sci.*, **31**, pp. 1500-1506.

BURSTEIN, C.Z. and MIRIN, A.A. 1970 Third-order difference methods for hyperbolic equations, *J. Comput. Phys.*, **5**, pp. 547-571.

CHRISTENSEN, O. and PRAHM, L.P. 1976 A pseudospectral model for dispersion of atmospheric pollutants, *J. Appl. Meteor.*, **15**, pp. 1284-1294.

CROWLEY, W.P. 1968 Numerical advection experiments, *Mon. Weath. Rev.*, **96**, pp. 1-11.

DAHLQUIST, G. 1963 A special stability problem for linear multistep methods, *BIT* **3**, pp. 27-43.

FORNBERG, B. 1975 On a Fourier method for the integration of hyperbolic equations, *SIAM J. Num. Anal.* **12**, pp. 509-528.

GEAR, C.W. 1971 Numerical initial value problems in ordinary differential equations, Prentice-Hall, Englewood Cliffs, N.J.

JELTSCH, R. and NEVANLINNA, O. 1981 Stability of explicit time discretizations for solving initial value problems, *Num. Math.*, **37**, pp. 61-91.

KREISS, H.O. and OLIGER, J. 1972 Comparison of accurate methods for the integration of hyperbolic equations, *Tellus* **XXIV**, pp. 199-215.

LAMBERT, J.D. 1973 Computational methods in ordinary differential equations, Wiley, New York - London.

LONG, P.E. (Jr) and PEPPER, D.W. 1981 An examination of some simple numerical schemes for calculating scalar advection, *J. Appl. Meteor.*, **20**, pp. 146-156.

MOLENKAMP, C.R. 1968 Accuracy of finite-difference methods applied to the advection equation, *J. Appl. Meteor.* **7**, pp. 160-167.

ORSZAG, S.A. 1971 Numerical simulation of incompressible flows within simple boundaries: accuracy, *J. Fluid Mech.* **49**, pp. 75-112.

PEPPER, D.W., KERN, C.D. and LONG, P.E. (Jr) 1979 Modeling the dispersion of atmospheric pollution using cubic splines and chapeau functions, *Atmos. Environ.* **13**, pp. 223-237.

STETTER, H.J. 1973 Analysis of discretization methods for ordinary differential equations, Springer, Berlin-Heidelberg-New York.

THOMSEN, P.G. and ZLATEV, Z. 1979 Two-parameters families of predictor-corrector methods for the solution of ordinary differential equations, *BIT* **19**, pp. 503-517.

ZLATEV, Z. 1978 Stability properties of variable stepsize variable formula methods, *Num. Math.* **31**, pp. 175-182.

ZLATEV, Z. 1981 Zero-stability properties of the three-ordinate variable stepsize variable formula methods, *Num. Math.* **37**, pp. 157-166.

ZLATEV, Z., BERKOWICZ, R. and PRAHM, L.P. 1981a Numerical treatment of the advection-diffusion equation: Part I Space discretization, Report MST LUFT-A47, Air Pollution Laboratory, National Agency of Environmental Protection, Risø National Laboratory, DK-4000 Roskilde, Denmark.

ZLATEV, Z., BERKOWICZ, R. and PRAHM, L.P. 1981b Numerical treatment of the advection - diffusion equation: Part II Test problems, Report MST LUFT-A49, Air Pollution Laboratory, National Agency of Environmental Protection, Risø National Laboratory, DK-4000 Roskilde, Denmark. To appear in Computers and Fluids.

ZLATEV, Z., BERKOWICZ, R. and PRAHM, L.P. 1981c Numerical treatment of the advection - diffusion equation: Part III Time integration, Report MST LUFT-A50, Air Pollution Laboratory, National Agency of Environmental Protection, Risø National Laboratory, DK-4000 Roskilde, Denmark. To appear in Journal of Computational Physics.

ZLATEV, Z., BERKOWICZ, R. and PRAHM, L.P. 1982a Numerical treatment of the advection - diffusion equation: Part IV Multilevel model, Report MST LUFT-A57, Air Pollution Laboratory, National

Agency of Environmental Protection, Risø National Laboratory, DK-4000 Roskilde, Denmark. To appear in Atmospheric Environment.

ZLATEV, Z., BERKOWICZ, R. and PRAHM, L.P. 1982b Numerical treatment of the advection - diffusion equation: Part V Documentation of ADM01, Report MST LUFT-A58, Air Pollution Laboratory, National Agency of Environmental Protection, Risø National Laboratory, DK-4000 Roskilde, Denmark.

ZLATEV, Z. and THOMSEN, P.G. 1979 Automatic solution of differential equations based on the use of linear multistep methods, *ACM Trans. Math. Software* **5** , pp. 401-414.

ZLATEV, Z. and THOMSEN, P.G. 1980 Differential integrators based on linear multistep methods. In Méthodes numérique dans les sciences de l'ingénieur - GAMN1 (Absi, E., Glowinski, R., Lascaux, P. and Veysseyre, H. eds.) Dunod, Paris, pp. 221-231.

FINITE DIFFERENCE METHODS FOR VISCOELASTIC FLOW

A.R. Davies

(University College of Wales, Aberystwyth)

1. INTRODUCTION

Viscoelastic fluids are neither purely viscous nor purely elastic. They are non-Newtonian since they do not obey the classical Newtonian constitutive relation; in particular their stress distribution depends on (or has a memory of) past deformation.

The understanding and prediction of how viscoelastic fluids behave in various flow situations is of considerable importance industrially, for example in the processing of foods, plastics, paints and soaps. Of particular relevance is the flow of highly elastic liquids in complex situations involving abrupt changes in geometry. In the numerical simulation of such flows the model constitutive relations must be capable of reflecting viscoelastic effects, and yet must be simple enough to allow numerical tractability. Recent investigations have been dominated by differential rheological models of Oldroyd/Maxwell type (Oldroyd, 1950). Both finite-difference and finite-element formulations have been used. The review article by Crochet and Walters (1983) contains a complete list of flow problems and geometries studied so far. All published work is restricted to steady incompressible flows in 2-dimensional planar or axisymmetric geometries; some viscoelastic free surface problems (particularly die-swell) have been solved by finite elements.

The aims of this paper are twofold: first, to outline the finite-difference methods developed by Davies et al. (1979) and Davies and Webster (1982) for complex viscoelastic flows; secondly, to shed light on one of the outstanding numerical problems in this area, namely the breakdown of convergence of algorithms at relatively low values of elasticity.

2. THE GOVERNING EQUATIONS

We concentrate attention on the so-called "upper-convected Maxwell" rheological model, characterized by a constant viscosity coefficient η and a constant relaxation time λ. In a rectangular Cartesian coordinate system x^j, the constitutive equations are

$$\tau^{jk} = -p\delta^{jk} + T^{jk}, \tag{2.1}$$

$$T^{jk} + \lambda \overset{\nabla}{T}{}^{jk} = 2\eta D^{jk}, \tag{2.2}$$

where τ^{jk} is the stress tensor, T^{jk} the extra-stress tensor, p the isotropic pressure, δ^{jk} the Kronecker delta, and D^{jk} the rate of deformation (rate of strain) tensor defined by

$$D^{jk} = \frac{1}{2}\left(\frac{\partial v^k}{\partial x^j} + \frac{\partial v^j}{\partial x^k}\right), \tag{2.3}$$

where v^k is the velocity. ∇ denotes the upper-convected derivative operator (Oldroyd, 1950) defined for any contravariant tensor E^{jk} by

$$\overset{\nabla}{E}{}^{jk} = \frac{\partial E^{jk}}{\partial t} + v^m \frac{\partial E^{jk}}{\partial x^m} - \frac{\partial v^j}{\partial x^m} E^{mk} - \frac{\partial v^k}{\partial x^m} E^{jm}, \tag{2.4}$$

where the summation convention is implied. Notice that (2.1) and (2.2) reduce to the classical Newtonian model when the elasticity parameter λ vanishes.

By way of illustration we consider steady incompressible 2-dimensional planar flow over a rectangular obstruction. The flow geomtry is shown in Fig. 1, where the entry AB and exit GH are assumed to be sufficiently far from the obstruction for the entry and exit flow to be regarded as "fully developed". The flow is generated by the movement of the plate AH in the direction indicated; the entry and exit flow is thus of Couette type.

It is useful to make the transition to non-dimensional variables

$$x^j \to \frac{x^j}{L}, \quad v^j \to \frac{v^j}{U}, \quad p \to \frac{pL}{\eta U}, \quad T^{jk} \to T^{jk}\frac{L}{\eta U} \qquad (2.5)$$

where the characteristic length L is the entry-length AB, and the characteristic velocity U is the plate velocity. We may introduce the two non-dimensional parameters

$$R = \frac{\rho UL}{\eta}, \quad W = \lambda\frac{U}{L}, \qquad (2.6)$$

where R is the conventional Reynolds number, and W is an elasticity parameter often called a Weissenberg number or Deborah number.

Taking Cartesian axes as shown, let u(x,y), v(x,y) denote the velocity components. In terms of the stream function ψ defined by

$$u = \frac{\partial\psi}{\partial y}, \quad v = -\frac{\partial\psi}{\partial x}, \qquad (2.7)$$

and the vorticity ω defined by

$$\omega = \frac{\partial v}{\partial x} - \frac{\partial u}{\partial y}, \qquad (2.8)$$

the non-dimensional equations which govern the flow of the Maxwell fluid are then

$$\nabla^2\psi = -\omega, \qquad (2.9)$$

$$R\left(\frac{\partial\psi}{\partial x}\frac{\partial\omega}{\partial y} - \frac{\partial\psi}{\partial y}\frac{\partial\omega}{\partial x}\right) = \frac{\partial^2 T^{xx}}{\partial x\partial y} - \frac{\partial^2 T^{yy}}{\partial x\partial y} + \frac{\partial^2 T^{xy}}{\partial y^2} - \frac{\partial^2 T^{xy}}{\partial x^2}, \qquad (2.10)$$

which are the (generalized) Navier-Stokes equations, and

$$\left.\begin{aligned}
&T^{xx}\left(1 - 2W\frac{\partial u}{\partial x}\right) + W\left(u\frac{\partial T^{xx}}{\partial x} + v\frac{\partial T^{xx}}{\partial y}\right) - 2WT^{xy}\frac{\partial u}{\partial y} = 2\frac{\partial u}{\partial x}\\
&T^{xy} + W\left(u\frac{\partial T^{xy}}{\partial x} + v\frac{\partial T^{xy}}{\partial y}\right) - WT^{xx}\frac{\partial v}{\partial x} - WT^{yy}\frac{\partial u}{\partial y} = \frac{\partial u}{\partial y} + \frac{\partial v}{\partial x}\\
&T^{yy}\left(1 - 2W\frac{\partial v}{\partial y}\right) + W\left(u\frac{\partial T^{yy}}{\partial x} + v\frac{\partial T^{yy}}{\partial y}\right) - 2WT^{xy}\frac{\partial v}{\partial x} = 2\frac{\partial v}{\partial y}
\end{aligned}\right\} \quad (2.11)$$

which are the constitutive relations (2.2).

Notice that whereas in the Newtonian problem ($W = 0$) we have only the two dependent variables ψ and ω, in the viscoelastic case we must solve for the additional three variables T^{xx}, T^{xy}, and T^{yy}, which are determined by the first-order hyperbolic system (2.11).

The substitution

$$T^{jk} = \bar{T}^{jk} + 2D^{jk} \quad (2.12)$$

where $\bar{T}^{jk}$ denotes the non-Newtonian contribution to the extra-stress tensor, reveals the elliptic nature of (2.10) in the form

$$\nabla^2\omega + R\left(\frac{\partial\psi}{\partial x}\frac{\partial\omega}{\partial y} - \frac{\partial\psi}{\partial y}\frac{\partial\omega}{\partial x}\right) = \frac{\partial^2}{\partial x\partial y}\left(\bar{T}^{xx} - \bar{T}^{yy}\right) + \left(\frac{\partial^2}{\partial y^2} - \frac{\partial^2}{\partial x^2}\right)\bar{T}^{xy} \quad (2.13)$$

It should be mentioned here that observable viscoelastic effects occur at fairly low Reynolds numbers, and in numerical simulations the left hand side of (2.13) is usually diffusion dominated. We therefore adopt (2.13) in preference to (2.10) because it is easier to solve for ω.

To summarize, the basic problem is to solve the coupled system of equations (2.9), (2.11) and (2.13) for the five variables ψ, ω, $\bar{T}^{xx}$, $\bar{T}^{xy}$ and $\bar{T}^{yy}$, subject to the appropriate boundary conditions.

2.1 Boundary conditions

It is convenient to use Dirichlet conditions for ψ and ω over the complete boundary. The no-slip condition on solid boundaries yields

$$\psi = \psi_0 \text{ on AH}; \quad \psi = \psi_1 \text{ on BCDEFG}; \tag{2.14}$$

where ψ_0 and ψ_1 are constants. The fully-developed flow condition on entry and exit has a linear velocity profile, yielding

$$\psi(y) = y(1 - \tfrac{1}{2}y) + \psi_0 \text{ on AB, GH}. \tag{2.15}$$

The vorticity boundary values must be found computationally, and this is discussed in section 3.1.

Because of the hyperbolic nature of (2.11) it is sufficient to specify T^{xx}, T^{xy} and T^{yy} at the entry AB only. We have

$$T^{xx} = 2W\left(\frac{\partial u}{\partial y}\right)^2 = 2W; \quad T^{xy} = \frac{\partial u}{\partial y} = -1; \quad T^{yy} = 0. \tag{2.16}$$

We note that the specification of extra-stress at entry must be consistent with the complete past history of deformation prior to entry, a feature not shared by the memoryless Newtonian fluid.

3. DISCRETIZATION

Consider a square grid of spacing h imposed on the flow region in Fig. 1. Let Ω_h and $\partial\Omega_h$ denote the sets of internal and boundary grid points, respectively. With h = 1/15, for example, Ω_h contains approximately 500 points, with correspondingly 2500 nodal variables; $\partial\Omega_h$ contains approximately 100 points.

The 5-point difference approximation for the Laplacian is used in (2.9) and (2.13). The discrete form of (2.9) is therefore

$$4\psi_{mn} - \psi_{m+1,n} - \psi_{m-1,m} - \psi_{m,n+1} - \psi_{m,n-1} = h^2\omega_{mn}, \quad (x_m, y_n) \in \Omega_h \tag{3.1}$$

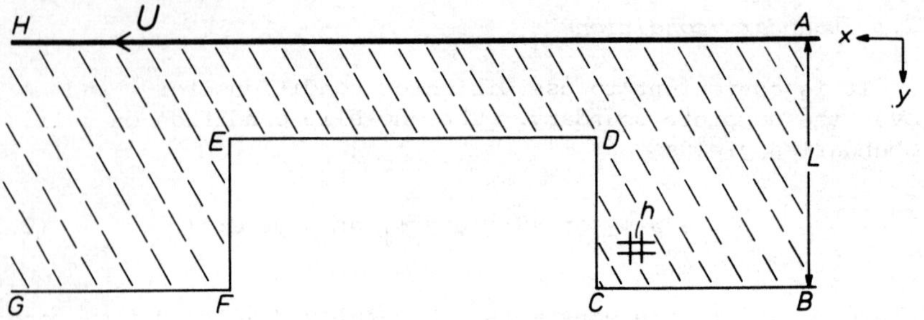

Fig. 1 Flow region.

with $$\psi_{mn} = \psi(x_m,y_n), \quad (x_m,y_n) \in \partial\Omega_h. \tag{3.2}$$

In (2.13) standard central-differences are used for the ψ-derivatives on the left, and also for the 2nd derivatives of $\bar{T}^{jk}$ on the right. Following Spalding (1972) the ω-derivatives are approximated by either central or upwind differences depending on the size of the maximum grid Reynolds number. A full discussion of the suitability of this differencing scheme is given by Davies and Webster (1982).

Defining the function

$$s(x) = \tfrac{1}{2}(|1 + x| + |1 - x|) \tag{3.3}$$

and the variables

$$\alpha_{mn} = \tfrac{1}{4}R(\psi_{m+1,n} - \psi_{m-1,n}), \quad \beta_{mn} = \tfrac{1}{4}R(\psi_{m,n+1} - \psi_{m,n-1}), \tag{3.4}$$

the discretized form of (2.13) may be written

$$K_0\omega_{mn} - K_1\omega_{m+1,n} - K_2\omega_{m-1,n} - K_3\omega_{m,n+1} - K_4\omega_{m,n-1} = -F_{mn},$$
$$(x_m,y_n) \in \Omega_h, \tag{3.5}$$

where $$K_0 = 2[s(\alpha_{mn}) + s(\beta_{mn})],$$

$$K_1 = s(\beta_{mn}) + \beta_{mn}, \quad K_2 = s(\beta_{mn}) - \beta_{mn},$$

$$K_3 = s(\alpha_{mn}) - \alpha_{mn}, \quad K_4 = s(\alpha_{mn}) + \alpha_{mn},$$

and $$F_{mn} = \tfrac{1}{4}(\bar{T}^{xx}_{m+1,n+1} - \bar{T}^{xx}_{m+1,n-1} - \bar{T}^{xx}_{m-1,n+1} + \bar{T}^{xx}_{m-1,n-1})$$

$$-\bar{T}^{yy}_{m+1,n+1} + \bar{T}^{yy}_{m+1,n-1} + \bar{T}^{yy}_{m-1,n+1} - \bar{T}^{yy}_{m-1,n-1})$$

$$- \bar{T}^{xy}_{m+1,n} - \bar{T}^{xy}_{m-1,n} + \bar{T}^{xy}_{m,n+1} + \bar{T}^{xy}_{m,n-1}. \quad (3.6)$$

3.1 Boundary vorticity approximation

In Fig. 2, let O denote a horizontal boundary grid-point, and 1,2 internal grid-points. A first order approximation for ω_0 may be obtained from the Taylor expansion

$$\psi_1 = \psi_0 + hu_0 - \tfrac{1}{2}h^2\omega_0 + O(h^3),$$

giving

$$\omega_0 = -(2/h^2)(\psi_1 - \psi_0 - hu_0) + O(h). \quad (3.7)$$

From the no-slip condition, $u_0 = 1$ on AH, and $u_0 = 0$ on BC, DE and FG. Similar approximations are valid on the vertical boundaries CD and EF. On AB and GH the fully-developed flow condition gives $\omega = 1$.

Several second order formulae for ω_0 are available (see, for example, (Roache, 1972), and (Gupta and Manohar, 1979)). Although, with the help of iterative damping, these formulae can lead to improved accuracy in Newtonian calculations, (Webster, 1979) and (Court, 1980) have found that in viscoelastic calculations no improvement in accuracy over formula (3.7) is achieved; moreover, serious convergence difficulties are experienced, irrespective of damping.

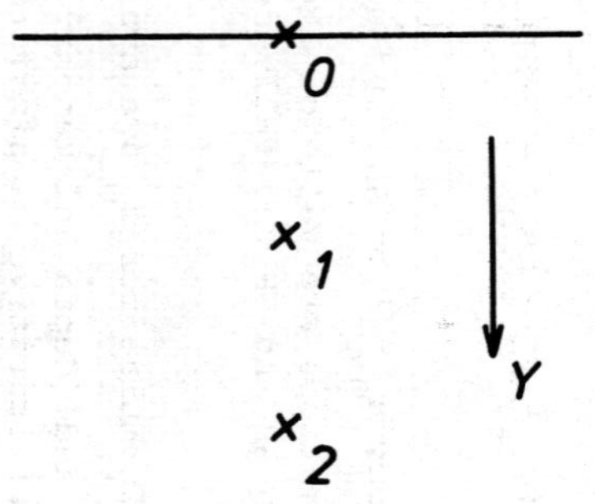

Fig. 2 Horizontal boundary.

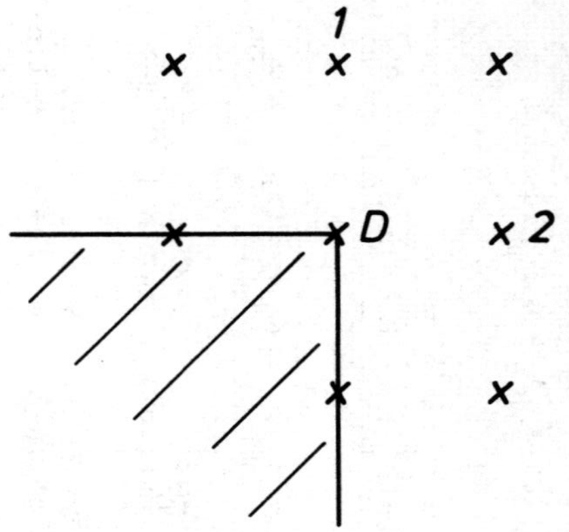

Fig. 3 Re-entrant corner.

At the re-entrant corners D and E, ω is singular, and the approximation (3.5) breaks down at grid-points immediately adjacent to D and E (see, for example, points 1 and 2, Fig. 3). For Newtonian problems the method of Kawaguti (1965) attempts to overcome this difficulty by assigning the fictitious value

$$\omega_D = -(2/h^2)(\psi_1 + \psi_2 - 2\psi_D), \tag{3.8}$$

(see Fig. 3). Similarly for E.

Holstein and Paddon (this conference) are able to justify Kawaguti's method by comparison with a rigorous asymptotic expansion method. For the Maxwell fluid, however, no asymptotic analysis yet exists for corner singularities. Despite obvious misgivings, therefore, we are forced to adopt a Newtonian approximation at corners, and we have accordingly used Kawaguti's method because of its simplicity. A fuller discussion of the use of this method in viscoelastic calculations is also given by Davies and Webster.

3.2 Discretization of the constitutive equations

Using the transformation (2.12), equations (2.11) may be written

$$\left.\begin{aligned} A_1\bar{T}^{xx} + WL\bar{T}^{xx} &= 2B\bar{T}^{xy} + F_1 \\ A_2\bar{T}^{xy} + WL\bar{T}^{xy} &= C\bar{T}^{xx} + B\bar{T}^{yy} + F_2 \\ A_3\bar{T}^{yy} + WL\bar{T}^{yy} &= 2C\bar{T}^{xy} + F_3 \end{aligned}\right\}, \tag{3.9}$$

where

$$\left.\begin{aligned}
L &\equiv \frac{\partial\psi}{\partial y}\frac{\partial}{\partial x} - \frac{\partial\psi}{\partial x}\frac{\partial}{\partial y}, \\
A_1 &= 1 - 2W\frac{\partial^2\psi}{\partial x\partial y}, \quad A_2 = 1, \quad A_3 = 1 + 2W\frac{\partial^2\psi}{\partial x\partial y}, \\
B &= W\frac{\partial^2\psi}{\partial y^2}, \quad C = -W\frac{\partial^2\psi}{\partial x^2}, \\
F_1 &= 2W\left[2\left(\frac{\partial^2\psi}{\partial x\partial y}\right)^2 - \frac{\partial\psi}{\partial y}\frac{\partial^3\psi}{\partial x^2\partial y} + \frac{\partial\psi}{\partial x}\frac{\partial^3\psi}{\partial x\partial y^2} + \frac{\partial^2\psi}{\partial y^2}\left(\frac{\partial^2\psi}{\partial y^2} - \frac{\partial^2\psi}{\partial x^2}\right)\right], \\
F_2 &= -W\left[2\frac{\partial^2\psi}{\partial x\partial y}\left(\frac{\partial^2\psi}{\partial x^2} + \frac{\partial^2\psi}{\partial y^2}\right) + \left(\frac{\partial\psi}{\partial y}\frac{\partial}{\partial x} - \frac{\partial\psi}{\partial x}\frac{\partial}{\partial y}\right)\left(\frac{\partial^2\psi}{\partial y^2} - \frac{\partial^2\psi}{\partial x^2}\right)\right], \\
F_3 &= 2W\left[2\left(\frac{\partial^2\psi}{\partial x\partial y}\right)^2 + \frac{\partial\psi}{\partial y}\frac{\partial^3\psi}{\partial x^2\partial y} - \frac{\partial\psi}{\partial x}\frac{\partial^3\psi}{\partial x\partial y^2} - \frac{\partial^2\psi}{\partial x^2}\left(\frac{\partial^2\psi}{\partial y^2} - \frac{\partial^2\psi}{\partial x^2}\right)\right].
\end{aligned}\right\} \quad (3.10)$$

In discretizing the operator L, care must again be taken to ensure stability of the resulting matrix systems. Diagonal dominance is achieved if upwind differencing is used when $A_k \geqslant 0$, with downwind differencing when $A_k < 0$.

Introducing the variables

$$\left.\begin{aligned}
\alpha^A_{mn} &= \tfrac{1}{4}W(\psi_{m+1,n} - \psi_{m-1,n})\,\mathrm{sign}(A) \\
\beta^A_{mn} &= \tfrac{1}{4}W(\psi_{m,n+1} - \psi_{m,n-1})\,\mathrm{sign}(A)
\end{aligned}\right\} \quad (3.11)$$

the discretized form of (3.9) is then

$$\left.\begin{aligned}
&K_0^{A_1}\bar{T}^{xx}_{mn} - K_1^{A_1}\bar{T}^{xx}_{m+1,n} - K_2^{A_1}\bar{T}^{xx}_{m-1,n} - K_3^{A_1}\bar{T}^{xx}_{m,n+1} - K_4^{A_1}\bar{T}^{xx}_{m,n-1} \\
&\qquad = h^2[2B_{mn}\bar{T}^{xy}_{mn} + F_{1,mn}]\mathrm{sign}(A_1), && (3.12a) \\
&K_0^{A_2}\bar{T}^{xy}_{mn} - K_1^{A_2}\bar{T}^{xy}_{m+1,n} - K_2^{A_2}\bar{T}^{xy}_{m-1,n} - K_3^{A_2}\bar{T}^{xy}_{m,n+1} - K_4^{A_2}\bar{T}^{xy}_{m,n-1} \\
&\qquad = h^2[C_{mn}\bar{T}^{xx}_{mn} + B_{mn}\bar{T}^{yy}_{mn} + F_{2,mn}], && (3.12b) \\
&K_0^{A_3}\bar{T}^{yy}_{mn} - K_1^{A_3}\bar{T}^{yy}_{m+1,n} - K_2^{A_3}\bar{T}^{yy}_{m-1,n} - K_3^{A_3}\bar{T}^{yy}_{m,n+1} - K_4^{A_3}\bar{T}^{yy}_{m,n-1} \\
&\qquad = h^2[2C_{mn}\bar{T}_{mn} + F_{3,mn}]\mathrm{sign}(A_3), && (3.12c)
\end{aligned}\right\}$$

where

$$\left.\begin{aligned}
&K_0^A = h^2|A_{mn}| + 2(|\alpha^A_{mn}| + |\beta^A_{mn}|), \\
&K_1^A = |\beta^A_{mn}| - \beta^A_{mn}, \quad K_2^A = |\beta^A_{mn}| + \beta^A_{mn}, \\
&K_3^A = |\alpha^A_{mn}| + \alpha^A_{mn}, \quad K_4^A = |\alpha^A_{mn}| - \alpha^A_{mn},
\end{aligned}\right\} \qquad (3.13)$$

and the coefficients A, B, C, and F may be computed from (3.10) by central-difference approximation.

3.3 Stress boundary values

These are needed only in evaluating the right-hand sides of (3.5). Except on AB they do not constitute boundary conditions, *per se*. They may therefore be found by substituting the appropriate boundary velocities in (3.9). On BCDEFG, where $u = v = 0$, the resulting equations are algebraic in the stresses and may be solved explicitly. On AH, however, where $u = 1$, $v = 0$, the stress values must be found implicitly by solving the appropriate difference equations. Fortunately the expressions in (3.9) simplify.

4. SOLUTION OF THE COUPLED MATRIX SYSTEMS

By far the simplest iterative method is to solve each matrix system in turn, evaluating each matrix and right-hand vector from current values of the variables. This is a Picard-type iteration amounting to linearization by decoupling.

Each individual matrix system may be solved by a variety of methods, direct or iterative. Perera and Walters (1977) and Davies et al. (1979) use classical SOR and Gauss-Seidel (GS) iteration, whereas Court et al. (1981) and Davies and Manero (1983) use pre-conditioned conjugate gradient methods.

A simple algorithm is as follows:

1. Guess an initial stream function vector $\underset{\sim}{\psi}_o$; set initial vectors $\underset{\sim}{\omega}_o$, $\underset{\sim}{\bar{T}}^{xx}_o$, $\underset{\sim}{\bar{T}}^{xy}_o$, $\underset{\sim}{\bar{T}}^{yy}_o$.

2. Perform r_1 sweeps of GS iteration for (3.12a) until $\|\underset{\sim}{\bar{T}}^{xx}_{r_1} - \underset{\sim}{\bar{T}}^{xx}_{r_1-1}\| < \varepsilon_1 \|\underset{\sim}{\bar{T}}^{xx}_{r_1-1}\|$, for some prescribed tolerance ε_1.

3. Similarly perform r_2, r_3 sweeps of GS iteration for (3.12b,c), respectively, with tolerances ε_2, ε_3.

4. Compute the remaining boundary stress values, and the boundary vorticity.

5. Perform r_4 sweeps of GS iteration for (3.5) until $\|\underset{\sim}{\omega}_{r_4} - \underset{\sim}{\omega}_{r_4-1}\| < \varepsilon_4 \|\underset{\sim}{\omega}_{r_4-1}\|$, for prescribed ε_4.

6. Perform r_5 sweeps of SOR iteration for (3.1) until $\|\underset{\sim}{\psi}_{r_5} - \underset{\sim}{\psi}_{r_5-1}\| < \varepsilon_5 \|\underset{\sim}{\psi}_{r_4-1}\|$, for prescribed ε_5.

7. If $\|\underset{\sim}{\psi}_{r_5} - \underset{\sim}{\psi}_o\|$ is sufficiently small, stop; otherwise let

   $\underset{\sim}{\bar{T}}^{xx}_o = \underset{\sim}{\bar{T}}^{xx}_{r_1}$, $\underset{\sim}{\bar{T}}^{xy}_o = \underset{\sim}{\bar{T}}^{xy}_{r_2}$, ... , $\underset{\sim}{\psi}_o = \underset{\sim}{\psi}_{r_5}$, and repeat steps 2-7.

There are two key factors which govern the successful convergence of this algorithm (Davies and Webster, 1982):

(a) The tolerances $\varepsilon_1, \ldots, \varepsilon_5$ must be carefully controlled. Crude values are needed at early stages, with reduced values as the algorithm progresses.

(b) The choice of initial vectors in step 1. Continuation with respect to the W-parameter is advisable. The solution of the Newtonian problem serves as an initial guess for the problem with $W = W_1 > 0$, the solution of which in turn serves as an initial guess for $W = W_2 > W_1$, etc.

5. BREAKDOWN OF CONVERGENCE

Results for the flow problem described in this paper may be found in Webster (1979) for W-values in the range $0 \leqslant W \leqslant 0.13$, ($W_1 = 0.04$, $W_2 = 0.1$). Above a critical value $W_{crit} \simeq 0.13$ the algorithm fails to converge.

Unfortunately this breakdown is not confined to any specific problem or algorithm. A limit on W is common to all published work on the numerical simulation of viscoelastic flows. It applies to finite-difference and finite-element techniques, to Picard-type and Newton-type iteration schemes, to differential and integral constitutive equations, and to flow problems with and without abrupt changes in geometry.

It is generally accepted that the value of W_{crit} depends on the constitutive model, the flow geometry, and the mesh size. Once these factors are fixed, the precise value of W_{crit} varies only slightly from treatment to treatment. As W approaches W_{crit} it is observed that the stress field becomes extremely sensitive to changes in the velocity field; spurious oscillations appear in the stress, yielding large and erroneous stress gradients. Furthermore, the steps in W in the continuation scheme become uneconomically small; the radius of convergence of the nonlinear iteration tends to zero.

Possible causes of the breakdown are discussed below. For completeness we assess the evidence from both finite-difference and finite-element calculations.

5.1 *Discretization error*

For mixed mesh size, it is clear that the discrete problem becomes ill-conditioned, at least in part, as W approaches W_{crit}. We must examine whether or not this ill-conditioning results from the choice of discretization. We note first that, with the exception of Mendelson et al. (1982), most authors find that a choice of cruder mesh size enables a larger value of W_{crit} to be reached (see, for example (Crochet and Keunings, 1982)).

Secondly, we consider the stress discretization. In finite-element contexts this is performed without upwinding, in contrast to the finite-difference approach in section 3.2. Although upwinding has a smoothing effect on the stresses, instabilities are observed in both cases. It would seem unlikely, therefore, that stress discretization is the basic source of ill-conditioning, although it would be enlightening to know the effect of stable, higher-order upwinding schemes.

Thirdly, we may rule out the numerical treatment of corner singularities as being the basic source of ill-conditioning. W_{crit} is not increased if sharp corners are rounded (Walters and Webster, 1982; Mendelson et al., 1982); moreover the breakdown still occurs in flow geometries without singularities (Crochet, 1982; Tiefenbruck and Leal, 1982).

5.2 Nonlinear iteration

Holstein (1981) and Tanner (1982) show that certain types of Picard iteration are only conditionally stable (dependent on W) for simple 1-dimensional viscoelastic flows. Their analyses, however, do not apply to Newton iteration, which would appear to be unconditionally stable for 1-dimensional flows. It is not easy to extend Holstein's or Tanner's analysis to 2-dimensional flows where it is known that both Picard and Newton iteration break down.

5.3 Bifurcation or loss of solution

Using continuation with respect to W, Mendelson et al. examine the behaviour of the Jacobian matrix J in Newton's method as W approaches W_{crit}. They find the determinant of J to be negative, and decreasing as W increases, thus giving no evidence of solution bifurcation. We note, however, that the value of the determinant is not always a safe indicator of (near) singularity; a clearer indication would be given by the eigenvalues of J.

5.4 Other possibilities

(1) The onset of time periodic flows. This could only be detected by time-dependent calculation, which is currently under investigation. Preliminary results for sufficiently large times confirm the steady flow solutions already calculated, with

similar breakdown at W_{crit}. There is as yet no indication of time-periodicity.

(2) 3-dimensional flows arising as bifurcations from the family of 2-dimensional flows already calculated. This could presumably be detected by using continuation methods on the full 3-dimensional flow equations. This is a formidable undertaking, and is probably not justified at present.

(3) Classical stability of the model fluid. Several authors (see, for example, (Chan Man Fong and Walters, 1965)) have investigated the temporal and spatial response of model rheological fluids to sinusoidal perturbation in velocity for simple flow fields. The steady Poiseuille and Couette flow of a Maxwell fluid become unstable however small the value of $W > 0$, for certain ranges of the sinusoidal frequency. Since, in computer calculations, the velocity increments will contain a range of Fourier modes, the onset of classical instability is a possibility worthy of exploration.

6. CONCLUSION

We have listed several possibilities which might explain the breakdown of convergence of algorithms for the numerical solution of viscoelastic flow. The problem is by no means resolved. The limit on elasticity is often so low that many of the important and dramatic experimental results which are available fall outside the scope of numerical simulation. To discover and overcome the fundamental cause of the breakdown is therefore of vital importance to this field of research at the present time.

REFERENCES

CHAN MAN FONG, C.F. and WALTERS, K. 1965 The solution of flow problems in the case of materials with memory II. The stability of plane Poiseuille flow of slightly viscoelastic liquids. *J. de Mécanique* **4**, 439.

COURT, H. 1980 Computational Rheological Fluid Mechanics. Ph.D Thesis, Univ. of Wales.

COURT, H., DAVIES, A.R. and WALTERS, K. 1981 Long-range memory effects in flows involving abrupt changes in geometry. Part 4. Numerical simulation using integral rheological models. *J. Non-Newtonian Fluid Mech.* **8**, 95.

CROCHET, M.J. 1982 The flow of a Maxwell fluid around a sphere. In Finite Elements in Fluids IV (R.H. Gallagher Ed.), New York: John Wiley.

CROCHET, M.J. and KEUNINGS, R. 1982 On numerical die-swell calculation. *J. Non-Newtonian Fluid Mech.* **10**, 85.

CROCHET, M.J. and WALTERS, K. 1983 Numerical methods in non-Newtonian fluid mechanics. To appear in Annual Reviews in Fluid Mechanics.

DAVIES, A.R., WALTERS, K. and WEBSTER, M.F. 1979 Long-range memory effects in flows involving abrupt changes in geometry. Part 3. Moving boundaries. *J. Non-Newtonian Fluid Mech.* **4**, 325.

DAVIES, A.R. and WEBSTER, M.F. 1982 The finite-difference solution of viscoelastic flows. I. Stream function and vorticity formulation. To be published.

DAVIES, A.R. and MANERO, O. 1983 The finite-difference solution of viscoelastic flows using pre-conditioned conjugate gradient methods. To be published.

GUPTA, M.M. and MANOHAR, R.P. 1979 Boundary approximations and accuracy in viscous flow computations. *J. Comput. Phys.* **31**, 265.

HOLSTEIN, H. 1981 The numerical solution of some rheological flow problems, Ph.D. Thesis, Univ. of Wales.

KAWAGUTI, M. 1965 Numerical solutions of the Navier-Stokes equations for the flow in a channel with a step. MRC TSR 574, Mathematics Research Center, Madison, Wisconsin.

MENDELSON, M.A., YEH, P.W., BROWN, R.A., ARMSTRONG, R.C. 1982 Approximation error in the finite element calculation of viscoelastic fluid flow. *J. Non-Newtonian Fluid Mech.* **10**, 31.

OLDROYD, J.G. 1950 On the formulation of rheological equations of state. *Proc. R. Soc.* **A200**, 523.

PERERA, M.G.N. and WALTERS, K. 1977 Long range memory effects in flows involving abrupt changes in geometry. Part 1. Flows associated with L-shaped and T-shaped geometries. *J. Non-Newtonian Fluid Mech.* **2**, 49.

ROACHE, P.J. 1972 Computational Fluid Dynamics. Albuquerque: Hermosa Publications.

SPALDING, D.B. 1972 A novel finite-difference formulation for differential expressions involving both first and second derivatives. *Int. J. Num. Meth. Eng.* **4**, 551.

TANNER, R.I. 1982 The stability of some numerical schemes for model viscoelastic fluids. *J. Non-Newtonian Fluid Mech.* **10**, 169.

TIEFENBRUCK, G. and LEAL, L.G. 1982 A numerical study of the motion of a viscoelastic fluid past rigid spheres and spherical bubbles. *J. Non-Newtonian Fluid Mech.* **10**, 115.

WALTERS, K. and WEBSTER, M.F. 1982 On dominating elastico-viscous response in some complex flows. *Phil. Trans. R. Soc.* (to appear).

WEBSTER, M.F. 1979 The Numerical Solution of Rheological Flow Problems, Ph.D. Thesis, Univ. of Wales.

A FINITE DIFFERENCE STRATEGY FOR RE-ENTRANT CORNER FLOW

H. Holstein

(Dept. of Computer Science, University College of Wales, Dyfed)

and

D.J. Paddon

(School of Mathematics, University of Bristol)

1. INTRODUCTION

It is well known that the no-slip boundary condition applied to the steady, plane, viscous, incompressible fluid flow over a rectangular re-entrant corner boundary leads to a singularity in the vorticity and pressure fields. A mathematical description of the singularity is given by Moffatt (1964).

The presence of the singularity implies that the stream function and vorticity do not have Taylor expansions in regions that include the corner. Consequently, the usual discretisation methods cannot be applied near the corner.

Ad hoc methods for circumventing the difficulties are discussed by Roache (1972). Such methods effectively impose extra conditions at the corner. They are derived in part through physical intuition and tacit application of Taylor expansions, with the aim of providing an effective value for the boundary vorticity at the corner. These methods lack mathematical rigour. We may therefore inquire whether (i) the calculated flow field is sensitive to the choice of corner method, and if so, (ii) which methods, if any, are to be preferred?

This paper is addressed to the above questions. The discussion makes use of the singular finite difference (SFD) corner treatment (Holstein and Paddon, 1981). The SFD method takes account of the corner singularity, and serves as a standard of comparison for the ad hoc methods. The comparisons are carried out for flows in contraction, expansion and cavity flow domains at Reynolds numbers in the range 0 to 30.

2. SENSITIVITY OF THE FLOW FIELD TO CORNER VELOCITY

The ad hoc methods lead to a considerable scatter in the estimated corner vorticity. It is therefore instructive to calculate the flow fields for a sequence of corner vorticities, arbitrarily imposed without reference to any particular corner strategy. Stream function plots from such a computational experiment are shown in Fig. 1. Qualitative changes are observed over a wide area. In answer to the first question raised above, we infer that the choice of corner method can indeed influence the calculated flow field at significant distances from the corner.

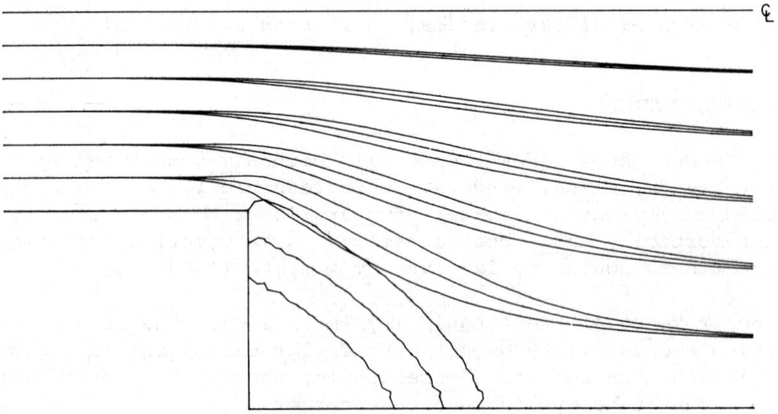

Fig. 1 Expansion, R=10, h=D/12, corner vorticities set to 0, 10, 20. The flow domain is described in section 5.

3. DISCRETISATION OF THE GOVERNING EQUATIONS

The governing equations for plane, steady, incompressible viscous flow may be expressed as

$$\frac{\partial^2 \psi}{\partial x^2} + \frac{\partial^2 \psi}{\partial y^2} = - \omega, \tag{3.1}$$

$$\frac{\partial^2 \omega}{\partial x^2} + \frac{\partial^2 \omega}{\partial y^2} = R\left(\frac{\partial \omega}{\partial x}\frac{\partial \psi}{\partial y} - \frac{\partial \omega}{\partial y}\frac{\partial \psi}{\partial x}\right), \tag{3.2}$$

where $\psi(x,y)$, $\omega(x,y)$ and R are the stream function, vorticity and Reynolds number, respectively. On rigid boundaries, the no-slip conditions

$$\psi = \psi_O, \qquad \partial\psi/\partial n = 0 \tag{3.3}$$

hold, where ψ_O is a constant and n is the normal direction.

We impose a rectangular grid on the flow domain such that the re-entrant corner is coincident with a grid point, and the re-entrant boundaries lie along grid lines. To simplify the discussion, we assume a square mesh with step length h.

Having discretised the governing equations over the interior of the flow domain, they may be solved by the usual direct or iterative methods, subject to stream function and vorticity boundary conditions. At regular boundary nodes, the vorticity is estimated by methods described by Roache(1972) e.g. Thom, Woods or Jensen. Procedures to be adopted at re-entrant corner points are discussed below.

4. SUMMARY OF CORNER METHODS

To describe the flow field in the corner region, we employ a polar coordinate system in which the corner is located at $r=0$ and the re-entrant boundaries are along $\theta = \pm 3\pi/4$. In this system, interior grid points of the rectangular grid adjacent to the corner have polar coordinates $(h, \pm\pi/4)$.

Let $\omega(+0, \theta_O)$ denote the vorticity calculated by a particular boundary method (e.g. Woods), when approaching the corner point $r \to +0$ along the boundary $\theta = \theta_O$, that is, the boundary method is applied at the corner using grid points along a line normal to the line $\theta = \theta_O$. In this notation, the corner vorticities defined by various ad hoc methods are given by

Corner method	Corner vorticity
Kawaguti (1965)	$-2(\psi(h,\pi/4) + \psi(h,-\pi/4) - 2\psi_O)/h^2$
Discontinuous	$\omega(+0, \pm 3\pi/4)$
Average	$(\omega(+0, 3\pi/4) + \omega(+0, -3\pi/4))/2$
Biased	$\max(\omega(+0, 3\pi/4), \omega(+0, -3\pi/4))$
45 degree wall	$\omega(+0, \pi/2)$
Continuous	0
Dennis and Smith (1980)	not defined.

The method of Dennis and Smith (1980) discretises equation (3.2) at $(h,\pm\pi/4)$ with respect to a diagonal grid of step length $h\sqrt{2}$, reference to the corner thereby being avoided. However, validity of a vorticity Taylor expansion in the corner neighbourhood is tacitly assumed by the method, and so we include it in the list of ad hoc methods.

The need for corner vorticities evidently arises from the use of central finite difference approximations for the vorticity derivatives at $r=h$, $\theta=\pm\pi/4$:

$$2h \left.\frac{\partial\omega}{\partial r}\right|_{h,\theta} = \omega(2h, \theta) - \omega^{(1)}(0, \theta), \tag{4.1}$$

$$h^2 \left.\frac{\partial^2\omega}{\partial r^2}\right|_{h,\theta} = \omega(2h, \theta) - 2\omega(h, \theta) + \omega^{(2)}(0, \theta), \tag{4.2}$$

where $\omega^{(1)}(0,\theta)$ and $\omega^{(2)}(0,\theta)$ are corner vorticities (not necessarily equal) appropriate to the first and second derivatives. Clearly, an independent estimate of the vorticity derivatives would obviate the need for estimating corner vorticities by ad hoc methods.

This approach is followed in the SFD method (Holstein and Paddon, 1981). It makes use of an argument by Moffatt (1964), leading to the conclusion that sufficiently near the corner the flow is Stokesian, satisfying

$$\psi(r, \pm\theta) = \psi_0 + A_1 r^{\Lambda_1} f_1(\theta) \pm A_2 r^{\Lambda_2} f_2(\theta) + O(r^{\Lambda_3}), \tag{4.3}$$

$$r^2\,\omega(r, \pm\theta) = A_1 r^{\Lambda_1} g_1(\theta) \pm A_2 r^{\Lambda_2} g_2(\theta) + O(r^{\Lambda_3}). \tag{4.4}$$

Here $\Lambda_1 \doteqdot 1.5444837$, $\Lambda_2 \doteqdot 1.9085292$, $\mathrm{Re}(\Lambda_3) \doteqdot 2.6$, f_1, f_2, g_1, g_2 are known functions of θ (see appendix), and A_1, A_2 are constants depending on the particular flow. Defining

$$\omega_1(r, \theta) = A_1 r^{\Lambda_1 - 2}\, g_1(\theta), \tag{4.5}$$

$$W(\alpha, r, \theta) = r^{\alpha}(\omega(r,\theta) - \omega_1(r,\theta)), \tag{4.6}$$

it is seen from (4.4) that the singularity at $r=0$ is removed from W, provided the constant α is chosen to exceed $2-\Lambda_2$.

Since W can now be defined at the corner, the radial derivatives of W at $(h,\pm\pi/4)$ may be approximated by central finite difference formulae. The expressions can be inverted to give the required vorticity derivatives in the form

$$2h \left.\frac{\partial\omega}{\partial r}\right|_{h,\theta} = c_{11}\ \omega(h,\theta) + c_{12}\ \omega(2h,\theta) + c_{13}\ \omega_1(h,\theta), \qquad (4.7)$$

$$h^2 \left.\frac{\partial^2\omega}{\partial r^2}\right|_{h,\theta} = c_{21}\ \omega(h,\theta) + c_{22}\ \omega(2h,\theta) + c_{23}\ \omega_1(h,\theta), \qquad (4.8)$$

where $\theta = \pm\pi/4$ and constants c_{ij} are given in the appendix. The value of $\omega_1(h,\pm\pi/4)$ is to be approximated from (4.3) and (4.5) by

$$\omega_1(h,\pm\pi/4) = \{g_1(\pi/4)/2f_1(\pi/4)\}\{\psi(h,\pi/4) + \psi(h,-\pi/4) - 2\psi_0\}/h^2. \qquad (4.9)$$

Thus the vorticity derivatives at $(h,\pm\pi/4)$ are expressed in terms of nodal values $\omega(h,\pm\pi/4)$, $\omega(2h,\pm\pi/4)$, $\psi(h,\pm\pi/4)$ and known constants, in a manner that takes account of the vorticity singularity at the corner. We call such expressions singular finite difference derivatives.

In deriving (4.9), we assumed that (4.3) describes the flow at distances $r = O(h)$ from the corner. It was shown in (Holstein and Paddon, 1981) that the expressions

$$R\left|\psi(h,\pm\pi/4) - \psi_0\right| \ll 1, \qquad \text{(local Stokes flow)} \qquad (4.10)$$

$$\left|\frac{\psi(h,\pi/4) - \psi(h,-\pi/4)}{\psi(h,\pi/4) + \psi(h,-\pi/4) - 2\psi_0}\right| \not> 1, \quad (A_1 \not< A_2) \qquad (4.11)$$

$$\frac{\psi(h_1,\pi/4) + \psi(h_1,-\pi/4) - 2\psi_0}{\psi(h_2,\pi/4) + \psi(h_2,-\pi/4) - 2\psi_0} \doteqdot \left(\frac{h_1}{h_2}\right)^{\Lambda_1}, \quad (A_1 \text{ independent of } h) \qquad (4.12)$$

provide a posteriori checks for this assumption. In (4.12), h_1 and h_2 correspond to different grid refinements. Failure of these tests indicates that smaller step lengths have to be used before the SFD theory becomes applicable.

5. COMPUTATIONAL EXPERIMENTS

We consider three rectangular flow domains, symmetric about their centre-lines, dimensioned as follows:

geometry of domain	length entry	length exit	length total	semi-width entry	semi-width exit	semi-width maximum
contraction	6D	4D	10D	2D	D	2D
expansion	4D	13D	17D	D	2D	2D
cavity	3D	3D	8D	D	D	3D

The basic unit of length, D, was non-dimensionalised to 0.5. Step lengths of D/6, D/12 and D/24 were used, and upwind differencing was not employed. The stream function difference between boundary and centreline was set at 0.5, and a Poiseuille flow was imposed at entry.

Visual inspection of computed stream function contours for contraction flows at R = 0, 1, 10 and 30 suggested three groupings within the ad hoc methods. With respect to increasing divergence from SFD, they were (i) Kawaguti, (ii) Discontinuous, Average, Biased, (iii) 45 degree wall, Continuous, D&S. Under mesh refinements, all methods moved closer to SFD. The grouping did not depend on mesh refinement or the choice of Thom, Woods or Jensen boundary methods. On this basis, we chose the Kawaguti and Discontinuous corner methods for further comparisons with SFD. Some results are also given for D&S. The Woods boundary method was adopted throughout, and Reynolds numbers were restricted to 0 and 30.

6. RESULTS

Stream function contour diagrams for the chosen flow domains are shown in Figs. 2 to 11. The contraction flow for R=30 is not shown because the small recirculating vortex is inadequately resolved on the presented scale. Contours for SFD and Kawaguti show a remarkable agreement. Spatial resolution of the differences is only just apparent at R=30, as shown in Figs. 5 and 10. Contours for the SFD and D&S methods are shown in Figs. 3, 6, 8 and 11. The spatial spread between the SFD and D&S contours is generally between 1.5 and 2 times the spread between the SFD and Discontinuous methods. This tendency was also observed for the other Reynolds numbers considered.

Values of the left hand side of inequality (4.10) are given in Table 1, for flow fields obtained with SFD at R=30. The

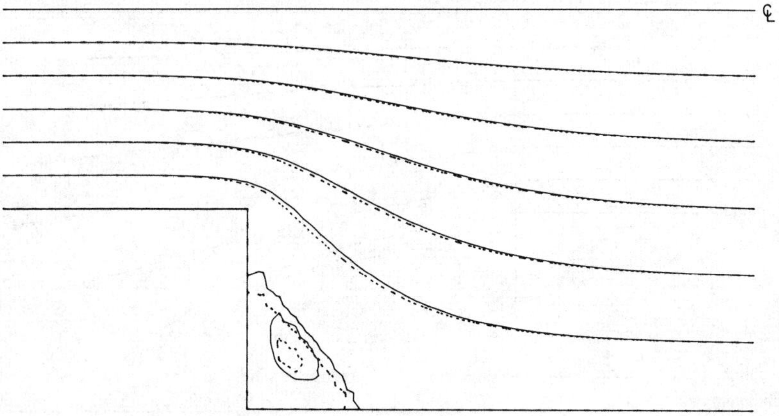

Fig. 2 Contraction or expansion, R=0, h=D/12, SFD and Kawaguti ————, Discontinuous ------------.

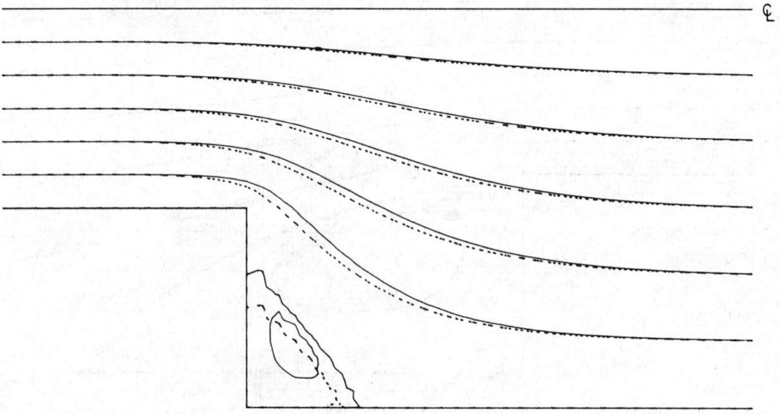

Fig. 3 Contraction or expansion, R=0, h=D/12, SFD and Kawaguti ————, D&S -------.

other corner methods yielded similar magnitudes. We deduce from Table 1 that inequality (4.10) is not satisfied when h=D/6. The SFD calculations therefore lack theoretical support at this discretisation.

Columns A,B and C of Table 2 represent values for the left hand side of inequality (4.11) at various step lengths. It is seen that (4.11) is satisfied in all cases. Columns E and F are obtained from the ratios A/B and B/C of the corresponding

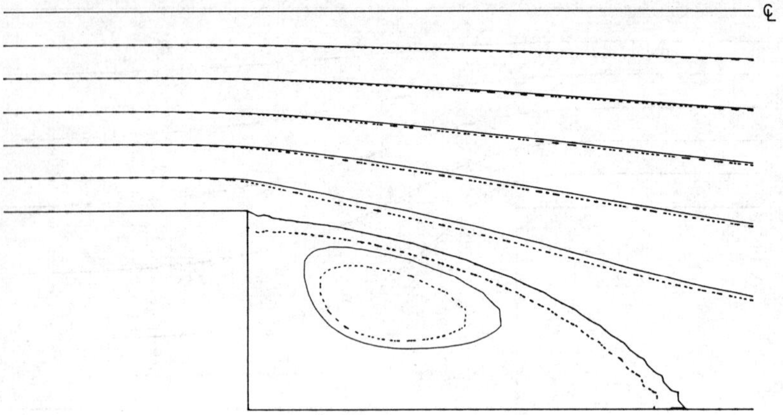

Fig. 4 Expansion, R=30, h=D/12, SFD ———, Discontinuous ---.

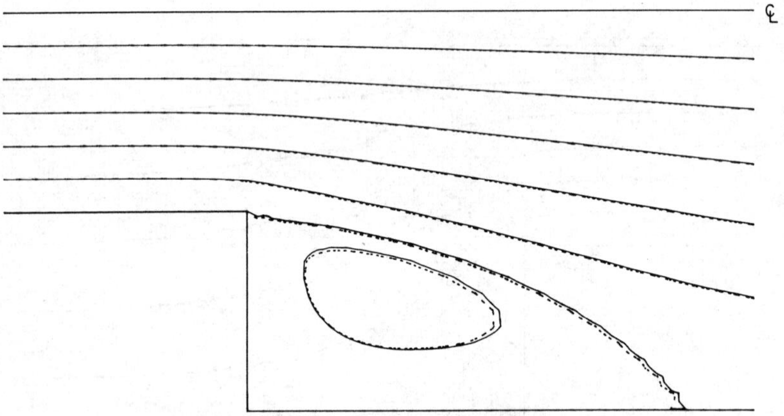

Fig. 5 Expansion R=30, h=D/12, SFD ———, Kawaguti ------.

entries in columns A, B and C. From equation (4.3), the theoretical limit of these ratios is $2^{(\Lambda_2-\Lambda_1)} \simeq 1.29$, as $h \to 0$. The approximation is well satisfied for the case R = 0, but less so for R=30. Finally, columns G and H, respectively, represent the left hand side of (4.12) for h_1=D/6, h_2=D/12 and h_1=D/12, h_2=D/24. They show good agreement with the theoretical limit $2^{\Lambda_1} \simeq 2.92$.

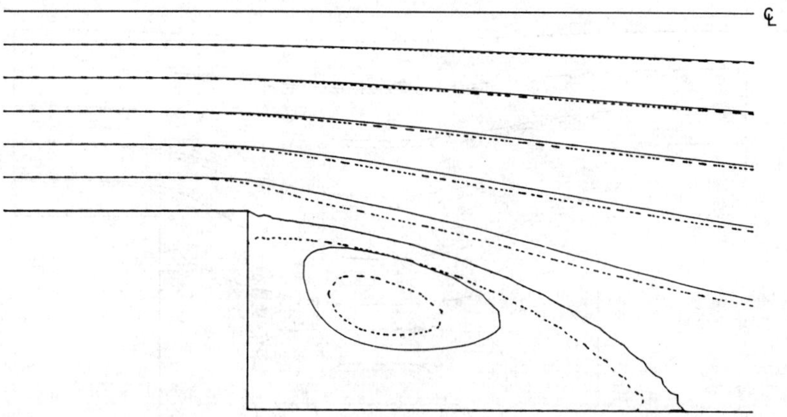

Fig. 6 Expansion, R=30, h=D/12, SFD ———, D&S -------.

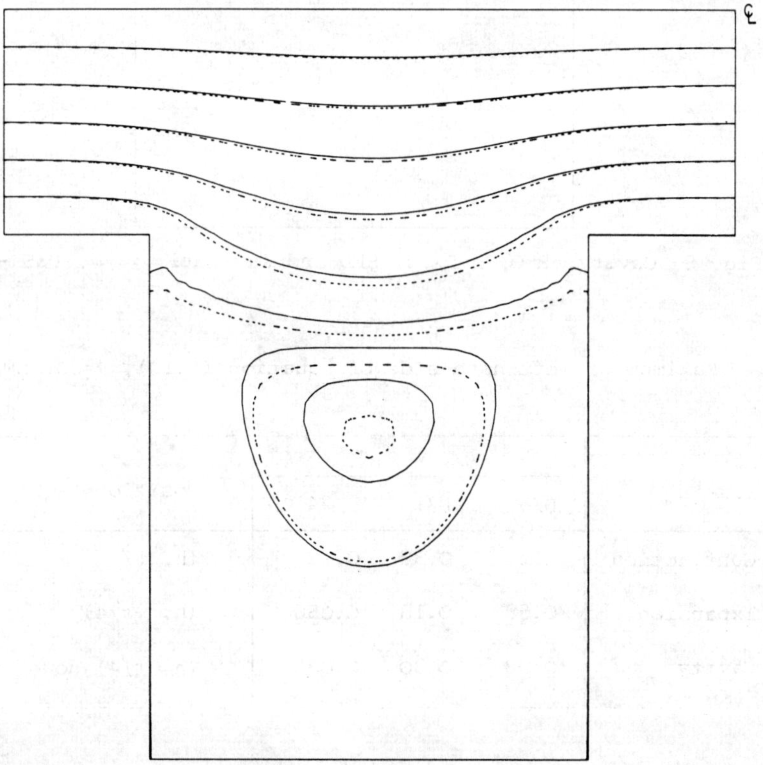

Fig. 7 Cavity, R=0, h=D/12, SFD and Kawaguti ————, Discontinuous -------.

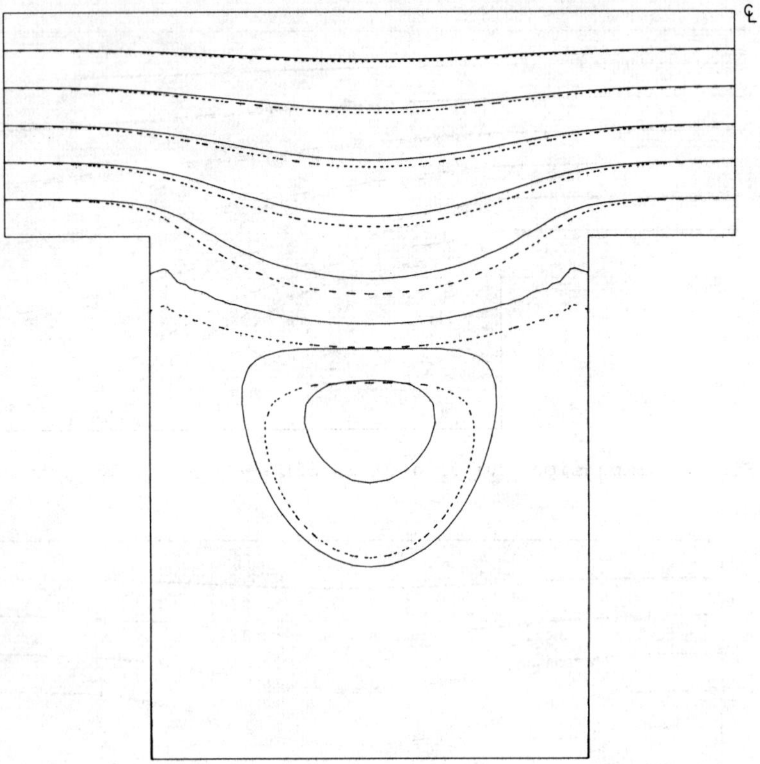

Fig. 8 Cavity, R=0, h=D/12, SFD and Kawaguti ———, D&S -----.

TABLE 1

Maximum of left hand side of equation (4.10), R=30, SFD.

Domain	h			Position
	D/6	D/12	D/24	
Contraction	1.2	0.38	0.12	(h, π/4)
Expansion	0.58	0.18	0.056	(h, -π/4)
Cavity	0.98	0.30	0.095	(h, π/4) downstream

TABLE 2

Left hand side of equations (4.11) and (4.12) - see section 7.

Domain	h=D/6 A	h=D/12 B	h=D/24 C	E (A/B)	F (B/C)	G	H	Reynolds Number
Contraction) Expansion)	0.440	0.352	0.276	1.25	1.28	2.95	2.93	0
Cavity	0.540	0.428	0.335	1.26	1.28	2.94	2.92	0
Contraction	0.087	0.067	0.051	1.30	1.30	2.80	2.83	30
Expansion	1.180	0.888	0.671	1.34	1.32	3.02	3.05	30
Cavity (upstream)	1.220	0.921	0.696	1.33	1.32	2.56	2.98	30
Cavity (downstream)	0.482	0.353	0.290	1.37	1.22	2.98	3.03	30

TABLE 3

Corner vorticity for the expansion domain, with reference to section 4.

Method	h=D/12		h=D/24		Reynolds Number
	$\omega^{(i)}(0,-\omega/4)$	$\omega^{(i)}(0,\pi/4)$	$\omega^{(i)}(0,-\pi/4)$	$\omega^{(i)}(0,\pi/4)$	
SFD(i=1)	14.82	12.09	19.58	16.81	0
SFD (i=2)	16.86	14.03	22.34	19.47	0
Kawaguti (i=1,2)	14.94	14.94	20.33	20.33	0
Discontinuous (i=1,2)	14.36	9.26	19.00	13.56	0
SFD (i=1)	10.17	6.72	12.43	8.88	30
SFD (i=2)	11.32	7.74	13.98	10.31	30
Kawaguti (i=1,2)	8.69	8.69	11.53	11.53	30
Discontinuous (i=1,2)	9.98	3.42	12.62	5.56	30

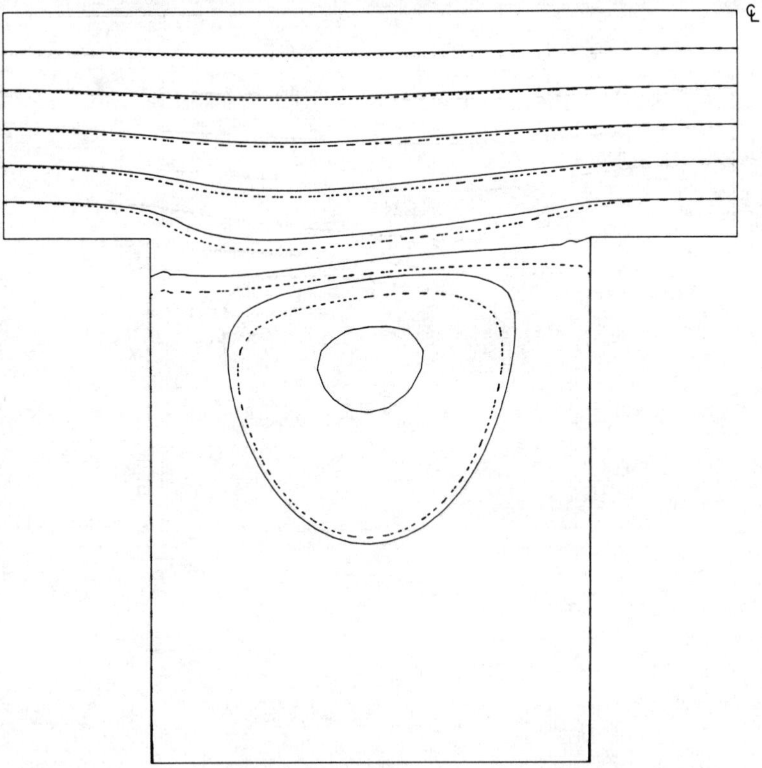

Fig. 9 Cavity, R=30. h=D/12, SFD ————, Discontinuous ------.

The SFD results for h=D/12, D/24 satisfy the checks of section 4, consistent with the assumptions for the method. We use it as a standard for corner methods. We must state by way of caution, however, that the ad hoc methods satisfy these checks remarkably well. This indicates that the ad hoc methods have convergence rates similar to SFD, as $h \to 0$.

The corner vorticities for the various methods may have a considerable scatter. Examples are shown in Table 3. For comparison, we include corner vorticities for the SFD method, obtained with the aid of (4.1) and (4.2).

7. DISCUSSION

The computational results suggest that the method of Kawaguti is the only ad hoc method to give little deviation from the SFD

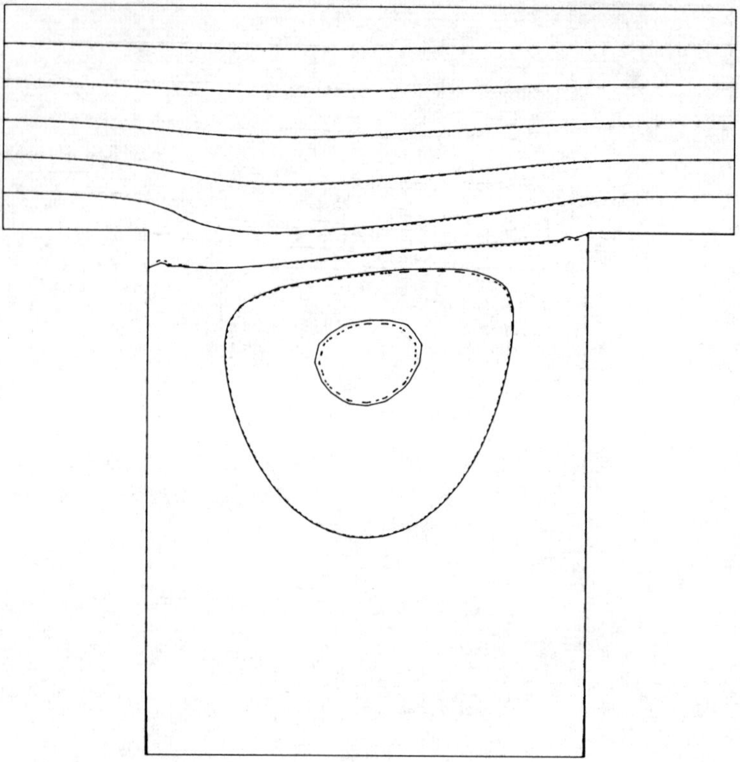

Fig. 10 Cavity, R=30, h=D/12, SFD ———, Kawaguti -------.

method, provided the mesh is sufficently fine for the SFD method to be applicable. Both methods therefore model the flow over a corner that is sharp with respect to the mesh length. This result is significant because of the relative ease with which the method of Kawaguti may be programmed.

A theoretical explanation of the observed ranking order of the corner methods can be given in the limiting case, in which h is chosen sufficiently small for terms $O(r^{\Lambda_1})$ to dominate other powers of r in expansions (4.3) and (4.4), when $r \lesssim h$. All corner flows (with $A_1 \neq 0$) then reduce to the same type of Stokes flow. The vorticity is then approximated by the first expansion term ω_1, given by (4.5) and (4.9). This vorticity can be substituted into both sides of (4.1) and (4.2) to give fictitious corner vorticities for the SFD method.

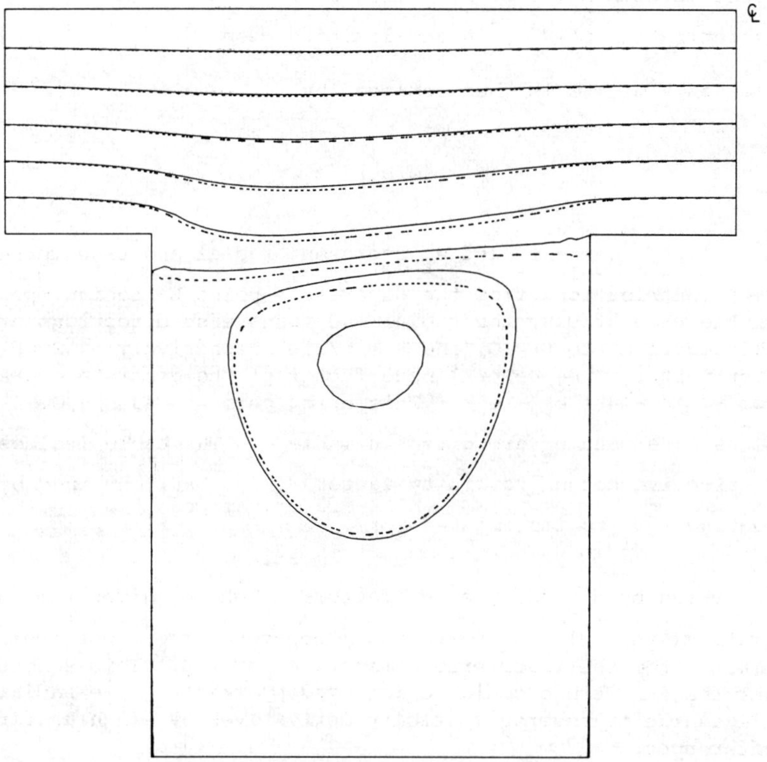

Fig. 11 Cavity, R=30, h=D/12, SFD ————, D&S -------.

In Stokes flow, we need only consider the second order derivatives. For all methods considered, the limiting corner vorticity $\omega^{(2)}(0,\pm\pi/4)$ appropriate to the second radial vorticity derivative takes the form

$$\omega^{(2)}(0, \pm \pi/4) = d_2\{\psi(h, \pi/4) + \psi(h,-\pi/4) - 2\psi_0\}/h^2, \qquad (7.1)$$

where the value of d_2 depends only on the corner method. In Table 4, the ad hoc methods are ordered by increasing deviation of d_2 from the SFD value. This ordering reproduces the computational ordering found earlier.

The methods may also be compared with respect to their discretisation of $\nabla^2\omega$, in the limiting case $\omega = \omega_1$. The exact relation $\nabla^2\omega = 0$ is then replaced by

$$\nabla_h^2 \omega = (\omega/h^2)\ (e_r + e_t) \tag{7.2}$$

at $(h, \pm\pi/4)$, where e_r and e_t represent radial and transverse error contributions from the discrete 5-point Laplacian operator. (In the case of D&S, the radial and transverse directions are taken parallel to $\theta = 0$, and $\theta = \pi/2$, respectively). Such an error perturbs the central node factor of the discrete operator from -4 to $-(4 + e_r + e_t)$. This error persists as $h \to 0$. Values of e_r and e_t are given in Table 4. The table includes an effective corner vorticity factor d_2 for D&S, obtained by equating $\nabla_h^2\omega_1$ to its rotated form.

Ranking by d_2 and by error factors e_r and e_t gives similar results (Table 4). We note that a non-zero error contribution remains from the transverse derivative for SFD. This suggests that the SFD method could be improved by treating the radial as well as the transverse vorticity derivatives by singular finite differences.

TABLE 4

Theoretical ranking - see section 7.

Corner Method	d_2	e_r	e_t
SFD	-1.75042	0	0.263
Kawaguti	-2.0	0.276	0.263
Discontinuous) Average) (with Woods Biased) Method)	-1.04741	-0.777	0.263
D & S	-0.48508	-0.170	-0.965

8. SUMMARY

Flow calculations employing singular finite difference derivatives (Holstein and Paddon, 1981) in the neighbourhood of a re-entrant corner are used as a standard of comparison for several ad hoc corner treatments. For steady flows through contraction, expansion and cavity flow domains at Reynolds numbers up to 30, it is found that the method of Kawaguti yields consistently close agreement with the singular finite difference method. Moreover, the method is easily programmed. Of the ad hoc methods considered, we recommend the method of Kawaguti for practical computation. A detailed report of the computational results summarised in this paper is in preparation (Paddon and Holstein, 1982).

9. ACKNOWLEDGEMENTS

We are grateful to Dr. I.P. Castro for drawing our attention to the work of Dennis and Smith, during the original presentation of this paper at the conference.

REFERENCES

DENNIS, S.R.C. and SMITH, F.T. 1980 Steady flow through a channel with a symmetrical constriction in the form of a step. *Proc. R. Soc.* A**372**, 393-414.

HOLSTEIN JR., H. and PADDON, D.J. 1981 A singular finite difference treatment of re-entrant corner flow. *J. Non-Newtonian Fluid Mech.* **8**, 81-93.

KAWAGUTI, M. 1965 Numerical solution of the Navier Stokes equation for the flow in a channel with a step. MRC TSR 574, Mathematical Research Center, Madison, Wisconsin.

MOFFATT, M.K. 1964 Viscous and resistive eddies near a sharp corner. *J. Fluid Mech.* **18**, 1-18.

PADDON, D.J. and HOLSTEIN JR., H. 1982 A survey of corner methods for plane, steady, viscous flows. University of Bristol Technical Report CS-82-05.

ROACHE, P.J. 1972 Computational Fluid Dynamics. Albuquerque: Hermosa.

APPENDIX

From (Holstein and Paddon, 1981) the functions f_1, f_2, g_1, g_2 in (4.3) and (4.4) are given by

$$f_1(\theta) = \cos(\Lambda_1-2)3\pi/4 \cos\Lambda_1 - \cos\Lambda_1 3\pi/4 \cos(\Lambda_1-2)\theta,$$

$$f_2(\theta) = \sin(\Lambda_2-2)3\pi/4 \sin\Lambda_2 - \sin\Lambda_2 3\pi/4 \sin(\Lambda_2-2) ,$$

$$g_1(\theta) = 4(\Lambda_1-1) \cos\Lambda_1 3\pi/4 \cos(\Lambda_1-2)\theta,$$

$$g_2(\theta) = 4(\Lambda_2-1) \sin\Lambda_2 3\pi/4 \sin(\Lambda_2-2)\theta.$$

Numerical values of the constants c_{ij} in (4.7), (4.8) are, for the choice $\alpha = 3 - \Lambda_2$ (c.f. (4.6)),

$$c_{11} = -2.18294, \quad c_{12} = 2.13091, \quad c_{13} = -0.282057,$$

$$c_{21} = 0.282779, \quad c_{22} = -0.194916, \quad c_{23} = 0.522375.$$

THE EFFECT OF PRESSURE APPROXIMATIONS ON FINITE ELEMENT CALCULATIONS OF INCOMPRESSIBLE FLOWS

D.F. Griffiths

(Dept. of Mathematical Sciences, University of Dundee)

1. INTRODUCTION

The most common mixed finite element formulations of incompressible flow problems make use of two spaces of piecewise polynomials - $\underline{X}^h$ and M^h - from which the velocity $\underline{u}$ and pressure p are approximated. With Ω a simply connected, bounded polygonal domain in $\mathbb{R}^2$ and with $H_0^1(\Omega)$ the usual Sobolev space of functions which have square integrable first derivatives and vanish on the boundary of Ω we shall, for the purposes of this paper, assume that $M^h \subset L_2(\Omega)$ and $\underline{X}^h \subset \underline{X} \equiv \{H_0^1(\Omega)\}^2$, though the latter inclusion may be weakened to allow nonconforming elements (Crouzeix and Raviart, 1973). The velocity in incompressible flow - governed, for example, by the Navier-Stokes equations - is constrained by the continuity equation to lie in the subspace $\underline{V} \subset \underline{X}$ defined by

$$\underline{V} = \{\underline{v} \in \underline{X},\ \text{div}\ \underline{v} = 0 \text{ in } \Omega\} \tag{1.1}$$

or equivalently,

$$\underline{V} = \{\underline{v} \in \underline{X},\ (q, \text{div}\ \underline{v}) = 0 \text{ for all } q \in L_2(\Omega)\} \tag{1.2}$$

where $\text{div}\ \underline{v} = \partial_x v_1 + \partial_y v_2$ for $\underline{v} = (v_1, v_2)^T$ and $(\cdot,\cdot)$ denotes the usual L_2 inner product.

Likewise, in mixed finite element formulations, the approximation $\underline{u}^h$ to the velocity $\underline{u}$ is constrained to be weakly solenoidal: it lies in the subspace $\underline{V}^h \subset \underline{X}^h$ defined by

$$\underline{V}^h = \{\underline{v}^h \in \underline{X}^h, (q^h, \text{div } \underline{v}^h) = 0 \text{ for all } q^h \in M^h\}. \quad (1.3)$$

An excellent account of the theoretical foundations of mixed methods is given by Girault and Raviart (1979). Briefly, the success or otherwise of these mixed methods depends on two properties. First of these is the compatibility of the spaces $\underline{X}^h$ and M^h and, secondly, there is the question of how well $(\underline{u},p) \in \underline{X}\times M$ can be approximated from $\underline{X}^h\times M^h$. A sufficient condition for compatibility is provided by the Babuska-Brezzi (BB) condition (Babuska, 1971 and Brezzi, 1974):

there exists a positive constant β, independent of h, such that

$$\sup_{\underline{v}^h \in \underline{X}^h} \frac{(q^h, \text{div } \underline{v}^h)}{\|\underline{v}^h\|_1} > \beta\|q^h\|_0, \text{ for all } q^h \in M^h, \|q^h\|_0 \neq 0 \quad (1.4)$$

where $\|\cdot\|$ is the usual L_2 norm, $\|\cdot\|_1$ is the norm on $\underline{X}$, i.e.

$$\|\underline{v}\|_1 = \{\|\underline{v}\|^2 + \|\text{grad } \underline{v}\|^2\}^{\frac{1}{2}}$$

and

$$\|q\|_0 = \inf_{c \in \mathbb{R}} \|q + c\|.$$

The norm $\|\cdot\|_0$ is designed to filter out the hydrostatic pressure q = constant. If $\underline{X}^h$ and M^h satisfy the BB condition and if the pair $(\underline{u}^h, p^h)$ denote the mixed finite element approximation to the solution pair $(\underline{u},p)$ of the Stokes equations (with body force and homogeneous boundary conditions) then it may be shown (Girault and Raviart, 1979) that

$$|\underline{u} - \underline{u}^h|_1 + \|p - p^h\| \leq C\{\inf_{\underline{v}^h \in \underline{X}^h} |\underline{u} - \underline{v}^h|_1 + \inf_{q^h \in M^h} \|p - q^h\|\} \quad (1.5)$$

where $|\underline{u}|_1 = \|\text{grad } \underline{u}\|$. This result shows that the assessment of how well $\underline{V}^h$ approximates $\underline{V}$ may be accomplished by estimating the standard error terms on the right of (1.5). Unfortunately, the verification in specific instances that the BB condition is satisfied is often a delicate exercise. Crouzeix and Raviart

(1973) avoided the BB condition by constructing interpolation operators $\Pi^h : \underline{X} \rightarrow \underline{X}^h$ which preserve divergence in the sense that

$$(q^h, \text{div } \Pi^h \underline{v}) = (q^h, \text{div } \underline{v}) \text{ for all } q^h \in M^h. \qquad (1.6)$$

It was subsequently shown by Fortin (1977) that the BB condition is satisfied whenever such an operator exists. Failure to satisfy (1.4), in the sense that the constant $\beta = 0$, is attributable to the linear dependence of the constraints imposed by M^h. That is, there exists at least one function $q^h \in M^h$, which is not the constant (hydrostatic) pressure, such that

$$(q^h, \text{div } \underline{v}^h) = 0 \text{ for all } \underline{v}^h \in \underline{X}^h. \qquad (1.7)$$

Functions q^h satisfying (1.7) are called spurious pressure modes (Sani et al., 1981a and b) and their existence means that the dimension of $\underline{V}^h$ is larger than expected. In earlier papers (Griffiths, 1979a, b, 1981) explicit bases were derived for a number of different velocity-pressure elements and this enabled, as a byproduct, the determination of the dimension of $\underline{V}^h$. By comparing the actual and expected dimensions of $\underline{V}^h$, the cases $\beta = 0$ and $\beta \neq 0$ may be distinguished although the latter does not rule out the possibility of β being dependent on h. For other aspects relating to the BB condition the interested reader is referred to Gustafson and Hartman (1981), Malkus (1981) and Walters and Carey (1981).

In subsequent sections we shall pursue the theme initiated by Griffiths (1979a, b and 1981) and describe bases for a variety of weakly solenoidal spaces generated by piecewise linear and quadratic representations, respectively, for pressure and velocity. Our primary interest here is not in the construction process per se but in elucidating the structure of $\underline{V}^h$ and the changes induced by changes in M^h. Some of the bases described may be of practical value but others do not conform to the definition of a finite element approximation because the representation of the velocity in one element depends on the geometric properties of neighbouring elements.

2. FINITE ELEMENT SPACES

The domain Ω is subdivided into triangular elements in the usual manner and $\underline{X}^h$ is taken to be the space of continuous, vector-valued, piecewise quadratic functions expressed in terms of the familiar Lagrange basis. We assume that Ω is simply connected, though this is not essential, and that the triangulation is such that each element has at most one edge lying on the boundary $\partial\Omega$ of $\bar{\Omega}$. To compute the dimension $\nu(\underline{X}^h)$ of $\underline{X}^h$, suppose that the triangulation is composed of t triangles, v interior vertices, s interior sides and b boundary sides. Thus, since functions in $\underline{X}^h$ vanish on $\partial\Omega$,

$$\nu(\underline{X}^h) = 2(s + v). \tag{2.1}$$

We now proceed to describe the structure of several weakly solenoidal spaces $\underline{V}^h$ which result from the imposition of different pressure spaces M^h on $\underline{X}^h$.

(i) $M_D^h = \text{div}\ \underline{X}^h \equiv \{q^h;\ q^h = \text{div}\ \underline{v}^h,\ \underline{v}^h \in \underline{X}^h\}$. This space consists of piecewise linear functions having no inter-element continuity. If $\underline{V}_D^h$ denotes the corresponding subspace of $\underline{X}^h$ defined by (1.3), then it is evident that functions in $\underline{V}_D^h$ have zero divergence: $\underline{V}_D^h \subset \underline{V}$, indeed $\underline{V}_D^h = \underline{X}^h \cap \underline{V}$. It then follows that, for each $\underline{v}^h \in \underline{V}_D^h$, there is a piecewise cubic function ψ^h, the discrete streamfunction, such that, for any two points P_1 and P_2 in Ω,

$$\psi^h(P_1) = \psi^h(P_2) + \int_\Gamma \underline{v}^h.\underline{n}\ ds$$

where $\underline{n}$ is the unit normal to an arbitrary path Γ connecting P_1 to P_2. The dimension of M_D^h is given by

$$\nu(M_D^h) = 3t$$

and therefore, taking account of the hydrostatic pressure,

$$\begin{aligned}\nu(\underline{V}_D^h) &= \nu(\underline{X}^h) - \nu(M_D^h) + 1 \\ &= 2v - b + 1,\end{aligned} \tag{2.2}$$

where we have made use of Euler's formulae

$$s = 3(v - 1) + b,$$
$$t = 2(v - 1) + b. \tag{2.3}$$

It follows from (2.1), (2.2) and (2.3) that $\nu(\underline{V}_D^h) < \frac{1}{4}\nu(\underline{X}^h)$ demonstrating that divergence-free functions constitute only a relatively small part of $\underline{X}^h$. Indeed, on grids with a small number of elements it can often occur that b > 2v in which case $\underline{X}^h$ is over-constrained (locked) and $\underline{V}_D^h = \{\underline{O}\}$. For instance, when Ω is the unit square divided into N^2 squares and then into $2N^2$ triangles, we have $v = (N-1)^2$ and $b = 4N$ giving $\nu(\underline{V}_D^h) = 2(N-2)^2 - 5$. Thus N = 4 (32 elements) is the smallest grid of this type that can sustain a non-trivial space $\underline{V}_D^h$.

Over-constraining of $\underline{X}^h$ can be avoided by treating each element as a macro-element wherein each element is subdivided into three sub-elements by the lines joining centroid to vertices. The dimensions of $\underline{X}^h$ and M_D^h are then increased to

$$\nu(\underline{X}_D^h) = 2(s + v + 4t), \qquad \nu(M_D^h) = 9t$$

and consequently,

$$\nu(\underline{V}_D^h) = s + 3v \tag{2.4}$$

where t, s and v refer to the macro- rather than the sub-elements of the grid. The expression (2.4) immediately suggests that $\underline{V}_D^h$ has three degrees of freedom associated with each interior vertex (streamfunction and two velocity components) and one degree of freedom (tangential velocity) at each midside node. This macro-element is equivalent to representing the stream-function ψ^h in terms of the well-known Clough-Tocher element and subsequently obtaining the velocity from $\underline{v} = \text{curl}\ \psi^h$. By explicitly solving the constraints in a manner suggested by Griffiths (1979b) a basis set for $\underline{V}_D^h$ may be constructed. Vector plots of typical members of this set are shown in Fig. 1 for a representative patch of elements. For a basis for the Clough-Tocher element we refer to Bernadou and Hassan (1981).

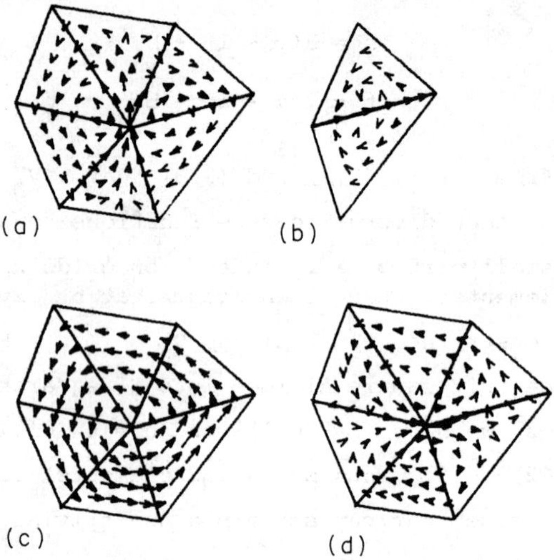

Fig. 1 Representative basis functions for $\underline{V}^h_D$. Note that each element shown is the union of three sub-elements.

We now turn our attention to the effects of choosing other pressure spaces. Since $M^h = \text{div}\ \underline{X}^h$ implies that $\text{div}\ \underline{v}^h = 0$ for $\underline{v}^h \in \underline{V}^h$, it is clear that taking a larger pressure space, in the sense that $M^h \supset \text{div}\ \underline{X}^h$, can have no effect on $\underline{V}^h$ - although in a Navier-Stokes problem it would undoubtedly lead to spurious pressure modes. We can therefore restrict ourselves to 'smaller' pressure spaces, though these may not be subspaces of $\text{div}\ \underline{X}^h$ (equal-order interpolation for example).

(ii) $M^h_C = \text{div}\ \underline{X}^h \cap C^0(\Omega)$, the space of continuous piecewise linear functions on Ω. M^h_C is provided with the usual basis of 'pyramid' functions giving one degree of freedom at each vertex of the grid (including boundary vertices). Thus,

$$\nu(M^h_C) = v + b$$

and

$$\nu(\underline{V}^h_C) = 2v + s + (t - b) \tag{2.5}$$

indicating that the weakly solenoidal space $\underline{V}_C^h$ has two degrees of freedom at each interior vertex, one at each midside node and one degree of freedom for each element centroid, excluding those elements having one edge on $\partial\Omega$ - recall that each element has at most one edge on $\partial\Omega$. A straightforward computation reveals that

$$(q^h, \text{div } \underline{\phi}_v) = 0, \quad \text{for all } q^h \in M_C^h, \tag{2.6}$$

where $\underline{\phi}_v$ is a typical basis function from $\underline{X}^h$ associated with a vertex node, i.e. $\underline{\phi}_v = (1,0)^T$ or $(0,1)^T$ at one vertex node and vanishes at all other nodes of the grid (Fig. 3a, b). These functions therefore also lie in $\underline{V}_C^h$ and contribute the 2v term in (2.5). To complete the specification of a basis for $\underline{V}_C^h$, consider the patch of elements shown in Fig. 2. With each midside node, typified by M_1, we associate a basis function $\underline{\phi}_{ms}$ which vanishes at all nodes except M_1, M_2 and M_4 where it has the values

$$\underline{\phi}_{ms} = \begin{cases} \alpha\, \underline{P_3P_4}/(A_1 + A_2) & \text{at } M_1 \\ \alpha\, \underline{P_4P_1}/(A_2 + A_5) & \text{at } M_4 \\ \alpha\, \underline{P_1P_3}/(A_1 + A_3) & \text{at } M_2 \end{cases} \tag{2.7}$$

A_j being the area of the jth element, $\underline{P_iP_j}$ representing the vector joining P_i to P_j and α being chosen such that $\underline{\phi}_{ms}$ has unit flux across $\underline{P_iP_j}$. Note that $\underline{\phi}_{ms}$ could also be defined on the nodes M_1, M_3 and M_5, which of these definitions is adopted is immaterial. Finally, with each element having no edge on $\partial\Omega$, typically element '1' in Fig. 2, we associate a basis function $\underline{\phi}_e$ which vanishes at all nodes except M_1, M_2 and M_3 where it has the values

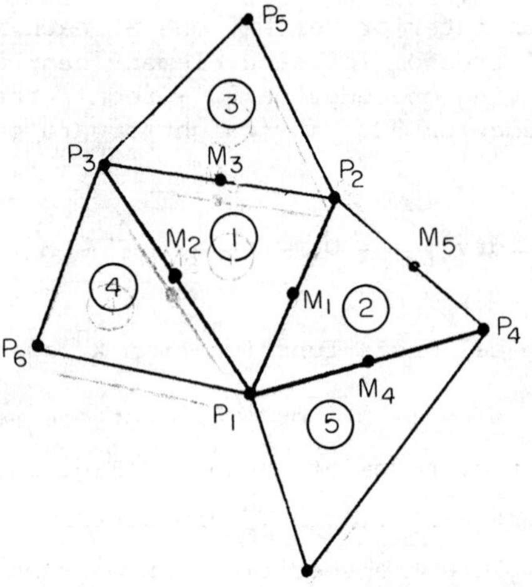

Fig. 2 A patch of elements for describing the basis functions for $\underline{V}_C^h$. The element numbers are circled.

$$\underline{\phi}_e = \begin{cases} \underline{P_1P_2}/(A_1 + A_2) & \text{at } M_1 \\ \underline{P_2P_3}/(A_1 + A_3) & \text{at } M_3 \\ \underline{P_3P_1}/(A_1 + A_4) & \text{at } M_2. \end{cases} \tag{2.8}$$

It may in fact be shown that div $\underline{\phi}_e$ = 0 inside element '1' though it does not vanish on the whole of Ω. Vector plots of representative basis functions are given in Fig. 3. From (2.7) and (2.8) it is seen that these basis functions violate one of the cardinal rules for a finite element basis, namely, that their definition on one element depends on the geometry of neighbouring elements. This precludes their implementation in conventional finite element packages.

The above basis set may have the merit of simplicity but it turns out not to be the most convenient for comparison with those of other spaces $\underline{V}^h$. From (2.5) and (2.3) it follows that

$$\nu(\underline{V}_C^h) = 5v + (t - b) + (b - 3) \tag{2.9}$$

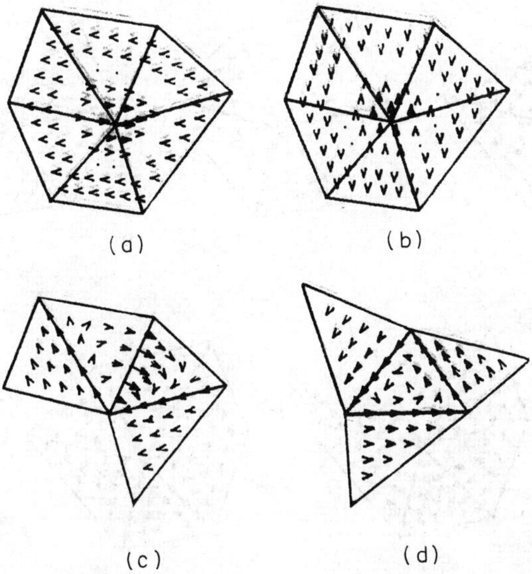

Fig. 3 Vector plots of typical basis functions from $\underline{V}_C^h$. The functions a) and b) interpolate the horizontal and vertical components of velocity whilst c) and d) represent the functions $\underline{\phi}_{ms}$ and $\underline{\phi}_e$ described by (2.7) and (2.8) respectively.

on which we base a partial description of a modified basis in which there are five degrees of freedom at each vertex node and no midside function. At a typical vertex, P_1 say in Fig. 2, we first add together all functions of the form (2.7) which are attached to edges emanating from P_1. The resulting function, denoted by $\hat{\underline{\phi}}_v^0$, is depicted in Fig. 4a. Next, the two vertex functions $\underline{\phi}_v$ of the original basis are modified by the addition of functions of the form (2.7), with suitable weights, to give $\hat{\underline{\phi}}_v^1$ and $\hat{\underline{\phi}}_v^2$ satisfying an element mass balance:

$$\int_e \operatorname{div} \hat{\underline{\phi}}_v^j \, de = 0, \qquad j = 1,2, \tag{2.10}$$

for each element $e \in \Omega$. Note that $\hat{\underline{\phi}}_v^0$ and $\underline{\phi}_e$ also satisfy a balance of the form (2.10). The structure of the two remaining

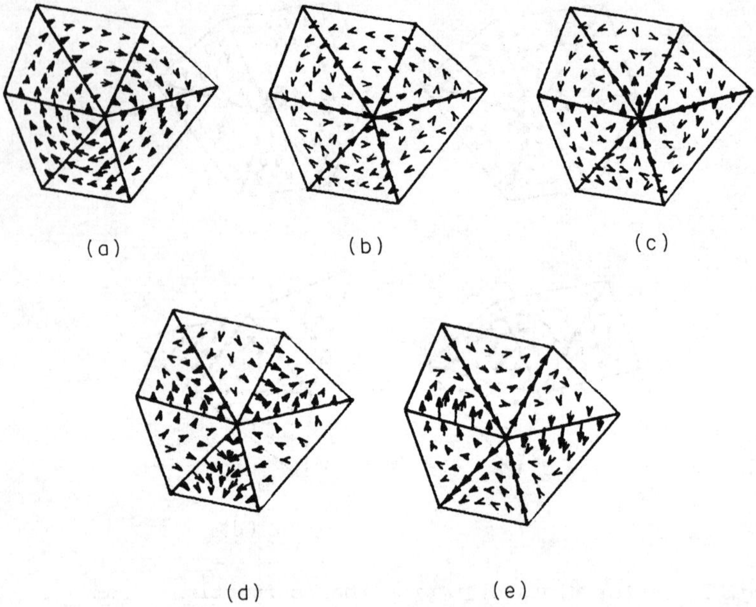

Fig. 4 The vector plots a),b)..,e) show, respectively, the modified basis functions $\hat{\underline{\phi}}_v^j$ (j = 0,1,..,4).

functions $\hat{\underline{\phi}}_v^3$ and $\hat{\underline{\phi}}_v^4$ associated with a vertex depends on how many elements meet at P_1. The only situation of interest here is when six triangles meet at each vertex. This being so, linear combinations of functions of the form (2.7), with weights ± 1, are formed and the results depicted in Fig. 4d,e. With the (t - b) functions of the form (2.8), there remain a further (b - 3) functions to be identified. Since these play no part in the subsequent discussion, they will be left unspecified.

The results of a convergence analysis of a mixed finite element solution of the Navier-Stokes equations based on $\underline{X}^h$ and M_C^h are given by Raviart (1979).

(iii) $M_{PC}^h = \{q^h;\ q^h|_e = \text{constant on each element } e \in \Omega\}$. If $\underline{V}_{PC}^h$ denotes the corresponding weakly solenoidal space, then

$$\nu(M^h_{PC}) = t, \qquad \nu(\underline{V}^h_{PC}) = 3v + s. \tag{2.11}$$

The construction of a possible basis for $\underline{V}^h_{PC}$ has been previously described by Griffiths (1979a,b). However, we again make modifications in order to facilitate comparison with the bases for other spaces $\underline{V}^h$. In fact, the three basis functions at each vertex may be chosen to be $\hat{\underline{\phi}}^0_v$, $\hat{\underline{\phi}}^1_v$, $\hat{\underline{\phi}}^2_v$ from $\underline{V}^h_C$ since they can satisfy (2.10). The remaining basis function, associated with a midside node, is displayed in Fig. 5; three such functions, with appropriate weights, combine to give $\underline{\phi}_e$ (Fig. 3d).

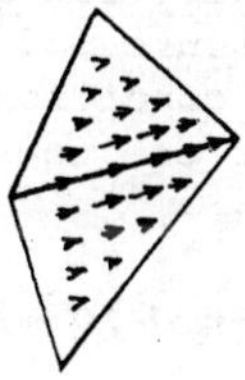

Fig. 5 A midside basis function from $\underline{V}^h_{PC}$. The remaining basis functions for this space are shown in Fig. 4a), b) and c).

The results of numerical experiments comparing the performance of mixed formulations based on $\underline{X}^h \times M^h_D$ and $\underline{X}^h \times M^h_{PC}$ are presented by Thompson (1975).

(iv) $M^h_{MC} = \{q^h;\ q^h|_e = \text{linear},\ q^h \text{ continuous at midside nodes}\}$.
In this case

$$\nu(M^h_{PC}) = s + b$$

and, following the course of the previous examples, one would expect $\nu(\underline{V}^h_{MC}) = 5v - 2$. However, this combination of spaces violates the BB condition: for certain regular grids there are three (rather than the single hydrostatic) pressure modes, i.e. three independent functions $q^h \in M^h_{MC}$ satisfying (1.7).

Consequently

$$\nu(\underline{V}^h_{MC}) = 5v \tag{2.12}$$

on these grids. One simple example of a grid which gives rise to these spurious pressure modes is given by the decomposition of Ω into equal parallelograms and then into triangles by diagonals parallel to a single direction. In the special case of a grid of equilateral triangles the structure of the pressure modes is quite elementary: The pressure takes the values 1,0,1,0,1,0 sequentially at the midside nodes of the edges emanating from any one vertex. There are three such modes and their sum gives the constant hydrostatic pressure. The two remaining modes are spurious and resemble the checkerboard (CB) modes described by Sani et al. (1981a,b). More generally, it may be shown that three pressure modes exist whenever the grid satisfies certain geometric conditions similar to those given by Sani et al. (1981a,b) for the four node velocity - piecewise constant pressure element.

On the assumption that the grid allows three pressure modes, the requisite number of basis functions for $\underline{V}^h_{MC}$ is provided by precisely the five (modified) vertex basis functions $\hat{\underline{\phi}}^j_v$ (j = 0,1,.,4) for $\underline{V}^h_C$ (see Fig. 4). It follows from the inclusion $M^h_{MC} \supset M^h_C$ that $\underline{V}^h_{MC} \subset \underline{V}^h_C$ and, therefore, a basis for $\underline{V}^h_{MC}$ can always be expressed in terms of the basis for $\underline{V}^h_C$, even on arbitrary grids.

The nature of the spurious pressure modes in this example is very similar to that encountered when using equal order interpolation, i.e. when $M^h = X^h$, the space of scalar piecewise quadratic functions. On a grid of equilateral triangles, for example, the pressure modes are actually identical.

For solutions $(\underline{u},p) \in \underline{V}\times M$ which are sufficiently smooth, the optimal order of the terms on the right of (1.5) is $O(h^2)$ as the grid is refined. This order is attainable with a piecewise linear approximation for the pressure and, most efficiently in mixed mode, with M^h_C. There is some evidence to suggest that the lack of an element mass balance of the form (2.10) can lead to non-physical numerical solutions on coarse grids (Gresho et al. 1980). This leads us to propose the following element.

(v) $M_*^h = M_C^h \cup M_{PC}^h$, the continuous piecewise linear and piecewise constant pressure fields are superposed. There are, therefore, two independent hydrostatic pressure modes and the pressure must be specified at two points (a vertex and a centroid) to ensure uniqueness. Now $\underline{V}_*^h = \underline{V}_C^h \cap \underline{V}_{PC}^h$,

$$\nu(\underline{V}_*^h) = 3v + (t - b) \tag{2.13}$$

and a basis is provided by the functions $\hat{\underline{\phi}}_v^0$, $\hat{\underline{\phi}}_v^1$, $\hat{\underline{\phi}}_v^2$ (Fig. 4 a, b,c) and $\underline{\phi}_e$ (Fig. 3d). $\hat{\underline{\phi}}_v^1$ and $\hat{\underline{\phi}}_v^2$ interpolate the horizontal and vertical components of velocity at a vertex whereas $\hat{\underline{\phi}}_v^0$ and $\underline{\phi}_e$ interpolate, respectively, (some multiple of) the streamfunction at a vertex and element centroid (the latter being some averaged value of the streamfunction over the element).

This element combines the approximating ability of M_C^h with the element mass balance of M_{PC}^h, though it could prove expensive to implement in a mixed finite element method since $\nu(M_*^h) \simeq 3\nu(M_C^h)$. One way around this problem might be to use the basis for $\underline{V}_{PC}^h$ described by Griffiths (1979a,b) and then impose the constraints implied by M_C^h in the usual manner by Lagrange multipliers. The total number of degrees of freedom in this system would be $\nu(M_C^h) + \nu(\underline{V}_{PC}^h) \simeq 7v$ compared with $\nu(M_C^h) + \nu(\underline{X}^h) \simeq 9v$ using $\underline{V}_C^h$ in mixed mode. If the basis described for $\underline{V}_*^h$ could be incorporated into a finite element package then the number of degrees of freedom required would be $\nu(\underline{V}_*^h) \simeq 5v$, with no additional constraints.

This device of superposing a piecewise constant pressure on an otherwise continuous field has been used to some effect by Gresho et al. (1980) for Boussinesq fluids; with a piecewise biquadratic velocity field, an underlying continuous bilinear pressure approximation was taken.

3. DISCUSSION

We have attempted to detail in this paper the interrelationships between the weakly solenoidal spaces generated by

$M_D^h = \text{div}\ \underline{X}^h$ and some of its subspaces. A number of the results follow from the simple observation that for two pressure spaces M_1^h and M_2^h with $M_1^h \cup M_2^h$, the corresponding weakly solenoidal spaces satisfy $\underline{V}_1^h \cap \underline{V}_2^h$. These results are summarised in the Venn diagram shown in Fig. 6. The dimensions of the various spaces $\underline{V}^h$ can be expressed, asymptotically, as multiples of v, the number of interior vertices in the grid, and therefore v is used as the horizontal scale on the rectangular sections of Fig. 6. Also indicated are the basis functions which span the various spaces.

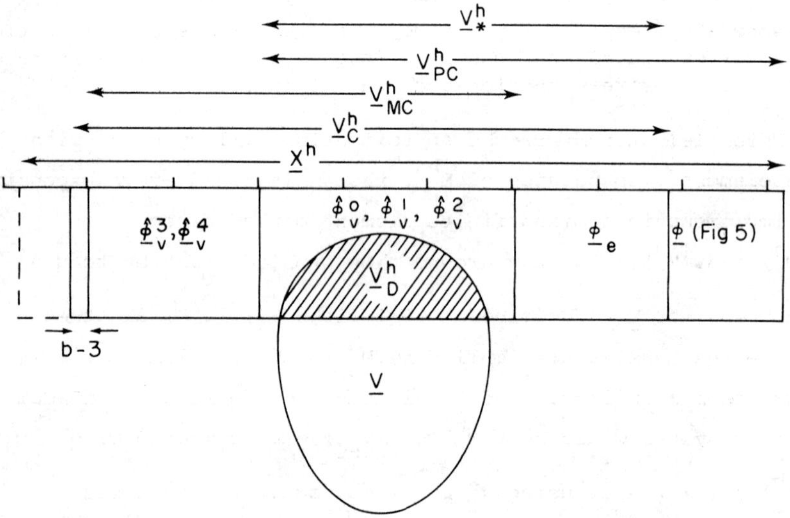

Fig. 6 Venn diagram illustrating the inter-relationship between $\underline{V}$, $\underline{X}^h$ and the various weakly solenoidal spaces $\underline{V}^h$. The shaded region represents $\underline{V}_D^h$.

The direct practical implications of this work is not clear but it is hoped that it will lead to a better understanding of mixed finite element methods for Stokes and Navier-Stokes flows.

REFERENCES

BABUSKA, I. 1971 Error bounds for finite element methods. *Num. Math.* **16**, 322-333.

BERNADOU, M. and HASSAN, K. 1981 Basis functions for the general Hsieh-Clough-Tocher triangles, complete or reduced. *Int. J. Num. Meth. Eng.* **17**, 784-789.

BREZZI, F. 1974 On the existence, uniqueness and approximations of saddle-point problems arising from Lagrange multipliers. R.A.I.R.O. 8-R2, 129,151.

CROUZEIX, M. and RAVIART, P-A. 1973 Conforming and non-conforming elements for solving the stationary Stokes equations I. R.A.I.R.O. R3, 33-76.

FORTIN, M. 1977 An analysis of the convergence of mixed finite element methods. R.A.I.R.O. 11-R3, 341-354.

FORTIN, M. 1981 Old and new finite elements for incompressible flows. *Int. J. Num. Meth. Fluids* **1**, 347-364.

GIRAULT, V. and RAVIART, P-A. 1979 Finite Element Approximation of the Navier-Stokes Equations. Lecture Notes in Mathematics, 749, New York:Springer-Verlag.

GRESHO, P.M., LEE, R.L., CHAN, S.T. and LEONE, J.M. Jr. 1980 A new finite element for Boussinesq fluids. In Proc. Third Int. Conf. on Finite Elements in Flow Problems.

GRIFFITHS, D.F. 1979a Finite elements for incompressible flows. *Math. Meth. in Appl. Sci.* **1**, 16-31.

GRIFFITHS, D.F. 1979b The construction of approximately divergence-free finite elements. In Mathematics of Finite Elements and Applications III (J.R. Whiteman, Ed.), London: Academic Press, 239-245.

GRIFFITHS, D.F. 1981 An approximately divergence-free 9-node velocity element (with variations) for incompressible flows. *Int. J. Num. Meth. Fluids* **1**, 323-346.

GUSTAFSON, K. and HARTMAN, R. 1981 Divergence-free bases for finite element schemes in hydrodynamics. Manuscript.

MALKUS, D.S. 1981 Eigenproblems associated with the discrete LBB condition for incompressible finite elements. *Int. J. Eng. Sci.* **19**, 1299-1310.

RAVIART, P-A. 1979 Finite element methods and Navier-Stokes equations. In Lecture Notes in Physics 91. New York: Springer-Verlag, 27-47.

SANI, R.L., GRESHO, P.M., LEE, R.L. and GRIFFITHS, D.F. 1981 The cause and cure(?) of the spurious pressures generated by certain FEM solutions of the incompressible Navier-Stokes equations, Part I. *Int. J. Num. Meth. Fluids* **1**, 17-43.

SANI, R.L., GRESHO, P.M., LEE, R.L., GRIFFITHS, D.F. and ENGELMAN, M. 1981 The cause and cure(?) of the spurious pressures generated by certain FEM solutions of the incompressible Navier-Stokes equations, Part II. *Int. J. Num. Meth. Fluids* **1** 171-204.

Thompson, E.G. 1975 Average and complete incompressibility in the finite element method. *Int. J. Num. Meth. Eng.* **9**, 925-932.

WALTERS, R.A. and CAREY, G.F. 1981 Analysis of spurious modes for the shallow water and Navier-Stokes equations, TICOM Report 81-3, Univ. of Texas at Austin.

ANALYSIS OF SOME LOW ORDER FINITE ELEMENT SCHEMES FOR THE NAVIER-STOKES EQUATIONS

M.J.P. Cullen

(Meteorological Office, Bracknell)

1. INTRODUCTION

This paper is motivated by an apparent gap between the very extensive theoretical studies of finite element methods applied to the Navier-Stokes equations, e.g. Temam (1977), Heywood and Rannacher (1982), Thomasset (1981), and the actual behaviour of the schemes in computations. The theoretical studies are concerned with obtaining "optimal" rates of convergence to the exact solution, where the "optimal" rate is the rate at which the piecewise polynomial approximation used converges to a general smooth function. It has been found, however, by Gresho et al. (1980, 1982), that there are marked differences among the quality of results obtained by schemes which should yield the same convergence rate. For instance, in two test problems, it was found that using a bilinear isoparametric element for velocity with the same bilinear approximation to the pressure gave much better results for the velocities than using piecewise constant pressures. The pressure fields were worse because of extra checkerboard modes. Both schemes were observed to give quadratic convergence in velocity as the mesh was refined, which is the optimal rate. However, the difference between the two results was more in line with the accuracy of the schemes interpreted as finite difference schemes, even though it is known that, in general, such an interpretation can be misleading for finite element algorithms. Finite element methods require the discrete representation to be obtained from the continuous data by a projection depending on the equation being solved. In the case of primitive variable approximations to the Navier-Stokes equations this projection involves the velocity and pressure simultaneously. It is not possible to state, for instance, that the discrete velocity field is a best fit to the continuous velocity field in some norm. This is why apparently large errors can occur, even though these errors then converge to zero at the predicted rate when the mesh is refined.

The results of Gresho et al. suggest that, for real applications on an affordable mesh, these errors may not be acceptable. The grid used to study flow past a cylinder there was quite well refined, and the grid required to obtain correct results with the bilinear velocity/piecewise constant pressure element had to be extremely smooth, Gresho et al. (1982), Brooks (1981). The mesh refinement needed to avoid errors in practical applications with more difficult boundary shapes may be prohibitive. It is therefore worth investigating alternatives. In this paper these errors are analysed and methods which avoid them are discussed. Much of the difficulty lies in the enforcement of the incompressibility condition, which is non-local. This condition has to be approximated by a discrete condition which is then imposed on the discrete approximation to the velocity field. The resulting approximation may no longer be close to a best fit to the velocity field, a fact that has caused serious difficulties even in proving the consistency of finite element schemes. The difficulty is equally present in finite difference velocity/pressure algorithms.

In the finite element context this difficulty can be resolved by approximating the velocity field using only functions which satisfy the discrete incompressibility condition. This has been extensively studied, by Griffiths (1981) and Fortin (1981), amongst others. In this case it should be possible to obtain a solution which is a best fit to the velocity by these divergence-free functions. Alternatively, we can seek a velocity field whose implied vorticity field is a best fit to the true vorticity. This type of method has been studied by Raviart (1977), and others.

In this paper we perform some analysis which illustrates these approaches, and suggests some advantages in the vorticity-based approach. Some experiments are carried out which confirm these results. It is possible to obtain correct solutions in general geometry without using any more than bilinear functions in the approximation scheme, an important advantage. A hybrid finite difference/finite element approach is also illustrated, which demonstrates how easily finite differences can be implemented on arbitrary grids.

2. THEORETICAL DISCUSSION

For the purposes of this paper we consider the two dimensional Navier-Stokes equations:

$$\left.\begin{aligned} \frac{\partial \underline{u}}{\partial t} + \underline{u}.\nabla\underline{u} + \frac{1}{\rho}\nabla p &= \nu\nabla^2\underline{u} + \underline{f} \\ \nabla.\underline{u} &= 0 \end{aligned}\right\} \text{ on } \Omega$$

$$\underline{u} = \underline{g} \quad \text{on } \partial\Omega. \qquad (1)$$

The three dimensional extensions to the results will be mentioned briefly. The standard velocity-pressure Galerkin approximation to (1) can be written, using C^0 basis functions $\underline{\chi}_n$ for velocity and C^{-1} functions θ_n for pressure, as

$$\int \left(\frac{\partial \underline{u}_h}{\partial t} + \underline{u}_h.\nabla\underline{u}_h\right).\underline{\chi}_n \, d\Omega - \frac{1}{\rho}\int p_h \nabla.\underline{\chi}_n \, d\Omega =$$

$$-\nu\int \nabla\underline{\chi}_n : \nabla\underline{u}_h \, d\Omega + \int \underline{f}_h.\underline{\chi}_n \, d\Omega \qquad \forall\, \chi_n \qquad (2)$$

$$\int \theta_n \nabla.\underline{u}_h \, d\Omega = 0. \qquad (3)$$

Equations (2) do not include equations involving $\underline{\chi}_n$ for boundary nodes since $\underline{u}_h$ is given on the boundary. In order to prove that the solution of (2) converges to the solution of (1) at "optimal" rate, three assumptions are required (e.g. Heywood and Rannacher, 1982).

(i) Regularity of the solution to the steady Stokes problem with the same boundary conditions as (1).

(ii) Convergence of a discrete approximation satisfying (3) to a non-divergent $\underline{u}$ at optimal rate.

(iii) Uniqueness of the discrete pressure derived from (2) and (3).

If all these conditions are met, then consistency in the solution of the time dependent problem follows by standard techniques if the initial data $(\underline{u}_0, p_0)$ are represented as $(\underline{u}_{0h}, p_{0h})$ satisfying

$$\int \Big[(\underline{u}_O.\nabla\underline{u}_O - \underline{u}_{Oh}.\nabla\underline{u}_{Oh} - \underline{f}_O + \underline{f}_{Oh}).\underline{\chi}_n + \frac{1}{\rho}(p_O - p_{Oh})\nabla.\underline{\chi}_n$$

$$+ \nu\,\nabla(\underline{u}_O - \underline{u}_{Oh}) : \nabla\underline{\chi}_n \Big]\, d\Omega = O \tag{4}$$

$$\int \nabla.(\underline{u}_O - \underline{u}_{Oh})\,\theta_n\, d\Omega = O.$$

It is clear that this projection will not give an exact best fit to $\underline{u}_O$ and p_O individually. The solution described by Gresho et al. (1982) to the steady Boussinesq flow over a step shows that even a zero velocity field may not be represented exactly. If the initial data are defined in another way, for instance as interpolated values, global convergence to the solution of (1) occurs because the inconsistency in the initial condition decays with time, and the final solution is a projection of the form (4) of the true solution.

The results of Gresho suggest that, on practical grids, the errors in the representation (4) may be too large to be acceptable. They also show that schemes which avoid these errors are consistent in a finite different sense, in other words are consistent when $\underline{u}_{Oh}$ and p_{Oh} are interpolants of $\underline{u}_O$ and p_O. These two facts suggest that we should seek finite element schemes which are consistent when $\underline{u}_{Oh}$ is a best fit to $\underline{u}_O$ in some norm, and p_{Oh} is a best fit to p_O. The experience of Gresho indicates that first order consistency in this sense would be enough to ensure accuracy on coarse grids, and that the "optimal" convergence rate would be observed after sufficient mesh refinement.

The difficulty with seeking such algorithms is that the non-local condition (3) has to be enforced on $\underline{u}_{Oh}$. If we wish to use a best fit in some sense to $\underline{u}_O$ this must use only trial functions satisfying (3), i.e., the approximately divergence-free functions of Griffiths (1981). Alternatively we can specify that the implied vorticity field is a best fit to the true vorticity, and use (3) to complete the definition of $\underline{u}_{Oh}$. Since a general velocity field is uniquely specified by its rotational and divergent parts, and the boundary conditions, this representation also gives accurate information about the velocity. In two dimensions the vorticity can be treated as a scalar ζ. Using basis functions ϕ_n to represent it, set

$$\int \phi_n \nabla \times \underline{u}_{Oh} \, d\Omega = \int \phi_n \nabla \times \underline{u}_O \, d\Omega \quad \forall \phi_n$$

$$\int \theta_n \nabla . \underline{u}_{Oh} \, d\Omega = 0 \qquad \forall \theta_n . \tag{5}$$

In three dimensions the first equation of (5) can still be used; it is now a vector equation. The second equation will not be satisfied, and there is therefore a clear distinction between a velocity and a vorticity based approach.

Even in two dimensions, the two representations are different. Compare them on a simple low order element. The standard bilinear velocity/piecewise constant pressure element is not suitable for this, since no local approximately divergence free basis can be constructed in general geometry, Griffiths (1981). We therefore use a different type of element; the lowest order case of those considered by Raviart (1977) and Fortin (1981). This element is illustrated in Figs. 1 and 2. The normal mass flux across each side of the quadrilateral element is specified as the mid-side nodal parameter. Within the element, in terms of the transformed coordinates (ξ,η) the flux across lines $\xi=$ const. is assumed linear in ξ and independent of η and vice versa. The fluxes across opposite sides of the quadrilaterals are thus connected. This construction can be carried out on any mesh even if it is not logically rectangular (Fig. 1). If we use piecewise constant functions for pressure, a basis element for velocity fields satisfying (3) is shown in Fig. 2. The total flux into each element is clearly zero. There is one basis function associated with each node, which can form an approximately divergence-free basis. We write these basis functions as χ_n.

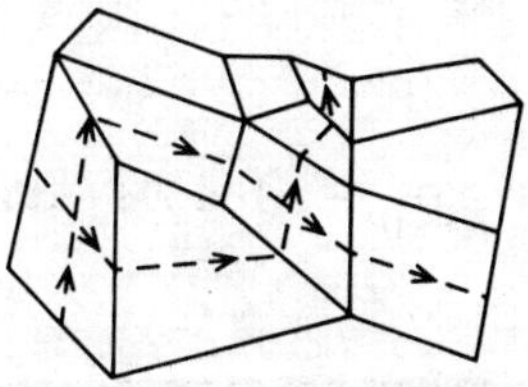

Fig. 1 Logical structure of elements using normal fluxes as unknowns.

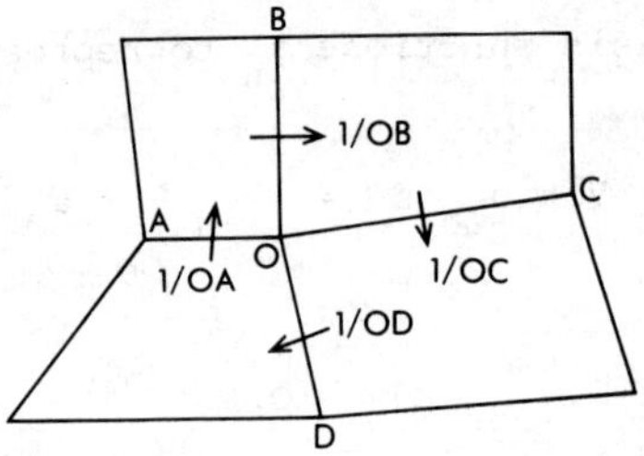

Fig. 2 Local approximately divergence-free basis.

It is known that this element does not yield consistent results with the velocity-pressure algorithm; we will show that it does if the projection (5) is used instead. This indicates an advantage of (5) over the divergence-free basis approach.

The difference can be shown by considering the following sub-problem derived from (1). Given $\underline{u}$ not satisfying $\nabla.\underline{u} = 0$ but satisfying the boundary conditions, suppose $P\underline{u}$ is such that

$$\begin{aligned} P\underline{u} &= \underline{u} + \nabla q \\ \nabla.(P\underline{u}) &= 0. \end{aligned} \tag{6}$$

This projection can be proved to be well defined (Chorin and Marsden (1979), p. 51). The problem then is: given $\underline{u}$, find $(P\underline{u})_h$ to satisfy (3) and a bound on $\|(P\underline{u})_h - P\underline{u}\|$ in an appropriate norm. The algorithms to be checked are:

(i) $$\int ((P\underline{u})_h - \underline{u}) \cdot \underline{\chi}_n \, d\Omega = 0 \qquad \forall\, \chi_n \tag{8}$$

where all $\underline{\chi}_n$ satisfy (3), i.e., the best fit to $\underline{u}$ in the divergence-free space;

(ii) $$B + \int ((P\underline{u})_h - \underline{u}) \times \nabla\phi_n \, d\Omega = 0 \qquad \forall\, \phi_n, \tag{9}$$

i.e., the best fit to the vorticity in the $\{\phi_n\}$ space. In each case

$$(P\underline{u})_h \in \text{span}\,(\underline{\chi}_n). \tag{10}$$

The boundary term B in equation (9) involves the integrated tangential velocity $\int_{\partial\Omega} \underline{u}.d\underline{\ell}$. Consistency of (9) is easy to prove, because this approximation to the velocity field can be derived from a bilinear streamfunction ψ by direct differentiation. Therefore (9) reduces to

$$B + \int (\hat{\underline{k}} \times \nabla(\psi_h - \psi)) \times \nabla\phi_n \, d\Omega = 0 \quad \forall \phi_n \tag{11}$$

where $\hat{\underline{k}}$ is a unit vector in the third direction. Equation (11) simply states that ψ_h is the Galerkin approximation to the solution of the Poisson equation

$$\nabla^2\psi = \nabla \times \underline{u}$$

using bilinear elements, and by standard theory

$$\|\psi_h - \psi\|_{L_2} = O(h^2)$$

and

$$\|\underline{u}_h - \underline{u}\|_{L_2} = O(h).$$

It is not possible however, to prove consistency of (8) on a general grid, though it is consistent on a rectangular grid. If we expand (9) in terms of bilinear functions $\theta(\xi,\eta)$ and write (u_1, u_2) to refer to the x and y components of $P\underline{u}_h - \underline{u}$ we obtain

$$B + \int \frac{1}{J}\left[(u_2x_\xi - u_1y_\xi)(\beta\theta_\eta - \alpha\theta_\xi) + (-u_2x_\eta + u_1y_\eta)(\gamma\theta_\eta - \beta\theta_\xi)\right] d\xi \, d\eta = 0 \tag{12}$$

where α, β, γ and J are the components of the metric tensor of the coordinate transformation given by

$$\alpha = x_\eta^2 + y_\eta^2, \quad \beta = x_\xi x_\eta + y_\xi y_\eta, \quad \gamma = x_\xi^2 + y_\xi^2, \quad J = x_\xi y_\eta - x_\eta y_\xi.$$

The use of (x,y) components in (12) avoids the need to distinguish covariant and contravariant components.

As can be seen from Fig. 2, each element of the divergence-free basis for $\underline{u}$ is associated with a bilinear basis function for the streamfunction. In terms of (x,y) components it can be written as

$$\left(\frac{1}{J}(-\theta_\xi x_\eta + \theta_\eta x_\xi), -\frac{1}{J}(\theta_\xi y_\eta - \theta_\eta y_\xi)\right)$$

and the orthogonal projection (8) reduces to

$$\int \left[u_1(\theta_\xi x_\eta - \theta_\eta x_\xi) - u_2(\theta_\xi y_\eta - \theta_\eta y_\xi\right] d\xi d\eta = 0. \quad (13)$$

This is different from (12) except on rectangular elements, for which $\alpha = y_\eta^2$, $\beta = 0$, $\gamma = x_\xi^2$, $J = x_\xi y_\eta$. There is thus no guarantee of convergence, and in fact it is known not to converge, though if the procedure is repeated with higher order basis functions it does converge. This indicates an advantage for the vorticity-based projection. This has to be set against the disadvantage that, in three dimensions, it is not known how to handle the boundary conditions and thus the correct inversion of the generalised version of (12), now a vector equation. In two dimensions, the vorticity-based approach appears to have some advantages at boundaries; for instance, the far field condition developed by Fornberg (1980) for flow past a cylinder could naturally be applied to (12).

This approach of considering a best fit to the velocity satisfying (3) is very restrictive and only allows consideration of a small number of elements. An alternative is not to enforce (3) immediately, but to analyse the consistency of the approximation to each equation in (1) separately. Thus if (2) and (3) are consistent approximations to the separate equations of (1), then over-all convergence follows, given the regularity assumption, by the same arguments as for finite differences. In terms of the sub-problem (6) this means that the approximation $\nabla_h p$ to ∇p must be consistent as well as (3) being a consistent approximation to $\nabla . u = 0$. This condition is not satisfied for the bilinear velocity/piecewise constant pressure element in general geometry; but is satisfied using bilinear pressures. In order to satisfy it using low order elements the Galerkin approximation to ∇p could be replaced by a finite difference approximation.

We illustrate this for the low order element already used for the other approximations. If we consider the unknown parameters to be simply the fluxes across each side of the quadrilaterals, then (3) states that the sum of the fluxes into each element is zero. Thus we solve (6) by setting

$$P\underline{u} = \underline{u} + \nabla_h p$$
$$\Sigma_{i=1}^{4} \ell_i (P\underline{u})_i = 0 \qquad (14)$$

around all elements. Here, ℓ_i is the length of side i and $\nabla_h p$ is a finite difference approximation to ∇p using values defined at element centroids. It has to be calculated in the direction normal to each side on the grid by interpolation or extrapolation (Fig. 3). The scheme shown uses four values of p to calculate $\nabla_h p$ across any side and is first order accurate. Convergence of (14) to the solution of (6) then follows by finite difference arguments.

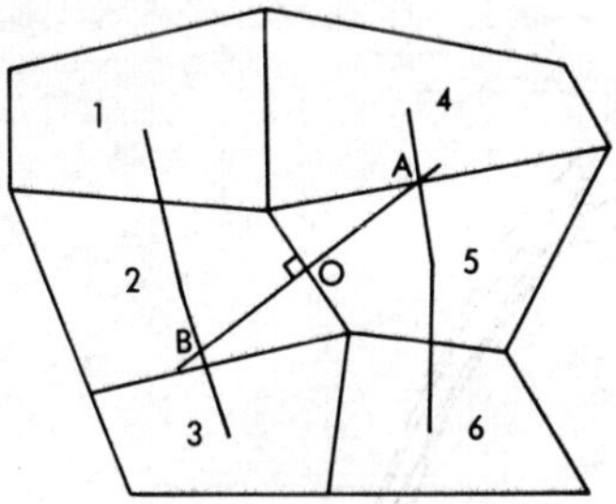

Fig. 3 Finite difference stencil for pressure gradient.

The algorithms (12), (13) and (14) can all be extended to algorithms for the full Navier-Stokes equations. The details are given in Cullen (1982) and are based on the methods described by Raviart (1977).

3. RESULTS

Results are shown for the standard test problem of flow past a circular cylinder. Fig. 4 shows the grid used, and Figs. 5 and 6 the steady state reached at Re = 40. Results are shown for very sensitive parameters, the vorticity and pressure on the cylinder surface, and for the following methods:

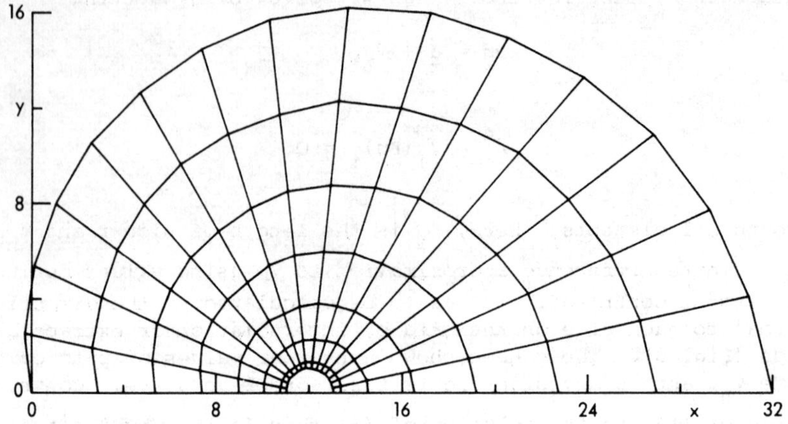

Fig. 4 Finite element mesh used for cylinder integrations.

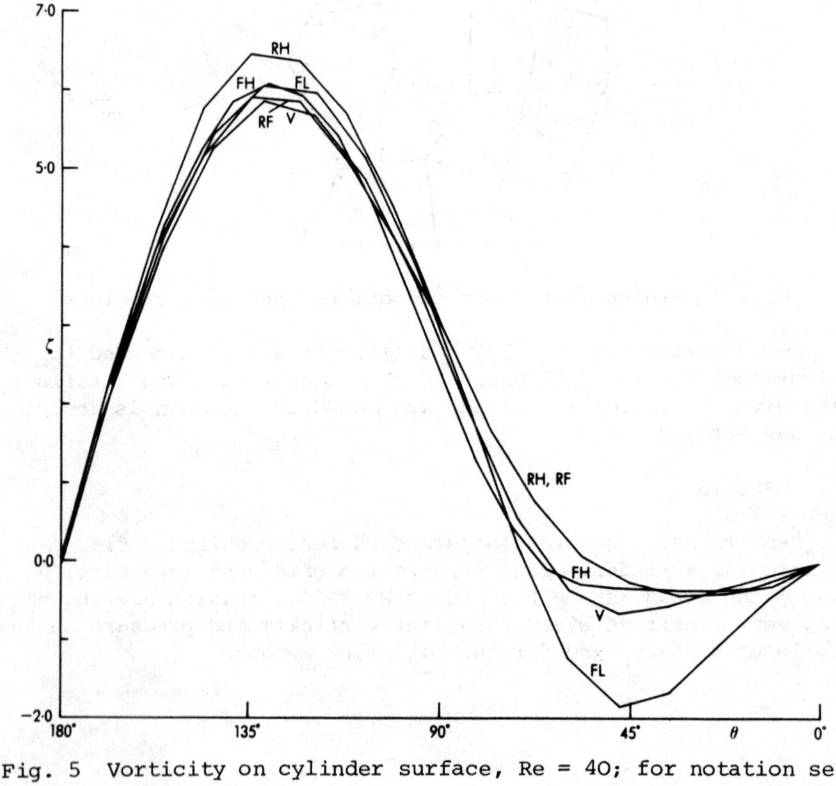

Fig. 5 Vorticity on cylinder surface, Re = 40; for notation see text.

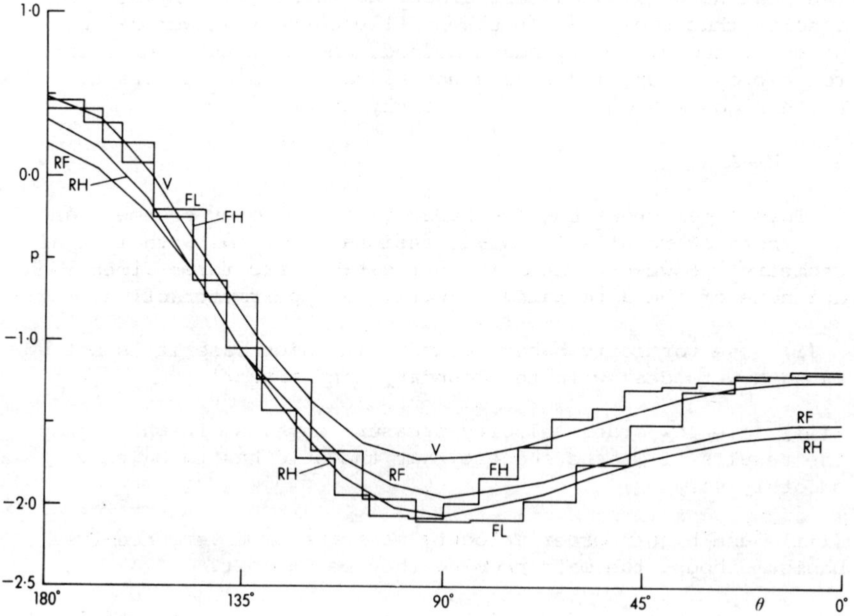

Fig. 6 Pressure on cylinder surface, Re = 40; for notation see text.

V: Vorticity based projection, (12), 15 x 7 grid.

FL: Finite difference projection (14); 15 x 7 grid.

FH: Finite difference projection, (14); 22 x 10 grid.

RH: Finite element streamfunction method (Olson and Tuann, (1976)) using 78 elements and C' basis functions.

RF: Finite difference solution (Dennis and Chang, (1970)).

The results show that FL does not appear to give the correct solution except on the finer mesh; while V does on the coarse mesh. There is some disagreement between these results and RH and RF in the wake region. This may be due to different specifications of the far-field boundary condition; a mixed condition was used here with the free stream specified at inflow points. The correct condition to use is discussed by Fornberg (1980). The results for the vorticity distribution quoted there use much higher resolution and may be regarded as reference solutions. The size of the region of positive vorticity is closer to that given by RF and RH; the magnitude of the positive vorticity and the peak negative value are closer to those given by V. These results thus show that an essentially correct answer can be given by the vorticity-based method, and that the finite difference projection, (14), does not appear to be as accurate. Further conclusions cannot be safely drawn.

4. SUMMARY

This paper shows how low order finite element schemes can give correct results in two-dimensional problems with irregular geometry. However, the main applications are three-dimensional and none of the alternatives available appears attractive:

(i) use vorticity based methods, in which case it is not known how to deal with the boundary conditions;

(ii) use low order velocity-pressure methods, in which case the results of Gresho suggest that the grid has to be very smoothly varying;

(iii) use higher order velocity-pressure or divergence-free basis methods, the main problem then being cost.

None of these difficulties is peculiar to the use of finite element methods, and the same comments could be equally applied to finite difference methods. Only much longer computational experience will suggest which alternative is the least unattractive.

ACKNOWLEDGEMENTS

This work was carried out while the author was visiting the Department of Mathematics, University of California, Berkeley. Much of it was carried out as joint work with Dr. P.M. Gresho and his group at the Lawrence Livermore Laboratory, with whom I had many stimulating discussions.

REFERENCES

BROOKS, A.M. 1981 A Petrov-Galerkin finite element formulation for convection dominated flows. Ph.D Thesis, California Institute of Technology.

CHORIN, A.J. and MARSDEN, J.E. 1979 A mathematical introduction to fluid mechanics. Springer-Verlag, N.Y.

CULLEN, M.J.P. 1982 Analysis and experiments with some low order finite element schemes for the Navier-Stokes equations. Submitted to *J. Comput. Phys*.

DENNIS, S.C.R. and CHANG, G.Z. 1970 Numerical solution for steady flow past a circular cylinder at Reynolds numbers up to 100. *J. Fluid Mech.* **42**, 471-489.

FORNBERG, B. 1980 A numerical study of steady viscous flow past a circular cylinder. *J. Fluid Mech.* **98**, 819-855.

FORTIN, M. 1981 Old and new finite elements for incompressible flows. *Int. J. Num. Meth. Fluids* **1**, 347-364.

GRESHO, P.M., LEE, R.L., CHAN, S.T. and SANI, R.L. 1982 In preparation.

GRESHO, P.M., LEE, R.L. and UPSON, C.D. 1980 F.E.M. solutions of the Navier-Stokes equations for vortex shedding behind a cylinder: experiments with the four node element In 3rd Int. Conf. on F.E.M. in Water Resources, 448-466.

GRIFFITHS, D.F. 1981 An approximately divergence-free 9-node velocity element (with variations) for incompressible flows. *Int. J. Num. Meth. Fluids* **1**, 323-346.

HEYWOOD, J.G. and RANNACHER, R. 1982 Finite element approximation of the non stationary Navier-Stokes problem, Part I: regularity of solutions and second order spatial discretisation. *SIAM J. Num. Analysis* to appear.

RAVIART, P.A. 1977 Finite element methods and Navier-Stokes equations. Springer Lect. Notes in Physics, **91**, 27-47.

TEMAM, R. 1977 Navier-Stokes equations. North-Holland, Amsterdam.

THOMASSET, F. 1981 Implementation of finite element methods for Navier-Stokes equations. Springer-Verlag, New York.

TUANN, S.Y. and OLSON, M.D. 1978 Numerical studies of the flow around a circular cylinder by a finite element method. *Computers and Fluids* **6**, 219-240.

A FINITE ELEMENT METHOD ON A FIXED MESH FOR THE STEFAN PROBLEM WITH CONVECTION IN A SATURATED POROUS MEDIUM

J.W. Barrett and C.M. Elliott

(Department of Mathematics, Imperial College, London)

1. INTRODUCTION

The artificial freezing of groundwater has important engineering applications (c.f. Sanger, 1968, and Proceedings of the 2nd International Symposium on Ground Freezing, NTH, Trondheim, Norway, 1980) particularly in ground stabilization at construction sites and the prevention of water leakage into tunnels or shafts. In this paper we consider a problem studied by Hashemi and Sliepcevich (1973), Goldstein and Reid (1978) and Frivik and Comini (1982): that of a freeze pipe inserted vertically into a water saturated soil. We make the following physical assumptions: (a) the problem is one of heat transport in a saturated homogeneous porous medium, (b) the temperature of the water is equal to that of the soil, (c) freezing takes place at a single temperature, (d) the fluid velocities are sufficiently small in the unfrozen soil to be governed by Darcy's law and in the frozen region are zero, (e) the groundwater is incompressible, (f) the viscous dissipation of energy may be neglected, (g) thermal gradients in the z-direction are negligible enabling the problem to be formulated in a horizontal cross-section of fixed vertical height. An account of the mathematical formulation of heat and mass transfer in a porous medium may be found in (Bear, 1972 p.641), which together with the above assumptions lead to the following mathematical model.

Let $\Omega \equiv \Omega^+(t) \cup \Gamma(t) \cup \Omega^-(t)$ be a fixed bounded region of $\mathbb{R}^2$ with boundary $\partial\Omega$, where for any time $t \in [0,T]$:

$$\Omega^+(t) \equiv \{\underline{x}\in\Omega: \quad \theta(\underline{x},t) > \theta_p\} \quad \text{the unfrozen region} \tag{1.1a}$$

$$\Omega^-(t) \equiv \{\underline{x}\in\Omega: \quad \theta(\underline{x},t) < \theta_p\} \quad \text{the frozen region} \tag{1.1b}$$

$$\Gamma(t) \equiv \overline{\Omega^+(t)} \cap \overline{\Omega^-(t)} \; . \tag{1.1c}$$

$\theta(\underline{x},t)$ is the temperature distribution ($\underline{x} \equiv (x,y)^T$) and θ_p is the phase change temperature. The transport of heat is governed by the equations:

$$C_e(\theta)\frac{\partial\theta}{\partial t} = \underline{\nabla}.(K(\theta)\underline{\nabla}\theta) + F \quad \text{in } \Omega^-(t) \tag{1.2a}$$

$$C_e(\theta)\frac{\partial\theta}{\partial t} = \underline{\nabla}.(K(\theta)\underline{\nabla}\theta) - C_w(\theta)\underline{V}.\underline{\nabla}\theta \quad \text{in } \Omega^+(t), \tag{1.2b}$$

where C_e and C_w are the effective volumetric heat capacities of the saturated porous medium and water, respectively, K is the effective thermal conductivity, $\underline{V}$ is the velocity of the water, $\underline{\nabla} \equiv \left(\frac{\partial}{\partial x}, \frac{\partial}{\partial y}\right)^T$ and F is the heat sink concentrated (i.e. non-zero) in a small area of Ω representing the freeze pipe. We will use the convention that F is extended to all of Ω by $F \equiv 0$ in $\Omega^+(t)$. This modelling of the freeze pipe is valid if the pipe's diameter is small compared to the diameter of $\Omega^-(t)$, otherwise we delete F in (1.2a) and impose on an internal closed boundary $\partial\Omega_F$, the pipe's surface, a suitable boundary condition.

On the moving boundary $\Gamma(t)$ we impose the following conditions: the temperature is continuous across $\Gamma(t)$ and equal to the phase change temperature,

$$\theta = \theta_p \quad \text{on } \Gamma(t) , \tag{1.3a}$$

there is no mass flow across $\Gamma(t)$,

$$\underline{V}.\underline{n} = 0 \text{ on } \Gamma(t) , \tag{1.3b}$$

and energy is conserved which, upon taking into account the latent heat, implies the Stefan condition

$$[K(\theta_p)\underline{\nabla}\theta]_-^+ . \underline{n} = - Lv_n \quad \text{on } \Gamma(t) , \tag{1.3c}$$

where $\underline{n}$ is the unit normal on $\Gamma(t)$ directed into $\Omega^+(t)$, v_n is the speed of the moving boundary in this direction, $[\ ]_-^+$ denotes a jump and L is the volumetric latent heat of fusion. We have also that the initial and boundary values of θ are prescribed:

$$\theta(\underline{x},0) = \theta_0(\underline{x}), \ \underline{x} \in \Omega; \ \theta(\underline{x},t) = \theta_B(\underline{x}), \ \underline{x} \in \partial\Omega. \qquad (1.4)$$

The velocity of the groundwater is given by Darcy's law in the unfrozen region,

$$\underline{V} = -[k/\mu]\underline{\nabla}p \ \text{in} \ \Omega^+(t), \qquad (1.5a)$$

and is zero in the frozen region, where k is the permeability of the porous medium, μ is the viscosity of the fluid and $p(\underline{x},t)$ is the pressure in the groundwater. In addition, we have that the water is incompressible,

$$\underline{\nabla}.\underline{V} = 0 \ \text{in} \ \Omega^+(t) \ , \qquad (1.5b)$$

and that either the velocity is specified in the far field,

$$\underline{V}(\underline{x},t) = \underline{V}_B(\underline{x}) \quad \underline{x} \in \partial\Omega \ , \qquad (1.5c)$$

or the pressure is specified,

$$p(\underline{x},t) = p_B(\underline{x}) \quad \underline{x} \in \partial\Omega. \qquad (1.5d)$$

Frivik and Comini (1982) obtained good agreement between experimental results obtained in a laboratory and numerical computations based on this model (although with the latent heat given out over a range of temperature).

The above equations can be made dimensionless by introducing new independent variables $\hat{t} = t/t_0$, $\hat{x} = x/x_0$ and new dependent variables $u = (\theta-\theta_p)/(\bar{\theta}_B-\theta_p)$, $P = -\bar{k}p/(\bar{p}_B x_0 \bar{\mu})$ where barred quantities are mean or representative values of the quantities $\{\theta_B, k, p_B, \mu\}$, x_0 is taken to be the diameter of the initial frozen region and the choice of t_0 will be discussed later. Setting $\bar{K}$ equal to the mean thermal conductivity we introduce:

$$a(u) = x_O^2 C_e(\theta)/(\bar{K}t_O),\ b(u) = x_O\bar{p}_B C_w(\theta)/\bar{K},$$

$$d(u) = K(\theta)/\bar{K},$$

$$\lambda = x_O^2 L/[\bar{K}t_O(\bar{\theta}_B-\theta_p)]\ ,\ f = F/[\bar{K}(\bar{\theta}_B-\theta_p)]$$

and $\kappa = \bar{\mu}k/(\bar{k}\mu)$. The resulting problem (writing t for $\hat{t}$ and x for $\hat{x}$) is: find $\{u(\underline{x},t),\ P(\underline{x},t),\ S(\underline{x},t)\}$ such that

$$a(u)\frac{\partial u}{\partial t} = \underline{\nabla}.(d(u)\underline{\nabla}u) + f \qquad \text{in } \Omega^-(t) \tag{1.6a}$$

$$a(u)\frac{\partial u}{\partial t} = \underline{\nabla}.(d(u)\underline{\nabla}u) - b(u)\underline{q}.\underline{\nabla}u \qquad \text{in } \Omega^+(t) \tag{1.6b}$$

$$\Gamma(t) \equiv \{\underline{x}:S(\underline{x},t)=0\} \text{ and sign } u(\underline{x},t) \equiv \text{sign } S(\underline{x},t) \tag{1.7a}$$

$$u = 0,\ [d(u)\underline{\nabla}u]_-^+ .\underline{\nabla}S = \lambda\frac{\partial S}{\partial t} \text{ and } \underline{q}.\underline{\nabla}S = 0 \text{ on } \Gamma(t) \tag{1.7b}$$

$$\underline{q} = \kappa\underline{\nabla}P \text{ and } \underline{\nabla}.(\kappa\underline{\nabla}P) = 0 \qquad \text{in } \Omega^+(t) \tag{1.8}$$

and

$$\begin{aligned} &P(\underline{x},t) = P_B(\underline{x}) \text{ and } u(\underline{x},t) = u_B(\underline{x}) \quad \underline{x} \in \partial\Omega, \\ &u(\underline{x},0) = u_O(\underline{x}) \quad \underline{x} \in \Omega. \end{aligned} \tag{1.9}$$

There are clearly two natural choices for t_O:

(i) $t_O = x_O^2 L/[\bar{K}(\bar{\theta}_B-\theta_p)]$, the order of the "growth" time of the frozen region due to conduction, yielding $\lambda = 1$ or
(ii) $t_O = x_O L/[\bar{p}_B\bar{C}_w(\bar{\theta}_B-\theta_p)]$, the order of the "growth" time due to convection, yielding $\lambda = x_O\bar{p}_B\bar{C}_w/\bar{K}$ or $\lambda = b$ if $C_w(\theta)$ is a constant.

In the case $a(u) \equiv 0$, $b(u)$ and $d(u)$ piecewise constant, discontinuous only at the phase change temperature, Goldstein and Reid (1978) transform the above system into a nonlinear integral differential equation which is then tackled numerically. There are various transformations of dependent and independent variables which can be applied to Stefan problems (1.6) - (1.9) so as to make the task of coping with the moving boundary more amenable numerically, see (Fox, 1979). We will consider one of the simplest and most effective ways, the enthalpy formulation of (1.6), which we describe in the next section. The problem is then seen to be one of solving a nonlinear time dependent diffusion-convection equation with discontinuous coefficients. In section 3 we present a finite element in space/finite difference in time discretisation of this formulation. For the "standard" Stefan problem, that is in the absence of convection, the key feature of a numerical scheme based on the enthalpy formulation is the avoidance of any front-tracking or mesh-adjustment. The aim of this paper is to show that the present problem, which contains the added difficulties of convection and the fact that the velocity field is dependent on the position of the moving boundary, can be solved simply and effectively on a fixed mesh. Finally in section 4 we present some numerical results.

2. THE ENTHALPY OR WEAK FORMULATION

An account of the enthalpy method in the absence of convection may be found in (Elliott and Ockendon, 1982). One can simplify equation (1.6) by using the Kirchoff transformation

$$U(u) = \int_0^u d(u')du', \tag{2.1}$$

since $\underline{\nabla}U = d(u)\underline{\nabla}u$. For ease of exposition we will now restrict ourselves to piecewise constant coefficients, discontinuous only at the phase change temperature, so that

$$a(u) = \begin{cases} a^+ & u > 0 \\ a^- & u < 0 \end{cases}, \quad d(u) = \begin{cases} d^+ & u > 0 \\ d^- & u < 0 \end{cases}, \tag{2.2}$$

$b(u) = db^+$ and $\kappa = 1$.

The development of the following formulation and numerical method for general temperature dependent coefficients is straightforward.

A generalised enthalpy or heat content $H(\underline{x},t)$ is defined by a mapping $\beta(.): \mathbb{R} \to \mathbb{R}$ so that

$$\beta(r) = \begin{cases} \alpha^+ r + \lambda & r > 0 \\ [0,\lambda] & r = 0 \\ \alpha^- r & r < 0 \end{cases} , \tag{2.3}$$

where $\alpha^+ = a^+/d^+$ and $\alpha^- = a^-/d^-$, and

$$H(\underline{x},t) \in \beta(U(\underline{x},t)). \tag{2.4}$$

As the enthalpy can take any value between 0 and λ at the phase change temperature $U = 0$, it is not defined uniquely in $L_2(\Omega)$ by the temperature distribution where there is a region of non-zero measure in which $U \equiv 0$. However, in our problem for any time t the temperature will vanish only on $\Gamma(t)$, a curve of measure zero in Ω. Also the inverse mapping $\beta^{-1}(.)$ is a unique Lipschitz continuous function so that the enthalpy value at a point always defines a unique temperature. The problem for U is then (1.6)-(1.9) with $\{a(u)u_t, d(u)\underline{\nabla}u\}$ being replaced by $\{H_t, \underline{\nabla}U\}$ and supplemented by (2.4).

If the triple of functions $\{H,U,P\}$ satisfy the equations (1.6)-(1.9) classically then they solve the equations:

$$\frac{\partial H}{\partial t} = \nabla^2 U - b\underline{q}.\underline{\nabla}U + f \quad \text{in } \Omega \tag{2.5a}$$

$$\underline{\nabla}.(\kappa(U)\underline{\nabla}P) = 0 \quad \text{in } \Omega \tag{2.5b}$$

$$\{\text{or } \nabla^2 P = 0 \text{ in } \Omega^+(t) \text{ with } \underline{\nabla}P.\underline{\nabla}S = 0 \text{ on } \Gamma(t)\}$$

$$\underline{q} = \kappa(U)\underline{\nabla}P \quad \text{in } \Omega, \qquad \kappa(U) = \begin{cases} 1 & U \geqslant 0 \\ 0 & U < 0 \end{cases} \tag{2.5c}$$

$$H(\underline{x},0) = H_0(\underline{x}) \equiv \beta(u_0(\underline{x})) \qquad \underline{x} \in \Omega , \tag{2.5d}$$

which holds in the sense of distributions on $\Omega_T \equiv \Omega\times(0,T)$.

Adopting the standard notation: $(\underline{w}_1,\underline{w}_2)_\Omega = \int_\Omega \underline{w}_1 . \underline{w}_2 d\underline{x}$, $L_2(\Omega)$ the Hilbert space consisting of square-integrable functions over Ω and $H^1(\Omega)$ the Hilbert space consisting of functions whose first derivatives are square-integrable over Ω, then more precisely the triple of functions {H,U,P} form a weak solution in the following sense.

Definition A weak solution is a triple of functions {H,U,P} with $H \in L_2(\Omega_T)$, $U \in C(\Omega_T) \cap L_2(0,T;H^1(\Omega))$, $P \in L_2(0,T;H^1(\Omega))$ and $U-U_B = P-P_B = 0$ on $\partial\Omega$ such that

$$\int_0^T [(H,w_t)_\Omega - (\underline{\nabla}U,\underline{\nabla}w)_\Omega - (b\underline{q}.\underline{\nabla}U-f,w)_\Omega]dt + (H_0,w(0))_\Omega = 0 \quad (2.6a)$$

for all test functions $w \in C^\infty(\bar{\Omega}_T)$ with $w = 0$ on $\partial\Omega\times(0,T)$ and $\Omega(T)$,

$$\int_0^T (\underline{\nabla}P.\underline{\nabla}w)_{\Omega^+(t)} dt = 0 \quad (2.6b)$$

for all test functions $w \in L_2(0,T;H^1(\Omega))$ with $w = 0$ on $\partial\Omega\times(0,T)$,

$$\Omega^+(t) = \{\underline{x}:U(\underline{x},t) > 0\}, \quad q = \begin{cases} \underline{\nabla}P & U \geq 0 \\ \underline{0} & U < 0 \end{cases} \quad (2.6c)$$

and

$$H(\underline{x},t) \in \beta(U(\underline{x},t)).$$

That (1.6)-(1.9) implies (2.6) is a straightforward matter of multiplying (1.6) and (1.8) by a test function and integrating by parts using the jump condition (1.7) and the boundary conditions to eliminate surface integrals. Similarly sufficiently smooth weak solutions with a smooth surface $S(\underline{x},t) = 0$ across which U changes sign are classical solutions. In the absence of convection it is known that there exist unique weak solutions (Elliott and Ockendon, 1982) but this is not known about the present Stefan problem with convection. (However see (Cannon, Di Benedetto and Knightly, 1980) who study the problem with the velocity field defined by the Navier-Stokes equations.)

The main feature of the enthalpy formulation from both the theoretical and numerical point of view is that the moving boundary and Stefan condition are contained implicitly in (2.5a) leading to a simple numerical algorithm. However, in the present problem $\Gamma(t)$ still appears explicitly in the pressure equation (2.5b).

Hashemi and Sliepcevich (1973) and Frivik and Comini (1982) study (2.5) with the enthalpy and permeability being smooth functions of temperature albeit rapidly changing in the neighbourhood of $U = 0$. In the absence of convection this is known to be a convergent approximation of (2.6). We wish to approximate directly by the finite element method the problem with discontinuous coefficients which in the absence of convection is also known to be a convergent approximation of (2.6), see (Elliott and Ockendon, 1982).

3. THE NUMERICAL METHOD

In this section we present a finite element in space/finite difference in time discretisation of (2.5) (or equivalently (2.6)). In the computational work we have taken Ω to be a square. For our finite element trial space, denoted by $S^h(\Omega)$, we have chosen continuous functions which are piecewise bilinear on square elements, having sides of length h, resulting from a uniform partition of Ω. If $V \in C(\Omega)$ then we denote by V^I the interpolate of V in $S^h(\Omega)$. We note that due to the lack of smoothness of the temperature U across the moving boundary there is no apparent benefit in using high order finite element spaces. To describe our numerical procedure we require the following sets of functions:

$$S_O^h(\Omega) = \{W \in S^h(\Omega) \,|\, W=0 \text{ on } \partial\Omega\}$$

and (3.1)

$$S_U^h(\Omega) = \{W \in S^h(\Omega) \,|\, W=U_B^I \text{ on } \partial\Omega\}.$$

With Δt denoting the fixed time step, an approximation from $S^h(\Omega)$ to U at time $n\Delta t$ is written as $U^n \equiv U^n(\underline{x}) = \Sigma\, U_j^{\,n}\phi_j(\underline{x})$, where the summation is performed over all vertices (mesh points) $\underline{x}_j$ of the partition and $\{\phi_j(\underline{x})\}$ is the standard basis for $S^h(\Omega)$ satisfying $\phi_j(\underline{x}_k) = \delta_{jk}$, the Kronecker delta. We now review the numerical approximation of the enthalpy formulation (2.5) in the absence of convection; for more details see (Elliott and Ockendon, 1982).

3.1 *A Galerkin Approximation in the Absence of Convection*

We consider the following standard Galerkin finite element approximation of (2.5), with $\underline{q} \equiv \underline{O}$ in Ω, using a θ method in time: find $U^{n+1} \in S_U^h(\Omega)$ and $H^{n+1} \in S^h(\Omega)$, with $H_j^{n+1} \in \beta(U_j^{n+1})\forall j$, such that

$$(H^{n+1}-H^n,\phi_i)_\Omega^h + \Delta t(\underline{\nabla} U^{n+\theta},\underline{\nabla}\phi_i)_\Omega = \Delta t(f,\phi_i)_\Omega \quad \forall\phi_i \in S_O^h(\Omega),$$

$$H^O = H_O^{\;I} \text{ and } U_j^{\;O} = \beta^{-1}(H_j^{\;O}). \tag{3.2}$$

In the above, the following notation has been adopted: $U^{n+\theta} = \theta U^{n+1} + (1-\theta)U^n$, $\theta \in [0,1]$ and $(.,.)_\Omega^h$ represents the use of numerical integration which results in the lumping of the mass matrix, that is $(W,\phi_i)_\Omega^h = h^2 W_i \;\forall W \in S^h(\Omega)$, $\phi_i \in S_O^h(\Omega)$. Denoting by $\underline{H}^n$ and $\underline{U}^n$ the vectors of interior nodal values of H^n and U^n, respectively, the approximation (3.2) gives rise to the following discrete system of equations:

$$\underline{H}^{n+1} + \theta\Delta t A\underline{U}^{n+1} = \underline{H}^n - (1-\theta)\Delta t\; A\underline{U}^n + \Delta t\;\underline{r} \tag{3.3}$$

where A is the matrix with entries $A_{ij} = h^{-2}(\underline{\nabla}\phi_j,\underline{\nabla}\phi_i)_\Omega$ and $\underline{r}$ is the vector with entries $r_i \equiv h^{-2}(f,\phi_i)_\Omega - B_i$; B_i representing the contribution of U_B^I, the given temperature on the boundary.

With an explicit time stepping procedure, that is $\theta = O$, H^{n+1} is defined explicitly by (3.3). From H_j^{n+1} we can define

U_j^{n+1} uniquely, as β is invertible, to obtain the temperature distribution. However, an explicit procedure has severe stability restrictions on the choice of Δt, that of $\Delta t = O(h^2)$. Thus one is led into using an implicit scheme, but this requires solving a nonlinear system of algebraic equations at the new time level. Elliott (1981) shows the existence and uniqueness of H^{n+1} and U^{n+1} satisfying (3.3) and moreover gives a modified version of the standard S.O.R. algorithm which provides a simple and effective method of solution. The algorithm is as follows: initialising ${}^{O}\underline{U}^{n+1} = \underline{U}^n$, on the ith S.O.R. iteration for node $\underline{x}_j$ one has to solve

$$ {}^{i}H_j^{n+1} + \theta\Delta t A_{jj}\, {}^{i}U_j^{n+1} = {}^{i}r_j^{n+1}(U^n, H^n, {}^{i}U^{n+1}, {}^{i-1}U^n) \quad (3.4)$$

for the update ${}^{i}U_j^{n+1}$. This is easily accomplished, due to the simple nature of the non-linearity, and we obtain

$$ {}^{i}U_j^{n+1} = \begin{cases} [{}^{i}r_j^{n+1} - \lambda]/[\alpha^+ + \theta\Delta t\, A_{jj}] & \text{for } {}^{i}r_j^{n+1} > \lambda \\ 0 & \text{for } {}^{i}r_j^{n+1} \in [0,\lambda] \\ {}^{i}r_j^{n+1}/[\alpha^- + \theta\Delta t\, A_{jj}] & \text{for } {}^{i}r_j^{n+1} < 0. \end{cases} \quad (3.5)$$

Performing a relaxation step we set

$$ {}^{i}U_j^{n+1} = \omega_j^{i}\, {}^{i}U_j^{n+1} + (1-\omega_j^{i})\, {}^{i-1}U_j^{n+1} $$

and

$$ {}^{i}H_j^{n+1} = [{}^{i}r_j^{n+1} - \theta\Delta t\, A_{jj}\, {}^{i}U_j^{n+1}]. \quad (3.6)$$

This algorithm is globally convergent provided one chooses $\omega_j^{i} = \omega \in (0,2)$ or $\omega_j^{i} = 1$ for a node j under-going phase change during the ith iteration. For $\theta \geq \frac{1}{2}$ it can be shown that (3.2) is unconditionally stable and convergent to the weak solution (2.6) with $\underline{q} \equiv \underline{0}$ in Ω. In practice we have chosen the fully implicit scheme as there are no advantages with $\theta = \frac{1}{2}$ due to the lack of regularity in time.

From the above one can see how simple the enthalpy method is to program. The position of the moving boundary can be recovered at time $n\Delta t$ from inspecting U^n or possibly H^n, and values at adjacent time levels. For one-dimensional problems there are various sophisticated recovery procedures to obtain more accurately the position of the moving boundary. However, many of these techniques do not generalise in a simple manner into higher dimensions. In our numerical work we have adopted the following procedure: along any element side the position of the moving boundary is determined from where the temperature U^n is zero. Then these points are connected by straight lines, so that one obtains a polygonal approximation Γ_h^n to $\Gamma(n\Delta t)$, which in turn defines approximations $\Omega_h^{-,n}$ and $\Omega_h^{+,n}$ to $\Omega^-(t)$ and $\Omega^+(t)$, respectively.

In the presence of convection, that is solving the full problem (2.5), there are two sources of difficulty to be overcome. First there is the calculation of the pressure which gives rise to the velocity field and secondly one has to approximate the convection term in the enthalpy/temperature equation (2.5a).

3.2 *The Calculation of the Pressure*

To obtain the velocity field one has to solve a Laplace equation for the pressure P in the polygonal domain $\Omega_h^{+,n}$. A standard finite element procedure would be to adjust the mesh at each time level so that the boundary Γ_h^n coincided with element boundaries, that is the mesh is adjusted to fit the domain $\Omega_h^{+,n}$. However, this adjustment leads to programming complications and one looses the main advantage of the enthalpy formulation: the use of a fixed mesh with no front tracking. One can overcome this difficulty by approximating the pressure on an unfitted finite element mesh, a technique proposed and analysed in (Barrett and Elliott, 1982). With $\Omega \equiv \overline{\Omega_h^{-,n}} \cup \overline{\Omega_h^{+,n}}$ partitioned into uniform square elements e, that is $\Omega = \cup e$, we define

$$D_h^n \equiv \bigcup_{e \in \beta_h^n} e \quad , \quad \beta_h^n \equiv \{e \in \Omega,\ e \cap \Omega_h^{+,n} \neq \{\phi\}\}$$

and (3.7)

$$S_P^h(D_h^n) = \{W \in S^h(D_h^n) \subseteq S^h(\Omega),\ W = P_B^I \text{ on } \partial\Omega\}.$$

The approximation to the pressure equation (2.5b) at time $n\Delta t$ is then: find $P^n \in S_P^h(D_h^n)$ such that

$$(\underline{\nabla}P^n, \underline{\nabla}\phi_i)_{\Omega_h^{+,n}} = 0, \qquad \forall \phi_i \in S_0^h(D_h^n). \tag{3.8}$$

Barrett and Elliott (1982) show that if one approximates a Poisson equation in a region with a curved boundary on which a Neumann condition is prescribed with unfitted bilinear elements and the curved boundary is replaced by its chord in each element of intersection, then the resulting approximation retains the optimal rate of convergence in the Dirichlet norm. Clearly, the use of an unfitted mesh requires integrals to be evaluated over partial elements, $e \cap \Omega_h^{+,n}$, as well as complete elements; but these can be easily calculated by splitting the region $e \cap \Omega_h^{+,n}$ into triangles and using simple quadrature rules, see (Barrett and Elliott, 1982). The resulting system of linear equations from (3.8) are solved iteratively using S.O.R., which is cost effective as one starts with a good initial guess from the value of the pressure at the previous time level. Thus the pressure can be simply and accurately calculated without the need for any mesh adjustment.

3.3 A Galerkin Approximation to the Enthalpy/Temperature Equation

At time $n\Delta t$, given the approximations U^n and H^n, and having calculated the pressure P^n, the next step is to integrate the enthalpy/temperature equation forward in time to obtain U^{n+1} and H^{n+1}. As stated previously, for stability reasons we have chosen a fully implicit approximation for the diffusive term. However, as we have the pressure at time $n\Delta t$ only it is natural to approximate the convection term explicitly. Thus with a standard Galerkin approximation to (2.5a), writing $\underline{q}.\underline{\nabla}U$ in conservative form, we have: find $U^{n+1} \in S_U^h(\Omega)$ and $H^{n+1} \in S^h(\Omega)$, with $H_j^{n+1} \in \beta(U_j^{n+1}) \forall j$, such that

$$(H^{n+1}-H^n,\phi_i)_\Omega^h + \Delta t(\underline{\nabla}U^{n+1},\underline{\nabla}\phi_i)_\Omega = b\Delta t([\underline{\nabla}P^n]U^n,\underline{\nabla}\phi_i)_{\Omega_h^{+,n}}$$

$$+ \Delta t(f,\phi_i)_\Omega \qquad \forall \phi_i \in S_0^h(\Omega), \tag{3.9}$$

$$H_0 = H_0^I \text{ and } U_j^0 = \beta^{-1}(H_j^0).$$

This approach has the advantage in that the convective term being explicit implies that only the right-hand side of the resulting system of nonlinear algebraic equations is affected and thus the modified S.O.R. algorithm described previously can be used with the existence and uniqueness of H^{n+1} and U^{n+1}, satisfying (3.9), guaranteed. Secondly in constructing the discrete equations corresponding to (3.9) the convective term requires integrals over $\Omega_h^{+,n}$ to be calculated, that is integrals over some partial elements. An efficient procedure is to perform these integrals simultaneously with those required for the pressure equation (3.8). For given U^n and hence $\Omega_h^{+,n}$ one can evaluate $(\underline{\nabla}\phi_j,\underline{\nabla}\phi_i)_{\Omega_h^{+,n}\cap e}$ and $(U^n\underline{\nabla}\phi_j,\underline{\nabla}\phi_i)_{\Omega_h^{+,n}\cap e}$ together element by element. Having solved for the pressure P^n satisfying (3.8) one can operate on it with the assembled matrix having entries $(U^n\nabla\phi_j,\underline{\nabla}\phi_i)_{\Omega_h^{+,n}}$ to obtain the contribution of the convection term in (3.9) with the minimum of effort.

However, the Galerkin approximation (3.9) has two major drawbacks. First, from a Von-Neumann analysis we see that a necessary condition for L_2 stability is that Δt has to be chosen such that $[b|\underline{\nabla}P^n|]^2\ \Delta t = O(\alpha^+)$; and secondly, as is well known in diffusion convection problems, as the mesh Peclet number $b|\underline{\nabla}P^n|h$ increases oscillations appear in the approximations U^n and H^n. This second phenomenon makes the task of determining $\Omega_h^{+,n}$ from U^n and H^n more difficult. Clearly the procedure proposed earlier would breakdown. As the convection term in (3.9) becomes more dominant both of the above limitations become more severe and one is led to use some form of upwind discretization for this term.

3.4 An Upwind Approximation to the Enthalpy/Temperature Equation

A simple procedure for removing the oscillations from the Galerkin formulation and improving the stability constraint on Δt is to replace the Galerkin approximation of the convection term, $(\underline{q}^n.\underline{\nabla}U^n,\phi_i)_\Omega$, by a directional upwind finite difference approximation, e.g. see (Tabata, 1977). In such a technique one approximates the convection term for a given node by a backward difference approximation in the direction of the velocity at that node. Thus in the present context of a uniform square mesh and if the velocity vector $\underline{q}^n \equiv (q_x^n,q_y^n)^T$ was given explicitly the above term would be replaced by

$$h\{q_x^n(\underline{x}_i)U^n(x_i,y_i)-[q_x^n(\underline{x}_i)-q_y^n(\underline{x}_i)]U^n(x_i-h,y_i)$$

$$- q_y^n(\underline{x}_i)U^n(x_i-h,y_i-h)\} \tag{3.10}$$

$$\text{for } q_x^n(\underline{x}_i) \geq q_y^n(\underline{x}_i) \geq 0;$$

with the obvious generalisations to other combinations of q_x^n and q_y^n. The difficulty of this approach for our problem is obtaining a representative velocity vector for each node from the pressure, especially for those nodes adjacent to the moving boundary. For fitted meshes a simple and accurate method to obtain velocities from the pressure is to sample the gradient of the pressure at the centroids of the element, which is a superconvergent procedure, see (Lesaint and Zlámal, 1979). We have adopted a generalisation of this approach for the present problem. For each element $e_i \equiv (x_i-h,x_i) \times (y_i-h,y_i)$ we assign a constant velocity vector, denoted by $\underline{Q}_i^n \equiv (Q_{xi}^n,Q_{yi}^n)^T$, using the following approximation of (2.5c):

$$\underline{Q}_i^n = h^{-2}\ \underline{m}(e_i \cap \Omega_h^{+,n})\ \underline{\nabla}P^n(x_i - \frac{h}{2}, y_i - \frac{h}{2}), \tag{3.11}$$

where $\underline{m}(D)$ denotes the measure of the domain D. In the problem of physical interest, that we report on in the next section, the imposed boundary conditions on the pressure imply that the x-velocity component of $\underline{q}$, q_x, is always non-negative. For ease of exposition of the next part of our procedure we will restrict ourselves to this case. First we impose this constraint on the x-component explicitly, that is set $Q^n_{xi} = \max(0, Q^n_{xi})$, $\forall i$. Now if the flow into the node $\underline{x}_i \equiv (x_i, y_i)$ is in one and only one direction that is $Q^n_{yi} . Q^n_{yj} \geq 0$ where $\underline{x}_j \equiv (x_i, y_i + h)$, then we set $\underline{Q}^n(\underline{x}_i)$ equal to the velocity approximation in the inflow element, that is the transfer of heat by convection in an element should only be affected by the velocity in that element; otherwise we average the two velocities. Thus summarizing, we have

(i) if $Q^n_{yi} \geq 0$ and $Q^n_{yj} \geq 0$, then $\underline{Q}^n(\underline{x}_i) = (Q^n_{xi}, Q^n_{yi})^T$,

(ii) if $Q^n_{yi} < 0$ and $Q^n_{yj} < 0$, then $\underline{Q}^n(\underline{x}_i) = (Q^n_{xj}, Q^n_{yj})^T$,

(iii) otherwise $\underline{Q}^n(\underline{x}_i) = (\tfrac{1}{2}[Q^n_{xi} + Q^n_{xj}], \quad \tfrac{1}{2}[Q^n_{yi} + Q^n_{yj}])^T$, (3.12)

The above can be extended to the general flow pattern in a straightforward way. Then we deine our upwind finite difference replacement to $(\underline{q}^n . \underline{\nabla} U^n, \phi_i)_\Omega$, which we will denote by $(\underline{Q}^n . \underline{\nabla} U^n, \phi_i)^U_\Omega$, to be (3.10) with $\underline{Q}^n(\underline{x}_i)$, defined by (3.12), replacing $\underline{q}^n(\underline{x}_i)$.

Our upwind approximation to (2.5a) is then: find $U^{n+1} \in S^h_U(\Omega)$ and $H^{n+1} \in S^h(\Omega)$, with $H^{n+1}_j \in \beta(U^{n+1}_j) \forall_j$, such that

$$(H^{n+1} - H^n, \phi_i)^h_\Omega + \Delta t(\underline{\nabla} U^{n+1}, \underline{\nabla}\phi_i)_\Omega = -b\Delta t(\underline{Q}^n . \underline{\nabla} U^n, \phi_i)^U_\Omega$$

$$+ \Delta t(f, \phi_i)_\Omega \qquad \forall \phi_i \in S^h_0(\Omega), \tag{3.13}$$

$$H^0 = H^I_0 \text{ and } U^0_j = \beta^{-1}(H^0_j).$$

This upwind approximation has the advantage that the necessary condition for L_2 stability is now vastly improved in that the time step Δt has to be chosen such that $b|\underset{\sim}{Q}^n| \frac{\Delta t}{\Delta x} = O(\alpha^+)$.

In some cases one may not be so interested in the evolution of the frozen region but only in the steady state, assuming it exists. In such cases one would like to use an unconditionally stable algorithm. This can be achieved by lagging the velocity field and making the upwind convective term fully implicit; that is: find $U^{n+1} \in S_U^h(\Omega)$ and $H^{n+1} \in S^h(\Omega)$, with $H_j^{n+1} \in \beta(U_j^{n+1})\forall j$, such that

$$(H^{n+1}-H^n,\phi_i)_\Omega^h + \Delta t(\underline{\nabla}U^{n+1},\underline{\nabla}\phi_i)_\Omega + b\Delta t(\underset{\sim}{Q}^n.\underline{\nabla}U^{n+1},\phi_i)_\Omega^U$$

$$= \Delta t(f,\phi_i)_\Omega \qquad \forall \phi_i \in S_0^h(\Omega), \tag{3.14}$$

$$H^0 = H_0^I \text{ and } U_j^0 = \beta^{-1}(H_j^0).$$

The approximation (3.14) gives rise to the following system of nonlinear algebraic equations:

$$\underline{H}^{n+1} + \Delta t\, A\underline{U}^{n+1} = \underline{H}^n + \Delta t\, \underline{r}, \tag{3.15}$$

where A is now the matrix with entries $A_{ij} = h^{-2}\,[(\underline{\nabla}\phi_j,\underline{\nabla}\phi_i)_\Omega + b(\underset{\sim}{Q}^n.\underline{\nabla}\phi_j,\phi_i)_\Omega^U]$. As the convective term is implicit, the matrix A is no longer symmetric positive definite and one has no guarantee concerning the existence and uniqueness of H^{n+1} and U^{n+1} and the convergence of the generalised S.O.R. algorithm. However, through the upwind approximation to the convective term, A is an M matrix and one may hope that this generalised S.O.R. algorithm may still converge. In practice the generalised S.O.R. algorithm does converge and the over-all approximation (3.14) is unconditionally stable with no oscillations.

4. NUMERICAL RESULTS

We now report on some numerical examples, each solving system (2.5) using the fully implicit upwind approximation (3.14). The domain Ω was taken to be the square (-4.4) x (-4,4) and the following parameters were fixed for all four examples:

$\alpha^+ \equiv \alpha^- \equiv 0.1$, $d^+ \equiv 1$ and $\lambda \equiv 1$. The freeze pipe was represented by a delta function at the origion of strength -2π, i.e., $f \equiv -2\pi\delta(\underline{x})$. The boundary conditions chosen for the temperature and pressure were $U_B \equiv 1$ and $P_B \equiv x$, that is in the far field the flow is parallel with the x-axis. The initial frozen region was taken to be the unit circle by specifying the initial enthalpy distribution $H_0(\underline{x})$ to be

$$H_0(\underline{x}) = \begin{cases} \|\underline{x}\|^2 - 1 & \|\underline{x}\| < 1 \\ 0.5 & \|\underline{x}\| = 1 \\ \min\{\|\underline{x}\|^2, 2\} & \|\underline{x}\| \geq 1. \end{cases} \qquad (4.1)$$

With the above data the solution is symmetrical about the x-axis and so one can solve the problem in half the region. Below we present results for four values of b: 0,1,2 and 5; so one can see how the presence and dominance of convection effects the growth or thaw of the frozen region.

Fig. 1 shows the position of the moving boundary at $t = 0, 2, 4$ and 6 in the case $b = 0$, that is no convection. As one can see the frozen region grows symmetrically (approaching a steady state of approximate radius 1.5). The square box in the bottom left-hand corner indicates the mesh size h which was chosen to be 0.25 for all the results presented here, for this run Δt was also chosen to be 0.25. Fig. 2 shows the position of the frozen region at $t = 0, 2, 4$ and 6 for $b = 1$ with $\Delta t = 0.25$. The presence of convection has almost completely halted any growth on the left-hand side, due to the inflow of warm water from the left, whilst the right side grows almost uneffected. The effect of increasing the dominance of convection, $b = 2$, is shown in Fig. 3. These results were obtained with $\Delta t = 0.125$ and from the position of the frozen region, shown at $t = 0, 1$ and 2, one can see that now thawing occurs on the left and the growth on the right is significantly less. As the convection term becomes dominant, $b = 5$, we see, from Fig. 4, that thawing occurs on the right-hand side as well as the left. These positions of the frozen region are at $t = 0, 0.5, 1$ and 1.5 with Δt chosen to be 0.05.

We have carried out various checks on the accuracy of our results: (i) comparison with different values of h and Δt, and (ii) in the case of $b = 1$ comparison with the Galerkin approximation (3.9), and by choosing Ω to be a larger square comparison with the steady state solution, where U_B and P_B are boundary

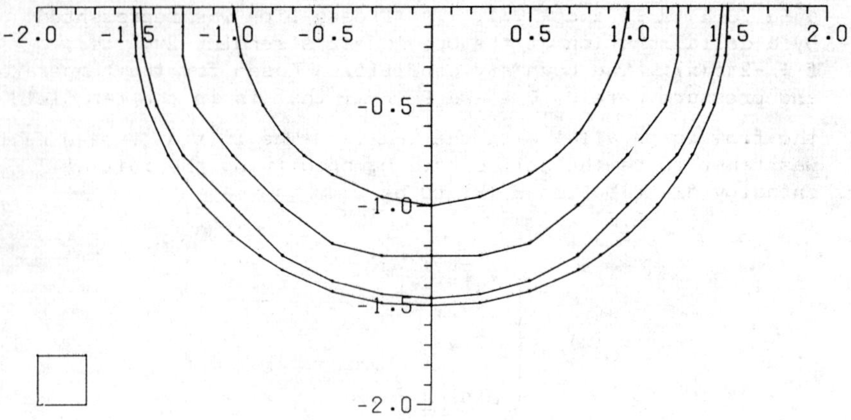

Fig. 1 Model problem with no convection, b = 0.

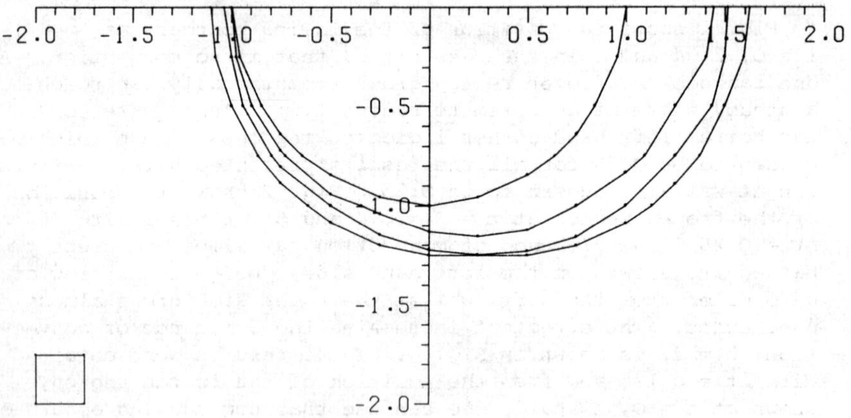

Fig. 2 Model problem with b = 1.

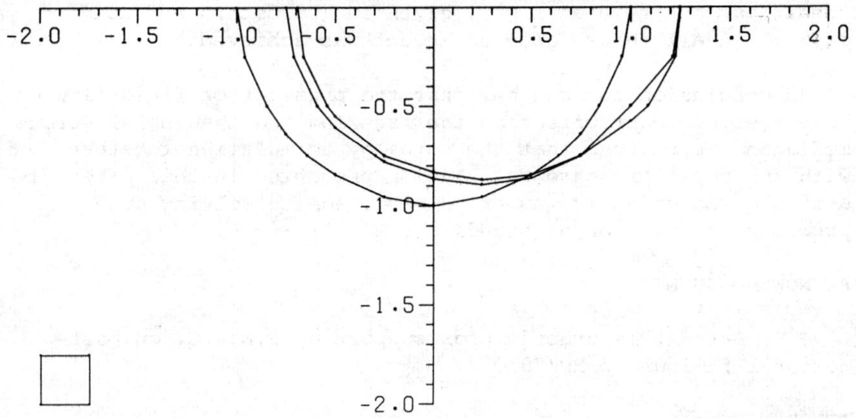

Fig. 3 Model problem with b = 2.

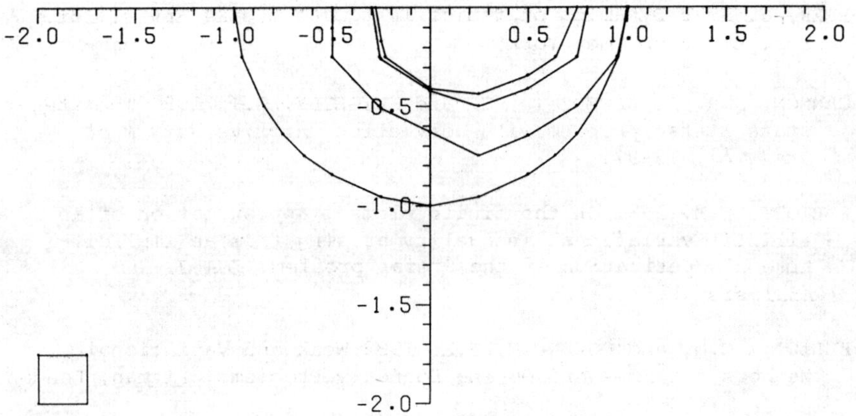

Fig. 4 Model problem with b = 5.

conditions achieved as $\underline{x} \to \infty$, given in (Goldstein and Reid, 1978). In all checks good agreement was achieved.

In conclusion one can see that the presence of fluid flow may have a significant effect on the freezing and thawing of porous media and we believe that the enthalpy formulation together with the novel pressure calculation presented in this paper give a simple and effective numerical approach to solving such problems to engineering standards.

ACKNOWLEDGEMENT

J.W. Barrett is grateful for support by S.E.R.C. on post-doctoral fellowship RF/5830.

REFERENCES

BARRETT, J.W. and ELLIOTT, C.M. 1982 A finite element method for solving elliptic equations with Neumann data on a curved boundary using unfitted meshes. Submitted to *IMA J. Num. Analysis*.

BEAR, J. 1972 Dynamics of fluids in porous media. Envir. Sci. Ser. Elsevier, New York.

CANNON, J.R., DI BENEDETTO, E. and KNIGHTLY, G.H. 1980 The steady state Stefan problem with convection. *Archive Rat. Mech. Anal.* **73**, 79-97.

ELLIOTT, C.M. 1981 On the finite element approximation of an elliptic variational inequality arising from an implicit time discretization of the Stefan problem. *IMA J. Num. Analysis,* **1**, 115-125.

ELLIOTT, C.M. and OCKENDON, J.R. 1982 Weak and Variational Methods for Free and Moving Boundary Problems. Pitman, London.

FOX, L. 1979 The Stefan problem: moving boundaries in parabolic equations. In A Survey of Numerical Methods for Partial Differential Equations (I. Gladwell and R. Wait, Eds.), Oxford Univ. Press, Oxford.

FRIVIK, P.E. and COMINI, G. 1982 Seepage and heat flow in soil freezing. *ASME J. Heat Transfer,* **104**, 323-328.

GOLDSTEIN, M.E. and REID, R.L. 1978 Effect of fluid flow on freezing and thawing of saturated porous media. *Proc. R. S. London A* **364**, 45-73.

HASHEMI, H.T. and SLIEPCEVICH, C.M. 1973 Effect of seepage stream on artificial soil freezing. *J. Soil Mech. Fdns Div. Am. Soc. Civ. Engrs.* **99**, (SM3), 267-287.

LESAINT, P. and ZLÁMAL, M. 1979 Superconvergence of the gradient of finite element solutions. R.A.I.R.O. Analyse Numerique, **13**, 139-166.

SANGER, F.J. 1968 Ground freezing in construction. *J. Soil Mech. Fdns. Div. Am. Soc. Civ. Engrs.* **94**, (SM1), 131-158.

TABATA, M. 1977 A finite element approximation corresponding to the upwind finite differencing. *Mem. Num. Maths.* **4**, 47-63.

VARIATIONAL INEQUALITIES APPLIED TO TIME-DEPENDENT FLOW IN A PHREATIC AQUIFER

A.W. Craig and W.L. Wood

(Department of Mathematics, University of Reading)

1. INTRODUCTION

This paper considers a moving boundary problem related to time-dependent flow of an incompressible fluid in a phreatic aquifer which forms a homogeneous dam separating two reservoirs. The problem is transformed into a variational inequality on a fixed region for which we can obtain existence, uniqueness and approximation results. The technique used is similar to that introduced by Baiocchi et al. (1973a,b). The numerical scheme used is much more efficient in practice than previous heuristic algorithms. Some results for a similar problem have previously been obtained by Torelli (1975) but without the storativity parameter included here.

Bear (1979) gives the equation for the flow of groundwater in a phreatic aquifer composed of an isotropic medium with constant hydraulic conductivity K and velocity potential u as

$$K\nabla^2 u = S_O \frac{\partial u}{\partial t} \tag{1.1}$$

where S_O is the specific volume storativity (dimension ℓ^{-1}).

The numerical model considered here is that of a dam separating two reservoirs whose levels vary with time (we shall consider later, in the numerical results, the particular case in which one of the reservoirs is tidal). Equation (1.1) holds in the fully saturated part of the dam. We suppose that the dam is sufficiently long that we need only consider the flow in a region which is two-dimensional in elevation. Thus x is the horizontal coordinate and y is the vertical and upwards as shown in Fig. 1.

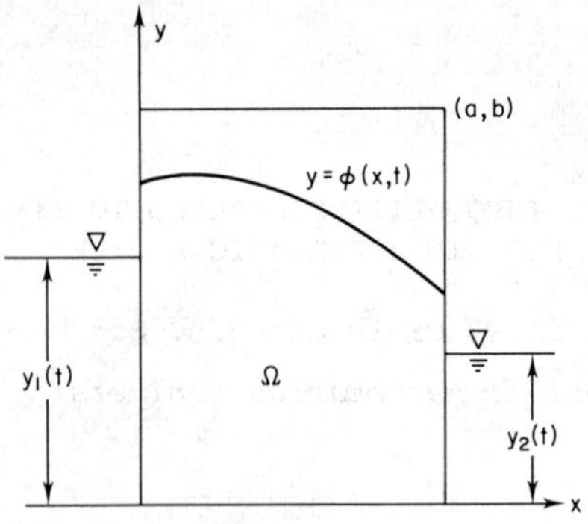

Fig. 1 Diagram of flow region.

The boundary conditions are given in detail later.

On the free surface, which is initially unknown, we have two boundary conditions

(i) pressure ≡ 0 which implies u = y

(ii) a nonlinear boundary condition determining the rate of change of potential across the free surface (Bear, 1979).

$$n_e \frac{\partial u}{\partial t} - K\left[\left(\frac{\partial u}{\partial x}\right)^2 + \left(\frac{\partial u}{\partial y}\right)^2 - \frac{\partial u}{\partial y}\right] = 0 \tag{1.2}$$

where n_e is the porosity of the medium. (We neglect any capillary fringe effect above the free surface.) We non-dimensionalise the equations by putting x = x'd;y = y'd; u = u'd (where d is a reference length), t = t'T (where t is a reference time),

$$\alpha = dn_e/TK; \tag{1.3}$$

then equation (1.1) becomes

$$\frac{\partial^2 u'}{\partial x'^2} + \frac{\partial^2 u'}{\partial y'^2} = S\frac{\partial u'}{\partial t'} \tag{1.4}$$

where

$$S = \frac{S_0 d^2}{KT}$$

and the non-linear boundary condition on the free surface given by equation (1.2) becomes

$$\alpha \frac{\partial u'}{\partial t'} = \left(\frac{\partial u'}{\partial x'}\right)^2 + \left(\frac{\partial u'}{\partial y'}\right)^2 - \frac{\partial u'}{\partial y'} \quad (1.5)$$

2. STATEMENT OF PROBLEM

We now drop primes and suppose that the following theory is applied to the non-dimensionalised equations.

We suppose that $a > 0$, $b > 0$ are the width and height of the dam; $y_1(t)$, $y_2(t)$ are the reservoir levels upstream and downstream, respectively, and that there is a horizontal impermeable base to the dam and the adjoining reservoirs. Further we suppose that

$$0 \le y_1(t) \le b \quad \text{and} \quad 0 \le y_2(t) \le b \quad \forall\, t \in [0,T] \quad (2.1)$$

where $[0,T]$: $T > 0$ is the time interval in which we seek the solution of the problem. We can then state the problem as follows.

<u>Problem 1</u>

$$\phi : [0,a] \times [0,T] \to [0,b] \quad (2.2)$$

$$y_1(t) \le \phi(0,t) \text{ and } y_2(t) \le \phi(a,t) \quad \forall t \in [0,T] \quad (2.3)$$

$$\phi(x,0) = \phi_0(x) \quad \forall x \in [0,a] \quad (2.4)$$

$$\Omega = \{x,y,t) : 0 < x < a,\ 0 < t < T,\ 0 < y < \phi(x,t)\} \quad (2.5)$$

and u, the velocity potential, is defined by

$$S \frac{\partial u}{\partial t} = \nabla^2 u \ , \ \forall (x,y,t) \in \Omega \quad (2.6)$$

with $u(0,y,t) = \max\{y, y_1(t)\}$ and

$$u(a,y,t) = \max\{y, y_2(t)\}, \ \forall t \in [0,T], \quad (2.7)$$

$$u_y(x,0,t) = 0 \quad (2.8)$$

$$u(x,y,0) = u_0(x,y). \quad (2.9)$$

If we define the surface Γ by the following relationship

$$\Gamma = \{(x,y,t) : 0 < x < a,\ 0 < t < T,\ y = \phi(x,t)\} \quad (2.10)$$

then the following relations hold:

$$u = y \text{ on } \Gamma \tag{2.11}$$

$$\alpha u_t = u_x^2 + u_y^2 - u_y \text{ on } \Gamma \tag{2.12}$$

We shall also suppose that we know $u(x,t)$, $\phi(x,t)$ $\forall t\in(-\infty,0]$.

The relationship $y = \phi(x,t)$ represents the free boundary Γ (note that we have assumed that ϕ is a single-valued function of x and t, therefore precluding any instability in the free surface). Ω is the flow region in which the velocity potential u must satisfy equation (2.6), fixed boundary conditions (2.7) - (2.9), and the two moving boundary conditions (2.11), (2.12).

3. WEAK FORMULATION OF PROBLEM

We now state the following two lemmas due to Torelli (1975). Set

$$B = \{(x,y,t): 0 < x < a,\ 0 < t < T,\ y = 0\}, \tag{3.1}$$

$$C_E^\infty(\bar{\Omega}) = \{\psi\in C^\infty(\bar{\Omega}): \psi = 0 \text{ in a neighbourhood of } \partial\Omega - (\Gamma\cup B)\}, \tag{3.2}$$

where $\partial\Omega$ is the boundary of the region Ω, $\bar{\Omega} = \Omega + \partial\Omega$. We then have

Lemma 1

$$\alpha\phi_t(x,t) = -\frac{\partial u}{\partial n}\sqrt{(1+\{\phi_x(x,t)\}^2)} \text{ on } \Gamma \tag{3.3}$$

Lemma 2

$$\alpha\phi_t(x,t) > -1 \tag{3.4}$$

Proofs See Torelli (1975).

Lemma 3

$$\int_\Omega (Su_t\psi + u_x\psi_x + u_y\psi_y)\, dxdydt = \alpha\int_\Omega \psi_t\, dxdydt \quad \forall\psi\in C_E^\infty(\Omega) \tag{3.5}$$

Proof See Craig (1982).

Note that equation (3.5) is a weak form of equation (2.6) with the term on the right hand side due to the non-linear moving boundary condition (2.12).

Now define

$$D = \{(x,y) : 0 < x < a, 0 < y < b\}, \tag{3.6}$$

$$Q = D \times (0,T). \tag{3.7}$$

We extend the function u, previously only defined in Ω to the region Q in the following manner

$$\tilde{u}(x,y,t) = \begin{cases} u(x,y,t) & \text{if } (x,y,t) \in \bar{\Omega} \\ y & \text{if } (x,y,t) \in \bar{Q} - \bar{\Omega} \end{cases} \tag{3.8}$$

We now have

<u>Lemma 3</u>

$$S \frac{\partial \tilde{u}}{\partial t} - \nabla^2 \tilde{u} = \left(\frac{\partial}{\partial y} - \alpha \frac{\partial}{\partial t}\right) \chi_\Omega \tag{3.9}$$

in the sense of distributions on Q, where

$$\chi_\Omega = \begin{cases} 1 & (x,y,t) \in \Omega \\ 0 & \text{elsewhere} \end{cases} \tag{3.10}$$

<u>Proof</u> See Craig (1982)

Let us now introduce the following transformation:

$$w(x,y,t) = \int_y^b \{\tilde{u}(x,\gamma,t-\alpha(\gamma-y)) - \gamma\} \, d\gamma. \tag{3.11}$$

This integration in the $(0,1,-\alpha)$ direction is suggested by equation (3.8) of Lemma 3. We also define

$$\begin{aligned} g(0,y,t) &= \int_y^b [y_1(t-\alpha(\gamma-y)) - \gamma]^+ \, d\gamma \\ g(a,y,t) &= \int_y^b [y_2(t-\alpha(\gamma-y)) - \gamma]^+ \, d\gamma \end{aligned} \tag{3.12}$$

$$g(x,b,t) = 0 \qquad (\text{where } f^+ = \tfrac{1}{2}(|f| + f)). \tag{3.13}$$

Theorem 1 The function w has the following properties

$$w \in C^1(Q) \tag{3.14}$$

$$w > 0 \text{ in } \Omega, \; w = 0 \text{ in } Q - \Omega \tag{3.15}$$

$$(\alpha\frac{\partial}{\partial t} - \frac{\partial}{\partial y})w = \tilde{u} - y \text{ in } Q \tag{3.16}$$

$$(S\frac{\partial}{\partial t} - \nabla^2)w = -\chi_\Omega \text{ in } Q \tag{3.17}$$

$$w|_{\partial D - B} = g \tag{3.18}$$

$$(\alpha\frac{\partial^2}{\partial t \partial y} + \frac{\partial^2}{\partial x^2} - S\frac{\partial}{\partial t})w|_B = 0 \tag{3.19}$$

Proof See Craig (1982)

We shall now make a weak formulation of the problem.

Let

$$\Gamma_n = \{(x,y) : 0 < x < a, \; y = 0\} \tag{3.20}$$

$$\Gamma_d = \partial D - \Gamma_n \tag{3.21}$$

be the Neumann and Dirichlet boundaries respectively of the region D and let $\Omega_{\bar{t}} = \Omega \cap \{t=\bar{t}\}$, $K(t) = \{v(x,y,t) \in H^1(D) : v = g$ on $\Gamma_d \; \forall \; t\}$ (3.22)

We now consider the following problem

Problem 2 Find $w(x,y,t) \in K(t)$ such that

$$w(x,y,0) = w_0(x,y) \tag{3.23}$$

$$S \int_D \frac{\partial w}{\partial t} (v-w)\, dxdy + \int_D \nabla w . \nabla (v-w)\, dxdy + \int_D (v^+ - w^+)\, dxdy$$

$$+ \int_{\Gamma_n} \frac{\partial w}{\partial y} (v-w)\, dx \geq 0 \quad \forall v \in K(t) \tag{3.24}$$

$$\left(\alpha \frac{\partial^2}{\partial t \partial y} + \frac{\partial^2}{\partial x^2} - S \frac{\partial}{\partial t}\right) w\Big|_{\Gamma_n} = 0 \ \forall \ t \tag{3.25}$$

<u>Theorem 2</u>. The function $w(x,y,t)$ defined in equation (3.11) is a solution of problem 2.

<u>Proof</u>. See Craig (1982).

We note that the boundary condition (3.25) is not a natural boundary condition for the variational formulation (3.24).

4. EXISTENCE AND UNIQUENESS RESULTS

The problem is solved by using a finite difference method to solve for approximate values of $w(x,y,t)$, but in order to establish existence and uniqueness a further transformation is introduced which enables us to use a result of Lions (1969).

We introduce a function $G(x,y,t)$ defined in Q whose trace on the appropriate boundaries gives the boundary conditions (3.18) and we define the following space:

$$V = \{v(x,y,t) \in H^1(D) \ : \ v = 0 \text{ on } \Gamma_d \ \forall t\} \tag{4.1}$$

We then have

<u>Problem 3</u> Given $w_0(x,y)$, $G(x,y,t)$ find $z(x,y,t)$ such that for z, z_t, $z_{tt} \in V$

$$z(x,y,0) = 0 \tag{4.2}$$

$$z_t(x,y,0) = w_0(x,y) - G(x,y,0) \tag{4.3}$$

$$(z_{tt}, v-z_t) + a(z_t, v-z_t) + b(z, v-z_t) + j(v) - j(z_t) \geq L(v-z_t)$$

$\forall t \in [0,T]$ and $v \in V$ where (4.4)

$$(u,v) = \alpha S \int_D uv\, dxdy \tag{4.5}$$

$$a(u,v) = \alpha \int_D \nabla u . \nabla v \, dxdy + S \int_{\Gamma_n} uv \, dx \tag{4.6}$$

$$b(u,v) = \int_{\Gamma_n} \frac{\partial u}{\partial x} \cdot \frac{\partial v}{\partial x} \, dx \tag{4.7}$$

$$j(u) = \alpha \int_D (u+G)^+ \, dxdy \tag{4.8}$$

$$L(u) = -(G_t, u) - a(G,u) + S \int_{\Gamma_n} w_O \, u \, dx \; -b\left(\int_O^t G(\tau)d\tau, \, u\right) - \alpha \int_{\Gamma_n} \frac{\partial w_O}{\partial y} u \, dx \tag{4.9}$$

<u>Theorem 3</u> (i) if w is a solution of Problem 2, then z defined by

$$z(x,y,t) = \int_O^t w(x,y,\tau) - G(x,y,\tau) \, d\tau \tag{4.10}$$

is a solution of Problem 3.

(ii) if z is a solution of Problem 3, then w defined by

$$w(x,y,t) = z_t(x,y,t) + G(x,y,t) \tag{4.11}$$

is a solution of Problem 2.

<u>Proof</u> See (Craig, 1982).

5. NUMERICAL PROCEDURES AND RESULTS

The approximation scheme used for the solution for the function w in Theorem 1, i.e., equations (3.15), (3.17), (3.18), (3.19) is the five-point finite difference for the Laplacian with Crank-Nicolson time stepping and a finite difference form for the boundary condition (3.19).

The solution of this resulting system of equations is by successive over-relaxation with projection (Cryer, 1970)(see also (Craig, 1982)).

The program has been tested for the following parameters: width of dam = 200m, height of dam = 200m, $n_e = 0.4$,

$K = 10^{-4}\ ms^{-1}$, $S_O = 10^{-5}$ m, giving $\alpha = 8 \times 10^5/T$, $S = 4 \times 10^3/T$ (values appropriate for sandstone).

If we are interested in a 12 hour period, this gives

$$\alpha = 18.5185\ ,\ S = 9.259 \times 10^{-2}$$

The region was divided into a fixed regular mesh with ten subintervals in the x and y directions giving $\Delta x = \Delta y = 20$m and the time period was divided into 60 subintervals giving $\Delta t = 720$ s.

The numerical example presented here is that of a dam with one stationary reservoir and one periodic reservoir. The left hand reservoir is held constant at y = 100 while the right hand reservoir starts at y = 200, falls linearly to y = 0 in 30 time steps (6 hours) and rises again linearly to y = 200 by 60 time steps (12 hours). It is then repeated with a 12 hour period. The solution is assumed to be steady initially, so we expect a running-in period and in fact it took 3 time periods to settle down to a completely periodic solution. As the right hand reservoir is oscillating about the position of the left hand one we also expect the flow direction to reverse and, for example, in the first period the direction has completely reversed by t = 4 hours, (note that the reservoir levels are equal at t = 3 hours), and then it reverts to its original direction by t = 10 hours (reservoir levels equal at t = 9 hours). We present a sequence of graphs representing the free boundary position at intervals over the first time period: Fig. 2a, b, c, d, e, f.

CONCLUSIONS

We have produced a method of solving a moving boundary problem in the theory of fluid flow through porous media which has both a sound theoretical basis and efficient numerical procedures. The cost of solving the problem is that of solving one matrix equation per time step.

For a more complete source of references to the application of variational inequalities in the theory of fluid flow through porous media see Bruch (1980), and for applications in other fields, see, e.g., Elliott (1981).

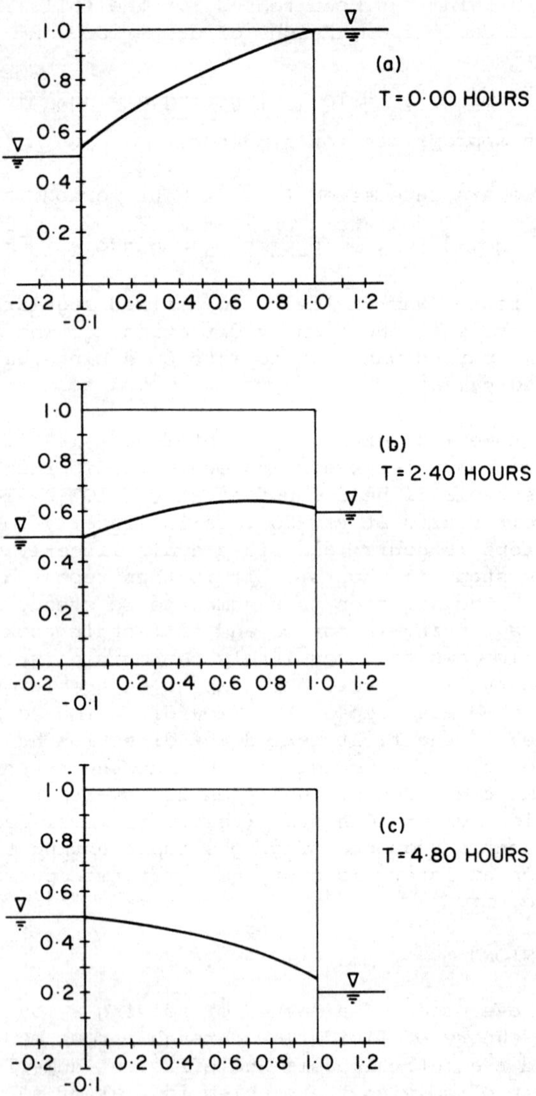
1·0
0·8
0·6
0·4
0·2
-0·2
-0·1
0·2 0·4 0·6 0·8 1·0 1·2
(a)
T = 0·00 HOURS
(b)
T = 2·40 HOURS
(c)
T = 4·80 HOURS

Fig. 2.

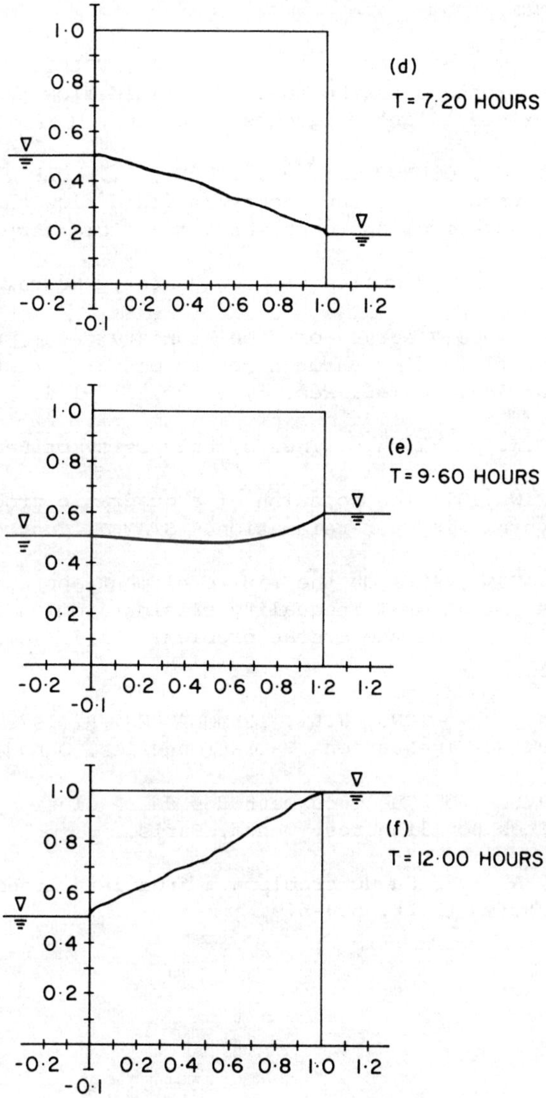

1·0
0·8
0·6
0·4
0·2
-0·2
-0·1
0·2 0·4 0·6 0·8 1·0 1·2
(d)
T = 7·20 HOURS
1·0
0·8
0·6
0·4
0·2
-0·2
-0·1
0·2 0·4 0·6 0·8 1·2 1·2
(e)
T = 9·60 HOURS
1·0
0·8
0·6
0·4
0·2
-0·2
-0·1
0·2 0·4 0·6 0·8 1·0 1·2
(f)
T = 12·00 HOURS

Fig. 2 contd.

REFERENCES

BAIOCCHI, C., MAGENES, E., POZZI, G.A., VOLPI, G. 1973 Free boundary problems in the theory of fluid flow through porous media: existence and uniqueness theorems. *Calcolo* **10**, 1-85.

BAIOCCHI, C., COMINCIOLI, V., GUERRI, L., VOLPI, G. 1973 Free boundary problems in the theory of fluid flow through porous media: a numerical approach. *Ann. Mat. Pura Appl.* **4**(97), 1-82.

BEAR, J. 1979 Hydraulics of Groundwater. McGraw Hill.

BRUCH, J. 1980 A survey of free boundary value problems in the theory of fluid flow through porous media: variational inequality approach. *Adv. in Wat. Res.* **3**, 65-80, 115-124.

CRAIG, A.W., 1982 Ph.D. Thesis, University of Reading.

CRYER, C.W. 1971 The solution of a quadratic programming problem using systematic over-relaxation. *SIAM J. Control.* **9**, 385-392.

ELLIOTT, C.M., 1981 On the finite element approximation of an elliptic variational inequality arising from an implicit time discretisation of the Stefan problem. *IMA J. Num. Analysis* **1**, 115-125.

GLOWINSKI, R., LIONS, J.L., TREMOLIERES, R. 1976 Analyse Numerique des Inequations Variationnelles. Dunod, Paris.

LIONS, J.L. 1969 Quelques methodes de résolution des problèmes aux limites non lineares. Dunod, Paris.

TORELLI, A. 1975 Su un problema a frontiera libera di evoluzioni. *Boll. U.M.I.* (4) **11**, 559-579.

THE USE OF STRONGLY STRETCHED GRIDS IN VISCOUS FLOW PROBLEMS

I.P. Jones

(AERE, Harwell, Oxon)

1. INTRODUCTION

One of the difficulties present in the numerical modelling of high Reynolds number fluid flow problems is the need to perform economical calculations with as few grid points as possible and yet to resolve the extremely small length scales found in boundary layers and shear layers. In some problems it is possible to use the ideas of matched asymptotic expansions, splitting the domain of interest into different regions, with simplified equations in these regions. One example of this is the classical treatment of flow over a slender aerofoil with an inviscid outer flow matched to a viscous boundary layer. Another well known example arises in the numerical simulation of turbulent channel flow. In this problem the fine scale flow near to a wall is usually modelled using a semi-empirical logarithmic profile. This profile is matched to a numerical solution of a turbulence model within the channel. In many cases however the approach of subdividing the region of interest breaks down and it becomes necessary to solve the full equations over the entire region. Internal flows, particularly ones with sharp corners, boundary layer separation and with body forces driving the flow often give rise to problems of this sort. An example which nicely demonstrates some of these difficulties is the compressible flow in a rapidly rotating cylinder. A set of standard parameters has been published (Scuricini, 1979) for use in comparing numerical results. Approximate analytic solutions at these parameter values for the end wall boundary layers show that these boundary layers, the Ekman layers, are 1/4mm to 1/2mm thick; yet the computations are carried out in a domain of 5 metres in extent. For this case the flow in the interior is driven by inflow/outflow from the end wall boundary layers and it is crucial, in numerical simulations, to resolve the flow in these boundary layers. On the side wall a temperature variation induces a "Stewartson $E^{1/3}$ boundary layer" which is elliptic in character and contains substantial recirculation. For this reason conventional boundary layer marching techniques will not work. Some of these features are illustrated in (Dickinson and Jones, 1981). Another example is high Rayleigh number laminar and turbulent buoyancy driven flow in a rectangular cavity with differential heating on the vertical walls. These flows exhibit thin boundary layers, recirculation within the boundary layer and the quantity of most practical importance, the Nusselt number is determined from the temperature gradient at the wall. Furthermore, for turbulent flow, the logarithmic law of the wall is not valid and hence the boundary layers need to be resolved in order to determine the temperature gradients at the walls.

In these and other problems it is therefore desirable to treat the domain as a single region and to solve the full equations within this domain, resolving all the features of interest. This paper considers some of the implications of using strong grid stretching so that the grid is fine where the solution changes rapidly and coarser where the changes are smaller. Two aspects are of particular interest in this study:

(a) the behaviour of standard iterative techniques such as successive over relaxation, SOR; and alternating direction implicit, ADI, applied to a simple diffusion equation;

and

(b) the accuracy of the finite difference solutions of simple model problems, a diffusion equation and a convection-diffusion equation.

2. MATHEMATICAL AND NUMERICAL FORMULATION

The idea underlying the difference formulae adopted is to transform the physical coordinates to a system where the variations in the solution are much less rapid (Blottner, 1975 and Kalnay de Rivas, 1972). These transformations are carried out in each coordinate direction only and may be applied after coordinate transformations to change an irregular geometry into a regular one. A uniform mesh is then used in the transformed coordinate together with conventional second order central differencing of the derivatives of the solution. This technique can be applied using either numerical or analytic differentiation of the transformation derivatives which now appear in the differential equations. In the present paper second order central differences are also used for the transformation derivatives. The difference formulae obtained may be found in (Jones and Thompson, 1980). This procedure for obtaining the difference formulae has the following advantages.

1. The formulae obtained are virtually identical to conventional non-uniform-grid difference formulae and hence can be implemented simply in many existing codes, e.g., ones based on a conservative control volume philosophy.

2. The transformations are used only to define the mesh in the physical coordinate system and numerical techniques may be used to obtain this mesh.

3. The results are formally second order accurate in $\Delta\xi$, the uniform mesh increment in the transformed coordinate. Hence there is no need to constrain mesh expansion ratios in order to obtain accurate results. Furthermore, techniques based on uniform grid methods, such as Richardson extrapolation, may be used to improve the accuracy of the solutions.

It is however important that the mesh does not expand rapidly in the boundary layers otherwise the truncation errors may be large.

Apart from simple analytic transformations, which do not have any degree of flexibility, three transformations have been used

$$\xi = x + \sum_{j=1}^{N_B} w_j \tan^{-1}[(x-\alpha_j)/\delta_j] \tag{1}$$

$$\xi = x + \sum_{j=1}^{N_B} w_j \tanh[(x-\alpha_j)/\delta_j] \tag{2}$$

and one given by an algorithm due to Kautsky and Nichols (1980). In these formulae ξ is the transformed coordinate, x the physical coordinate, N_B the number of boundary and interior layers, α_j the positions where greater resolution is required, δ_j the thickness of the region in which resolution is required and w_j a weighting function. The first two of these are based upon a suggestion by Orsag and Israeli (1974). In practice equation (1) is used for practical computations as the tanh function employed in (2) has too sharp a cutoff at the edge of a boundary layer. For theoretical studies however equation (2) has many advantages because of this sharp cutoff. Principally, for small δ_j, it gives a uniform grid outside the boundary layers with the number of points in each region being in proportion to the weights w_j. For example, with N_B=1, δ_1<<1, w_1=1.0, and the position of the boundary layer at $\alpha_1=0.0$, half of the points form a uniform grid with $\delta x = 2/(N-1)$, where δx is this uniform grid interval and N the total number of points, i.e., N-1 intervals. The other half falls into the first interval between x=0.0 and $x=\delta x$. This behaviour makes it much easier to understand many of the results obtained although the sharp nature of the transition often gives rise to large errors near to the transition points. The algorithm of Kautsky and Nichols has a great deal of flexibility and enables meshes to be chosen to satisfy constraints on the mesh expansion ratios. These points, and the application of the methods to practical problems, are discussed in more detail in (Jones and Thompson, 1980 and Jones, 1981). Most of the results presented in this paper were obtained using equation (2) to define the mesh although the results themselves do not depend upon the exact transformation used. The results are valid whether analytic or numerical differentiation of the transformation is performed and essentially only depend upon the distribution of the grid points in the physical coordinate.

3. RATE OF CONVERGENCE OF ITERATIVE METHODS

The theory underlying the use of successive over-relaxation is well understood for finite difference equations where the coefficient matrix is symmetric and positive definite (or similar to one), see e.g. (Wachspress, 1966). In this case the rate of convergence is dominated by the eigenvalue of largest modulus, the spectral radius, of the iteration matrix. Furthermore the optimum rate of convergence and the corresponding value of the relaxation parameter may be calculated from the spectral radius of the Jacobi iteration matrix. These optimum rates of convergence have been investigated for finite difference approximations of the simple one-dimensional diffusion equation

$$\frac{d}{dx}\left(k(x)\,\frac{dy}{dx}\right) = f(x) \qquad (3)$$

where k(x) and f(x) are specified functions and y is specified on the boundaries x=0 and 1. Three cases have been studied:

1. grid compression near to the boundaries at x=0 and 1 with k(x)=1,

2. grid compression about the centre of the domain x=1/2, with k(x)=1,

3. varying k so that it is low near the walls and high in the centre. The aim is to try and understand the behaviour when a diffusion coefficient appropriate for turbulent flow is adopted.

The results for these cases are only strictly applicable to the one-dimensional case although they may be applied to two-dimensional cases which have the same grid structure and diffusion coefficient in each direction. Furthermore the results may be applied to stationary ADI, i.e., with the optimum constant-time steps or iteration parameters, since it may be shown, asymptotically, that it converges at the same rate as optimum SOR. The methods used to obtain the spectral radii and the results obtained are described in detail in (Jones, 1981). The results show that for the first case, boundary grid compression, the rate of convergence increases dramatically as the grid is redistributed to put more points near to the boundaries. For example, the number of iterations required to gain one significant figure of accuracy is roughly half of that required for the corresponding uniform grid with the same number of grid points. Conversely, for grid compression in the centre to resolve an interior layer the rate of convergence decreases rapidly as the grid is compressed. To illustrate this, for a case with a minimum step size δx_{min} of 10^{-4} in the centre and 16 intervals over-all the results indicate that asymptotically about 100 iterations are required to gain 1 significant figure, compared to about 6 for a regular grid with 16 intervals. From the transformation viewpoint a diffusion coefficient which increases in the centre is similar to a transformation which compresses the points in the centre. It can be expected therefore that the iterative methods will degrade in this case. For example with a diffusion coefficient which increases from 1 on the boundary to 1000 in the centre and 16 uniform intervals, the results show an increase in the number of iterations needed to gain 1 significant figure, asymptotically, from about 6 for constant k to about 21 for the simple assumed profile for k(x).

4. ACCURACY ON MODEL PROBLEMS

The numerical solution of two simple model problems is discussed here. The first is a simple diffusion equation given by

$$\frac{d^2y}{dx^2} = f(x) \qquad (4)$$

with y=0 at x=0 and 1, where f(x) is chosen so that the solution is

$$y = 1 - \exp\{-x(1-x)/\varepsilon\} + \sin(\pi x). \quad (5)$$

For small ε this has strong boundary layers at $x=0$ and 1 and also, because of the sin term, has a non-trivial structure in the centre. Space does not permit a detailed discussion of the results so only a brief summary is presented here. Further details may be found in Hunter and Jones (1981). The results obtained demonstrate the over-all second order accuracy of the difference approximations with the errors being divided by 4 as the number of grid intervals doubled. Furthermore results could be obtained which were to some extent invariant of the value of ε provided the grid was chosen carefully with values of δ_j in equation (2) equal to 10 ε. This meant the grid expanded very rapidly outside the boundary layers. Values of $\delta_j=5\varepsilon$ gave lower errors but since the grid was expanding at the edge of the boundary layer the asymptotic nature of the errors, e.g., monotonicity as the grid is refined, was not as good. For this reason care should be taken when using extrapolation techniques. Further results not presented by Hunter and Jones (1981) indicate lower errors generally for the $\tan^{-1}$ transformation, equation (1), with $\delta_j = \varepsilon$. Some experiments have also been performed using a piecewise linear finite element approximation. The difference approximations remain the same, the only difference being to the form of the right-hand side. Assuming piecewise linear behaviour of the right-hand side a Simpson's like approximation is obtained instead of a point value. The pointwise errors of the finite difference approximation were generally slightly lower than for the piecewise linear right-hand side. To compensate for this however the form of the errors in the piecewise linear case was much nicer and had better asymptotic properties. This was particularly noticeable when a bad distribution of grid points was taken. The finite difference results then did not always exhibit the second order nature of the approximation whereas the finite element results did. This is illustrated in Fig. 1 which shows a comparison of the distribution of errors for a case where $\varepsilon=10$, i.e., no boundary layers. The grid was chosen however to concentrate some of the grid points near to the walls and a third of the points formed a uniform grid in the centre. The graph shows the errors plotted against the transformed coordinate so that each grid point is equally spaced from the next. The important feature to note is the nice second order form of the results in the centre for the piecewise linear approximation. The finite difference errors are lower but exhibit sharp spikes and non-asymptotic behaviour.

The second example is a much studied one, the convection-diffusion equation

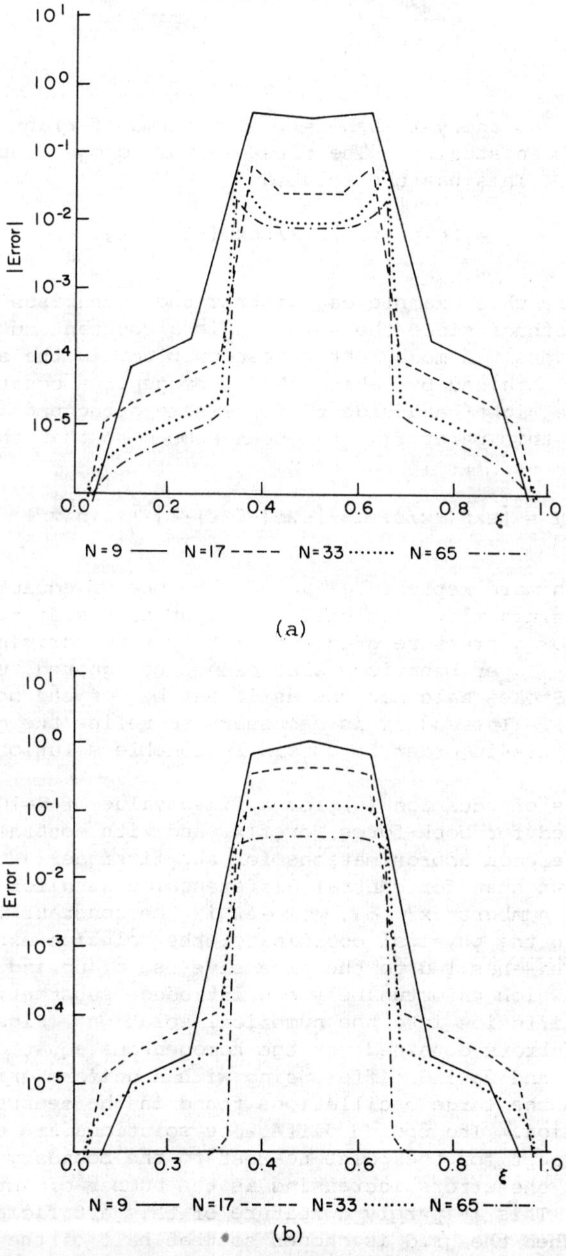

Fig. 1 Graph of absolute value of the errors for the diffusion equation against the transformed coordinates, (a) finite difference approximation, (b) piecewise linear right-hand side.

$$\varepsilon \frac{d^2y}{dx^2} + \frac{dy}{dx} = f(x) \tag{6}$$

with y=0 at x=0 and y=1 when x=1. Two forms of right-hand side, f(x), have been studied. The first is the homogeneous equation with f(x)=0. This has the solution

$$y = [\exp(-x/\varepsilon)-1]/[\exp(-1/\varepsilon)-1]. \tag{7}$$

Unfortunately this example can distort the comparison of numerical methods since the solution is a constant outside the boundary layers. A much better test problem, in the author's opinion, is obtained by taking the inhomogeneous equation and choosing the right-hand side to force some structure into the centre. In this paper f(x) has been chosen so that the solution is

$$y = [\exp(-x/\varepsilon)-1]/[\exp(-1/\varepsilon)-1] + \sin(\pi x) \tag{8}$$

This is much more representative of the type of equation which represents fluid flow problems, the right-hand side representing the effect of a pressure gradient or body force driving the flow. The boundary layer behaviour also cannot be ignored, usually, for Navier-Stokes calculations as it may be for the homogeneous equation (6). Instead it is necessary to refine the grid in the boundary layers in order to obtain reasonable solutions.

Solutions of equation (6) for a fixed value of $\varepsilon=10^{-4}$ have been obtained for both forms for f(x) and with central and upwind difference approximations for the first derivative. It is well known that for central differences on a uniform mesh and mesh Peclet numbers $\delta x/\varepsilon > 2$, where δx is the constant mesh increment in the physical coordinate, the solution exhibits wiggles. This has led to the extensive use of upwind difference techniques which unfortunately can introduce substantial numerical diffusion into the numerical solution. Figs. 2a and b show the errors obtained for the homogeneous equation (6), for central and upwind differencing with a uniform grid. These demonstrate the large oscillations found in the central difference solution. The upwind difference solutions are extremely accurate except for the point nearest to the boundary layer which shows the errors increasing as the number of intervals increases. This is purely a feature of this artificial test problem. When the grid is chosen so that half of the grid points are in the boundary layer and the other half form a uniform grid outside the situation is somewhat different. The central difference results in Fig. 2c show the expected

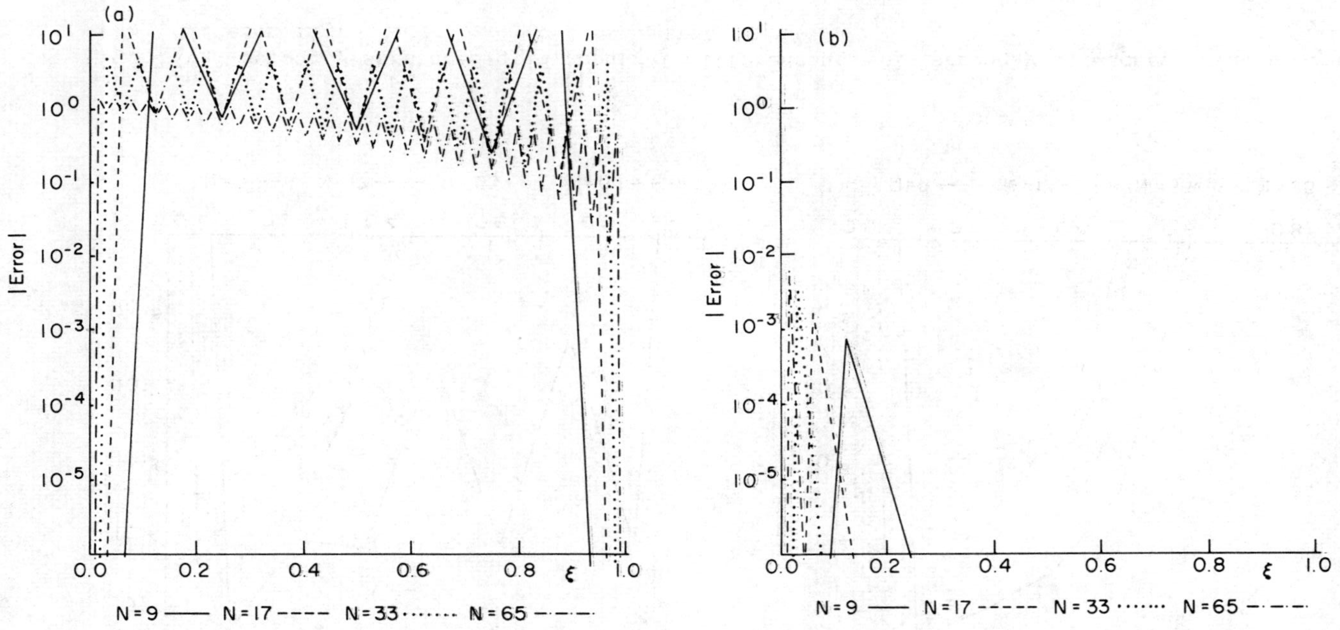

Fig. 2 Graph of absolute values of the errors for the homogeneous convection diffusion equation, $\varepsilon=10^{-4}$, against the transformed coordinate, (a) uniform grid, central differencing, (b) uniform grid, upwind differencing.

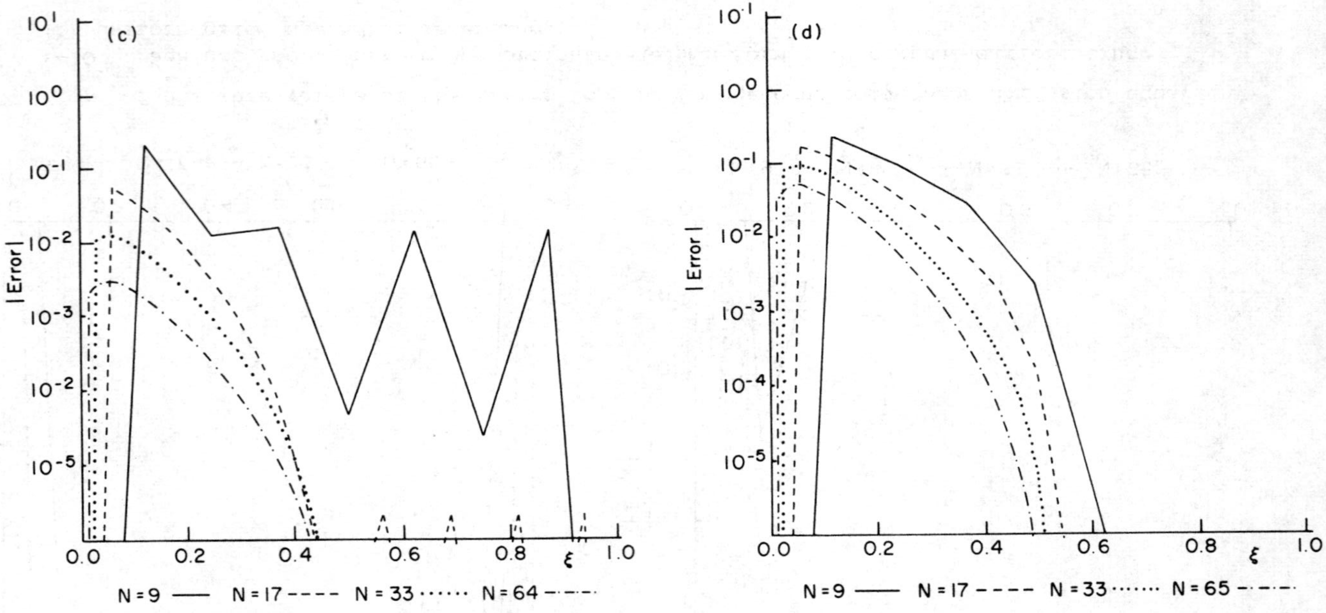

Fig. 2 (c) strongly stretched grid, central differencing, (d) strongly stretched grid, upwind differencing.

oscillations where the mesh Peclet number is large, but, apart from the case with 8 intervals these oscillations are occurring in, at most, the fifth significant figure of the solution. The results also exhibit second order accuracy within the boundary layer. For upwind differencing the results in Fig. 2d now show first order behaviour of the errors within the boundary layer and again are accurate outside the boundary layer where the solution is a constant.

A much more realistic view of these difference formulae is found for the inhomogeneous case. The uniform grid results again show oscillations for central differencing, Fig. 3a, and the upwind difference results, Fig. 3b, now do not give particularly accurate results. When half the points are chosen to lie within the boundary layer even worse errors are found, in Fig. 3d, for upwind differencing. The accuracy for 64 intervals is now just under 10%. The central differencing results, Fig. 3c, hardly show any signs of the oscillations as they are swamped by the errors in approximating the structure. For 32 intervals the largest errors are about 1%, far smaller than the upwind differencing results with twice as many points. Extrapolating the results from Fig. 6c we find that over 500 grid intervals would be required to obtain errors of about 1%, compared to the 32 intervals for central differencing. It is also important to realise that outside the boundary layers the mesh Peclet numbers for the central differencing results are comparatively large, about 600 for the case of 32 intervals.

After presenting this paper at the IMA Conference on numerical methods in fluid dynamics the author's attention was drawn to the work of Segal (1980) who has explored similar ideas to these on homogeneous and non-homogeneous convection-diffusion equations.

5. CONCLUDING REMARKS

This paper has examined some of the implications of the use of strong grid stretching in order to resolve boundary and interior layers. The results have indicated that in certain cases, particularly interior refinement and a varying diffusion coefficient, standard iterative techniques may need a large number of iterations in order to obtain convergence. The results also showed improved rates of convergence for boundary grid compression. The author has carried out some preliminary numerical experiments on laminar natural convection flows to investigate the convergence rates. The experiments indicate enhanced rates of convergence on diffusion dominated Stokes type flows for boundary grid compression. Unfortunately, for this type of flow, strong grid stretching is not required. For convection dominated high Rayleigh number laminar flows however

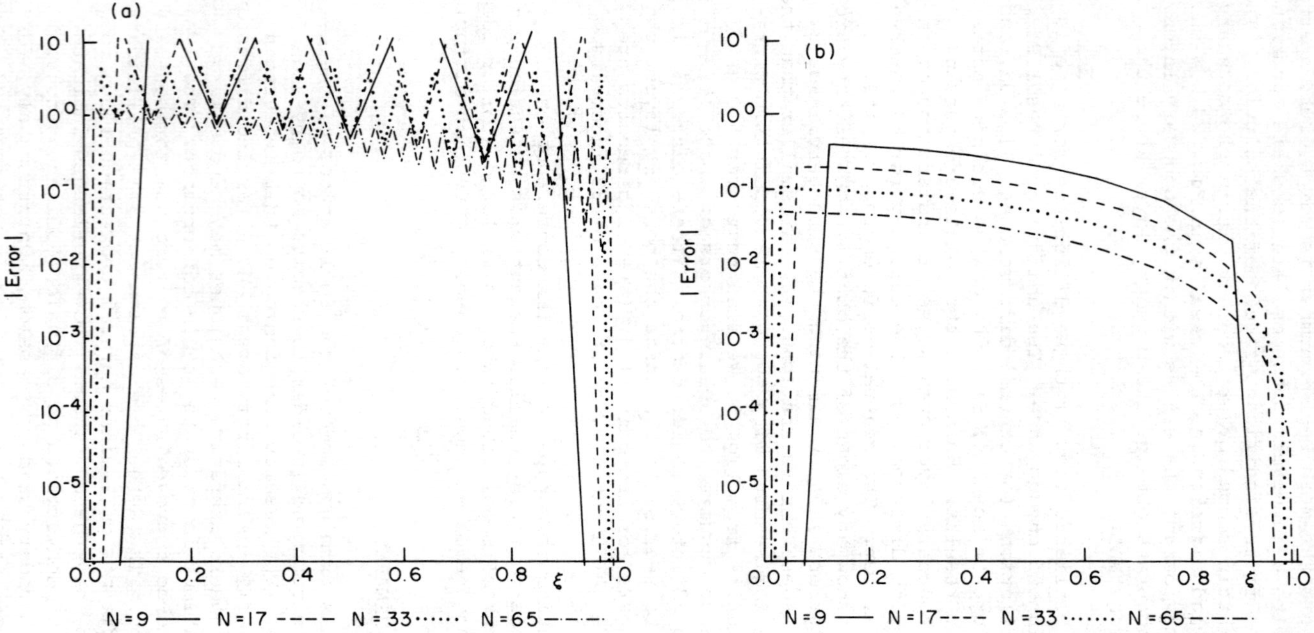

Fig. 3 Graph of absolute value of the errors for the inhomogeneous convection diffusion equation, $\varepsilon=10^{-4}$, against the transformed coordinate, (a) uniform grid, central differencing, (b) uniform grid, upwind differencing.

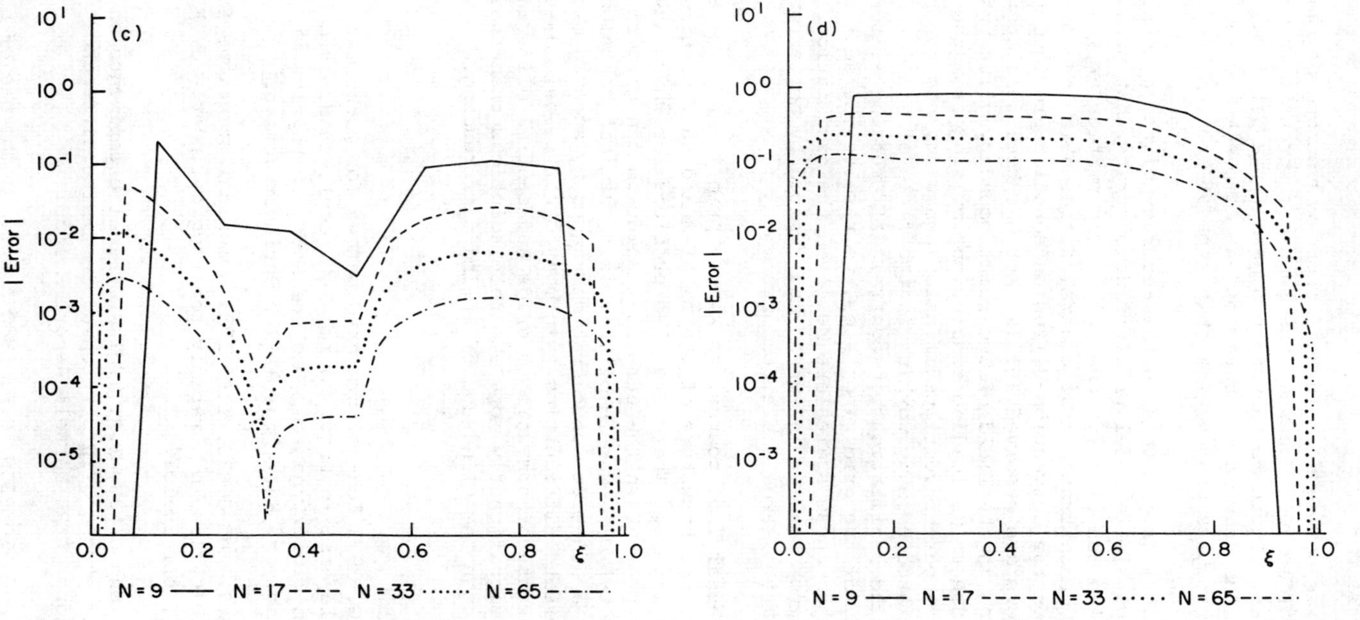

Fig. 3

(c) strongly stretched grid, central differencing, (d) strongly stretched grid, upwind differencing.

this effect was not found. Instead there was strong degradation of the rate of convergence when the mesh was stretched towards the boundaries. There is therefore a need for a detailed analysis for the convergence rates of the complete coupled iterative strategies adopted for this type of problem.

The numerical experiments on the accuracy of the numerical results on two simple model problems demonstrate that strong grid stretching can be very beneficial. In particular the results have demonstrated that central differencing can be used effectively on the convection-diffusion equation provided the boundary layers are resolved. This fact has been well known to people working in the field of two point boundary value problems but, partly due to misleading conclusions arising from the use of the homogeneous convection-diffusion equation, has not been fully appreciated by many working in the field of fluid mechanics. The oscillations in central differencing arise from the inability of the grid, in places, to resolve the scales of the flow. When these scales are resolved the oscillations become very much smaller in extent and swamped by the errors in approximating the large scale features.

It is necessary, of course, to know a priori, or to be able to determine, the structure of the problem and the location of the boundary layers. There are also problems for which any oscillations present in the numerical solution can give rise to problems. A typical example of this is a chemical reaction system where a negative concentration caused by such an oscillation could create difficulties. In many other cases, however, the judicious use of non-uniform grids and central differencing can provide significantly more accurate solutions than those obtained through the blind use of upwind differencing.

REFERENCES

BLOTTNER, F.G. 1975 Non-uniform grid method for turbulent boundary layers. In Proc. Fourth Int. Conf. on Numerical Methods in Fluid Dynamics. Lecture Notes in Physics **35**, (R.D. Richtmyer, Ed.), pp. 91-97, Springer-Verlag.

DICKINSON, G.J. and JONES, I.P. 1981 Numerical solutions for the compressible flow in a rapidly rotating cylinder. *J. Fluid Mech.* **107**, pp. 89-107.

HUNTER, I.C. and JONES, I.P. 1981 Numerical experiments on the effects of strong grid stretching in finite difference calculations. AERE Harwell Report R 10301.

JONES, I.P. and THOMPSON, C.P. 1980 A note on the use of non-uniform grids in finite difference calculations. In

Boundary and interior layers, computational and assymptotic methods. (J.J.H. Miller, Ed.), pp. 332-341, Boole Press.

JONES, I.P. 1981 Asymptotic rates of convergence of iterative methods applied to finite difference calculations on non-uniform grids. AERE Harwell Report R 10302.

KALNAY DE RIVAS, E. 1972 On the use of non-uniform grids in finite difference equations. *J.Comput. Phys.* **10**, pp. 202-210.

KAUTSKY, J. and NICHOLS, N.K. 1980 Equidistributing meshes with constraints. *SIAM J. Sci. Stat. Computing* **1**, pp. 499-511.

ORSZAG, S.A. and ISRAELI, M. 1974 Numerical simulation of viscous incompressible flows. Annual Reviews in Fluid Mechanics **6**, pp. 281-318.

SCURICINI, G.B. 1979 Proceedings of the Third Workshop on Gases under Strong Rotation, Panel discussion. CNEN, Viale Regina Margherita 125, Rome.

SEGAL, A. 1980 On the need for upwind differencing for elliptic singular boundary value problems, Part 1, Report NA-27, Part 2, Report NA-35. Technical University Delft.

WACHSPRESS, E.L. 1966 Iterative solution of elliptic systems. Prentice Hall.

FINITE DIFFERENCE SCHEMES AND SOLUTION METHODS FOR THE FLUID DYNAMICS EQUATIONS USED IN THE ROD BUNDLE CODE SABRE

J.N. Lillington and J.D. Macdougall

(UKAEA, AEE Winfrith)

1. INTRODUCTION

The SABRE code has been developed for calculating the flow through nuclear reactor fuel elements, which consist of a bundle of fuel pins inside an outer sheath or "wrapper", cooled by coolant (water, sodium or gas) flowing primarily along the fuel pins. In particular the code is designed to provide a calculational tool that can be used for the nonstandard configurations that occur in safety analysis, and thus deals not only with the flow in unperturbed fuel pin bundles, but also with operational problems such as rod bowing, and with accident situations in which such events as blockages, coolant boiling and natural convection might occur. Whereas under standard operating conditions the flow is generally along the direction of the fuel pins, for many of the situations in which we are interested strong crossflows and possibly regions of recirculating flow may occur. This necessitates the use of a solution method for the hydrodynamic equations which, while it may be biased to a particular flow direction, must be satisfactory for general flow regimes, including recirculating wakes.

In the SABRE code we have alternative methods of solution available, a fully implicit method based on the SIMPLE method, developed at Imperial College (Patankar and Spalding (1972)) and a semi-implicit method based on the ICE method developed at Los Alamos by Harlow and Amsden (1971). These methods have been implemented in SABRE using a subchannel discretization of the geometry and a staggered mesh in which the velocity nodes lie midway between the pressure nodes, as shown in Fig. 1. We have chosen to use a subchannel representation as it gives the most accurate practical representation of the pin bundle. To obtain greater detail a reasonably fine mesh in every subchannel would be required, and this becomes impractical as it would involve orders of magnitude more computation. The use of a porous body representation requires as data average quantities over length scales equal to or greater than the pin pitch, and

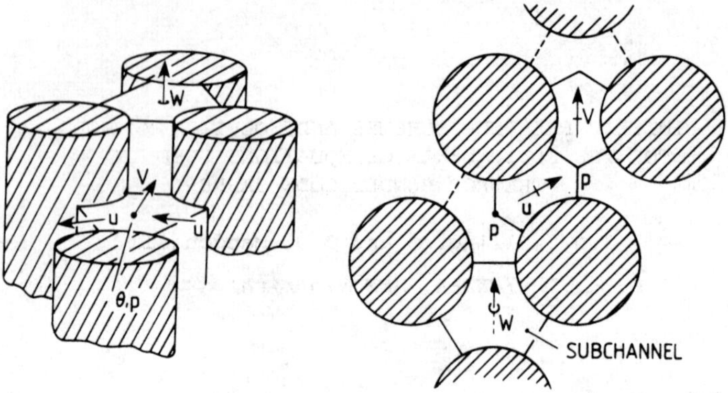

Fig. 1 Method (geometry).

thus, cannot calculate accurately details of the flow on the subchannel scale. The resulting difference equations in subchannel geometry differ little in general character from the equations for a continuum, but two points are worth noting. First the convection of transverse momentum across the pin array needs to be treated carefully: in SABRE by making assumptions about the crossflow we are able to decouple the transverse momentum equations in the transverse direction; this is discussed in Potter (1979) and, as it has little bearing on the points we wish to discuss in this paper, it will not be considered further here. The second point is that in deriving the difference equations the wall shear stress gives rise to a resistance term which is represented using a friction factor correlation and this term tends to improve the stability of the equations compared with those for free flow.

We wish to discuss in this paper three particular topics which have arisen in the development of the code; these topics are not specific to the subchannel geometry of SABRE but are general problems involved in solving the thermohydraulics equations. We start by summarising the methods used in SABRE to solve the equations to obtain both steady state and transient solutions, and comment on the relative merits of alternative iteration and time integration procedures. This is followed by a discussion of spatial differencing and the problem of numerical (false) diffusion. Finally we consider some problems arising in the implementation of a homogeneous boiling model, where discontinuities in the gradient of the density present problems.

2. SOLUTION METHODS

The thermohydraulic equations may be written:

Conservation of mass $\frac{\partial \rho}{\partial t} = -\underline{\nabla}.\rho\underline{u}$

Conservation of momentum $\frac{\partial \rho \underline{u}}{\partial t} = -\underline{\nabla}.\rho\underline{u}\underline{u} + \mu\nabla^2\underline{u} - \underline{\nabla}p + \underline{R}$

Conservation of energy $\frac{\partial \rho I}{\partial t} = -\underline{\nabla}.\rho\underline{u}I + \Gamma\nabla^2 I - p\underline{\nabla}.\underline{u} + \underline{S}$

Equation of state $\rho = \rho\ (I,p)$

These equations are to be solved by finite difference techniques, and we have considered (single step) finite difference schemes in which the time derivatives $\partial\phi/\partial t$ are represented by $(\phi^{(n)} - \phi^{(n-1)})/\partial t$ where $\phi^{(n)}$ is the value of ϕ at $t = t^{(n)}$. When the right hand sides of the conservation equations are evaluated at the old time $t^{(n-1)}$ we have an explicit solution scheme and when they are evaluated at the new time $t^{(n)}$ we have an implicit scheme; we shall refer to schemes in which a mixture of new and old time values are used as semi-implicit. Now, as is well known, for an explicit scheme the time step must satisfy the sound speed Courant condition, that is, the time step must be shorter than the minimum transit time of a sound wave across a spatial mesh, whereas no such type of restriction exists for an implicit scheme. However, the implicit scheme requires much more work per timestep to evaluate the values of the variables at the new time. Thus in choosing a temporal finite difference scheme a balance must be chosen between implicitness and explicitness, and this will largely be controlled by the time scale of the transients being studied. For the nuclear reactor fuel bundles in which we are interested, typical time constants for transients would be of the order of seconds, ranging down to perhaps milliseconds for some very fast transients; whereas the sound speed Courant condition would restrict time steps typically to the order of tens of microseconds, resulting in explicit calculations requiring an excessive number of timesteps.

We have thus confined ourselves to semi-implicit and implicit methods for our work with SABRE, basing our methods on the ICE and SIMPLE algorithms. For the semi-implicit ICE method we solve the equations in the form:

$$\frac{\partial \rho^{(n-1)}}{\partial p} \cdot \frac{p^{(n)} - p^{(n-1)}}{\delta t} + \frac{\partial \rho^{(n-1)}}{\partial I} \cdot \frac{I^{(n)} - I^{(n-1)}}{\delta t} = -\underline{\nabla}.\rho\underline{u}^{(n)}$$

$$\frac{\rho\underline{u}^{(n)} - \rho\underline{u}^{(n-1)}}{\delta t} = -\underline{\nabla}.\rho\underline{u}\underline{u}^{(n-1)} + \mu\nabla^2\underline{u}^{(n-1)} - \underline{\nabla}p^{(n)} + \underline{R}^{(n-1)}$$

$$\frac{I^{(n)} - I^{(n-1)}}{\delta t} = -\underline{u}^{(n-1)}.\nabla I^{(n)} + \frac{\Gamma}{\rho^{(n-1)}}\nabla^2 I^{(n)} - \frac{p}{\rho}\underline{\nabla}.\underline{u}^{(n-1)} + \frac{1}{\rho}S^{(n-1)}$$

$$\rho^{(n)} = \rho(I^{(n)}, p^{(n)})$$

It has been shown (Hirt (1968)) that these equations are stable for a time step satisfying the flow Courant condition ($u\delta t < \delta x$), which for our fuel bundle problem gives a time step limitation of the order of tens of milliseconds. This gives quite satisfactory time steps for the faster transients in which we are interested, but is still restrictive for many of the slower transients. For these slower transients we may use a fully implicit solution of the conservation equations.

The semi-implicit equations for ICE as used in SABRE require the solution of spatial linear differential equations for the energy and pressure, and an explicit evaluation of the velocity and density. The procedure for advancing a time step is thus fairly straightforward and, as we find in practice that a single predictor corrector pass gives adequate convergence of the energy equations, while only a few iterations are required to solve the pressure equation to an adequate accuracy, it is also quite fast.

The implicit solution of the equations is based on the application of the following iterative scheme,

$$\frac{\rho^{(n,m*)} - \rho^{(n)}}{\delta t} = -\underline{\nabla}.\rho^{(n,m*)}\underline{u}^{(n,m)}$$

$$\frac{\rho^{(n,m-1)}[\underline{u}^{(n,m)} - \underline{u}^{(n,m-1)}]}{\delta t'} + \frac{\rho\,\underline{u}^{(n,m-1)} - \rho\underline{u}^{(n-1)}}{\delta t} =$$

$$-\underline{\nabla}.\rho\underline{u}^{(n,m-1)}\underline{u}^{(n,m*)} - \mu\nabla^2\underline{u}^{(n,m*)} - \underline{\nabla}p^{(n,m)} + \underline{R}^{(n)}$$

$$\frac{I^{(n,m)} - I^{(n,m-1)}}{\delta t'} + \frac{I^{(n,m-1)} - I^{(n-1)}}{\delta t} = -\underline{u}^{(n,m^*)} . \underline{\nabla} I^{(n,m)} +$$

$$\frac{1}{\rho^{(n,m-1)}} [\Gamma \nabla^2 I^{(n,m)} - p \underline{\nabla} . \underline{u}^{(n,m-1)} + S^{(n)}]$$

$$\rho^{(n,m)} = \rho(I^{(n,m)}, p^{(n,m)})$$

where $$\frac{\rho^{(n,m-1)} (\underline{u}^{(n,m^*)} - \underline{u}^{(n,m-1)})}{\delta t'} + \frac{\rho \underline{u}^{(n,m-1)} - \rho u^{(n-1)}}{\delta t} =$$

$$-\underline{\nabla} . \rho \underline{u}^{(n,m-1)} \underline{u}^{(n,m^*)} - \mu \underline{\nabla}^2 \underline{u}^{(n,m^*)} - \underline{\nabla} p^{(n,m-1)} + \underline{R}^{(n)}$$

$$\rho^{(n,m^*)} = \rho(I^{(n,m)}, p^{(n,m-1)})$$

where δt is the time step, and $\delta t' \leqslant \delta t$ may be chosen by the user.

In this scheme the momentum equations with the old pressure are first solved for $\underline{u}^{(n,m^*)}$ and then the energy equation is solved for $I^{(n,m)}$, enabling $\rho^{(n,m^*)}$ to be determined.

Taking the divergence of the momentum equations with the new pressure and substituting in the mass equation, a Poisson type equation for $p^{(n,m)}$ is then obtained which is then solved for the pressure, enabling $\underline{u}^{(n,m)}$ to be determined. Finally $\rho^{(n,m)}$ is calculated, completing the iteration cycle.

As the mth iterate values of $\underline{u}$, I and p are merely intermediate values required on the path to the final solution, these values are not usually converged, the usual strategy being to make a single iteration of $\underline{u}$ and I and several iterations on p.

In SABRE we have a fair amount of flexibility in the control of the iterative cycle, and indeed alternative solution sequences are available (e.g. the energy may be calculated at the end of the cycle rather than the middle). In addition we are able to relax the values of each variable, and also to select a suitable value of $\delta t' < \delta t$. We have in general found a value of $\delta t' \sim 0.01$ suitable for most of our problems, combined with no relaxation, except perhaps on p for very slow flows (natural recirculation). For two phase problems we find a significant underrelaxation on the density (~ 0.1) is required; we are working on improving convergence here.

Our experience with the relative merits of the semi-implicit and implicit solution methods has been that the implicit method shows a gain in computer efficiency over the semi-implicit method for slow transients where time steps greater than about 5 - 10 times the flow Courant time step are acceptable on accuracy grounds. We would point out, however, that we have found in practice that the amount of work, the number of iterations per time step, can increase significantly if too large a change is requested in a time step when using the implicit method. This problem has been particularly troublesome in two-phase calculations.

The steady state method in SABRE corresponds to the implicit method with $1/\delta t = 0$. Comparing calculations using this (SIMPLE) method, with the asymptotic behaviour of the semi-implicit ICE method has not shown any significant difference in the work required to obtain a converged solution.

3. SPATIAL DIFFERENCING AND FALSE DIFFUSION

To illustrate the problem we consider the convection term in the energy conservation equation in two dimensions, for the case when the velocity field is known. This term is

$$\underline{\nabla}.\rho\underline{u}I$$

when we use the conservation form of the energy equation. If we integrate this term over the "control volume" surrounding the current energy node (see Fig. 2), we obtain

$$J \equiv \frac{1}{V}\int_V \underline{\nabla}.\rho\underline{u}\, I\, dV = \frac{1}{V}\int_S \rho\underline{u}\, I.\, \underline{dS}\ .$$

Approximating the integrals over the faces by the values at the centres of the faces gives

$$J \simeq (-\rho u I_{i-\frac{1}{2}j} + \rho u I_{i+\frac{1}{2}j})/\delta x + (-\rho v I_{ij-\frac{1}{2}} + \rho v I_{ij+\frac{1}{2}})/\delta y\ .$$

We note that using this control volume analysis ensures that the finite difference representation of the convection terms is conservative. Using the staggered mesh representation of the variables as in SABRE, the velocities u and v are defined at the nodes at which they are required, but the energy is not. The natural representation would be to take the approximation $I_{i-\frac{1}{2}j} = \frac{1}{2}(I_{i-1j} + I_{ij})$, giving rise to

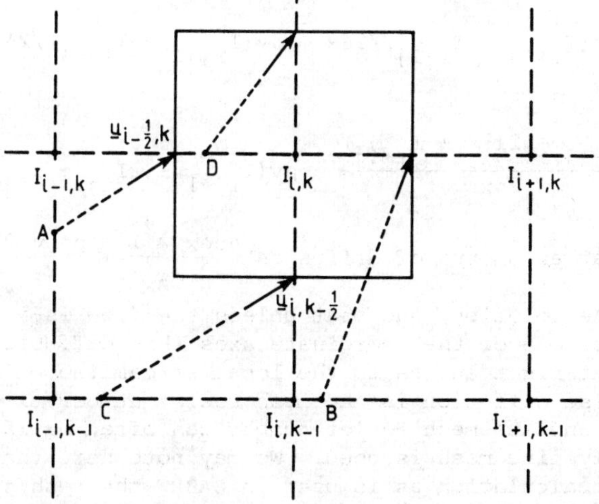

Fig. 2 VUDS location of upwind values.

$$J \simeq \{\rho u_{i+\frac{1}{2}j} I_{i+1j} - (\rho u_{i+\frac{1}{2}j} - \rho u_{i-\frac{1}{2}j}) I_{ij} - \rho u_{i-\frac{1}{2}j} I_{i-1j}\}/2\delta x$$

$$+ \{\rho v_{ij+\frac{1}{2}} I_{ij+1} - (\rho v_{ij+\frac{1}{2}} - \rho v_{ij-\frac{1}{2}}) I_{ij} - \rho v_{ij-\frac{1}{2}} I_{ij-1}\}/2\delta y$$

For a constant velocity field this reduces to the central difference form

$$\rho u(I_{i+1j} - I_{i-1j})/2\delta x + \rho v(I_{ij+1} - I_{ij-1})/2\delta y$$

and we observe that only alternate energy nodes are coupled by this term, thus giving rise to the possibility of chequerboard oscillations in the solution.

A common way of overcoming this problem, and one that we have used in SABRE, is to use upwind (or donor) differencing. Taking the velocity components u and v as positive, we find that for this approximation

$$J \simeq (\rho u_{i+\frac{1}{2}j} I_{ij} - \rho u_{i-\frac{1}{2}j} I_{i-1j})/\delta x + (\rho v_{ij+\frac{1}{2}} I_{ij} - \rho v_{ij-\frac{1}{2}} I_{ij-1})/\delta y$$

which reduces to

$$\rho u(I_{ij} - I_{i-1j})/\delta x + \rho v(I_{ij} - I_{ij-1})/\delta y$$

for a constant velocity field. But this may be written as

$$\rho u(I_{i+1j} - I_{i-1j})/2\delta x + \rho v(I_{ij+1} - I_{ij-1})/2\delta y$$

$$- \frac{\rho u(I_{i-1j} - 2I_{ij} + I_{i+1j})}{2\delta x} - \rho v(I_{ij-1} - 2I_{ij} + I_{ij+1})/2\delta y$$

showing that an amount of diffusion $\frac{\rho u\delta x}{2}\frac{\partial^2 I}{\partial x^2} + \frac{\rho v\delta y}{2}\frac{\partial^2 I}{\partial y^2}$ has been added to the solution, and that unless the flow is in the direction of one of the coordinate axes this diffusion acts both along and at right angles to the local streamline. The magnitude of this false diffusion is the mesh Peclet number times the true diffusion, and the mesh Peclet number can often be large unless an extremely fine mesh is used. We may note here that with a subchannel calculation as is used in SABRE the mesh size is fixed and a finer mesh cannot be used to remove these false diffusion effects.

Now a large stream Peclet number means that convection processes dominate diffusion processes, and so in such cases the effects of true diffusion will generally have a small effect on the answer, but this is not always the case. In particular consider a uniformly rotating body of fluid containing a heat source and surrounded by a constant temperature boundary. In this case there is no net transport of energy along the stream lines, all the energy transport is by diffusion normal to the streamlines. A solution to this problem using upwind differences and a finite difference mesh resulting in a significant Peclet number would clearly give meaningless results.

To overcome this difficulty, and at the same time avoid the problem of oscillatory solutions through decoupling of the equations, the vector upwind difference scheme has been developed. In such schemes, the one we use being illustrated in Fig. 2, instead of using the nodal energy value obtained by moving upwind normal to the face, an interpolated value obtained by moving upwind in the direction of flow at the face is used. The scheme illustrated in Fig. 2 still results in false diffusion, but for an equal mesh this is now a magnitude

$$\frac{\rho\delta}{2}\left[(u + v)\frac{\partial^2 I}{\partial s^2} + (v - u)\frac{\partial^2 I}{\partial s\partial n}\right],$$

where s and n indicate coordinates along and perpendicular to the flow vector. We thus observe that with this scheme there is no false diffusion normal to the streamline. There is however a cross term; it is possible to reduce the error term to pure diffusion along the streamline by an alternative choice of upwind interpolation, but in practice the results of alternative schemes do not differ greatly from the one presented here.

Our experience using this scheme has been very satisfactory, but it is possible for pathological cases to arise in which an energy node becomes decoupled; this appears to happen only if all outflows from a control volume are very oblique to the control volume surfaces.

A further improvement in the truncation error is obtained for the case of a spatially variable source by replacing the vector upstream value, I_A say, by

$$I_A + \frac{S_{i-1k}\,\delta x}{\rho|u|_{i-\frac{1}{2}k}} .$$

This correction is based on the argument that $\rho u \partial I/\partial s \approx S$.

In SABRE we have also applied the VUDS scheme to momentum transport, but to date our experience is rather limited. However we have found significantly more difficulty in converging VUDS than UDS calculations. A further point of interest we have observed concerns the flow entering a sudden contraction; with upwind differences we calculate that the flow adheres to the side wall after the contraction, whereas with the VUDS scheme we find the flow separates and a recirculating region forms adjacent to the wall. Further discussion of the VUDS scheme is given in Lillington (1981).

4. BOILING MODELLING

In SABRE, boiling is represented using a homogeneous model with slip. This model was chosen as it was believed to give an adequate physical representation of the boiling phenomena, and could be implemented using the existing structure of the code. With this model the thermohydraulic equations can be expressed in a similar form to the single phase equations, and can be solved using the same procedures as for the single phase case. As the modelling problems discussed here are not connected with slip, we will consider the problem of homogeneous boiling without slip, in which the mass, momentum and energy equations are unaltered from the single phase fluid case, but in which the equation of state becomes

$$\rho = \rho_L(I,p) \qquad I \le I_L$$

$$\rho = \frac{\rho_{VS}\rho_{LS}(I_V - I_L)}{\rho_{LS}(I - I_L) + \rho_{VS}(I_V - I)} \qquad I_L \le I \le I_V$$

$$\rho = \rho_V(I,p) \qquad I_V \le I$$

where $I_L = I_L(p)$ and $I_V = I_V(p)$ are the local liquid and vapour saturation energies, and $\rho_{VS} = \rho_V(I_V,p)$, $\rho_{LS} = \rho_L(I_L,p)$. The density defined by this equation of state is a continuous function of I and p, but it has a large discontinuity of gradient at $I = I_L(p)$ and $I = I_V(p)$, the boundaries between single and two phase flow (see Fig. 3a); at normal (a few bars) pressures, the vapour density ρ_V will typically be a factor 10^3 small than ρ_L.

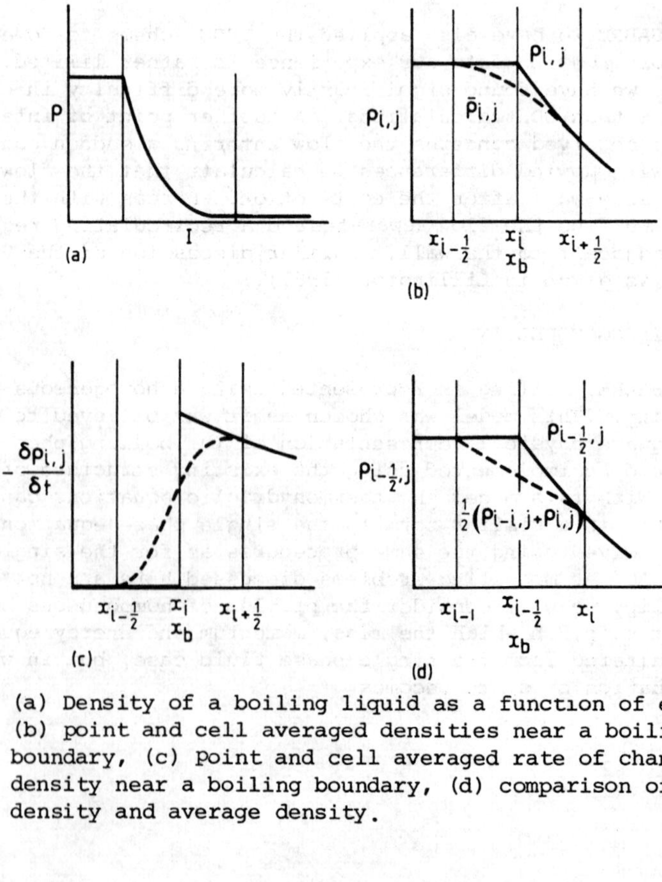

Fig. 3 (a) Density of a boiling liquid as a function of energy, (b) point and cell averaged densities near a boiling boundary, (c) Point and cell averaged rate of change of density near a boiling boundary, (d) comparison of density and average density.

To illustrate the problems that arise with this boiling model, consider first the mass conservation equation

$$\frac{\partial \rho}{\partial t} = -\underline{\nabla}.\rho\underline{u} .$$

We obtain a conservative form for the difference equations by integrating this equation over a control volume, and in two dimensions we obtain

$$\frac{\partial \rho_{ij}}{\partial t} = \rho u_{i-\frac{1}{2}j} - \rho u_{i+\frac{1}{2}j} + \rho v_{ij-\frac{1}{2}} - \rho v_{ij+\frac{1}{2}} .$$

When all the variables and their first derivatives are continuous this is a reasonably accurate approximation. However in the boiling case where the spatial derivatives of the density are discontinuous at the two phase boundaries, this approximation gives rise to problems which manifest themselves in the solution at best as discontinuities in the time rates of change of the variables, and at worst result in a discontinuous solution in time even if a solution can be obtained; these discontinuities appear whenever the boiling boundary passes a node.

Let us consider the problems in detail. The term $\frac{\partial \rho_{ij}}{\partial t}$ is an approximation to $\frac{1}{V}\frac{\partial}{\partial t}\int \rho dV$. We consider what happens if a discontinuity in the density gradient parallel to the y axis moves across the control volume at velocity u. Then the nodal value at the centre of the control volume is given by

$$\rho_{ij} = \begin{cases} \rho_o & t < t_o + \dfrac{x_i - x_{i-\frac{1}{2}}}{u} \\[2ex] \rho_o + \alpha[x_i - x_{i-\frac{1}{2}} - u(t - t_o)] & t > t_o + \dfrac{x_i - x_{i-\frac{1}{2}}}{u} \end{cases}$$

if the density gradient $\partial\rho/\partial x$ is zero ahead of the discontinuity and α behind it. However the control volume averaged value is given by

$$\bar{\rho}_{ij} \equiv \frac{1}{V}\int \frac{\partial \rho}{\partial t}\, dV = \begin{cases} \rho_o & t < t_o \\[2ex] \rho_o - \dfrac{\alpha u^2 (t-t_o)^2}{2(x_{i+\frac{1}{2}} - x_{i-\frac{1}{2}})} & t_o < t < t_o + \dfrac{x_{i+\frac{1}{2}} - x_{i-\frac{1}{2}}}{u} \\[2ex] \rho_o + \alpha[\frac{1}{2}(x_{i+\frac{1}{2}} - x_{i-\frac{1}{2}}) - u(t-t_o)] & t_o + \dfrac{x_{i+\frac{1}{2}} - x_{i-\frac{1}{2}}}{u} < t. \end{cases}$$

Figs. 3b and 3c illustrate the behaviour of ρ_{ij} and $\bar{\rho}_{ij}$ and their time derivatives. In SABRE we estimate the position of the boiling boundary along the axis of the subchannel and make a correction to $\partial\rho_{ij}/\partial t$ allowing for the axial variation of density only. In this manner we remove the discontinuity from $\partial\rho_{ij}/\partial t$ and thus alleviate the numerical problems considerably. We believe that removing this discontinuity from the finite difference representation of the rate of change of density is the most important single improvement that can be made to the finite difference equations; if it is not done, no flow out of a control volume can take place until half the control volume is boiling.

There could be a case for making similar corrections to the time rates of change of momenta and energy, but it is believed this would have a much smaller effect, although further investigation of this point is required.

Having dealt with the time derivative, let us consider the remaining terms. Here two points arise, first the representation of the surface integral by the centre node value, and secondly the definition of the density at the central node. We have done nothing about the former, but have found it necessary to take care with the definition of the density. For a single phase calculation it is quite adequate to define the density at a velocity node as an average density, for example

$$\rho_{i-\frac{1}{2}j} = \tfrac{1}{2}(\rho_{i-1j} + \rho_{ij}) \ .$$

However as we see in Fig. 3d this gives the incorrect behaviour of $\rho_{i-\frac{1}{2}j}$ as the boiling boundary moves from (i-1j) to (i,j). We have overcome this by taking

$$\rho_{i-\frac{1}{2}j} = \rho(I_{i-\frac{1}{2}j},\ p_{i-\frac{1}{2}j})$$

where the energy and pressure values are obtained by interpolation: thus

$$I_{i-\frac{1}{2}j} = \tfrac{1}{2}(I_{i-1j} + I_{ij}),\ p_{i-\frac{1}{2}j} = \tfrac{1}{2}(p_{i-1j} + p_{ij})$$

in a region of constant heat source, but for a spatially variable heat source this variation is allowed for in calculating $I_{i-\frac{1}{2}j}$. This gives an approximation to $\rho_{i-\frac{1}{2}j}$ which has the correct qualitative behaviour, and is used consistently for all the difference equations in SABRE.

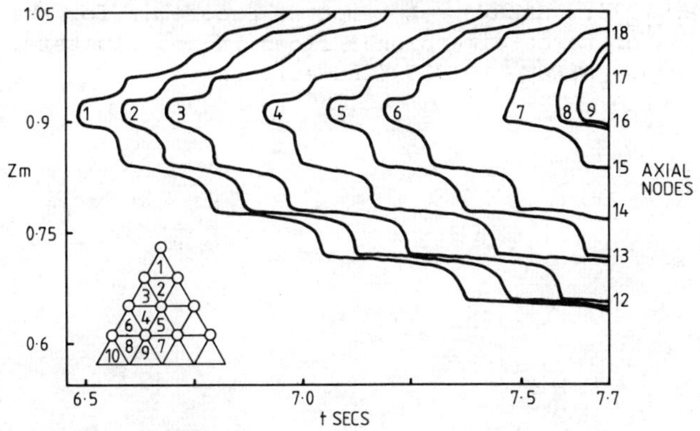

Fig. 4 Development of boiling region in 37 pin fuel element.

Fig. 4 shows the development of a boiling region in a 60° sector of a 37 pin bundle as calculated by SABRE. We observe we have obtained a continuous solution but there is a tendency for the boiling boundary to move faster through one half of the control volume than through the other half. Work is continuing to attempt to improve the representation.

As a final comment, we note that the above calculation has been carried out using an equilibrium thermodynamics model. The use of a non-equilibrium model, which is probably more correct physically, improves the stability of the boiling calculations. In SABRE we have so far only investigated a simple time constant non-equilibrium model, and this has been implemented in an Eulerian frame, rather than in the more physical Lagrangian frame.

5. REFERENCES

HARLOW, F.H. & AMSDEN, A.A. 1971. A numerical fluid dynamics calculation method for all flow speeds. *J. Comput. Phys.* **8**, 197.

HIRT, C.W. 1968. Heuristic stability theory for finite difference equations. *J. Comput. Phys.* **2**, 339.

LILLINGTON, J.N. 1981. An assessment of vector upstream differencing scheme predictions for the transport of a scalar in unidirectional and recirculating flow regimes. AEE Winfrith Rep. AEEW - M 1833.

PATANKAR, S.V. & SPALDING, D.B. 1972. A calculation procedure for heat mass and momentum transfer - three dimensional parabolic flows. *Int. J. Heat and Mass Trans.*, **15**, 1787.

POTTER, R. 1979. SABRE1 - A computer programme for the calculation of three-dimensional flows in rod clusters. AEE Winfrith Rep. AEEW - R 1057 (2nd edition).

FINITE ELEMENT METHOD WITH CLEBSCH REPRESENTATION

E. Detyna

(Department of Mathematics, University of Reading)

1. INTRODUCTION

The aim of this paper is to describe a Finite Element Method (FEM) for solving ideal compressible inviscid fluid flows. Developing any such method involves three parts:

(a) choosing a variational principle from which flow equations can be derived;

(b) choosing the space of trial functions to approximate the solution;

(c) adopting a method of solving the resulting non-linear equations.

These three parts are not entirely independent since the choice of variational principle to some extent predetermines the choice of finite elements and the method of solving the non-linear equations thus obtained.

A non-linear set of equations may have discontinuous solutions (weak solutions), and in general these weak solutions are not unique but depend on the particular formulation of the equations. The solutions on either side of a discontinuity have to be joined by Rankine-Hugoniot conditions which are obtained by integrating the conservation laws over a slim volume containing the discontinuity. It is clear that one particular form of equations, namely conservation form, is most convenient. Of course, in regions where solutions are smooth any equivalent form of equations will produce a unique solution. But since the conservation form of equations has to be used near discontinuities it may as well be used in the whole region. This is the reason why most numerical schemes for solving fluid flows use Euler's equations in conservative form. The finite difference methods take as a starting point the differential equations and discretise them in an appropriate manner. Hence it is natural to discretise the equations in the conservative form.

The finite element methods differ from finite difference methods in that they require some form of variational principle as a starting point. Now we have a choice of variational principles: one method is to seek weak solutions by the Petrov-Galerkin method, with as many variational principles as there are differential equations. The disadvantage of this method is that the choice of a test space for non-self adjoint or non-linear equations is not clear.

A more satisfactory approach when it is possible is as follows: a physical problem, like perfect fluid flow, has a physical variational principle and all the equations of motion are derivable from the one principle. These equations are not necessarily in conservation form, but in addition to the equations of motion the conservation laws are also derivable from the variational principle with the aid of Noether's theorem. Thus the equations of motion can be used in regions of smooth flow while Rankine-Hugoniot conditions obtained from conservation laws can be used for "joining" discontinuities. The disadvantage of this method is that discontinuities have to be dealt with separately from the rest of the solution. But, on the other hand, the equations of motion are in a simple canonical form which should be easier to solve than the conservation form.

In order to demonstrate the method an FEM for one-dimensional steady flow through a pipe with varying diameter is described in this paper. The main feature of the method is the use of the same equations for both supersonic and subsonic flow without addition of any numerical viscosity.

2. VARIATIONAL PRINCIPLE

It has been shown, Detyna (1981), that the steady flow of a perfect fluid without spin can be described by the variational principle

$$\delta A = \delta \int_{\Omega} d_3x \ L = 0 \tag{2.1}$$

where the Lagrangian L is

$$L = \rho[H - \tfrac{1}{2}(\nabla\phi + S\nabla\lambda + H\nabla\beta)^2 - E(\rho,S)]: \tag{2.2}$$

here H is the total enthalpy and $E(\rho, S)$ is the internal energy; ρ and S are mass density and entropy, respectively. The functions ϕ, λ and β are potentials without any particular physical meaning, but they define the velocity

$$\underline{v} = -(\nabla\phi + S\nabla\lambda + H\nabla\beta). \tag{2.3}$$

The equations of motion follow from (2.1) by taking variations with respect to each variable in turn:

$$\delta\phi : \quad \nabla\cdot(\rho\underline{v}) = 0, \tag{2.4}$$

$$\delta\lambda : \quad \nabla\cdot(\rho S\underline{v}) = 0, \tag{2.5}$$

$$\delta\beta : \quad \nabla\cdot(\rho H\underline{v}) = 0, \tag{2.6}$$

$$\delta\rho : \quad \tfrac{1}{2}\underline{v}^2 + h = H, \tag{2.7}$$

$$\delta S : \quad \underline{v}\cdot\nabla\lambda = T, \tag{2.8}$$

$$\delta H : \quad \underline{v}\cdot\nabla\beta = 0, \tag{2.9}$$

where $h = \partial(\rho E)/\partial\rho$ and $T = \partial E/\partial S$ are specific enthalpy and temperature, respectively. The first three equations describe the conservation of mass, entropy and energy.

Since the Lagrangian (2.2) is invariant with respect to translation of the coordinates, the conservation of momentum is also induced, Detyna (1981):

$$\nabla\cdot(\rho\underline{v}\underline{v} + \underline{\underline{I}}p) = 0 \tag{2.10}$$

where $p \equiv L$ is pressure.

The equations (2.4) and (2.9) are used in the regions of smooth flow while conservation laws (2.4), (2.6) and (2.10) determine conditions at discontinuities, where entropy S is no longer conserved; hence equation (2.5) is replaced by a physical requirement of non-reversibility

$$\Delta S \geq 0$$

across the discontinuity. (It should be noted that there are only five independent physical variables ρ, S and $\underline{v}$ and hence the six conservation equations including (2.5) are overdetermined.)

In order to show how the FEM can be applied to this problem, one-dimensional flow through a pipe with varying diameter is studied. Let us suppose the flow is in the x-direction: then the integration in (2.1) over y- and z-coordinates can be performed since no functions depend on y or z. Furthermore, the one-dimensional flow is necessarily homoenergic $H = H_0$ and the variable β can be transformed out by replacing ϕ with $\phi - H_0\beta$.

Thus the variational principle (2.1) now reads

$$\delta A = \delta \int dx \; [H_O - \tfrac{1}{2}(\phi_x + S\lambda_x)^2 - E(\rho,S)]\rho R = 0 \tag{2.11}$$

where $R = R(x)$ is the cross-section of the pipe at point x.

This variational principle differs from the previous one in that it is explicitly dependent on the x-coordinate through the function $R(x)$. Consequently, the momentum is no longer conserved and we have instead an equation

$$(\rho R v^2 + Rp)_x = pR_x \tag{2.12}$$

which describes the rate of change of momentum with pdR/dx as its source.

The equations of motion can be obtained from the variational principle (2.11) in a standard way and are not given here.

3. FINITE ELEMENTS

The Lagrangian in (2.11) depends on functions ρ and S and first derivatives of ϕ and λ: hence the simplest trial functions are piecewise constant for ρ and S and piecewise linear for ϕ and λ. Let us divide the computational region (x_O, x_N) into N elements with nodal values at $x_O, x_1, \ldots, x_N$. The nodal values of ϕ and λ are denoted by ϕ_i and λ_i, $i=0,1,\ldots N$; the constant values of ρ and S on the element i defined as (x_{i-1}, x_1) are denoted by ρ_i, S_i, $i=1,2,\ldots,N$.

With such trial functions the action A in (2.11) is

$$A = \sum_{i=1}^{N} \rho_i R_i \left\{ H_O - \tfrac{1}{2} \left(\frac{\phi_i - \phi_{i-1} + S_i(\lambda_i - \lambda_{i-1})}{\Delta_i} \right)^2 - E_i \right\} \Delta_i \tag{3.1}$$

where $\Delta_i = x_i - x_{i-1}$ and R_i is the integral

$$R_i = \Delta_i^{-1} \int_{x_{i-1}}^{x_i} R(x) \, dx$$

The approximate equations are now obtained by minimizing action A with respect to all the nodal values ϕ_i, λ_i and constants ρ_i, S_i:

$$\frac{\partial A}{\partial \rho_i} = [H_O - \tfrac{1}{2}v_i^2 - h_i]R_i\Delta_i = 0 \tag{3.2}$$

$$\frac{\partial A}{\partial S_i} = [v_i(\lambda_i - \lambda_{i-1}) - T_i\Delta_i]R_i\rho_i = 0 \tag{3.3}$$

$$\frac{\partial A}{\partial \phi_i} = \rho_i R_i v_i - \rho_{i+1}R_{i+1}v_{i+1} = 0 \tag{3.4}$$

$$\frac{\partial A}{\partial \lambda_i} = \rho_i R_i S_i v_i - \rho_{i+1}R_{i+1}S_{i+1}v_{i+1} = 0, \tag{3.5}$$

where $v_i = (\phi_i - \phi_{i-1} + S_i(\lambda_i - \lambda_{i-1}))/\Delta_i$ is the velocity of the fluid in element i.

The equations (3.2) to (3.5) are non-linear and we propose to solve them by iteration. It has been noted, Buneman (1981), Detyna (1981, 1982), that the variables ρ and ϕ, ρS and λ are conjugate in the Hamiltonian sense. Therefore it is natural to solve the above equations in appropriate pairs - (3.2) with (3.4) and (3.3) with (3.5). Let us deal with the first pair of equations. The equation (3.2) depends on a single value ρ_i (through h_i) and two values of ϕ: ϕ_{i-1} and ϕ_i (through v_i). On the other hand, the equation (3.4) depends on three values of ϕ: ϕ_{i-1}, ϕ_i and ϕ_{i+1}, and two values of density: ρ_i and ρ_{i+1}.

Therefore, in the subsonic region, we can regard equation (3.2) as a definition of ρ_i, while equation (3.4) is

$$-\frac{\rho_i R_i}{\Delta_i}\phi^{i-1} + \left(\frac{\rho_i R_i}{\Delta_i} + \frac{\rho_{i+1}R_{i+1}}{\Delta_{i+1}}\right)\phi^i - \frac{\rho_{i+1}R_{i+1}}{\Delta_{i+1}}\phi^{i+1} = -\frac{\rho_i R_i T_i S_i}{v_i} + \frac{\rho_{i+1}R_{i+1}T_{i+1}S_{i+1}}{v_{i+1}}, \tag{3.6}$$

an elliptic "three-point" formula for ϕ^i, where a superscript denotes a new iterative value.

In the supersonic region, the equations (3.2) and (3.4) are regarded as equations for ϕ and ρ, respectively:

$$\delta\phi^{i} = \delta\phi^{i-1} + (\tfrac{1}{2}v_i^{\,2} + h_i - H_O)\,\Delta_i/v_i \tag{3.7}$$

$$\rho^{i+1} = \rho^{i}(R_i v_i/R_{i+1}v_{i+1}), \tag{3.8}$$

where equation (3.7) was obtained by assuming a new approximation for ϕ_i to be $\phi^i = \phi_i + \delta\phi^i$ and linearising (3.2) with respect to $\delta\phi^i$. The equations (3.7) and (3.8) are "marching algorithms". The two methods of solution of equations (3.2) and (3.4) are well suited for the boundary conditions: for subsonic flow equation (3.6) requires Dirichlet or Neumann conditions while equations (3.7) and (3.8) for supersonic flow require Cauchy conditions.

The second pair of equations, (3.3) and (3.5), are different in that the only boundary conditions that can be given are the Cauchy ones, say λ_O and S_1. This is because it is physically inappropriate to prescribe values of entropy at both ends of the flow. Hence the equations are solved by the marching algorithm in an obvious manner. In the case of transonic flow, equation (3.5) is replaced at the point of the shock by a formula giving the increase of entropy from the conservation laws.

Let us suppose the inflow is supersonic while outflow is subsonic with full boundary conditions (ρ_1, v_1 and S_1) at the inflow and one other boundary condition (say v_N or p_N) at the outflow. These boundary conditions overdetermine the flow if all the equations (3.2) to (3.5) are used. But a careful study of the above described method of solution shows that (3.8) for $i = j$, where x_j is the shock point, cannot be used since the element $j + 1$ is already in the subsonic region and ρ_{j+1} should be calculated from equation (3.2) for $i = j + 1$. That means the mass is not conserved across the shock unless it is moved to a new position - a procedure equivalent to a calculation of shock speed in time dependent problems. A linear approximation to a new nodal position x^j which preserves mass gives

$$x^{j} = x_j + s\delta\tau \tag{3.9}$$

where

$$\delta\tau = \begin{cases} \Delta_j/v_j & \text{for } s > 0 \\ \Delta_{j+1}/v_{j+1} & \text{for } s < 0 \end{cases}$$

and $$s = (\rho_{j+1}v_{j+1}R_{j+1} - \rho_j v_j R_j)/(\rho_{j+1}R_{j+1} - \rho_j R_j).$$

It is seen that the equation (3.9) can be interpreted as the "movement" of the shock with speed s in time $\delta\tau$. This interpretation is indeed justifiable since, say equation (3.7), can be derived from time dependent upwind equations with variable time step $\Delta t_i = \Delta_i/v_i$ as in (3.9). Notwithstanding the interpretation, the equation (3.9) can be used either for calculation of a new nodal position $x = x^j$ of the shock (shock fitting) or for indicating the element in which the shock occurs without recal-recalculating the new nodal values (shock capture).

4. NUMERICAL RESULTS

In order to test the method, flow in various pipes with different boundary conditions was studied. As an example we show in detail the solution of the flow in the pipe shown in Fig. 1 with inlet conditions $v_1 = 1.4$, $\rho_1 = 1.0$ and $S_1 = 0.0$; at the outlet the one condition is $v_N = 0.4$ or $p_N = 1.7$. (The above values are in the standard units normalised to the sound velocity at the inlet.) According to the exact solution, the condition at the outlet results from a shock at $x_s = 3.8619$.

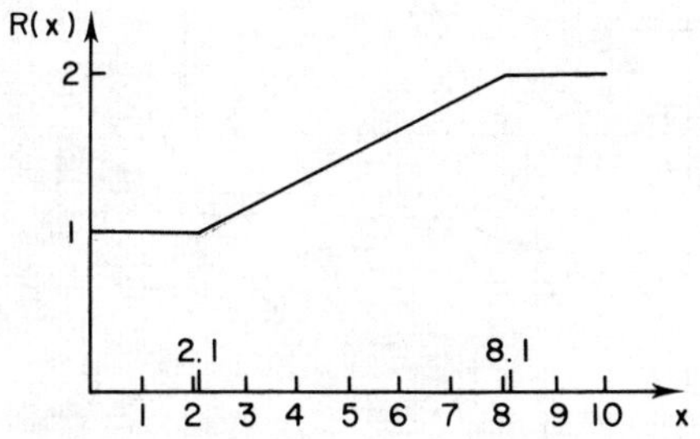

Fig. 1 Cross-section R(x) of the pipe studied numerically.

We note that the method requires a boundary condition for ϕ (or velocity) at the outlet rather than pressure, but these two values are connected through the formula

$$p = \rho(H_0 - \tfrac{1}{2}v^2 - E)$$

stating that the Lagrangian is the pressure. In order to start the iteration procedure the following distributions of density, velocity and entropy were chosen:

$$\rho_i = \begin{cases} 1.0 \text{ for } x_i < 7.0 \\ 1.8 \text{ for } x_i > 7.0 \end{cases}$$

$$v_i = \begin{cases} 1.4/\rho_i R_i & \text{for } i < N \\ 0.4 & \text{for } i = N \end{cases}$$

$$S_i = 0.0$$

as shown in Fig. 2 together with the exact solution. Such a trial solution ensures the shock position at the beginning of iterations is far removed from its exact location.

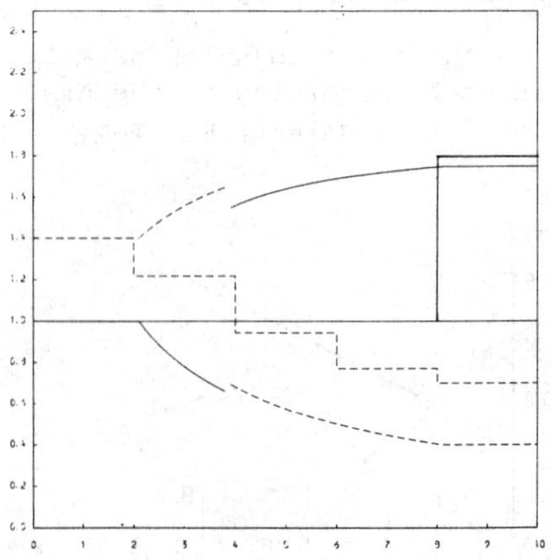

Fig. 2 The trial and exact solutions for density ρ (continuous line) and velocity v (broken line). The trial solution is the step function.

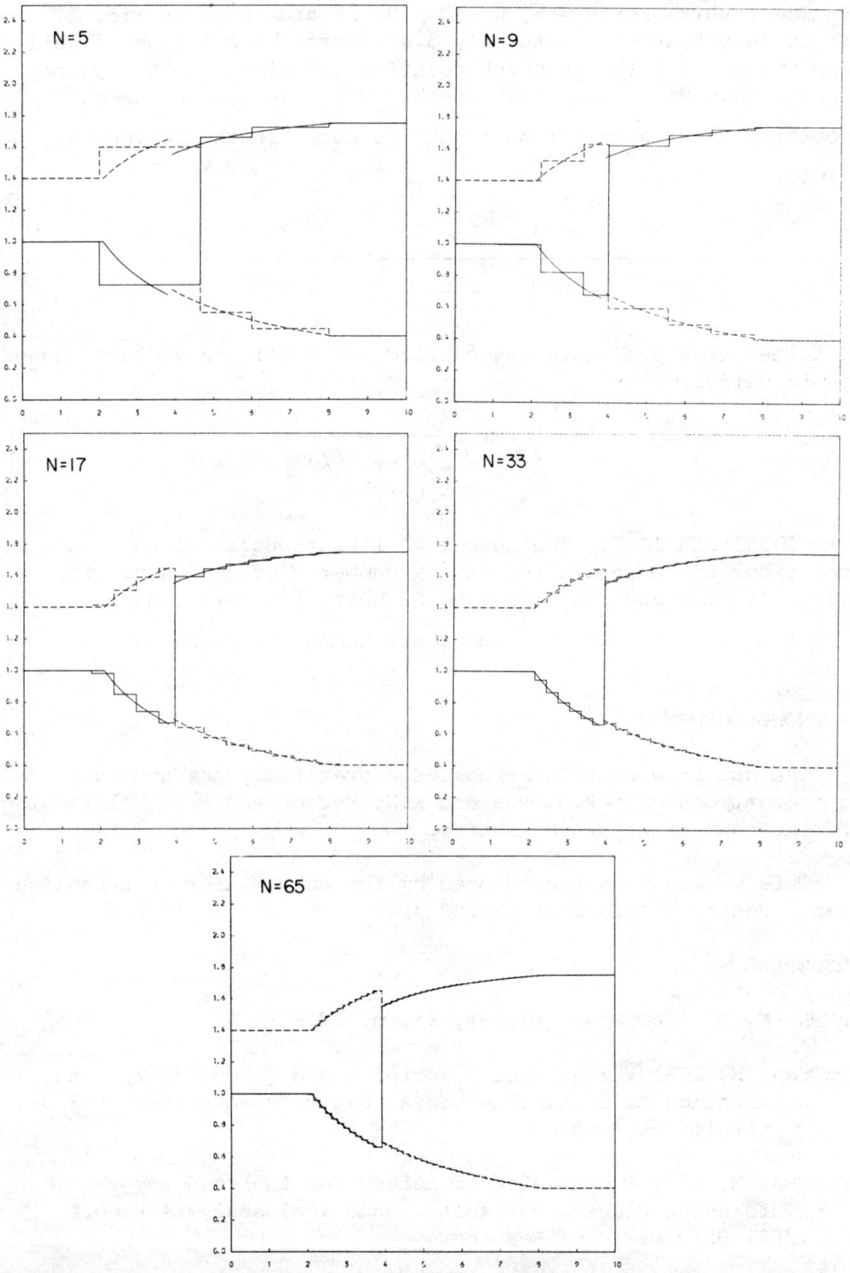

Fig. 3 The approximate and exact solutions for density ρ (continuous line) and velocity v (broken line). The approximate solution is the step function. The number N of elements used is shown on each figure.

The results for N = 5, 9, 17, 33, 65 are shown in Fig. 3. It is seen that even with only 5 elements the solution is well approximated, with the shock position calculated with accuracy better than 8% of the tube length. The error in the shock position $\delta x = |x_s - x_s^E|$/tube length is shown in the following table:

N	5	9	17	33	85
δx	0.078	0.012	0.0093	0.0068	0.0053

The iterative procedure was carried out until the average change δv in velocity

$$\delta v = \sqrt{\sum_i |v^i - v_i|^2 / N}$$

was less than 10^{-8}. The number of iterations required was of the order of 20 regardless of the number N of elements. No significant change in solutions is noted if iterations are terminated when $\delta v \leq 10^{-6}$ but their number decreases by a quarter.

ACKNOWLEDGEMENTS

The author wishes to acknowledge gratefully the numerous conversations with P.L. Roe and K.W. Morton and critical reading of the manuscript by the latter.

This work has been sponsored by the Royal Aircraft Establishment, Research Agreement AT/2035/042.

REFERENCES

BUNEMAN, O. 1980 *Phys. Fluids*, **23**, p. 1716.

DETYNA, E. 1981 Variational principles and gauge theory. An application to continuous media, Fluid Dynamics Report 2/81, University of Reading

DETYNA, E. 1982 Finite element method for inviscid compressible fluid using Clebsch variables, Numerical Analysis Report 1/82, University of Reading.

A GENERAL PURPOSE PROGRAM FOR THE ANALYSIS OF FLUID FLOW PROBLEMS

N.C. Markatos, N. Rhodes and D.G. Tatchell

(Concentration Heat and Momentum Limited, London)

1. INTRODUCTION

1.1 The General Approach to Fluid-Flow Modelling

Fluid-flow and heat and mass transfer play a large part in engineering, particularly in power-production equipment chemical-process plant, aircraft, ship and automobile design, nuclear plant and environmental problems. The engineering designer requires more detailed information than ever before on the fluid motion and thermal conditions of his equipment in order to meet increasingly stringent specification of efficiency, safety and economy.

The present paper reports on a general computer program which can be used to predict the flow and heat transfer in a wide range of engineering situations, and its application to two specific problems. The general computer program, PHOENICS (Spalding, 1981a), is designed to solve the differential equations for three-dimensional, transient, compressible, two-phase turbulent flows and heat transfer: namely, continuity, momenta, energy, chemical species concentration, turbulence energy, and turbulence dissipation rate. The simultaneous solution of these equations by means of a finite-domain solution algorithm yields the values of the variables at all internal grid nodes. Full account is taken of interphase interaction processes.

The general-purpose code utilises a core-program which is common to all applications, and a user-input section which allows the program to be easily adapted to particular problems. The advantages of the generalised approach to fluid-flow modelling are as follows:

- The general code provides a facility for building complex models on a good foundation which, being in constant use, is thoroughly tested. The engineer can therefore concentrate on the physical realism of his model, and does not need to provide suitable solution algorithms for each problem considered.

- The code is arranged so that new problems can be set up quickly, and solutions rapidly obtained, particularly if the simplest modelling assumptions are used in the first instance. Modelling details can be refined as experience is gained with a new problem.

- Internal aspects of the code can be more easily maintained and improved than would be the case if a number of specific codes were used.

- New developments or improvements are immediately accessible to all users.

1.2 Purpose of the Paper

The purpose of the paper is to provide a brief description of the main features of the PHOENICS program, and to describe two recent applications, namely:

- Explosion Containment in a Fast-Breeder Reactor: A three-dimensional, two-phase problem (the phases being separated), where as a consequence of an accident leading to overheating, boiling occurs, and the reactor coolant is abruptly set in motion. Forces on the vessel and its internal structures and the behaviour of the expanding gas-bubble are determined.

- Icing of an air-launched torpedo: The so-called Stefan problem occurs in a number of engineering situations. The code has been used to predict the results of an experimental simulation of the torpedo icing problem in which water freezes on a cold plate and subsequently melts as the plate increases in temperature.

1.3 Outline

Of the remaining four sections of the paper, the next section (Section 2) describes the general features of the code, the equations and solution method. Sections 3 and 4 describe the two applications of the code and concluding remarks are made in Section 5.

2. THE COMPUTER CODE

2.1 General Features

The general-purpose, fluid-dynamics computer code system PHOENICS (Parabolic, Hyperbolic or Elliptic Numerical Integration

Code Series) has a common central core program capable of simulating flow processes which are:

- one-, two- or three-dimensional;
- one- or two-phase;
- steady or transient;
- elliptic or parabolic;
- laminar or turbulent;
- subsonic, supersonic or transonic; and
- chemically reacting or inert.

The central core-program - named EARTH - is the main equation solver. The EARTH program is used in all flow simulations and therefore to ensure that generality and reliability can be maintained, it is not accessible to the user.

In order that the engineer can build a model of the flow in question, by activating the capabilities of the EARTH-program, a user-SATELLITE-program is provided. The SATELLITE is divided into two sections, the first section includes a BLOCK DATA subroutine, in which the necessary variables to define the problem are inserted; a PORDAT subroutine, which is required if non-regular geometries are to be specified by blockage factors; and a FLDDAT subroutine, where initial fields are specified. The data contained in this section are transmitted to the EARTH program at the beginning of each run.

The second section of the 'SATELLITE' program is required if additional information is to be provided which requires interaction with the iterative calculation. This 'GROUND' subroutine is linked to EARTH and called at various points in the computational cycle. Inputs provided through GROUND might include, for example, particular physical property formulations, alternative turbulence modelling data, flow-regime selection criteria or correlations for interphase heat and mass transfer.

2.2 Variables

Dependent and independent variables are specified by the user in the BLOCK DATA subroutine. The independent variables are x, y and z if a cartesian coordinate system is used, x, r and θ if polar coordinates are required. For unsteady flows, time, t, is also an independent variable.

There are twenty-five dependent variables which can be selected by the user:

- pressure p and the pressure correction p';
- three first-phase velocities, u_1, v_1, w_1;
- three second-phase velocities, u_2, v_2, w_2;
- phase volume fractions r_1, r_2 (and r_2^*);
- turbulence energy and dissipation rate, k and ε;
- first- and second-phase enthalpies, h_1 and h_2;
- enthalpies of a third fluid, h_3;
- four concentrations, C_1, C_2, C_3, C_4;
- three radiation-flux sums, R_x, R_y, R_z; and
- two spares.

Solution is obtained only for those variables selected in BLOCK DATA.

2.3 Governing Equations

The equations for all variables have the common form:

$$\frac{\partial}{\partial t}(r\rho\phi) + \text{div}\,(r\rho\vec{V}_\phi - r\Gamma_\phi \text{grad}\phi) = rS_\phi$$

transient convection diffusion source

where: r $\equiv$ phase volume fraction; $\rho \equiv$ density;

$\phi \equiv$ dependent variable; $\vec{V} \equiv$ velocity vector;

$\Gamma_\phi \equiv$ exchange coefficient (laminar or turbulent); and

$S_\phi \equiv$ source or sink term.

Integration of the equations over a staggered grid gives a set of finite-domain equations of the general form:

$$\phi_P = \frac{\Sigma_i\, a_i\, \phi_i + b}{\Sigma_i\, a_i - c}$$

where ϕ is the dependent variable and Σ_i denotes summation over the neighbouring grid points. The coefficients a_i represent the effects of convection and diffusion and $(b+c\phi_p)$ contains the transient term, and, for the momentum equations, the pressure term and integrated source term for the cell. The use of upwind differencing in evaluating the convection terms ensures that the a_i coefficients are always positive, so ensuring numerical stability.

2.4 Solution Methods

The finite-domain equations are solved using an iterative guess-and-correct procedure which is fully-implicit in transient cases, and based on the SIMPLE algorithm (Patankar and Spalding, 1972). For two-phase-flow calculations the IPSA algorithm of Spalding (1980, 1981b) is used.

A novel feature of the program, and one which has a significant influence on its ability to handle fine-grid problems, is the order of visitation which is selected when variables at cell-centres are up-dated.

The order chosen involves what are called 'repeated z-direction sweeps' through the integration domain. The whole set of cells is regarded as consisting of one-cell-thick 'slabs', extending in the x and y directions, and piled one on top of the other in the z-direction.

A single sweep therefore starts with attention being paid to the bottom (low-z) slab of cells. The finite-domain equations are solved for all the cells in this slab, the values of ϕ's at the next-higher slab being regarded as known. Attention then passes to the second slab, the ϕ-values there being adjusted by reference to those in the slabs both above and below. Then the next-higher slab is attended to; and so on, until the adjustment sweep has been completed.

At the end of a sweep, of course, the final solution has not yet been achieved; so the process must be repeated until a reasonable degree of convergence is achieved.

When the flow is a transient one, the same iterative-sweep procedure is employed. However, the number of sweeps needed to achieve convergence at a particular time-step reduces greatly with the magnitude of the time interval; when this is very small, a single sweep is all that is required.

As well as the full, elliptic, iterative solution procedure, EARTH contains as a user-activated option a marching integration, parabolic solution procedure. This proceeds in the slab-by-slab manner described above, except that the boundary-layer form of the equations is solved, so links to the high-z slab are absent; and, the w source term involves a z-direction gradient in mean, rather than local, pressure.

As a consequence of the absence of high-z links, solution is achieved in a single sweep. Solution at each slab is achieved in the same way as the elliptic solution, and uses largely the same coding sequences. Iteration may be performed at each slab, if required.

Because only a single sweep is required, calculated values at upstream slabs need not be retained as the solution moves downstream. Thus, a three-dimensional parabolic solution requires variables to be stored only two-dimensionally, in the xy plane. All this is arranged automatically by EARTH when the parabolic mode of solution is selected.

As is usual in special-purpose parabolic or boundary-layer programs, EARTH contains built-in options for curvature of the z-direction, and for expansion of the domain of solution in either the x or y direction, or both, in order to accommodate a spreading boundary-layer. Both confined and unconfined flows can be handled.

3. APPLICATION 1: EXPLOSION CONTAINMENT IN A FAST-BREEDER REACTOR

3.1 *Introduction*

The containment vessel of a fast-breeder reactor is designed to withstand forces which might result from an accident situation when, as a consequence of overheating, boiling occurs, a gas bubble forms, and the coolant liquid is abruptly set in motion. The forces which result from fluid impact on the vessel are important to the designer. Extensive reactor safety studies have been carried out in which computer codes have been used to predict the forces on the vessel in these hypothetical situations, see, for example, Hoskin and Herrmann (1982). In the codes employed, equations are solved only within the liquid. In the gaseous regions the pressures are assumed to be uniform and are calculated from pressure-time or pressure-volume relationships. Shear forces acting at the interfaces are neglected.

The codes are also restricted to two-dimensional axisymmetric situations. In typical fast-breeder reactor designs, however, the presence within the sodium pool of primary coolant pumps and heat-exchangers make the problem three-dimensional. The present calculations were performed to demonstrate that three-dimensional explosion containment calculations are practicable.

3.2 Problem Definition

Calculations have been performed to predict the fluid motions and forces on the structure of an idealised containment vessel, similar to those used in the COVA series of experiments (Kendall et al., 1979). A cylindrical vessel of radius 350mm and height 560mm is considered, with internal structures as shown in Fig. 1. Three-dimensional calculations have been performed for a segmented cylindrical internal structure, Fig. 1, Case 2. These are compared with a two-dimensional calculation where the internal structure was a continuous cylinder. An initially-spherical, high-pressure gas-bubble is assumed to form within the closed vessel, which is filled with liquid up to a level close to the top, and above this with a cover-gas.

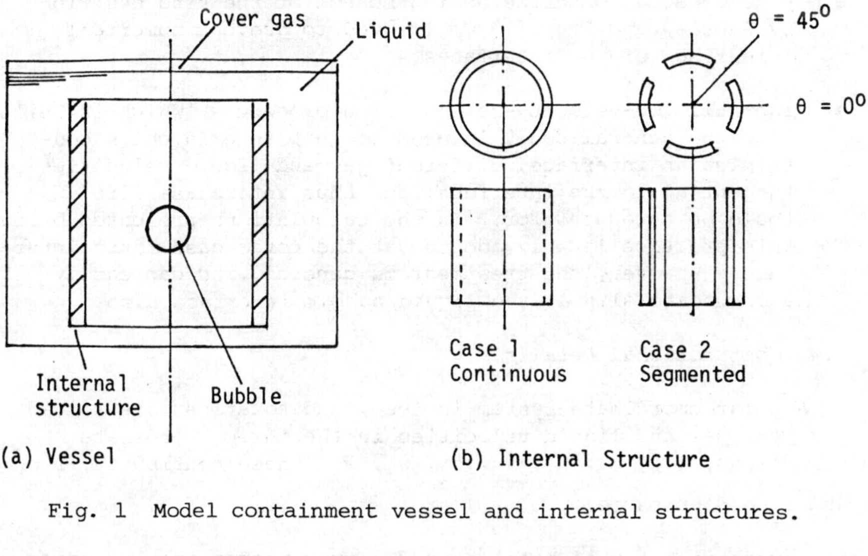

Fig. 1 Model containment vessel and internal structures.

The bubble is at a height of 280mm, its radius is 34.5mm and its initial pressure is 10^8 N/m^2. Elsewhere the pressure is 1.013×10^5 N/m^2. The upper liquid interface is 25mm from the top of the vessel, and the fluids are presumed to be initially at rest.

3.3 Modelling Approach

The modelling approach adopted for the calculations is as follows:

- The calculation domain extends over the whole vessel, including the bubble and cover-gas so that account can be taken of the gas-liquid interactions at the interfaces. This is particularly important for the cover-gas region, the behaviour of which affects the roof-impact pressure.

- Gas/liquid interface positions are deduced from solution of built-in phase-conservation equations. This method is adopted in preference to interface tracking using marker particles, which is normally used in containment codes, because it ensures that, in a converged solution, mass continuity for each phase is perfectly conserved; and extension to three-dimensional cases is perfectly straightforward.

- A donor-acceptor differencing scheme of the kind described by Ramshaw and Trapp (1976) is used to prevent numerical 'smearing' of the interfaces.

- The full two-velocity-set solution procedure, which is built into the general code, is used so that in grid cells containing an interface, different gas and liquid velocities tangential to the interface, and thus interphase slip, could be accounted for. In the calculations presented below, this practice is only adopted at the cover-gas/liquid interface. However, the treatment is general, and can easily accommodate slip at the liquid/bubble interface also.

3.4 Computational Details

A polar-coordinate system is used, and equations are solved for: the gas and liquid velocities in the three coordinate directions, i.e. u_g, u_ℓ, v_g, v_ℓ, w_g, w_ℓ; phase continuity, r_ℓ and r_g and pressure, p.

Compressibility of both gas and liquid phases are included, and the friction coefficient at the cover gas liquid interface is determined from the velocity difference tangential to the interface, the area, and a friction factor, for which a typical value of 0.003 has been chosen.

The vessel surfaces and internal structures are impervious to flow and rigid, and viscous effects and wall shear stress are neglected.

3.5 *Presentation of Results*

Typical results are shown in Fig. 2 in the form of velocity-vectors, and contours of pressure and volume-fraction at 0.5ms and 1.5ms; and Figs. 3 and 4 show the variation of pressure on the vessel lid during the transient, and the vertical variation of pressure on the outer wall and inner structure of the vessel at times when the largest forces occur. The pressure plots also include the results for a two-dimensional axisymmetric calculation in which the inner structure was taken to be continuous rather than segmented.

Fig. 2 shows the velocity vectors, pressure and volume fraction contours at 0.5ms and 1.5ms. Two plots are shown at each time. Figs. 2(a) and (c) give the results at the symmetry plane which passes through the structure; Figs. 2(b) and (d) give the results in the plane through the gap; $\theta=0^\circ$ and $\theta=45^\circ$ respectively in Fig. 1.

At 0.5ms the velocity vectors (shown in the left-hand halves of the Figures) indicate a predominant flow upwards out of the cylinder as liquid is displaced by the expansion of the bubble. In the plane of the gap, there is a large radial flow component which turns upwards, the cover-gas being easily compressed. At 1.5ms the velocities at the top of the vessel are beginning to move outwards since the cushioning effect of the now completely compressed cover-gas has been lost.

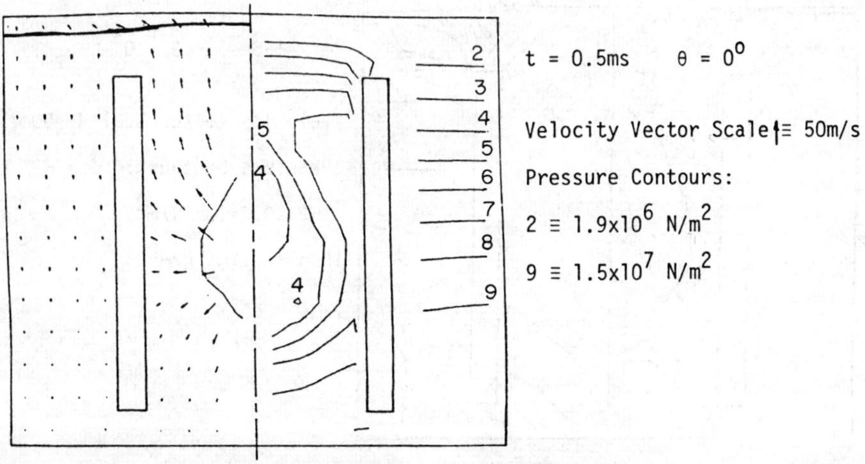

Fig. 2(a) Case 2 - Segmented structure: velocity vectors pressure contours and R_ℓ = 0.5 contours.

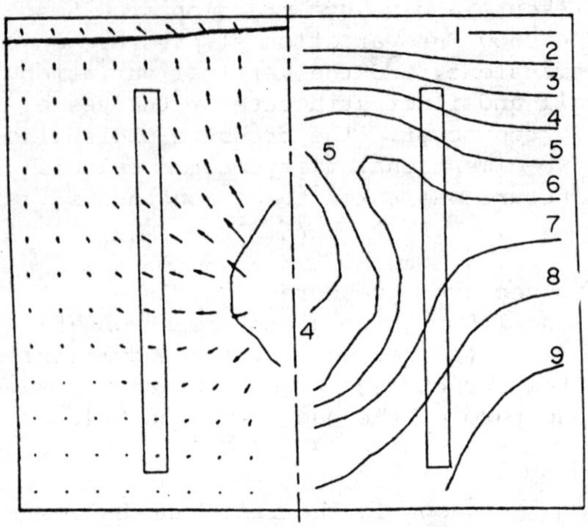

t = 0.5ms θ = 45°

Velocity Vector Scale ↑ ≡ 50m/s

Pressure Contours:

1 ≡ 1x10^5 N/m^2

9 ≡ 1.6x10^7 N/m^2

Fig. 2(b) Case 2 - continued.

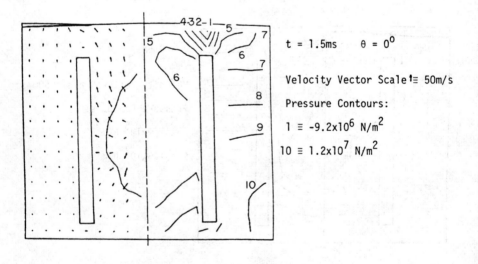

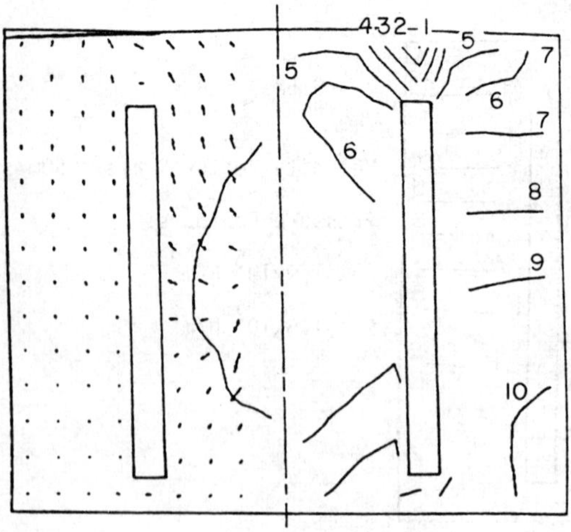

t = 1.5ms θ = 0°

Velocity Vector Scale ↑ ≡ 50m/s

Pressure Contours:

1 ≡ -9.2x10^6 N/m^2

10 ≡ 1.2x10^7 N/m^2

Fig. 2(c) Case 2 - continued.

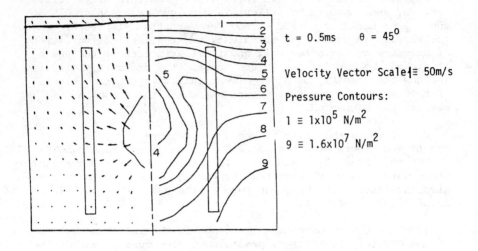

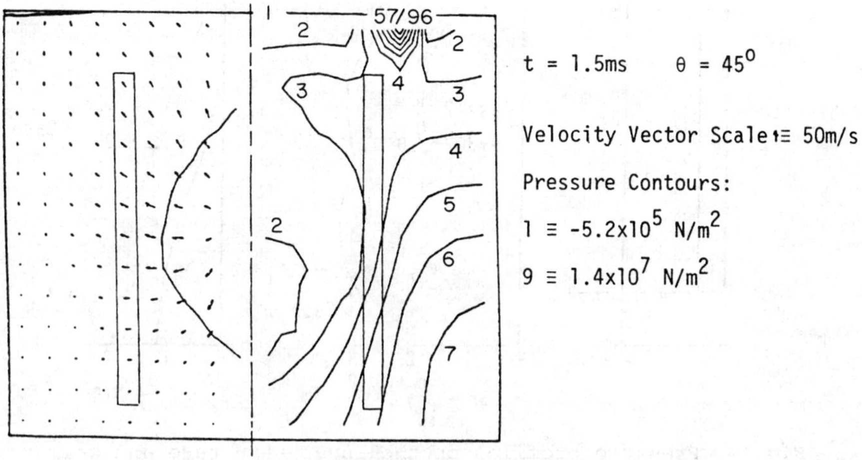

Fig. 2(d) Case 2 - continued.

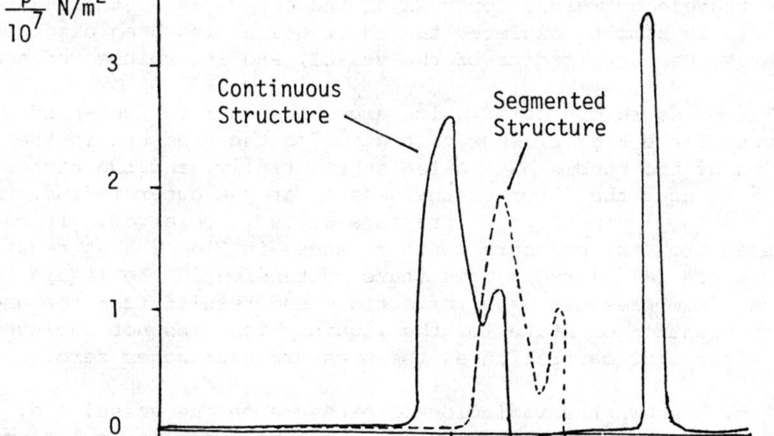

Fig. 3 Variation of pressure at the centre of the lid.

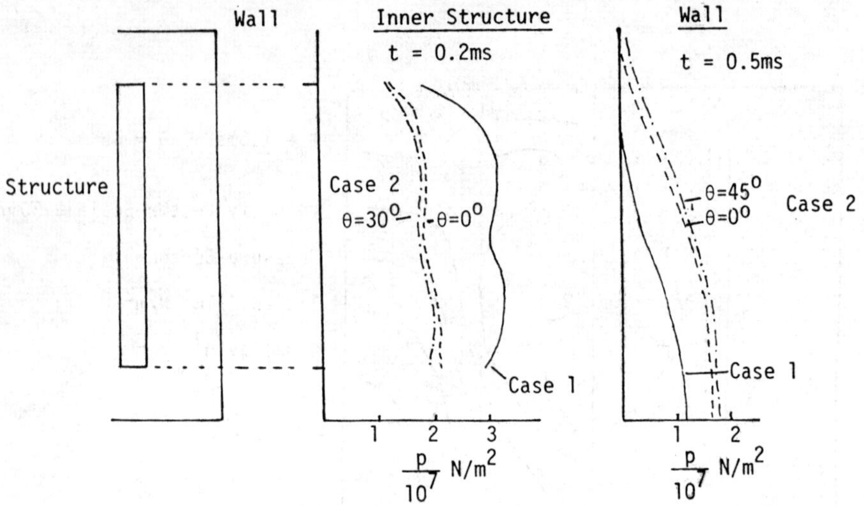

Fig. 4 Pressure profiles on the inner structure and wall.

The volume-fraction contours are also shown on the left-hand halves of the figures. The bubble is relatively small at 0.5ms, but at 1.5ms has expanded considerably, particularly in the vertical direction and in the plane of the gap.

The cover-gas interface is seen to rise above the bubble at 0.5ms. The liquid impacts on the roof of the vessel, and a wave travels outwards, compressing the cover-gas. At 1.5ms this process is almost complete, the cover-gas having been displaced towards the outer radius of the vessel, and its volume reduced.

The pressure contours (which are shown on the right-hand halves of the Figure) show that at 0.5ms the pressure in the region of the bubble has fallen substantially, and the highest value is near the floor of the vessel, at the outer radius. At 1.5ms a complex pattern of pressure-waves is observed. It can be seen from the pressure contour ranges in Fig. 2 that negative values are calculated. This state of tension in the liquid arises from pressure wave reflections and results from the use of an equation of state for the liquid which does not include the effects of cavitation as the pressure approaches zero.

Fig. 3 shows the variation of pressure on the vessel lid, at a location above the bubble. The two-dimensional calculation performed for a continuous structure shows a higher peak pressure than for the segmented structure, and there is a delay in the impact associated with the latter. The second peak in roof pressure which is observed for the continuous structure case arises from wave reflections. These have not occurred within the timescale of the three-dimensional calculations, and indeed may not take the same form in view of the altered geometry.

The pressures acting on the structures are radically changed by alteration of the internal arrangements. For the continuous structure, the highest pressures are exerted on the inner structure and on the centres of the lid and floor. In the segmented design the pressures are much greater at the outer walls of the vessel, and the pressures on the inner structure are reduced. This is shown in Fig. 4 by the distributions of pressure along the inner and outer walls at times when peak pressure occurs. These are 0.2ms for the inner structure, and 0.5ms for the outer wall.

4. APPLICATION 2: ICING OF AN AIR-LAUNCHED TORPEDO

4.1 Introduction

Certain types of torpedo are designed for helicopter launching, and in cold weather conditions the temperature of the torpedo hull can become lower than sea-water temperature. As a consequence of this, a layer of ice can form on the surface which might affect the operation of device temporarily.

Numerical studies were carried out to investigate the timescale of the icing-deicing process, a validation exercise being carried out initially during which predictions were made for an experimental situation investigated by Marconi Space and Defence Systems Limited. The results of this validation exercise are described below.

4.2 Problem Definition

The experimental situation is illustrated in Fig. 5. A metal plate was cooled to a temperature of -26°C and attached to a water-tunnel. Water of known salinity and at a temperature of -1°C was passed over the plate. The freezing point of the water was -1.9°C.

The thickness of ice which formed on the plate was measured for particular plate temperatures. The ice thickness increased initially, but reduced as the plate warmed.

4.3 Modelling Approach

The initial approach to modelling this problem considered a longitudinal 'slice' of the plate, as shown in Fig. 6a. This two-dimensional model included representations of the plate, the ice and the water, and hence equations were solved for two

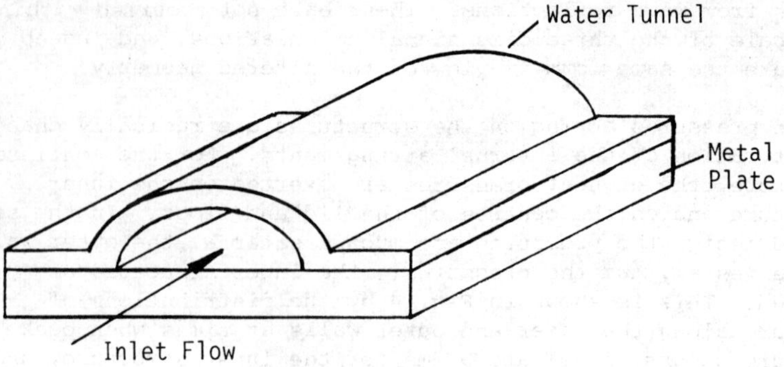

Fig. 5 Icing of an air launched torpedo: experimental situation.

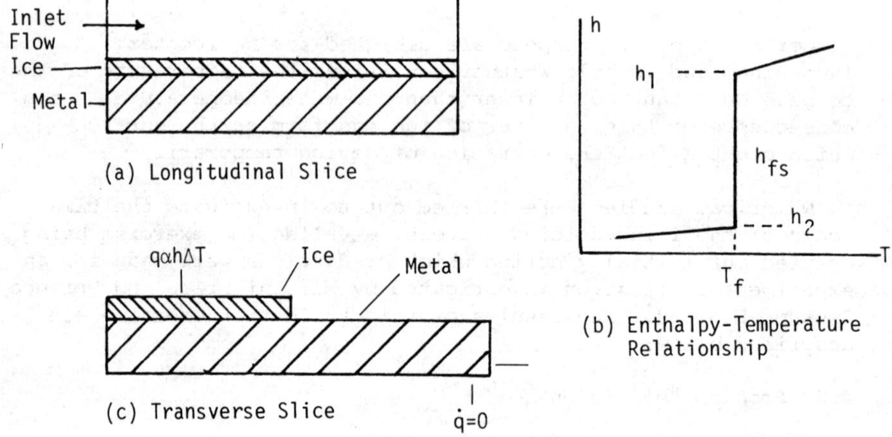

Fig. 6 Icing of an air launched torpedo: modelling approaches.

velocities, u and v, enthalpy, h and pressure, p. Where the solution domain extended into the plate, the velocities were set to zero and the thermal conductivity appropriate for the plate material was prescribed.

The freezing process itself was modelled as follows:

- The latent heat of fusion of ice was taken into account by expressing the diffusive flux vector in the enthalpy equation in terms of temperature. The temperature, T, is related to the enthalpy, h, by the relationship shown in Fig. 6b, in which the freezing temperature, T_f, and the enthalpies h_1 and h_2 are defined as:

$$h_1 = C_p\, T_f$$

$$h_2 = h_1 - h_{fs}$$

where: C_p = specific heat of fluid; and

h_{fs} = latent heat of fusion.

The temperature can then be expressed in terms of enthalpy by:

$$T = h/C_p \qquad h > h_1$$

$$T = T_f \qquad h_1 \geqslant h > h_2$$

$$T = (h+h_{fs})/C_p \qquad h_2 \geqslant h.$$

- The formation of the ice was modelled by assuming an infinite resistance to flow at the freezing temperature.

This stage of the study showed that the thermal and velocity boundary layers whilst being displaced by the ice, maintained the same profile, and hence the fluid-to-ice heat transfer, was not affected by ice formation. It was also shown that there was no longitudinal variation in ice thickness. The temperature equalisation within the metal showed that the sections of plate not in contact with the water would constitute a large heat sink, and should be included in the model.

The second stage of the study therefore considered a transverse slice of the plate as shown in Fig. 6c. The problem thus reduces to one of heat conduction within the plate, and only an equation for enthalpy is solved. The flow model is replaced by a heat source in the region where ice is formed which is proportional to a fixed heat transfer coefficient, obtained from the Dittus-Boelter correlation, and the temperature difference.

A zero heat-flux boundary condition was presumed elsewhere.

4.4 Presentation of Results

Figs. 7 and 8 show the variation of average ice depth with time and plate temperature respectively. Fig. 7 shows that the

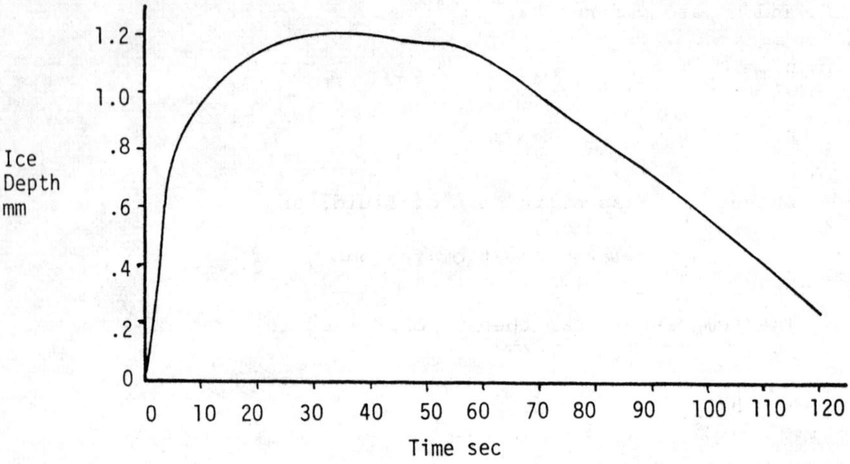

Fig. 7 Variation of ice depth versus time.

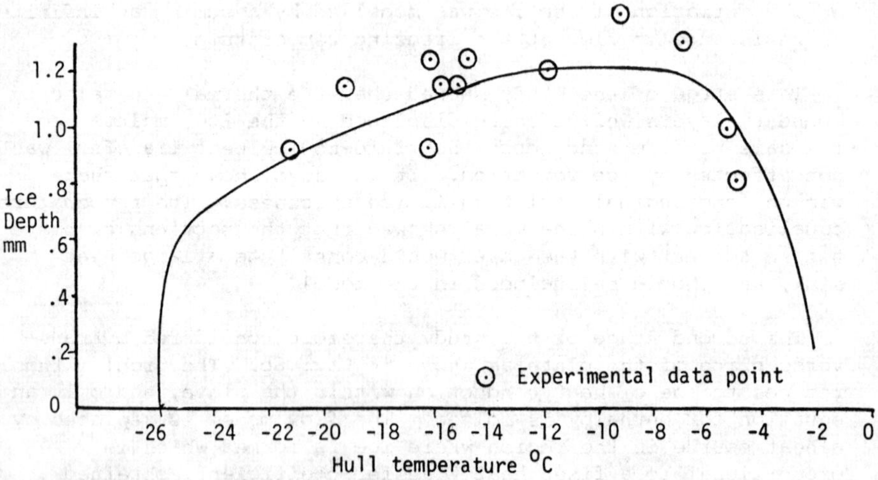

Fig. 8 Variation of ice depth versus hull temperature.

main thickness of ice forms to a depth of about 1.2mm within the first 20 seconds of the transient, and then reduces more slowly.

Fig. 8 compares the predictions with experimental measurements. It can be seen that the prediction agrees very well with the data.

5. CONCLUDING REMARKS

The foregoing sections have provided a brief description of the general-purpose flow program, PHOENICS, and its use in the solution of two problems.

Application of the code to a three-dimensional fast-breeder reactor containment problem shows that such calculations are practicable, and that extension from a two-dimensional axisymmetric case was straightforward. Before the results could be taken as quantitatively correct, however, it would be necessary to investigate the importance of grid and time-step and of physical effects such as viscous stresses at the walls and slip at the bubble/liquid interface.

In the case of the icing problem, all the physical effects are well represented, the results are grid and time-step independent, and also agree with experimental data.

The two problems which have been described represent typical applications of the code. The program is also currently being used for a wide range of problems involving nuclear reactors and associated equipment, gun ballistics, rockets, ships and submarines.

6. ACKNOWLEDGEMENTS

The authors wish to express their thanks to Marconi Space and Defence Systems Limited for permission to publish the results of the icing-problem, and to C. Aldham and J. Ludwig who performed the PHOENICS calculations reported in this paper.

7. REFERENCES

HOSKIN, N. E. and HERRMANN, W. 1982 A Survey of Several Finite Difference Codes Used in Two-Dimensional Unsteady Flow Problems. These proceedings.

KENDALL, K.C., ARNOLD, L.A., BROADHOUSE, B.S., JONES, A., YERKESS, A. and BENUZZI, A. 1979 Experimental Validation of the Containment Codes ASTARTE and SEURBNUK. In 5th Int. Conference on Structural Mechanics in Reactor Technology, Berlin, August 1979.

RAMSHAW, J.D. and TRAPP, J.A. 1976 A Numerical Treatment for Low-Speed Homogeneous Two-Phase Flows with Sharp Interfaces. *J.Comput. Phys.* Vol. **21**, pp. 438-453.

PATANKAR, V. and SPALDING, D.B. 1972 A Calculation Procedure for Heat, Mass and Momentum Transfer in Three-Dimensional

Parabolic Flows. *Int. J. of Heat and Mass Transfer* Vol. **15**, pp. 1787-1806.

SPALDING, D.B. 1980 Numerical Computation of Multi-Phase Fluid Flow and Heat Transfer. In Recent Advances in Numerical Methods in Fluids Vol. **1**, (C. Taylor and K. Morgan, Eds.), Pineridge Press, Swansea, 1980.

SPALDING, D.B. 1981a A General Purpose Computer Program for Multi-Dimensional One- and Two-Phase Flow. In Mathematical and Computers in Simulation, North-Holland, Vol. XXIII, pp. 267-276.

SPALDING, D.B. 1981b Developments in the Procedure for Numerical Computation of Multi-Phase-Flow Phenomena with Interphase Slip, Unequal Temperature, etc. In 2nd Int. Conference on Numerical Methods, Maryland University, September 1981.

THE METHOD OF CHARACTERISTICS APPLIED TO PROBLEMS IN COSMICAL GAS DYNAMICS

S.A.E.G. Falle

(Department of Mathematics, University of Leeds)

1. INTRODUCTION

Cosmical gas dynamics presents some computational problems which are not encountered in terrestrial gas dynamics. Firstly there are often very strong shocks (Mach number greater than 1000). Secondly there are radiative loss terms which can vary from being negligible in one part of the flow to being very large in another. And finally there are body forces such as gravity or radiation pressure which, under certain circumstances can dominate the flow.

The combination of strong shocks and a large radiative loss term leads to very steep temperature gradients. This means that schemes with substantial numerical diffusion cannot be used. Also, the nature of the radiative loss term is such that it is often necessary to refine the mesh near shocks. Under these conditions shock fitting offers a number of advantages over other methods.

This paper describes the application of a one dimensional Lagrangian scheme to some of these Astrophysical problems. The scheme is a modified version of Hartree's (1958) method and is based on differencing the equations along the characteristics. Shocks are treated by shock fitting, and it is possible to use different meshes and time steps in different parts of the flow.

Interstellar Gas

In many situations, the motion of the interstellar gas is described by the following equations:-

$$\rho\left(\frac{\partial}{\partial t} + \underline{u}.\nabla\right)\underline{u} = -\nabla P + \underline{F} \ , \qquad (1.1)$$

$$\frac{\partial \rho}{\partial t} + \nabla.(\rho\underline{u}) = 0, \qquad (1.2)$$

$$\rho\left(\frac{\partial\rho}{\partial t} + \underline{u}.\nabla\right)E = \underline{u}.\underline{F} - P\nabla.\underline{u} - \Lambda, \tag{1.3}$$

$$E = \frac{P}{(\gamma - 1)\rho} . \tag{1.4}$$

Here $\underline{u}$, P, ρ, and E are the velocity, pressure, density and internal energy per unit mass. $\underline{F}$ is a body force. The possible forms of $\underline{F}$ are:-

(a) $\underline{F} = 0$. (This is the case for explosions and other high energy phenomena.)

(b) $\underline{F} = -\nabla\psi$. (ψ is a gravitational potential. Typical problems are accretion and star formation.)

(c) $\underline{F} = \underline{K}\rho^2$. ($\underline{K}$ is a constant vector. This occurs in quasars and stellar winds.)

Λ is a radiative energy loss rate per unit mass. The possibilities are:-

(a) $\Lambda = \rho\phi(T)$. (This is radiative cooling in an optically thin gas.)

(b) $P = c_0^2\rho$ (Isothermal). (This is the case if $\Lambda(T_0) = 0$, and $t_c = E/\Lambda << t_d$ where t_d is the dynamical time. Here $c_0 = \sqrt{(P/\rho)}$ is the isothermal sound speed at the equilibrium temperature T_0.)

For temperatures greater than 10^4K, φ(T) is of the form shown in Fig. 1. For such a loss function, the cooling time becomes very long at high temperatures and then decreases rapidly at lower temperatures. This means that most high energy flows start off adiabatically, but radiative cooling eventually becomes important when the gas has cooled sufficiently by expansion.

It can easily be shown that the gas is thermally unstable if $d\log\phi/d\log T < 3$. As can be seen this is the case for $T > 10^5$ K. This means that gas which is heated to temperatures higher than 10^5 K is unstable. In many cases the cooling time becomes shorter than the sound crossing time so that shocks are formed.

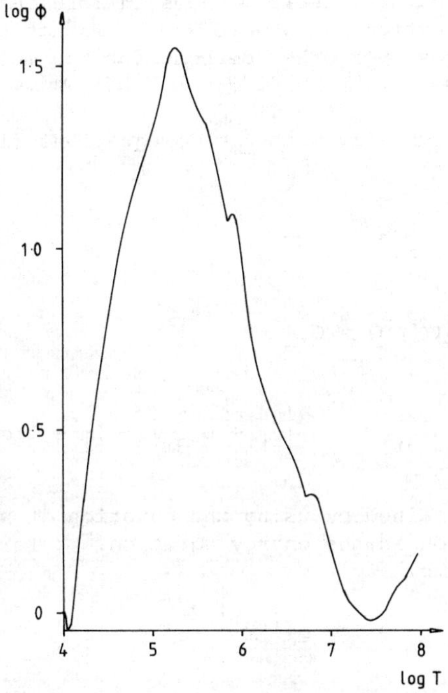

Fig. 1 Radiative cooling law for $T \geqslant 10^4$ K.

If artificial viscosity methods are used to study these types of problems, then either the mesh near shocks must be sufficiently fine for the Courant-Friedrichs time to be shorter than the cooling time, or the cooling must be set to zero inside a shock. Otherwise, since it takes about a Courant-Friedrichs time for a fluid element to pass through the shock, gas which is initially at 10^4 K would never be heated to high temperatures by a shock. It is also often necessary to use a finer mesh behind shocks because of the rapid cooling there. This means that the programming convenience, which is the main advantage of these methods over shock fitting, is largely lost.

Since the dominant effect in these flows is the large radiative loss term Λ, it is essential to use a Lagrangian scheme. Otherwise the errors which an Eulerian scheme inevitably introduces in the entropy can be magnified by the thermal instabilities. For this reason we use a modified version of Hartree's (1958) method which involves differencing the characteristic equations on a Lagrangian mesh. To use characteristic coordinates would involve interpolation for the energy equation. This would be unsound for the same reason as an Eulerian scheme.

We will only consider one space dimension. Although it is possible to use characteristic methods in more than one space dimension, it is extremely awkward and it is not clear that they have any advantages over other methods for the multidimensional case.

For spherical symmetry with $\underline{F} = 0$, equations (1.1) - (1.3) reduce to

$$\frac{\partial u}{\partial t} + u\frac{\partial u}{\partial r} = -\frac{1}{\rho}\frac{\partial P}{\partial r}, \tag{1.5}$$

$$\frac{1}{r^2}\frac{\partial}{\partial r}(r^2\rho u) = 0, \tag{1.6}$$

$$\left(\frac{\partial}{\partial t} + u\frac{\partial}{\partial r}\right)P = c^2\left(\frac{\partial}{\partial t} + u\frac{\partial}{\partial r}\right)\rho - (\gamma - 1)\rho\Lambda. \tag{1.7}$$

(1.7) has been obtained by using the equation of state and the continuity equation in the energy equation. c is the adiabatic sound speed given by

$$c = \left(\frac{\gamma P}{\rho}\right)^{1/2} . \tag{1.8}$$

Define a Lagrangian variable m by

$$dm = r^2\rho dr . \tag{1.9}$$

We can then write (1.5) - (1.7) in characteristic form

$$\frac{\partial u}{\partial t} \pm \frac{1}{\rho c}\frac{\partial P}{\partial t} = \mp\frac{2uc}{r} \mp \frac{(\gamma - 1)\Lambda}{c} \text{ along } C^+, C^-, \tag{1.10}$$

$$\frac{\partial P}{\partial t} - c^2\frac{\partial \rho}{\partial t} = -(\gamma - 1)\rho\Lambda \text{ along } C^0, \tag{1.11}$$

$$\frac{\partial r}{\partial t} = u. \tag{1.12}$$

The characteristics are given by

$$\frac{dm}{dt} = \pm r^2\rho c \qquad \text{for } C^+, C^-, \tag{1.13}$$

and

$$\frac{dm}{dt} = 0 \qquad \text{for } C^0. \tag{1.14}$$

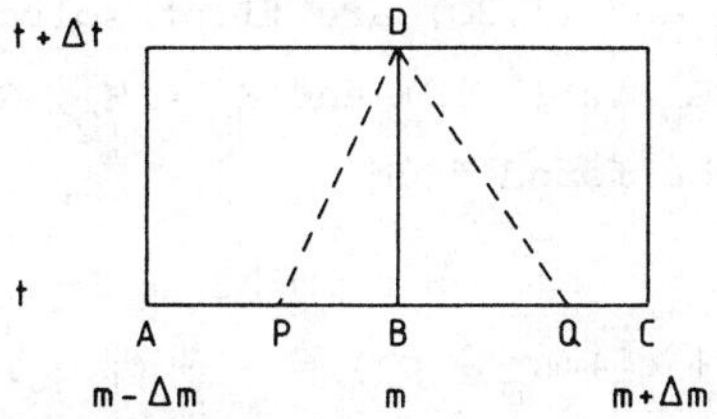

Fig. 2 Computational method for an ordinary mesh point.

These equations are now differenced as follows. In Fig. 2 PD and QD are approximations to the backward C^+ and C^- characteristics through D. If $m+p\Delta m$, $m+q\Delta m$ are the positions of P and Q, then p and q are found from

$$\begin{aligned} p\,\frac{\Delta m}{\Delta t} &= -\frac{1}{2}(F_P + F_D), \\ q\,\frac{\Delta m}{\Delta t} &= \frac{1}{2}(F_Q + F_D), \end{aligned} \tag{1.15}$$

where

$$F = r^2 \rho c . \tag{1.16}$$

Here subscripts denote the values at the corresponding points. The values of F at P and Q are obtained by linear interpolation on AB and BC, respectively.

Now write

$$H = \frac{1}{\rho c}, \quad G = \frac{2uc}{r} + \frac{(\gamma - 1)\Lambda}{c} . \tag{1.17}$$

(1.10) can then be differenced as follows

$$u_D - u_P + \frac{1}{2}(H_P + H_D)(P_D - P_P) = -\frac{1}{2}(G_P + G_D)\Delta t. \tag{1.19}$$

$$u_D - u_Q - \frac{1}{2}(H_Q + H_D)(P_D - P_Q) = \frac{1}{2}(G_Q + G_D)\Delta t. \tag{1.20}$$

In (1.19) and (1.20) u_P, u_Q, P_P, and P_Q are found from quadratic interpolation on ABC. H_P, H_Q, G_P, G_Q need only be calculated with linear interpolation.

Equations (1.19) and (1.20) are first solved for u_D, and P_D with $H_D = H_P$, $G_D = G_P$ in (1.19) and $H_D = H_Q$, $G_D = G_Q$ in (1.20). ρ_D and r_D can then be found from

$$P_D - P_B - \frac{1}{2}(c_D^2 + c_B^2)(\rho_D - \rho_B) = -\frac{1}{2}(\gamma - 1)(L_D + L_B)\Delta t. \tag{1.21}$$

$$r_D - r_B = \frac{1}{2}(u_D + u_B)\Delta t \tag{1.22}$$

where

$$L = \rho\Lambda. \tag{1.23}$$

Again in (1.21) and (1.22) the values of c and L at D are initially set equal to those at B. Once provisional values of the quantities at D have been obtained, then they can be used to calculate H_D, G_D, c_D and L_D. These are then used in (1.19) - (1.22) to calculate the final values of the quantities at D. The system is then second order accurate in both space and time.

The scheme is stable provided

$$|p| \leqslant 1, \ |q| \leqslant 1,$$

i.e. if

$$(\Delta t/\Delta m) \leqslant r^2 \rho c. \tag{1.24}$$

This is simply the usual Courant-Friedrichs condition.

2. SHOCKS

Fig. 3 shows the way in which a shock moving to the right is treated. CD is the path of the shock, and PD is the backward C^+ characteristic through D. The position $m+p\Delta m$ of P is found from

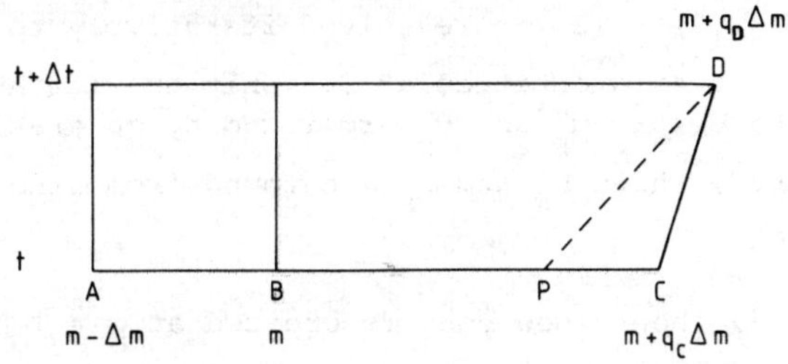

Fig. 3 Computational method for a shock moving to the right when the flow ahead of the shock is known.

$$(q_D - p)\frac{\Delta m}{\Delta t} = \frac{1}{2}(F_P + F_D) \tag{2.1}$$

Again F_P is found from linear interpolation on BC.

If ρ_O, P_O, u_O are the quantities ahead of the shock and V_D is the shock velocity, then we have the shock relations

$$\rho_D(V_D - u_D) = \rho_O(V_D - u_O), \tag{2.2}$$

$$P_D + \rho_D(V_D - u_D)^2 = P_O + \rho_O(V_D - u_O)^2 \tag{2.3}$$

$$\frac{\gamma P_D}{(\gamma - 1)\rho_D} + \frac{1}{2}(V_D - u_D)^2 = \frac{\gamma P_O}{(\gamma - 1)\rho_O} + \frac{1}{2}(V_D - u_O)^2. \tag{2.4}$$

The relation along the characteristic PD is

$$u_D - u_P + \frac{1}{2}(H_P + H_D)(P_D - P_P) = -\frac{1}{2}(G_P + G_D)\Delta t. \tag{2.5}$$

The position of the shock in Lagrangian and Eulerian coordinates is found from

$$(q_D - q_C)(\Delta m/\Delta t) = \frac{1}{2}\rho_O\{(r^2V)_D + (r^2V)_C\}, \tag{2.6}$$

$$r_D - r_C = \frac{1}{2}(V_D + V_C)\Delta t. \tag{2.7}$$

Equations (2.2) - (2.7) are solved iteratively to find V_D and hence all other quantities at D. Only one iteration is necessary. As before u_P and P_P are found by quadratic interpolation on ABC, while H_P and G_P are found from linear interpolation on BC.

Once $q_D \geqslant 2$, then a new zone is created at $q = 1$ by quadratic interpolation.

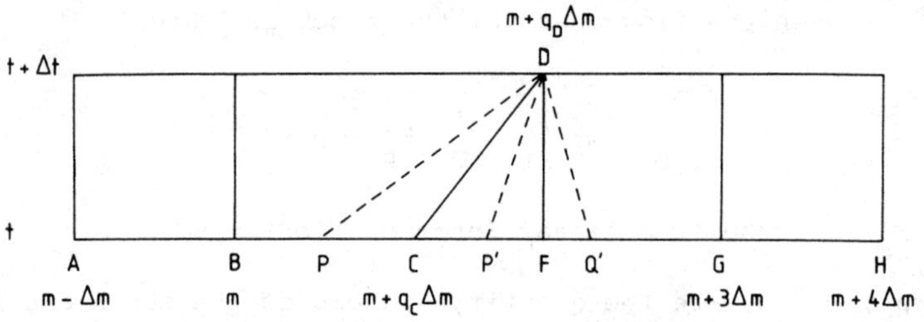

Fig. 4 Computational method for a shock moving to the right when the flow ahead of the shock is unknown

If the quantities ahead of the shock are not known, then they must be calculated as shown in Fig. 4. The pre-shock values at D are obtained from the relations along P'D, Q'D (C^+ and C^-) and along FD (C^0). The values at P', Q' and F are obtained by interpolation on CGH using the pre-shock values at C.

The equations giving the pre-shock values at D are iterated together with (2.2) - (2.7). Again only one iteration is needed. Once $q_C \geqslant 2$, then we create a new zone at $q_C = 1$ and remove the zone at G.

3. NON UNIFORM TIME STEP

The mesh can be divided into regions with different values of Δm and Δt in each. In order to obtain the solution along lines of constant t, Δt must be of the form $i^k \Delta t_b$ where Δt_b is a basic time step, i is a fixed integer and k is an integer which is different in each region. In practice we only use $i = 2$.

At a boundary where the mesh changes the situation is as shown in Fig. 5. The values at the base of the characteristics are obtained by interpolation at the latest time which has been calculated. Fig. 5 shows the case when the time step changes by

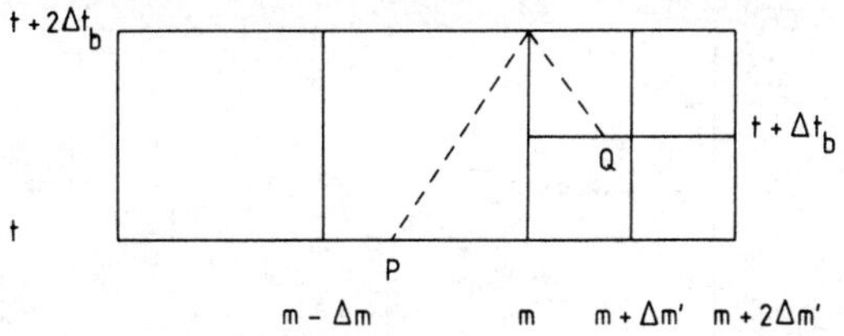

Fig. 5 Computational method at a boundary where the mesh changes

a factor of 2. In this case the quantities at P are from interpolation at t_n, while those at Q are from interpolation at $t_n + \Delta t_b$.

4. CONTACT DISCONTINUITIES

These are treated in the same way as a normal zone, except that we need two equations along C^O. It is often convenient to combine a contact discontinuity with a change in Δm and Δt.

5. FORMATION OF SHOCKS

If shocks form then there is a small dissipation due to the interpolation, but since the equations are not in conservation form, the entropy change is wrong. This is not serious as long as the shocks are weak. Fig. 6 shows the pressure profile for a shock with a Mach number of 2.

To decide whether a shock should be treated by shock fitting, we look at the quantity R defined by

$$R = - \frac{\rho c \Delta m}{P} \frac{\partial u}{\partial m} \quad . \tag{5.1}$$

If $R \simeq 1$, the shock is treated by shock fitting. The values two zones ahead of the maximum value of R are taken to be the pre-shock values, while those two zones behind are taken to be the post shock values. This results in some errors when the shock is set up, but as long as the shock is weak these errors are small.

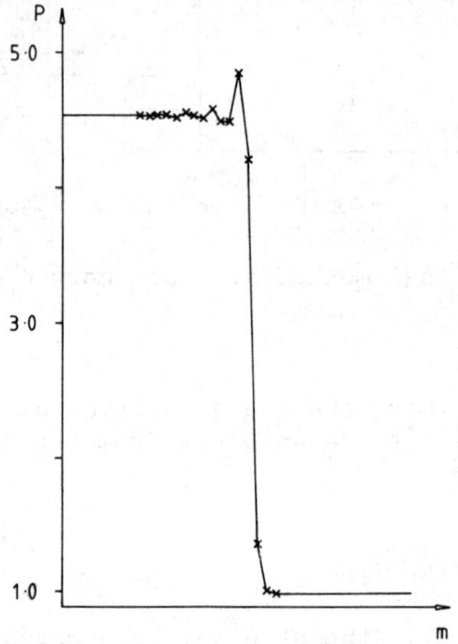

Fig. 6 Pressure as a function of the Lagrangian variable m for a shock with Mach number 2.

RESULTS

Fig. 7 shows a comparison between the Sedov similarity solution for a strong explosion and the numerical results after 1000 time steps (Falle, 1975). It can be seen that the agreement is excellent for all the variables. This is true even near the centre where the velocity gradients in the Lagrangian frame are large. For this calculation four different mesh spacings were used with the finer mesh near the centre.

In Fig. 8 the numerical results are compared with the first term of an asymptotic expansion for a continuous mass loss problem (Dyson et al., 1980). The error in the asymptotic solution is $O(1/R^2)$ where R is the dimensionless radius of an expanding shock. Again the numerical results are in excellent agreement with the asymptotic expansion.

The comparisons in Figs. 7 and 8 are for problems with no radiative losses or body force. In Fig. 9 we show the results for a plane parallel isothermal problem with a body force of the form $F = K\rho^2$ (Falle et al., 1981).

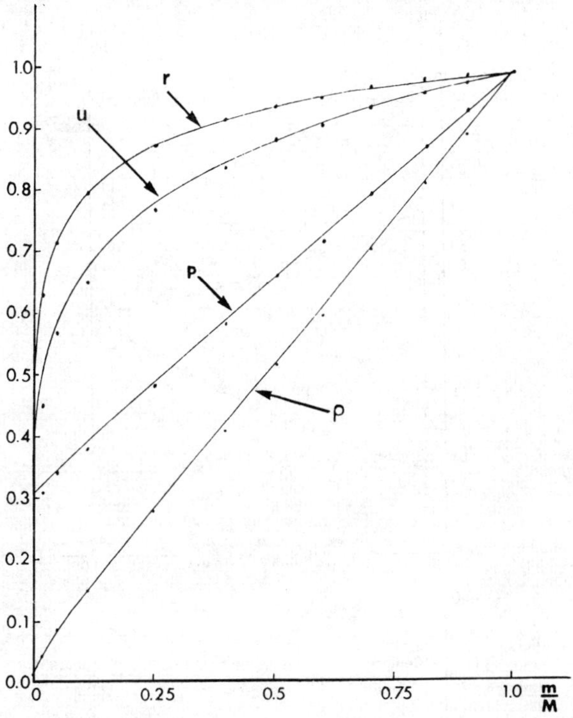

Fig. 7 Comparison between a Sedov similarity solution and the numerical results after 1000 time steps. The curves are from the similarity solution and the points are from the computations (Falle, 1975).

Again the numerical results are compared with the first term of an asymptotic expansion. The asymptotic expansion is singular and the dimensionless density is shown as a function of the dimensionless distance in both the inner and outer region. In this case, the small parameter in the asymptotic expansion is about 0.5 and so most of all the small discrepancy between it and the numerics is due to errors in the asymptotic expansion.

Finally in Fig. 10 we show the results for a supernova explosion with a cooling law of the form shown in Fig. 1 (Falle, 1981). The initial state is a Sedov solution as shown in Fig. 10(a). The radiative term is initially neglible, but as the temperature falls due to expansion, the cooling time decreases until it is much shorter than the dynamical time. This leads to a sudden drop in the pressure of the gas and the subsequent

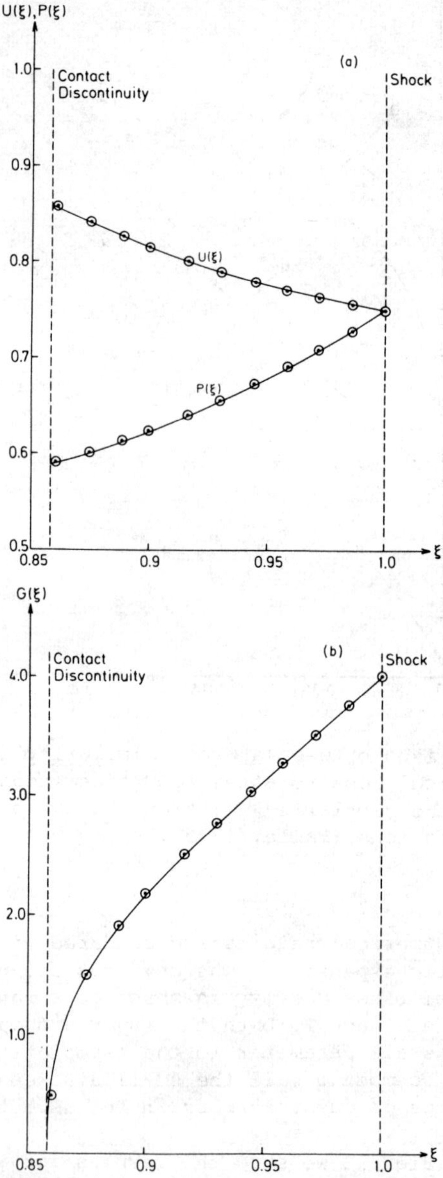

Fig. 8 Comparison between an asymptotic expansion and the computations for a continuous mass loss problem. Curves are for the asymptotic expansion and the points are from the computations. U, P and G are the dimensionless velocity, pressure and density respectively (Dyson et al., 1980).

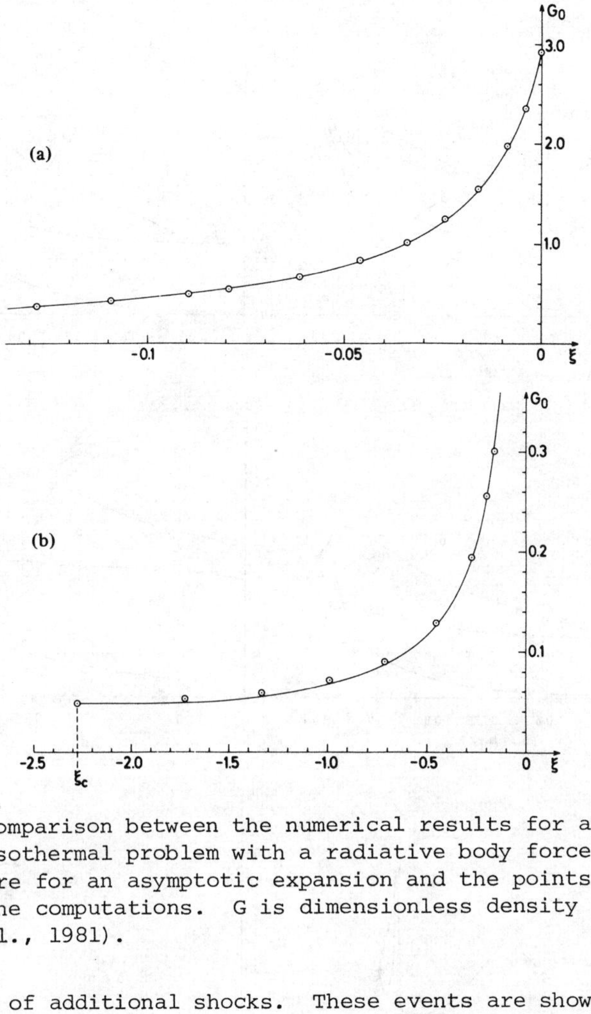

Fig. 9 Comparison between the numerical results for an isothermal problem with a radiative body force. Curves are for an asymptotic expansion and the points are from the computations. G is dimensionless density (Falle, et al., 1981).

formation of additional shocks. These events are shown in Figs. 10(d)-(g). The final stage has two shocks, one bounding the remnant, and another inward facing shock in which fast moving gas from the interior is decelerated by the slower cool gas. This shock is at $r/r_s = 0.8$ in Fig. 10(h). Notice that in Fig. 10(i) there is a sharp change in the density at $r/r_s \simeq 0.85$ with no corresponding change in pressure. This is due to the rapid cooling of the gas. It is this kind of feature which makes the use of a non diffusive scheme necessary.

In the supernova problem the cooling time behind the shocks becomes very short. It is therefore necessary to refine the

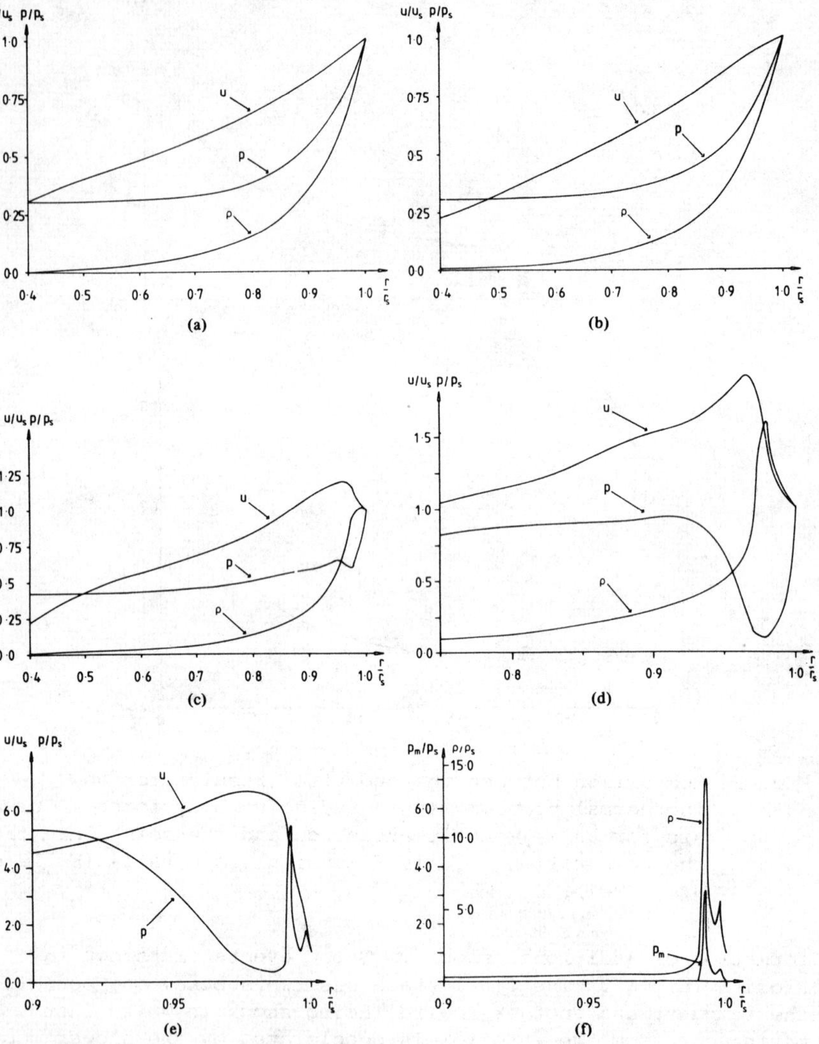

Fig. 10 Numerical results for a spherical supernova explosion. (a) Sedov solution for $\gamma=5/3$. (b) $\tau=1.04$. (c) $\tau=1.45$. (d) $\tau=1.59$. (e) $\tau=1.76$. (f) $\tau=1.76$. (g) $\tau=1.83$. (h) $\tau=4.46$. (i) $\tau=4.46$. Here τ is a dimensionless time (Falle, 1981).

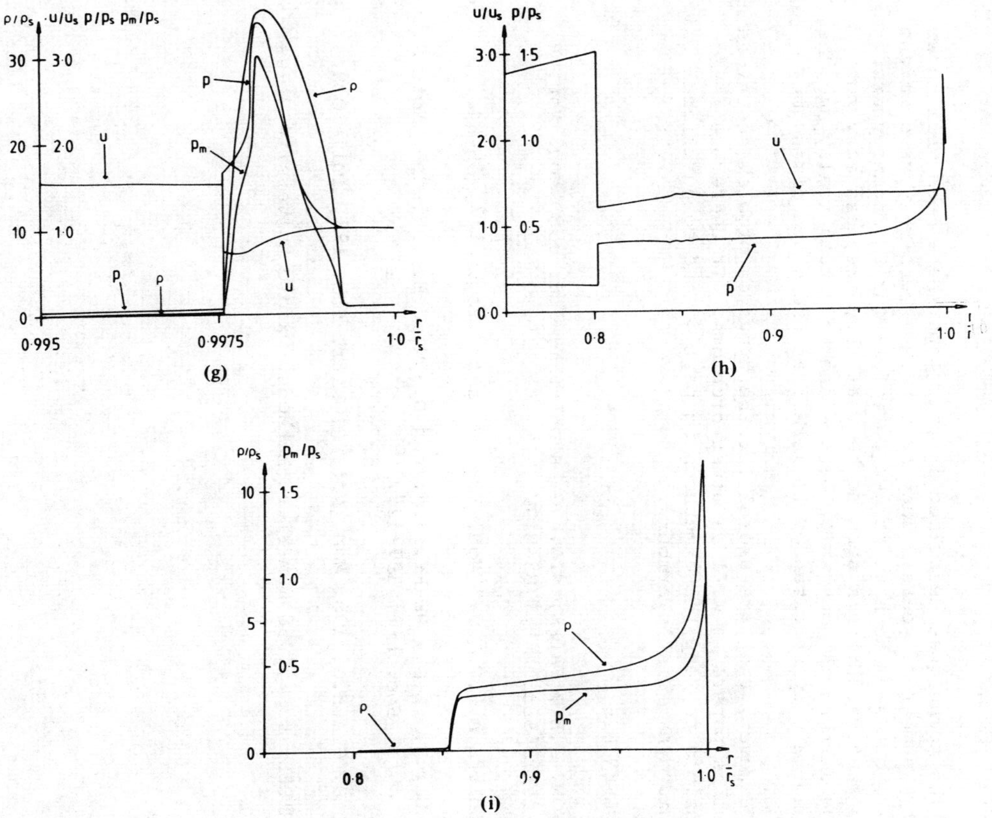

Fig. 10 (continued)

mesh behind these shocks. It is for this reason that shock fitting is so suitable for this type of problem.

CONCLUSION

The techniques described in this paper have been well known for a number of years. They have been extensively used in the past, but seem to have fallen out of favour in recent years. The reasons for this seem to be that they are less easy to program than methods which handle shocks automatically and they cannot easily be extended to more than one space dimension.

However, they are useful for the types of problem described in this paper, as well as other problems. It therefore seems sensible to keep these methods as part of the computational fluid dynamicist's armoury.

REFERENCES

DYSON, J.E., FALLE, S.A.E.G. and PERRY, J.J. 1980 *Mon. Not. R. astr. Soc.* **191**, 785-819.

FALLE, S.A.E.G. 1975 *Mon. Not. R. astr. Soc.* **172**, 55-84.

FALLE, S.A.E.G., PERRY, J.J. and DYSON, J.E. 1981 *Mon. Not. R. astr. Soc.* **195**, 397-427.

FALLE, S.A.E.G. 1981 *Mon. Not. R. astr. Soc.* **195**, 1011-1028.

HARTREE, D.R. 1958 Numerical Methods. Oxford University Press.

DONOR CELL, FCT-SHASTA AND FLUX SPLITTING METHOD: THREE FINITE DIFFERENCE TECHNIQUES APPLIED TO ASTROPHYSICAL SHOCK-CLOUD INTERACTIONS

J. Nittmann

(Department of Applied Mathematical Studies, University of Leeds)

1. INTRODUCTION

In this paper one first order and two second order finite difference techniques are applied to one dimensional test problems. In particular we demonstrate their applicability to model complex gas flows, by considering a shock-cloud interaction problem. The numerical solutions are compared with the exact solutions.

The study of the interaction of a strong interstellar shock with a high density gas cloud has been of particular interest in astrophysics in connection with

(i) whether a shock can trigger the gravitational collapse of the cloud owing to the additional compression produced by the shock,

(ii) whether a gravitationally stable cloud can be accelerated by the shock to reach velocities comparable with the shock velocity in the ambient gas.

The large compression of a gas cloud by an interstellar shock will lead to the gravitational collapse of the cloud if the density of the shocked gas exceeds the critical density for collapse. This process has been suggested as a mechanism which is capable of triggering star formation on a large scale in the arms of spiral galaxies. The initial shock is driven by a spiral density wave. Newly born stars of high luminosity can drive shocks which could possibly trigger star formation on a small scale. The strongest shocks are associated with supernova explosions. High velocity features in absorption line studies within old supernova remnants lead to the suggestion that gas clouds have been accelerated by the SN shock.

In order to model the evolution of a system containing shocks and density discontinuities in more than one dimension we have employed in the past (i.e. Nittmann (1981), Nittmann et al. (1982)) the Fluid In Cell technique (FLIC), which is based on the donor cell transport algorithm, and FCT-Shasta. Owing to the lack of exact solutions to complex multidimensional systems one is forced to employ at least two independent techniques to solve the associated set of non-linear partial differential equations, modelling such problems, in order to corroborate results.

In this paper we compare these two well established methods with a recently developed technique based on a flux vector splitting method. In order to test the accuracy of all three techniques we compare the numerical solutions with the exact solution for a one dimensional problem by considering the impact of a shock with a contact discontinuity. The remainder of the paper is as follows: In Section 2 we introduce the main characteristics of the FLIC technique. It is well known that the donor cell mass transport algorithm causes large amplitude errors in subsonic regions of the flow. In order to correct numerical diffusion errors we extend the FLIC technique by adding a flux limiting antidiffusion stage. In Section 3 we derive the principle of the FCT-algorithm. In particular, we show that the amount of numerical diffusion is adjustable. In Section 4 we introduce a flux splitting algorithm (FS2) suggested recently by van Albada et al. (1981). In Section 5 we compare solutions obtained by the numerical techniques with the analytic solutions of a non-trivial one dimensional problem. Finally, concluding remarks are made in Section 6.

2. FLUID IN CELL TECHNIQUE

The FLIC algorithm developed by Gentry et al. (1966) is used to solve the Euler equations

$$\frac{\partial U}{\partial t} + \frac{\partial F(U)}{\partial x} = 0 \tag{2.1}$$

with

$$U = \begin{pmatrix} \rho \\ \rho u \\ E \end{pmatrix} \quad \text{and} \quad F(U) = \begin{pmatrix} \rho u \\ \rho u^2 + p \\ (E + p)u \end{pmatrix} \tag{2.2}$$

where ρ, u, p and E represent the density, velocity, pressure and total energy. The algorithm proceeds in two stages.

(i) Lagrangian stage

During the first stage provisional values for the velocity $\tilde{u}$ and internal energy $\tilde{I}$ are found by solving

$$\frac{\partial \tilde{u}}{\partial t} = -\frac{1}{\rho}\frac{\partial (p+q)}{\partial x} \tag{2.3}$$

$$\frac{\partial \tilde{I}}{\partial t} = -\frac{(p+q)}{\rho}\frac{\partial u}{\partial x} \tag{2.4}$$

Since internal energy is not conserved in a shock an artificial viscous pressure term q is introduced, which is of the form

$$q = \begin{cases} k\,\rho\, c\, \delta u \ , & \text{if } \delta u < 0 \\ 0 & \text{if } \delta u \geqslant 0 \end{cases} \tag{2.5}$$

k is a constant and c is the sound speed ($c = \sqrt{(\gamma p/\rho)}$). In addition q is necessary to maintain stability for subsonic flows.

(ii) Transport stage (Donor cell differencing)

In order to calculate the mass flow between two cells one assumes that the transported mass is directly proportional to the density of the cell from which it flows with velocity $\tilde{u}$. A new value for the cell densities can be obtained by applying the law of conservation of mass.

$$\rho^{n+1} = \rho^{n} - \frac{\Delta t}{\Delta x}\cdot \begin{cases} (\tilde{u}^{n}_{i+\frac{1}{2}}\cdot\rho^{n}_{i} - \tilde{u}^{n}_{i-\frac{1}{2}}\cdot\rho^{n}_{i-1}) \text{ for } \tilde{u}^{n}_{i+\frac{1}{2}}>0 \text{ and } \tilde{u}^{n}_{i-\frac{1}{2}}>0 \\ (\tilde{u}^{n}_{i+\frac{1}{2}}\rho^{n}_{i+1} - \tilde{u}^{n}_{i-\frac{1}{2}}\rho^{n}_{i}) \text{ for } \tilde{u}^{n}_{i+\frac{1}{2}}<0 \text{ and } \tilde{u}^{n}_{i-\frac{1}{2}}<0 \end{cases} \tag{2.6}$$

Donor cell differencing prevents the formation of negative densities. In order to obtain the transported values for the momentum and the total energy one assumes that the mass which crosses the boundaries carries with it momentum and total energy of the donor cell.

$$\frac{\partial(\rho u)}{\partial t} = -\frac{\partial(\rho\tilde{u}\tilde{u})}{\partial x} \tag{2.7}$$

$$\frac{\partial E}{\partial t} = -\frac{\partial(\tilde{E}\tilde{u})}{\partial x}\ , \text{ where } \tilde{E} = \frac{\rho\tilde{u}^{2}}{2} + \tilde{I}\rho \tag{2.8}$$

One obvious advantage of the FLIC technique lies in its simplicity, which allows easy programming. In addition it is an easy task to extend this algorithm to treat a more complex hyperbolic system i.e. one including magnetic fields (Nittmann, 1981). As we shall see later, the FLIC algorithm treats supersonic problems reasonably well, but large amplitude errors develop in subsonic regions of the flow. Fig. 1 shows the propagation of a density profile in a uniform flow field. Large amplitude errors are seen to develop. In order to improve the numerical solutions for subsonic flow fields van Leer has suggested a modified transport stage. This is demonstrated in D. Young's paper in these proceedings. However, we employ a method suggested by Book et al. (1975) to improve the behaviour of diffusive first and second order techniques. This requires the introduction of an antidiffusion step at the end of the transport stage.

(iii) Antidiffusion Step

In order to establish the appropriate amount of antidiffusion one has to work out how much numerical diffusion is introduced during the transport stage owing to finite differencing. This can be done by investigating the continuity equation (2.6)

$$\rho_i^{n+1} = \rho_i^n - (\tilde{u}_{i+\frac{1}{2}}^n \rho_i^n - \tilde{u}_{i-\frac{1}{2}}^n \rho_{i-1}^n) \cdot \frac{\Delta t}{\Delta x}$$

We assume a uniform flow field and study the propagation of a particular fourier mode

$$\rho_i^n = \rho^n e^{-ikj\Delta x}, \text{ where } i = \sqrt{(-1)}. \tag{2.9}$$

For the amplification coefficient $A = \rho^{n+1}/\rho^n$ we obtain from Equation 2.6

$$A = 1 - \varepsilon(1 - e^{-ik\Delta x}) = 1 - \varepsilon(1 - \cos k\Delta x) - i\varepsilon \sin k\Delta x \tag{2.10}$$

where $\varepsilon = \tilde{u}(\Delta t/\Delta x)$. Amplification coefficients will be complex in general for a difference scheme. To obtain a real expression we define the amplification factor $|A|^2 = A \cdot A^*$, where A^* is the complex conjugate of A. We obtain for the donor cell algorithm,

$$|A|^2 = 1 - 2(\varepsilon - \varepsilon^2)(1 - \cos k\Delta x) \tag{2.11}$$

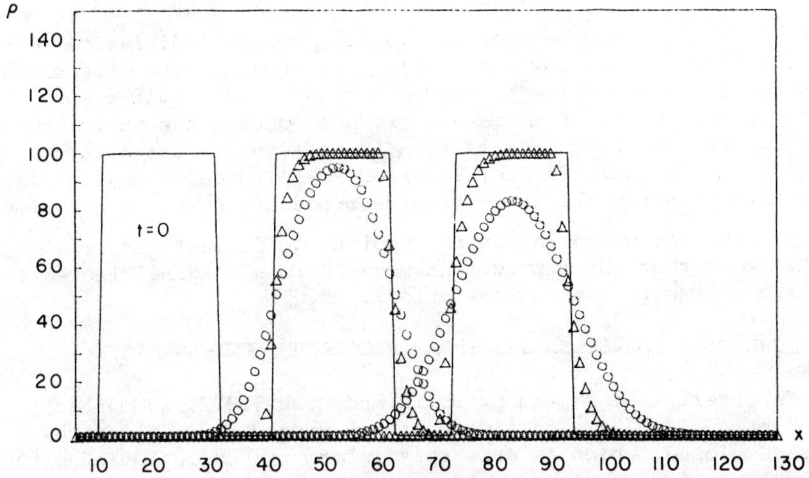

Fig. 1 Numerical results obtained using FLIC (circles) and antidiffused FLIC (triangles) in modelling the uniform propagation of a square density wave for two time intervals. The solid line represents the exact solution.

Stability requires $|A|^2 \leqslant 1$. But ε is always smaller than one (in order to enable a stable solution at the Lagrangian stage), so $|A|^2$ will be considerably smaller than 1. The error in the amplitude of the physical quantity is referred to as numerical diffusion.

Book et al., suggest an explicit antidiffusion function

$$\rho_i^{AD} = \rho_i^{n+1} - \eta(\rho_{i+1}^{n+1} - 2\rho_i^{n+1} + \rho_{i-1}^{n+1}) \qquad (2.12)$$

to correct the error. The antidiffusion coefficient η can be obtained by calculating the total amplification factor $(|A|^2)^T$ for the antidiffused transport stage:

$$(|A|^2)^T = |A|^2 \cdot |A^{AD}|^2 = [1-2(\varepsilon-\varepsilon^2)(1-\cos k\Delta x)].[1+2\eta(1-\cos k\Delta x)]^2$$
$$= 1-[2(\varepsilon-\varepsilon^2)-4\eta](1-\cos k\Delta x)-[8\eta(\varepsilon-\varepsilon^2)-4\eta^2](1-\cos k\Delta x)^2+\ldots$$
$$(2.13)$$

To remove the $(1-\cos k\Delta x)$ term the appropriate choice is $\eta = \frac{1}{2}(\varepsilon-\varepsilon^2)$. The greatly improved result for a propagating density profile can be seen in Fig. 1. Residual diffusion, due to the higher terms in $(1-\cos k\Delta x)$, will remain. The steepening of the diffused profile is controlled by a flux limiter which ensures that no new extrema are produced during the antidiffusion stage. We will come back to this in Section 3. One point worth noting is: The structure of Equation 2.11 indicates that it is possible to apply the antidiffusion procedure every second time step only, for which the antidiffusion coefficient $\eta = \varepsilon-\varepsilon^2$. This means that the improvement results in a marginal increase of c.p.u time in the regions of 40%.

3. THE FLUX CORRECTED TRANSPORT ALGORITHM: FCT-SHASTA

In a series of papers by Boris and Book (1973, 1976) and Book et al. (1975) a multidimensional second order algorithm was developed, which is able to simulate the evolution of both supersonic and subsonic gas flows without the loss of important information as a result of numerical diffusion. The algorithm can be divided into three successive steps:

(i) Computation of a transported diffused solution

At the first step the Shasta transport operator is applied which relates ρ^{TD} to ρ^n by

$$\rho^{TD} = \frac{1}{2} Q_-^2(\rho^n_{i-1} - \rho^n_i) + \frac{1}{2} Q_+(\rho^n_{i+1} - \rho^n_i) + (Q_+ + Q_-)\rho^n_i ,$$

$$Q_\pm = \left(\frac{1}{2} \mp u_i^{n+\frac{1}{2}} \frac{\Delta t}{\Delta x}\right) / \left[1 \pm \left(u^{n+\frac{1}{2}}_{i\pm 1} - u_i^{n+\frac{1}{2}}\right) \frac{\Delta t}{\Delta x}\right] \tag{3.1}$$

For a uniform velocity field this reduces to a simpler form,

$$\rho_i^{TD} = \rho_i^n - \frac{\varepsilon}{2}(\rho^n_{i+1} - \rho^n_{i-1}) + (\mu + \frac{\varepsilon^2}{2})(\rho^n_{i+1} - 2\rho^n_i + \rho^n_{i-1}) \tag{3.2}$$

with $\mu = \frac{1}{8}$ and $\varepsilon = u(\Delta t/\Delta x)$. This formula is a simple two sided differencing of the convection term plus a strong diffusion. The strong diffusion is a main feature of FCT. Book et al. show that numerical diffusion during the transport stage can limit phase errors which may produce undesirable overshoots and undershoots at steep gradients. The amplification factor for Equation 3.2 is

$$|A|^2 = [1 - 2\mu(1-\cos k\Delta x)]^2 - \frac{\varepsilon^2}{2} (1 - 2\varepsilon^2)(1-\cos k\Delta x)^2 \quad (3.3)$$

and $\varepsilon^2 \leq \frac{1}{2}$ is required for stability.

(ii) Antidiffusion step

To remove the large diffusion introduced during the first step a negative diffusion term is added. Three forms of antidiffusion are available:

(a) explicit AD: $$\rho_i^{AD} = \rho_i^{TD} - \eta(\rho_{i+1}^{TD} - 2\rho_i^{TD} + \rho_{i-1}^{TD}) \quad (3.4)$$

(b) implicit AD: $$\rho_i^{AD} = \rho_i^{TD} - \eta(\rho_{i+1}^{AD} - 2\rho_i^{AD} + \rho_{i-1}^{AD}) \quad (3.5)$$

(c) phoenical AD: $$\rho_i^{AD} = \rho_i^{TD} - \eta(\rho_{i+1}^{T} - 2\rho_i^{T} + \rho_{i-1}^{T}) \quad (3.6)$$

where ρ^T is the transported solution which can be obtained by subtracting the diffused (velocity independent diffusion) solution from ρ^{TD}. The implicit and phoenical antidiffusion are superior to the explicit one. Phoenical antidiffusion, owing to its explicitness, is less expensive in terms of computer time and has been employed in our computations. To see how the antidiffusion coefficient η influences the solution, we derive the amplification factor for phoenical Shasta which is,

$$|A|^2 = [1-\varepsilon^2(1-\cos k\Delta x) - 2\mu\varepsilon^2(1-\cos k\Delta x)^2]^2$$
$$+ \varepsilon^2[2(1-\cos k\Delta x) - (1-\cos k\Delta x)^2]\,[1+2\eta(1-\cos k\Delta x)]^2 \quad (3.7)$$

For the linear and quadratic terms of $(1-\cos k\Delta x)$ we obtain

$$|A|^2 = 1-2(\varepsilon^2-\varepsilon^2)(1-\cos k\Delta x) + (\varepsilon^4-4\mu\varepsilon^2-\varepsilon^2+8\eta\varepsilon^2)(1-\cos k\Delta x)^2$$

with $|\varepsilon| = \frac{1}{2}$ the choice of $\eta = \mu = \frac{1}{8}$ limits residual diffusion. If we choose η slightly smaller than $\frac{1}{8}$, then a more diffused solution will be obtained. This will be shown in Section 5.

The antidiffusion formula (3.4) - (3.6) will steepen up any gradients and may lead to the formation of overshoots and undershoots. To prevent the creation and growth of unphysical extrema a flux limiter is introduced in step three.

(iii) Flux limitation (flux correction)

The antidiffusion fluxes have to be modified (corrected) to prevent the growth of unphysical extrema. The flux limitation formula takes the form:

$$f^{C}_{i+\frac{1}{2}} = \text{sign}\,(f_{i+\frac{1}{2}}).\max\left[\begin{array}{l} 0 \\ \min\left\{\begin{array}{l} \text{sign}\,(f_{i+\frac{1}{2}})(\rho^{TD}_{i}-\rho^{TD}_{i-1}) \\ |f_{i+\frac{1}{2}}| \\ \text{sign}\,(f_{i+\frac{1}{2}})(\rho^{TD}_{i+2}-\rho^{TD}_{i+1}) \end{array}\right\} \end{array}\right] \tag{3.8}$$

where the final value of ρ_i^{n+1} is given by

$$\rho_i^{n+1} = \rho^{TD} - f^{C}_{i+\frac{1}{2}} + f^{C}_{i-\frac{1}{2}}$$

A similar expression (3.8) exists for $f^{C}_{i-\frac{1}{2}}$. Unfortunately this procedure limits the width of allowed maxima to three cells. Extrema consisting of less than three cells are cut off. This effect is referred to in the literature as the clipping effect. Zalesak (1979) has proposed a modified formulation to (3.8), which improves the solutions for propagating peak profiles. Fig. 2 illustrates the numerical solution for a propagating density profile using FCT-Shasta.

4. SECOND ORDER FLUX SPLITTING METHOD (FS2)

The principle of the method of flux vector splitting has been put forward by Steger and Warming (1981) to enable the use of upwind difference schemes for both supersonic and subsonic regions of the flow. One sided upwind schemes are in principle unstable in flows where characteristic velocities are smaller than the sound speed. To illustrate the procedure we write the Euler equations (2.1) as a quasi-linear system

$$\frac{\partial U}{\partial t} + A(U)\,\frac{\partial U}{\partial x} = 0 \tag{4.1}$$

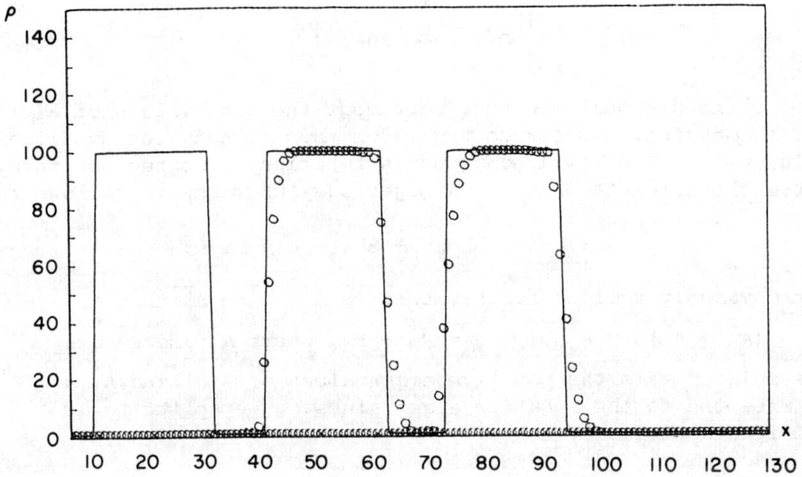

Fig. 2 As in Fig. 1 but results obtained using FCT-Shasta (circles).

where A(U) is the Jacobian matrix $\partial F/\partial U$. With $U = (\rho, m, E)^T$, and $p = (\gamma-1)[E - m^2/(2\rho)]$

$$F(U) = \begin{bmatrix} m \\ (\gamma-1)E + (3-\gamma)\dfrac{m^2}{2\rho} \\ \gamma\dfrac{Em}{\rho} - (\gamma-1)m^3/(2\rho^2) \end{bmatrix} \tag{4.2}$$

and

$$A(U) = \begin{bmatrix} 0 & 1 & 0 \\ (\gamma-3)\dfrac{u^2}{2} & (3-\gamma)u & \gamma-1 \\ (\gamma-1)u^3-\gamma\dfrac{Eu}{\rho} & \gamma\dfrac{E}{\rho} - 3(\gamma-1)\dfrac{u^2}{2} & \gamma u \end{bmatrix} \tag{4.3}$$

A(U) can be diagonalized by the transformation

$$Q^{-1} A\, Q = \Lambda \quad \text{with} \tag{4.4}$$

$$\Lambda = \begin{bmatrix} u & 0 & 0 \\ 0 & u+c & 0 \\ 0 & 0 & u-c \end{bmatrix} \tag{4.5}$$

where the diagonal elements represent the eigenvalues of A(U). It is of great importance that, for this method, the non-linear flux vector F(U) is a homogeneous function of degree one in U. From the Euler theorem for homogeneous functions it follows that

$$F = AU \tag{4.6}$$

It obviously follows further that if $A = A^+ + A^-$ then $F = (A^+ + A^-)U = A^+U + A^-U = F^+ + F^-$, where A^+ and F^+ are associated with the positive eigenvalues of A(U) and A^- and F^- correspond to the negative eigenvalues. The splitting is derived as follows:

By virtue of (4.4), (4.5) and (4.6)

$$F = A\,U = Q\Lambda\,Q^{-1}U \tag{4.7}$$

The eigenvalues λ_i (4.5) can be expressed as

$$\lambda_i = \lambda_i^+ + \lambda_i^- \tag{4.8}$$

where

$$\lambda_i^+ = \tfrac{1}{2}(\lambda_i + |\lambda_i|), \quad \lambda_i^- = \tfrac{1}{2}(\lambda_i - |\lambda_i|) \tag{4.9}$$

so that if $\lambda_i \geq 0$, then $\lambda_i^+ = \lambda_i$, $\lambda_i^- = 0$, with the converse result for $\lambda_i < 0$.

Using the above formula, we split the diagonal matrix

$$\Lambda = \Lambda^+ + \Lambda^- \tag{4.10}$$

where Λ^+ and Λ^- have as diagonal elements λ_i^+ and λ_i^-, respectively. Equation (4.7) can be rewritten as

$$\begin{aligned} F &= Q(\Lambda^+ + \Lambda^-)Q^{-1}U = (A^+ + A^-)U \\ &= F^+ + F^- \end{aligned} \tag{4.11}$$

The eigenvalues given by Equ. (4.5) are split according to Equs. (4.8) and (4.9) into

$$\begin{aligned} \lambda_1^+ &= \tfrac{1}{2}(u+|u|) & \lambda_1^- &= \tfrac{1}{2}(u-|u|) \\ \lambda_2^+ &= \tfrac{1}{2}(u+c+|u+c|) & \lambda_2^- &= \tfrac{1}{2}(u+c-|u+c|) \\ \lambda_3^+ &= \tfrac{1}{2}(u-c+|u-c|) & \lambda_3^- &= \tfrac{1}{2}(u-c-|u-c|) \end{aligned} \tag{4.12}$$

The corresponding subvectors F^+ and F^- can be obtained from the generalized fluxvector (see Steger and Warming, 1981). The authors suggest a Predictor-Corrector algorithm which employs the split flux vectors to solve the model equations. However we applied an algorithm put forward by van Albada et al. (1981).

The cell averages U_i^n are advanced to half timestep $t^{n+\frac{1}{2}}$ by

$$U_i^{n+\frac{1}{2}} = U_i^n - \frac{\Delta t}{2\Delta x}\left[F\left(U^n_{(i+\frac{1}{2})-}\right) - F\left(U^n_{(i-\frac{1}{2})+}\right)\right] \tag{4.13}$$

where the particular interface values $U^n_{(i\pm\frac{1}{2})\mp}$ are given by

$$U^n_{(i\pm\frac{1}{2})\mp} = U_i^n \pm \tfrac{1}{2}(\delta U)_i^n \tag{4.14}$$

$$(\delta U)_i^n = \mathrm{ave}(U_{i+1}^n - U_i^n,\ U_i^n - U_{i-1}^n) \tag{4.15}$$

The function ave(a,b) is given by

$$\mathrm{ave}(a,b) = \frac{(b^2+c^2).a + (a^2+c^2).b}{a^2 + b^2 + 2c^2} \tag{4.16}$$

where c^2 is small non-vanishing bias of the order $(\Delta x)^3$. The time centred fluxes $\phi^{n+\frac{1}{2}}_{i\pm\frac{1}{2}}$ which satisfy the equation

$$\frac{U_i^{n+1} - U_i^n}{\Delta t} + \frac{\phi^{n+\frac{1}{2}}_{i+\frac{1}{2}} - \phi^{n+\frac{1}{2}}_{i-\frac{1}{2}}}{\Delta x} = 0, \tag{4.17}$$

can now be obtained by

$$\phi_{i+\frac{1}{2}}^{n+\frac{1}{2}} = F^{+}(U_{(i+\frac{1}{2})-}^{n+\frac{1}{2}}) + F^{-}(U_{(i+\frac{1}{2})+}^{n+\frac{1}{2}}) \tag{4.18}$$

with F^{+} and F^{-} as the split fluxes. The interface values at $t^{n+\frac{1}{2}}$ are

$$U_{(i+\frac{1}{2})\mp}^{n+\frac{1}{2}} = U_{i}^{n+\frac{1}{2}} \pm \tfrac{1}{2}(\delta U)_{i}^{n} \tag{4.19}$$

Fig. 3 compares the numerical solution for a propagating density profile using FS2 with the exact solution. Fig. 4 compares the established profile of a moving shock for FLIC (a), FCT-Shasta (b) and FS2 (c). FLIC shows a relatively smooth profile extending over approximately 3-4 cells. FCT and FS2 show a similar shock profile which covers only 2 cells.

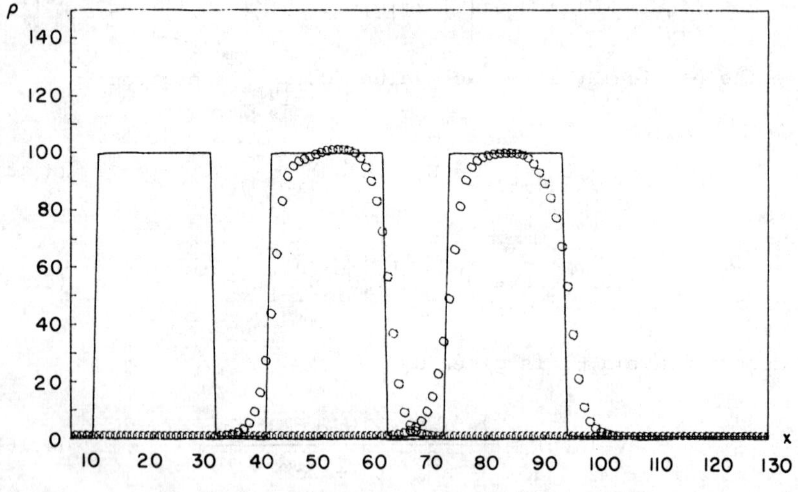

Fig. 3 As in Fig. 1 but results obtained using FS2 (circles).

Comparing the results for the propagating density profiles (Figs. 1-3) FCT certainly gives the best result. The anti-diffused FLIC algorithm shows a fairly good result too, while FS2 develops a slightly more diffused solution than FCT. This simple test example only gives a moderate insight into how well the techniques can be applied to complex flow problems. In order to apply the numerical techniques to a non-trivial problem

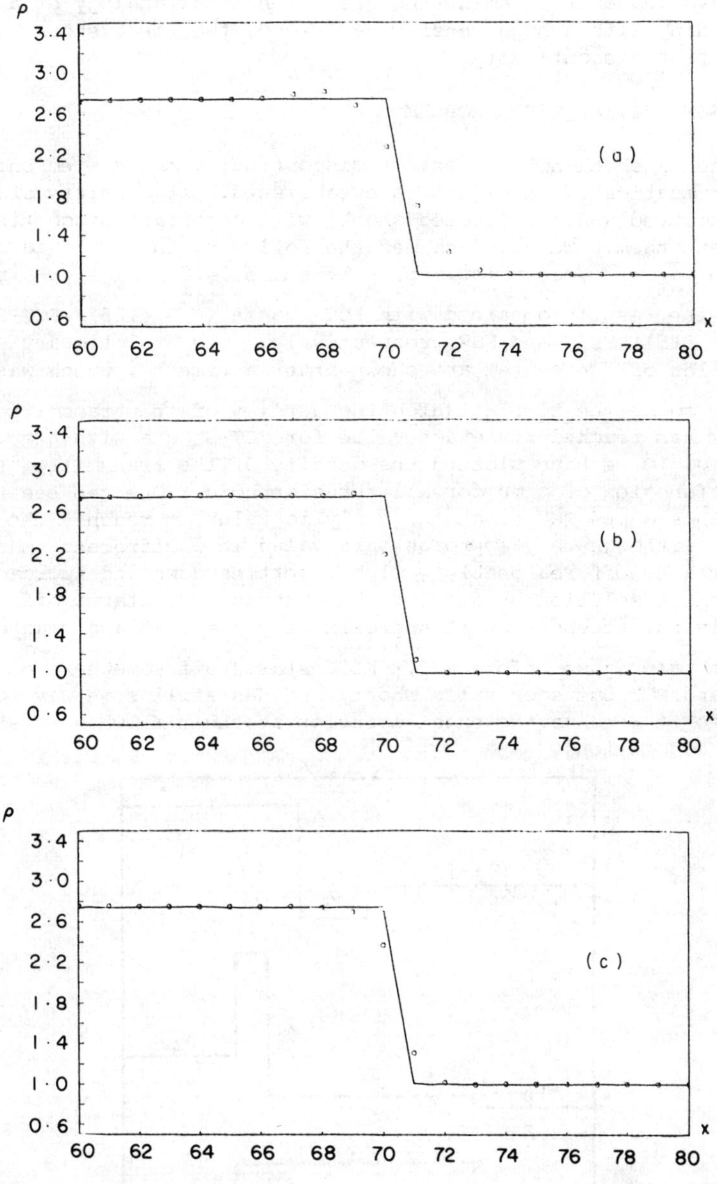

Fig. 4 Numerical results for a uniformly propagating shock (circles) for FLIC (a), FCT-Shasta (b) and FS2 (c). The solid lines represent the exact solutions.

we have chosen to examine the one dimensional analogy of a shock colliding with a high density gas cloud, that is a shock hitting a contact discontinuity.

5. NON-TRIVIAL TEST PROBLEM

When a shock hits a contact discontinuity the system shown diagrammatically in Fig. 5 is established. It consists of a transmitted and a reflected shock, with a contact discontinuity between them. We have assumed the following initial data at $t = 0$, $\rho_1 = 1.$, $\rho_4 = 100.$, $p_1 = 1.$ and $p_2 = 8.$ Figs. 6-9 represent the results obtained with FCT-Shasta ($\eta = .125$), FCT-Shasta ($\eta = .115$), FLIC and FS2, respectively. The evolutionary profiles of the system are shown after a time t_α, which was found to be the time at which the density of the transmitted shock has reached its exact value for FCT-Shasta with $\eta = .125$. In Fig. 10 we have plotted the density of the transmitted shock as a function of time for all three methods. One can see that FCT-Shasta and FS2 reach the analytic value at roughly the same time, although they approach this value in a different manner. FS2 starts off reasonably well but settles down and approaches the exact solution gradually. In contrast FCT starts off rather slowly but steepens up at approximately $t = t_\alpha/2$ and reaches the exact value just before FS2. FLIC starts off somewhere between FS2 and FCT but soon falls short, and has still some way to go before it reaches the exact solution at the time when we stopped the computations.

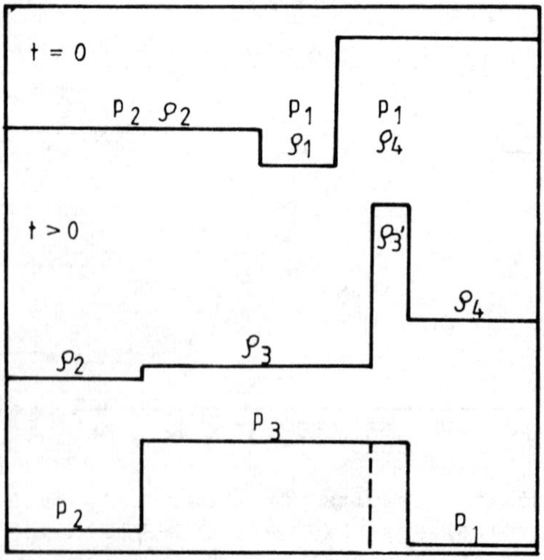

Fig. 5 Diagrammatic representation of the pressure and density distribution, before and after the shock has hit the interface.

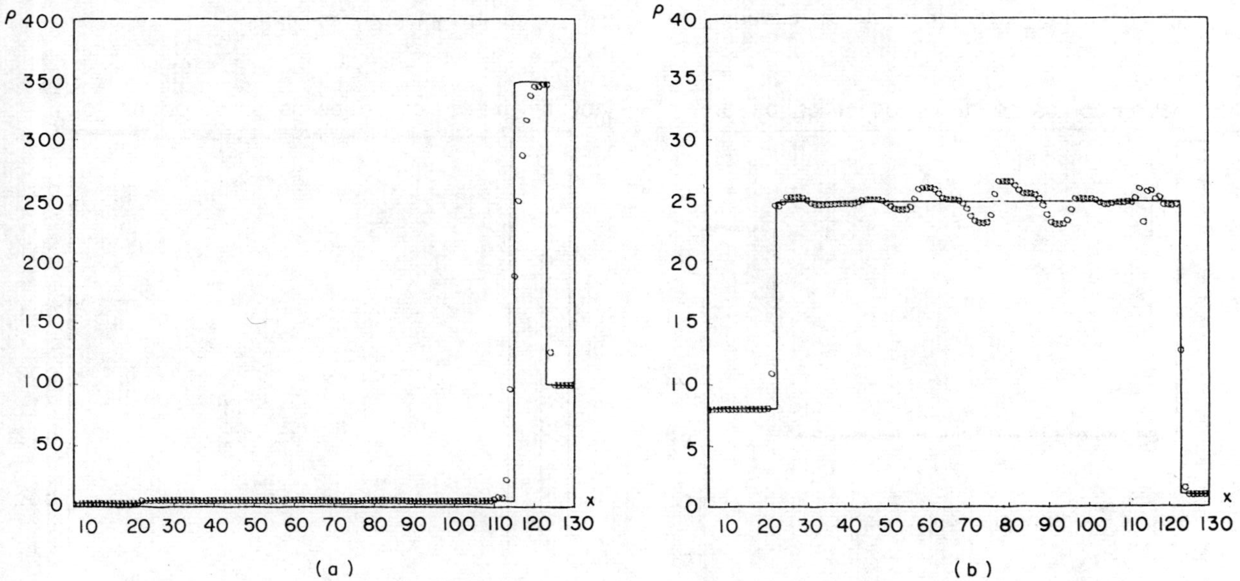

Fig. 6 Numerical results (circles) for the density (a) and pressure distribution (b) using FCT-Shasta ($\eta = .125$). The solid lines represent the exact solutions.

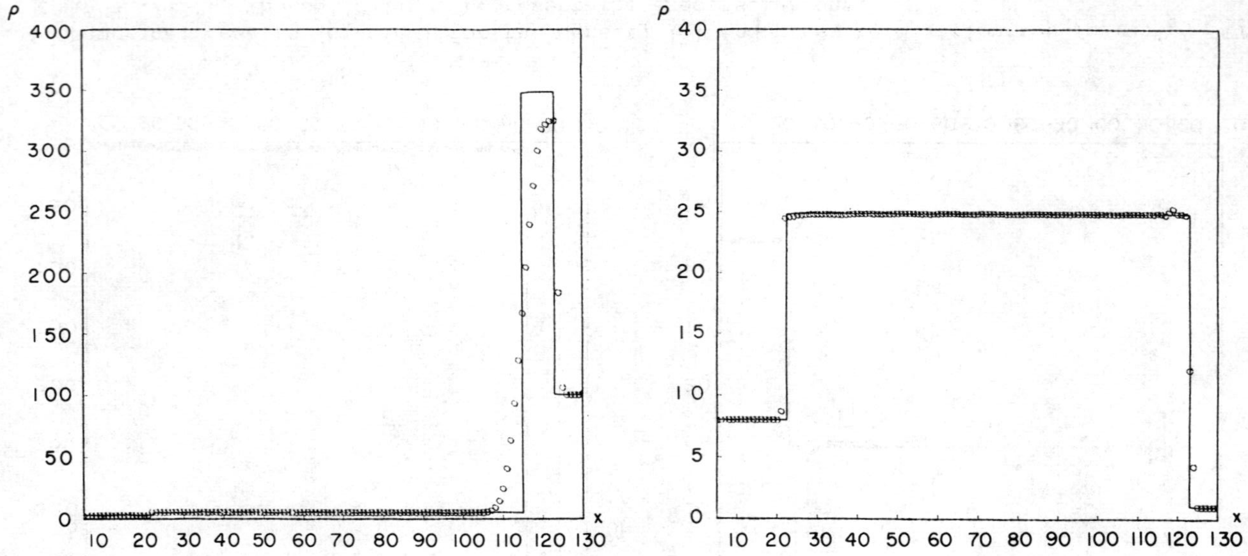

(a) (b)

Fig. 7 As in Fig. 6 but using FCT-Shasta ($\eta = .115$).

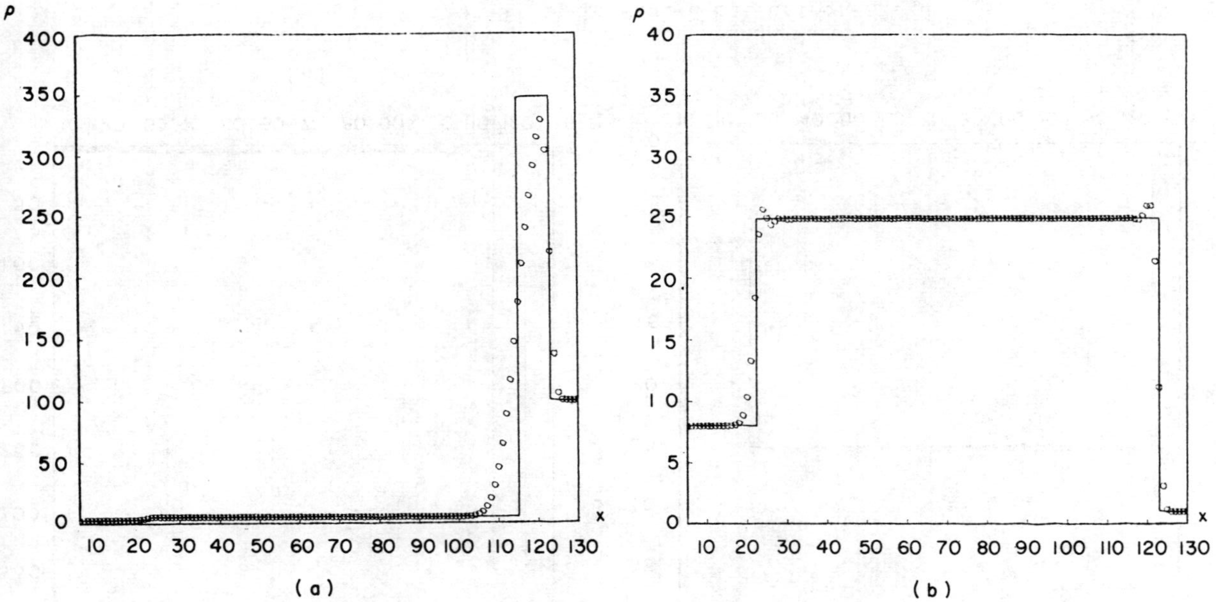

Fig. 8 As in Fig. 6 but using FLIC.

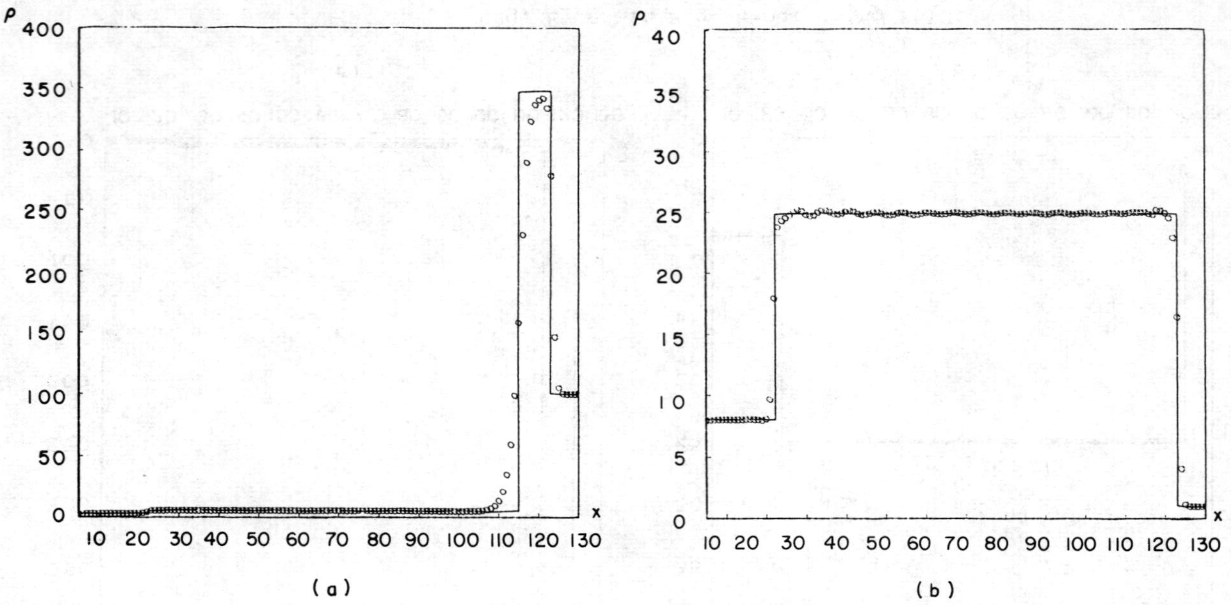

Fig. 9 As in Fig. 6 but using FS2.

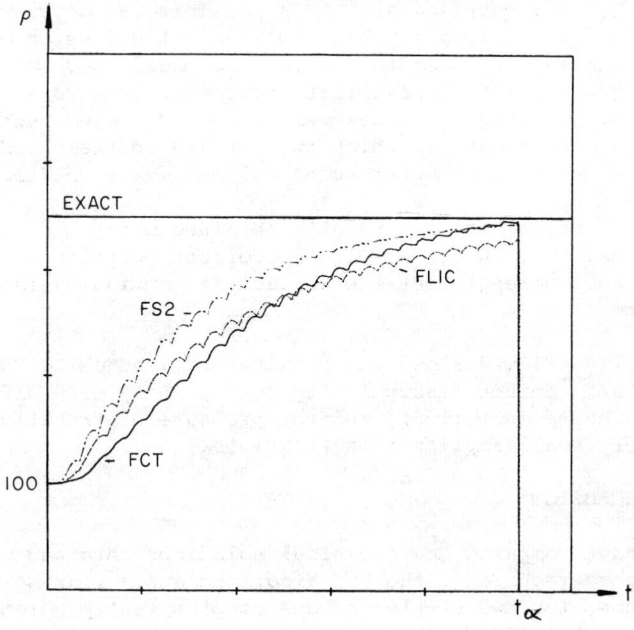

Fig. 10 Density of the transmitted shock versus time for FCT-Shasta (solid line), FLIC (dashed line) and FS2 (hyphenated line). The straight solid line represents the exact solution.

The time t_α can be derived from the formula

$$t_\alpha = \frac{\alpha \Delta x}{u_{ST} - u_{CD}} \tag{5.1}$$

where u_{ST} and u_{CD} are the velocities of the transmitted shock and the contact discontinuity, respectively. We found α to be 8 for FCT and FS2 and > 10 for FLIC. It is interesting to see that the small oscillations, which are superimposed on all three curves in Fig. 10, have a frequency which is one over the time it takes the transmitted shock to traverse one cell.

Comparing the exact solutions with the numerical ones of Figs. 6-9 we find that FCT-Shasta shows the best result for the density distribution. However, large pressure oscillations in the gas behind the reflected shock are present in Fig. 6b. A detailed study showed that these oscillations occur also in the other physical quantities. It is possible, as discussed in Section 3, to apply a slightly smaller antidiffusion coefficient in FCT-Shasta, in order to obtain a more diffused solution. Choosing $\eta = 0.115$ the oscillations can be removed as illustrated in Fig. 7b. However, a more diffused density distribution results in which the density in the cloud falls, giving a deviation between numerical and exact solution.

Fig. 8 represents the results obtained using the FLIC algorithm. The shocks are in the correct position and are of the correct strength but the contact discontinuity is very much diffused.

Finally, Fig. 9 shows the results obtained using FS2. The shocks and contact discontinuity are slightly more diffused than in FCT-Shasta ($\eta = .125$), and the pressure distribution shows only very small amplitude oscillations.

6. CONCLUSION

We have compared the numerical solutions obtained by a donor cell based technique, the FCT algorithm and a flux splitting technique, for two simple and one complex test problem, with the exact solutions. Assuming a uniform grid and identical initial values, we find that FCT-Shasta gives the best results for simple subsonic and supersonic flow problems. However, for the complex flow problem of a shock hitting a contact discontinuity numerical oscillations develop in all physical quantities. This effect has been reported by Nittmann et al. (1982) before. These numerical oscillations can be removed by considering slightly more diffused solutions, but this increases amplitude errors. The results obtained using FS2 indicate that the flux splitting method is superior to the flux corrected transport method. It is obvious from the plots that the first order uncorrected, donor cell based method (FLIC), is certainly less accurate and the numerical solutions obtained cannot stand comparison with the numerical solutions of the other two methods. A further discussion on donor cell techniques, in particular one which is able to handle two or more fluids, and FCT techniques is given in (Gaskell and Nittmann, 1982). The two fluid option is being considered as a possibility of inhibiting numerical diffusion at interfaces.

Sod (1978) and van Albada et al. (1981) have performed similar investigations on several techniques including a FCT algorithm. FS2 only is studied in the latter. Sod uses a shock tube flow as a test problem and acknowledges the improvements associated with the FCT-antidiffusion stage. Van Albada et al. compare several numerical techniques and how well they cope with steady state solutions. They conclude that FCT-Shasta is highly unsuitable for their problem. The FCT-Shasta solution did not converge to the steady state solution. In contrast they obtained the best result with FS2, but it should be noted that their flux splitting procedure differs somewhat from the one described in Section 4.

We believe that further investigations, particularly in two dimensions are necessary. FCT-Shasta has been used to study the evolution of a shocked axisymmetric gas cloud in two dimensions by Nittmann et al. (1982) and it is a straightforward task to extend the algorithm to investigate MHD-Shocks (Nittmann and Gaskell, 1982). The flux splitting technique on the other hand becomes considerably more complex if applied to a two dimensional MHD problem and its applicability has still to be shown.

ACKNOWLEDGEMENT

I am much indebted to Dr. P.H. Gaskell for his collaboration, in particular on the FCT-Shasta program and for contributing to the manuscript.

REFERENCES

VAN ALBADA, G.D., VAN LEER, B. and ROBERTS, JR., W.W. 1981 ICASE Report No. 81-24.

BOOK, D.L., BORIS, J.P. and HAIN, K.H. 1975 *J. Comput. Phys.* **18**, 248.

BORIS, J.P. and BOOK, D.L. 1973 *J. Comput. Phys.* **11**, 38.

BORIS, J.P. and BOOK, D.L. 1976 *J. Comput. Phys.* **20**, 397.

GASKELL, P.H. and NITTMANN, J. 1982 Departmental Report, in preparation

GENTRY, R.A., MARTIN, R.E. and DALY, B.J. 1966 *J. Comput. Phys.* **1**, 87.

NITTMANN, J. 1981 *Mon. Not. R. astr. Soc.* **197**, 699.

NITTMANN, J., FALLE, S.A.E.G. and GASKELL, P.H. 1982 *Mon. Not. R. astr. Soc.* **201**, in print.

NITTMANN, J. and GASKELL, P.H. 1982 in preparation.

SOD, G.A. 1978 *J. Comput. Phys.* **27**, 1.

STEGER, J.L. and WARMING, R.F. 1981 *J. Comput. Phys.* **40**, 263.

ZALESAK, S.T. 1979 *J. Comput. Phys.* **31**, 335.